Basic Neuroscience

Anatomy & Physiology

Second Edition

Basic Neuroscience

Anatomy & Physiology

Arthur C. Guyton, M.D.

Professor Emeritus
Department of Physiology and Biophysics
University of Mississippi Medical Center
Jackson, Mississippi

W.B. SAUNDERS COMPANY

Harcourt Brace Jovanovich, Inc.

Philadelphia · London · Toronto · Montreal · Sydney · Tokyo

W. B. SAUNDERS COMPANY
Harcourt Brace Jovanovich, Inc.

The Curtis Center
Independence Square West
Philadelphia, PA 19106

Library of Congress Cataloging in Publication Data

Guyton, Arthur C.
    Basic neuroscience / Arthur C. Guyton. — 2nd ed.
        p.   cm.
    Includes bibliographical references and index.
    ISBN 0-7216-3993-3
    1. Neurophysiology.   2. Neuroanatomy.   I. Title.
    [DNLM: 1. Nervous System — anatomy & histology.   2. Nervous System
— physiology.   WL 101 G992b]
    QP355.2.G89   1992
    612.8 — dc20
    DNLM/DLC
    for Library of Congress   91-25780
                                                                CIP

Portions of this book including both text and illustrations have appeared previously in the
Textbook of Medical Physiology by Dr. Arthur C. Guyton published by W. B. Saunders, 1991.

Editor: Martin J. Wonsiewicz
Designer: Karen O'Keefe
Cover Designer: Michelle Maloney
Production Manager: Peter Faber
Manuscript Editor: Holly Lakens
Illustration Specialist: Brett MacNaughton
Indexer: Roger Wall

Basic Neuroscience, Second Edition   ISBN 0-7216-3993-3

Printed in the United States of America.

Last digit is the print number: 9   8   7   6   5   4   3

Preface

In the past, neuroanatomy and neurophysiology have mostly been taught as separate subjects. However, every course in neuroanatomy almost invariably includes major amounts of neurophysiology, and it is impossible to teach a course in neurophysiology without discussing neuroanatomy. Even so, most attempts to present both neuroanatomy and neurophysiology in the same textbook have produced multiauthor books that usually encompass far too much for the student to study in the time that is available.

The first edition of this text, entitled *Basic Neuroscience: Anatomy and Physiology,* was designed to present the basics of both of these subjects as a single integrated discipline. This second edition follows the same principles as the first. However, except for the first few chapters on gross anatomy, the text has been extensively revised. The reason is simple: rapid strides in our understanding of many neural mechanisms have revealed what had remained a mystery until recently. Our new understanding requires much updating of discussions of microfunctional anatomy and the physiology and chemistry of nervous function.

Basic Neuroscience, 2nd Edition, is intended for a wide variety of students who want to know how the nervous system works, be they medical or dental students, students of basic neuroanatomy and physiology, students of psychology, students of biology, or others with similar interests.

The text begins with chapters on the gross anatomy of the nervous system, with color pictures showing virtually every important neuroanatomical structure. These chapters provide a background for understanding of the gross organization of the nervous system while helping the student begin to conceive its function.

The remainder of the text discusses the basic functional anatomy and physiology of each part of the nervous system. For instance, throughout the text are diagrams of the many nervous tracts that carry information from one neural locus to another. Also discussed are the anatomy and physiology of each sensory organ — the eyes, the ears, the vestibular apparatus, and the organs of somatic sensation (smell, taste, and so forth). Likewise, the anatomy and function of the neural effector organs — the skeletal muscles, the heart, and the glands — are described. Also, the chemical anatomies of the nerve cell membrane, the neuronal cell body, the nerve endings, and the synapses are discussed as a basis for understanding the processing of nerve signals.

The ultimate goal is to present a composite picture of neural function, so that the student can understand how the nervous system controls most bodily activities and at the same time functions as a sensing, feeling, acting, and thinking organ.

My own deep interest in the nervous system began when I was still a student. In fact, at that time I put together my first book by photographically enlarging a great many microscopic cross-sectional slides of the nervous system and collecting them into a large personal atlas of neuroanatomy. Later, when I was a resident in neurosurgery, the study of neuroanatomy once again became of paramount importance in my work. More recently, my research carried me deeply into various aspects of neural control of the body, especially control of the circulation and, to a lesser extent, control of respiration and endocrine functions. From these varied interests, I have learned to respect the nervous system as a masterpiece of design, with all those magical properties that make life meaningful and exciting.

I wish to thank many people who have made this text possible, particularly Ivadelle Osberg Heidke and Gwendolyn Robbins for their excellent secretarial services, and Tomiko Mita, Michael Schenk, Tina Burnham, Diane Flemming, Iris Nichols, and Patricia Johnson for the artwork. I am also indebted to the staff of the W.B. Saunders Company for their continued excellence in the publication of this book, especially to Martin J. Wonsiewicz, Amy Norwitz, Brett MacNaughton, Karen O'Keefe, and Peter Faber, whose editorial and technical help have been invaluable.

ARTHUR C. GUYTON

Contents

INTRODUCTION

1

Introduction: Structural and Functional Highlights of the Nervous System

The nervous system is the sensing, thinking, and controlling system of our body. To perform these functions, it collects sensory information from all of the body—from a myriad of special sensory nerve endings in the skin, from the deep tissues, from the eyes, the ears, the equilibrium apparatus, and other sensors—and transmits this information through nerves into the spinal cord and brain. The cord and the brain may react immediately to this sensory information and send signals to the muscles or internal organs of the body to cause some response, called a *motor response.* Or, under other conditions, no immediate reaction might occur at all; instead the sensory information is stored in one of the brain's memory banks. There it is compared with other memories already stored; it is combined with other information; and from the various combinations, new thoughts are achieved. Then, perhaps a few minutes later, a month later, or even several years later, this extensive processing of information might at last lead to some motor response, maybe a very simple one or maybe very complex, such as building a house or piloting a space craft. Also, nervous activation of internal organs of the body, such as increasing the heart rate or increasing peristalsis in the intestines, may also be part of a motor response.

Thus, the nervous system is said to subserve three principal functions: (1) *sensory function;* (2) *integrative function,* which includes the memory and thinking processes; and (3) *motor function.*

■ THE MAJOR DIVISIONS OF THE NERVOUS SYSTEM

Figure 1–1 illustrates the two major divisions of the nervous system: (1) the *central nervous system,* which in turn is composed of the *brain* and the *spinal cord,* and (2) the *peripheral nervous system.*

The brain is the principal integrative area of the ner-

vous system—the place where memories are stored, thoughts are conceived, emotions are generated, and other functions related to our psyche and to complex control of our body are performed. To perform these complex activities, the brain itself is divided into many separate functional parts, which we shall begin discussing in the following chapter.

The spinal cord serves two functions. First, it serves as a conduit for many nervous pathways to and from the brain. Second, it serves as an integrative area for coordinating many subconscious nervous activities, such as reflex withdrawal of a part of the body from a painful stimulus, reflex stiffening of the legs when a person stands on his feet, and even crude reflex walking movements. Thus, the spinal cord is much more than simply a large peripheral nerve.

The peripheral nervous system is illustrated to the left in Figure 1–1, showing it to be a branching network of nerves that is so extensive that hardly a single cubic millimeter of tissue anywhere in the body is without nerve fibers. These fibers are of two functional types: *afferent fibers* for transmission of sensory information into the spinal cord and brain and *efferent fibers* for transmitting motor signals back from the central nervous system to the periphery, especially to the skeletal muscles (see Chapter 24). Some of the peripheral nerves arise directly from the brain itself and supply mainly the head region of the body. These, called *cranial nerves,* are not illustrated in Figure 1–1, but they will be discussed later. The remainder of the peripheral nerves are *spinal nerves,* one of which leaves each side of the spinal cord through an intervertebral foramen at each vertebral level of the cord.

NERVOUS TISSUE

Nervous tissue, whether it be in the brain, the spinal cord, or the peripheral nerves, contains two basic types of cells:

1. *Neurons* conduct the signals in the nervous system. There are probably at least 100 billion of these in the entire nervous system.

2. *Supporting* and *insulating cells* hold the neurons in place and prevent signals from spreading between the neurons where this is not desired. In the central nervous system these supporting and insulating cells are collectively called the *neuroglia*. In the peripheral nervous system they are the *Schwann cells*.

The Central Nervous System Neuron

Figure 1–2 illustrates a typical neuron of the brain or spinal cord. Its principal parts are the following:

1. *Cell body.* It is from this that other parts of the neuron grow. Also the cell body provides much of the nourishment that is required for maintaining the life of the entire neuron.

2. *Dendrites.* These are multiple branching outgrowths from the cell body. They are the main receptor portions of the neuron. That is, most signals that are to be transmitted by the neuron enter by way of the den-drites, although some enter also through the surface of the cell body. The dendrites of each neuron usually receive signals from literally thousands of contact points with other neurons, points called *synapses*, as we shall discuss later.

3. *Axon.* Each neuron has one axon leaving the cell body. This is the portion of the neuron that is usually called the *nerve fiber*. It may extend only a few millimeters, as is the case for the axons of many small neurons within the brain, or it may be as long as a meter in the case of the axons (nerve fibers) that leave the spinal cord to innervate the feet. The axons carry the nerve signals to the next nerve cell in the brain or spinal cord or to muscles and glands in peripheral parts of the body.

4. *Axon terminals and synapses.* All axons branch near their ends many times, often thousands of times. At the end of each of these branches is a specialized axon terminal that, in the central nervous system, is called a *presynaptic terminal, a synaptic knob,* or a *bouton* because of its knoblike appearance. The presynaptic terminal lies on the membrane surface of a dendrite or cell body of another neuron, thus providing a con-

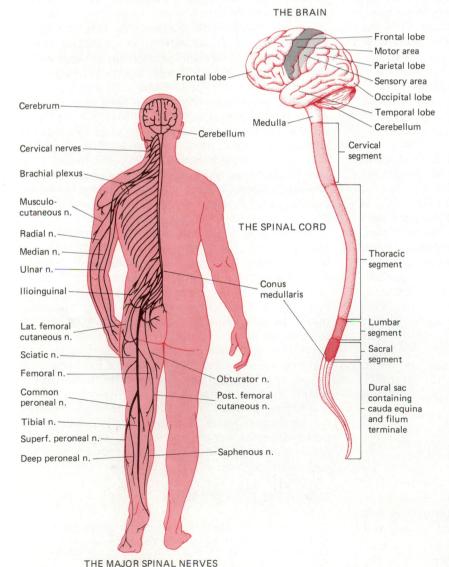

THE BRAIN

Frontal lobe
Motor area
Parietal lobe
Sensory area
Occipital lobe
Temporal lobe
Cerebellum

Frontal lobe

Cerebrum
Cervical nerves
Cerebellum
Brachial plexus
Medulla
Musculo-cutaneous n.
Radial n.
Median n.
Ulnar n.
Ilioinguinal
Lat. femoral cutaneous n.
Sciatic n.
Femoral n.
Common peroneal n.
Tibial n.
Superf. peroneal n.
Deep peroneal n.

Conus medullaris

Obturator n.
Post. femoral cutaneous n.

Saphenous n.

THE SPINAL CORD

Cervical segment
Thoracic segment
Lumbar segment
Sacral segment
Dural sac containing cauda equina and filum terminale

THE MAJOR SPINAL NERVES

Figure 1–1. The principal anatomical parts of the nervous system.

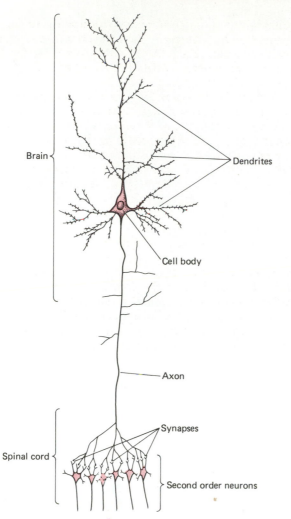

Figure 1–2. Structure of a large neuron of the brain, showing its important functional parts.

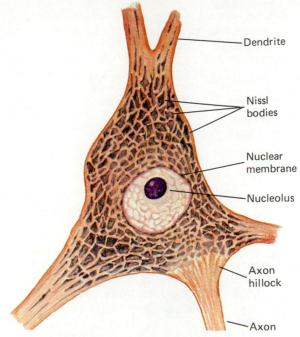

Figure 1–3. The neuronal cell body.

tact point called a *synapse* through which signals can be transmitted from one neuron to the next. When stimulated, the presynaptic terminal releases a minute quantity of a hormone called a *transmitter substance* into the space between the terminal and the membrane of the neuron, and the transmitter substance then stimulates this neuron as well.

Figure 1–3 illustrates more details of the neuronal cell body. It depicts a typical *nucleus* with a very prominent *nucleolus*. Also shown are *Nissl bodies,* which are parts of a specialized endoplasmic reticulum that synthesizes substances required to keep the neuron alive. These substances are transported into the axon and dendrites through a system of microtubules called *neurofibrils.* Finally, note in Figure 1–3 that the axon arises from a conical pole of the cell body called the *axon hillock.*

The Neuroglia

Figure 1–4 illustrates a large neuron of the spinal cord surrounded by its supporting tissue, the neuroglia. The cells in the neuroglia are called *glial cells.* Many of them function similarly to the fibroblasts of connective

tissue; that is, they form fibers that hold the tissue together. But others serve the same function as the Schwann cells of the peripheral nerves to wrap *myelin sheaths* around the larger nerve fibers, thus providing typical *myelinated nerve fibers* that transmit signals at velocities as great as 100 m a second, the same as in peripheral nerves as will be described in Chapter 6. The very small nerve fibers do not have myelin sheaths and therefore are called *unmyelinated fibers,* but even these are insulated from each other by interposition of glial cells between the fibers, much the same way that Schwann cells insulate the unmyelinated nerve fibers from each other in the peripheral nerves.

RECAPITULATION OF THE FUNCTIONAL STRUCTURE OF THE NERVOUS SYSTEM

Figure 1–5 gives a composite view of the structural highlights of the complete nervous system. To the left

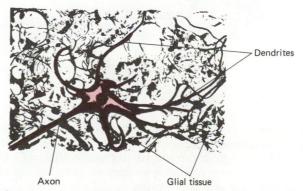

Figure 1–4. A large neuron of the spinal cord surrounded by its supporting tissue, called neuroglia.

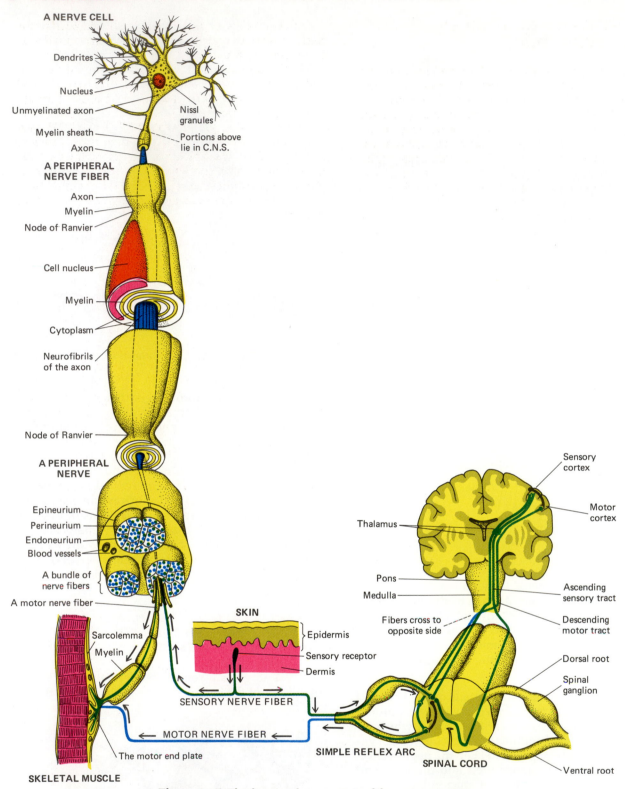

A NERVE CELL

Dendrites

Nucleus

Unmyelinated axon

Nissl granules

Myelin sheath

Portions above lie in C.N.S.

Axon

A PERIPHERAL NERVE FIBER

Axon

Myelin

Node of Ranvier

Cell nucleus

Myelin

Cytoplasm

Neurofibrils of the axon

Node of Ranvier

A PERIPHERAL NERVE

Epineurium

Perineurium

Endoneurium

Blood vessels

A bundle of nerve fibers

A motor nerve fiber

Sarcolemma

Myelin

SKIN

Epidermis

Sensory receptor

Dermis

SENSORY NERVE FIBER

The motor end plate

MOTOR NERVE FIBER

SKELETAL MUSCLE

SIMPLE REFLEX ARC

SPINAL CORD

Sensory cortex

Motor cortex

Thalamus

Pons

Medulla

Fibers cross to opposite side

Ascending sensory tract

Descending motor tract

Dorsal root

Spinal ganglion

Ventral root

Figure 1–5. The functional components of the nervous system.

is a typical neuron, a motor neuron located in the spinal cord that sends a large myelinated nerve fiber to a muscle through a peripheral nerve. To the right is shown the course of both a sensory and a motor nerve fiber entering and leaving the spinal cord through a peripheral nerve. Also shown is an ascending sensory tract in the cord that carries millions of sensory nerve fibers upward from the cord to the brain and a descending motor tract that carries millions of motor fibers downward.

■ ORGANIZATIONAL PLAN FOR THIS TEXT

The ultimate goal of this text is to explain how the nervous system performs its multiple roles as a sensing, thinking, and controlling system. Obviously, this requires detailed knowledge of the structure of each part of the nervous system as well as an understanding of how each part functions. Therefore, we will begin in Section II (Chapters 2 through 4) with a survey of the *gross anatomy* of the nervous system, principally to provide the basic terminology and interrelationships needed for later discussion. Then, throughout the remainder of the text, the *functional anatomy* of each organ system will be presented, along with its *physiology.*

In Section III (Chapters 5 and 6), we will present the basic biophysics and physiology of signal transmission in nerves and muscle.

The functional anatomy and physiology of the central nervous system and its peripheral sensory and motor connections will be given in Sections IV through VI (Chapters 7 through 23).

Finally, Section VII (Chapters 24 through 29) will discuss the many ways in which the nervous system controls most of the functional systems of the nonnervous portions of the body, such as the circulation, respiration, gastrointestinal function, body temperature, most hormonal secretion, and sexual and reproductive functions.

GROSS ANATOMY

Gross Anatomy of the Nervous System:

I. General Divisions of the Brain; the Cerebrum; the Diencephalon

■ THE BRAIN AND ITS DIVISIONS

The brain is the portion of the nervous system that is located in the cranial cavity. Figure 2–1 illustrates a lateral view of the brain, Figure 2–2 an inferior view (its ventral surface), and Figure 2–3 a sagittal view as seen in the median plane of the brain. Unfortunately, several different terminologies are used to describe the different parts of the brain, three of which are listed in Table 2–1. The most widely used terminology in medical circles is given in the right-hand column of the table in which the brain is divided into six separate parts: (1) the *cerebrum,* (2) the *diencephalon,* (3) the *mesencephalon,* (4) the *cerebellum,* (5) the *pons,* and (6) the *medulla oblongata,* usually called simply the "medulla."

It is important to recognize the relationship of this widely used terminology to the classical terminology given in the left-hand column of the table and also to several anglicized classical terms given in the middle column. The cerebrum is the same as the *telencephalon,* and the telencephalon and diencephalon together constitute the *prosencephalon,* or *forebrain,* which is the large massive portion of the brain filling the anterior and superior three fourths of the cranial cavity.

The mesencephalon, also called the *midbrain,* is a minute portion of the brain located at the base of the forebrain, as illustrated in Figures 2–2 and 2–3. Yet, despite its small size, it is the only connecting link between the forebrain and all the lower portions of the brain and spinal cord.

The cerebellum, pons, and medulla all lie in the posterior fossa of the cranial cavity, and together they constitute the *rhombencephalon,* or *hindbrain.*

One can see from the figures that the major mass of the brain is the cerebrum, and it is also clear that the next largest portion of the brain is the cerebellum. This might make one think that the other four parts of the brain—the diencephalon, the mesencephalon, the pons, and the medulla—are of relatively little importance. We shall see later that these parts are absolutely crucial to the maintenance of nervous function, indeed, far more so than any equivalent mass of the cerebrum or cerebellum.

THE CEREBRUM

The Cerebral Hemispheres and the Corpus Callosum. For the next few moments let us study the external views of the cerebrum in Figures 2–1, 2–2, and 2–3 and also a horizontal section of the cerebrum shown in Figure 2–4. The first notable feature that one observes about the cerebrum is that it is composed of two large bilateral masses, the *cerebral hemispheres,* illustrated in Figures 2–1, 2–2, and 2–3. These two hemispheres are connected with each other through several bundles of nerve fibers, the two most important of which are discussed here:

1. The *corpus callosum* is illustrated in sagittal section in Figure 2–3 and in horizontal section in Figure 2–4. Note that the corpus callosum is a broad band of fibers extending almost half the length of the cerebral hemisphere. Its importance is made apparent by its extremely large number of fibers, about 20 million.

2. The *anterior commissure* is also shown in Figure 2–3. This is a much smaller bundle, probably no more than a million fibers. It is located several centimeters below the anterior third of the corpus callosum, and it interconnects mainly the anterior and medial portions of the two temporal lobes.

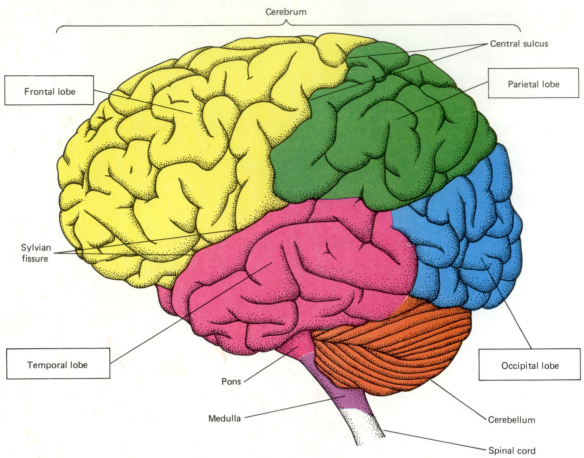

Figure 2–1. Left lateral view of the brain, showing the principal divisions of the brain and the four major lobes of the cerebrum.

Corresponding points in almost all areas of the two hemispheres interconnect with each other in both directions via the fibers in these two bundles, which allows continuous communication between the two hemispheres. When the corpus callosum and anterior commissure are destroyed, each of the two hemispheres functions as a separate brain, even thinking separate thoughts and causing separate reactions in the two sides of the body.

Cerebral Convolutions, Fissures, and Sulci. The next distinctive feature of the cerebrum is the folds in its surface. These are called *cerebral convolutions,* and each convolution is called a *gyrus.* The grooves between the gyri are called either *fissures* or *sulci,* the larger and deeper ones generally being called fissures whereas the great majority, less deep, are the sulci. Four of the principal fissures or sulci are illustrated in Figures 2–1 and 2–2. They are listed here:

1. The *longitudinal fissure* separates the two cerebral hemispheres from each other in the midsagittal plane of the basin.
2. The *central sulcus* extends in an approximate inferosuperior direction on the lateral side of each hemisphere and divides the cerebrum approximately into an anterior half and a posterior half.
3. The *lateral fissure,* also called the *Sylvian fissure,* extends along the lateral aspect of each cerebral hemisphere for about half its length.

4. The *parieto-occipital sulcus* originates from the side of the longitudinal fissure about one quarter the distance anterior to the posterior pole of the hemisphere and then extends laterally and anteriorly for about 5 cm.

To some extent, these fissures and sulci demarcate separate functional parts of the cerebrum, as we shall discuss.

The Lobes of the Cerebrum. Figures 2–1 through 2–4 illustrate that the cerebrum is divided into four major *lobes* and a fifth minor one. The major lobes are (1) the *frontal lobe,* (2) the *parietal lobe,* (3) the *occipital lobe,* and (4) the *temporal lobe;* the minor lobe is (5) the *insula.*

The central sulcus separates the frontal lobe from the parietal lobe. The lateral fissure demarcates the frontal lobe and the anterior portion of the parietal lobe from the temporal lobe. And the parieto-occipital sulcus separates the superior part of the parietal lobe from the occipital lobe. The separation between the temporal lobe and the occipital lobe is less distinct. We shall see later that the area where the parietal, temporal, and occipital lobes all meet is the major area of the brain where integration of sensory information occurs, with sensory information from the body feeding into this area through the parietal lobe, visual information through the occipital lobe, and auditory information through the temporal lobe. In contrast, we

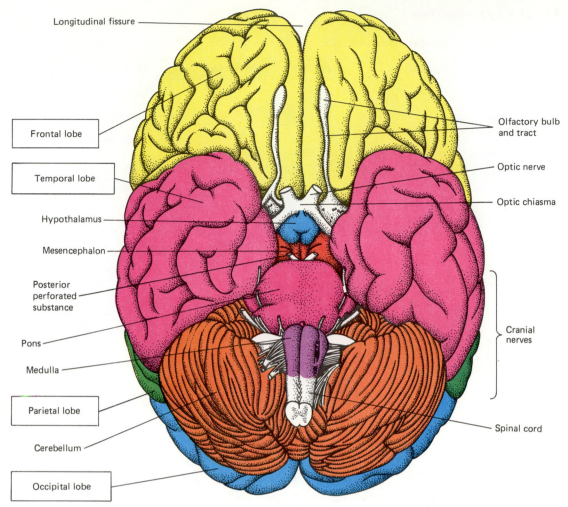

Longitudinal fissure

Frontal lobe

Temporal lobe

Hypothalamus

Mesencephalon

Posterior
perforated
substance

Pons

Medulla

Parietal lobe

Cerebellum

Occipital lobe

Olfactory bulb
and tract

Optic nerve

Optic chiasma

Cranial
nerves

Spinal cord

Figure 2 – 2. Basal view of the brain.

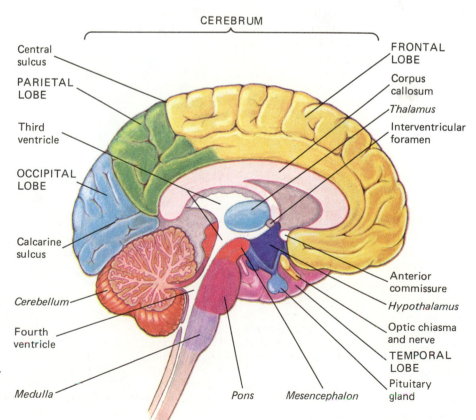

CEREBRUM

Central
sulcus

PARIETAL
LOBE

Third
ventricle

OCCIPITAL
LOBE

Calcarine
sulcus

Cerebellum

Fourth
ventricle

Medulla

FRONTAL
LOBE

Corpus
callosum

Thalamus

Interventricular
foramen

Anterior
commissure

Hypothalamus

Optic chiasma
and nerve

TEMPORAL
LOBE

Pituitary
gland

Pons   *Mesencephalon*

Figure 2 – 3. Medial view of the left half
of the brain, showing especially the relationship of the cerebrum to the brain
stem and cerebellum.

TABLE 2–1 Divisions of the Brain

Classical Terminology	Anglicized Terminology	Most Widely Used Terminology
Encephalon	Brain	Brain
Prosencephalon		
Telencephalon		Cerebrum
Diencephalon		Diencephalon (or thalamus, hypothalamus, and surroundings)
Mesencephalon	Midbrain	Mesencephalon
Rhombencephalon	Hindbrain	
Metencephalon		
Cerebellum		Cerebellum
Pons		Pons
Myelencephalon		
Medulla oblongata		Medulla (or medulla oblongata)

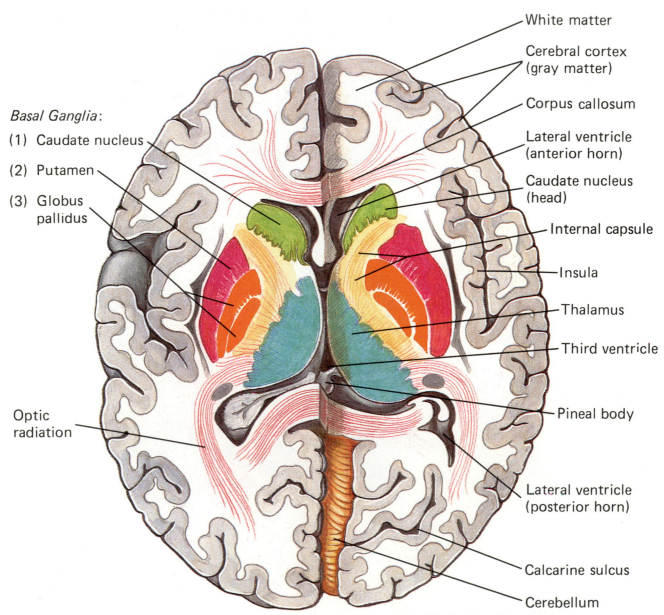

Basal Ganglia:

(1) Caudate nucleus

(2) Putamen

(3) Globus pallidus

White matter

Cerebral cortex (gray matter)

Corpus callosum

Lateral ventricle (anterior horn)

Caudate nucleus (head)

Internal capsule

Insula

Thalamus

Third ventricle

Pineal body

Optic radiation

Lateral ventricle (posterior horn)

Calcarine sulcus

Cerebellum

Figure 2–4. A horizontal section through the cerebrum at the level of the basal ganglia and thalamus.

shall see that the frontal lobe is concerned primarily with control of muscle movement and also with certain types of thinking processes.

The insula cannot be seen from the surface of the cerebrum. Instead, it lies deep in the lateral fissure. The horizontal section of the brain in Figure 2–4 shows that the lateral sulcus has a broad flat bottom covered by overhanging lips from the frontal, parietal, and temporal lobes. This flat bottom is the *insula,* and the lips are called *opercula* of the other lobes. Unfortunately, we know little about the function of the insula except that it probably acts as part of the limbic system (which will be discussed later) to help control behavior.

Table 2–2 summarizes the structures of the cerebrum and also gives its functional parts, which will be discussed further in the following sections of this chapter.

THE CEREBRAL CORTEX—GRAY MATTER AND WHITE MATTER

Now let us look inside the cerebrum to see how its internal structure is organized. Figure 2–4 shows a horizontal section through the cerebrum. It is composed of areas that appear gray to the naked eye, called *gray matter,* and other areas that appear white, called *white matter.* The gray matter is collections of great numbers of neuronal cell bodies that all together give it its grayish hue. The white matter is composed of great bundles of nerve fibers leading to or from the nerve cells in the gray matter; its white appearance is caused by the brilliant white color of the myelin sheaths of nerve fibers.

Figure 2–4 shows especially that a thin shell of gray matter covers the entire surface of the cerebrum, including the fissures and sulci. This is the *cerebral cortex.* One of the principal advantages of having the many fissures and sulci is that they triple the total area of the cerebral cortex; the exposed surface area of the brain is only about two thirds of 1 ft^2 or 600 cm^2, but the total area of the cerebral cortex is about 2 ft^2 or 1800 cm^2.

The cerebral cortex is the portion of the brain most frequently associated with the thinking process, even though it cannot provide thinking without simultaneous action of most deep structures of the brain. Yet, the cerebral cortex is the portion of the brain in which essentially all of our memories are stored, and it is also the area most responsible for our ability to acquire our many muscle skills. We still do not know the basic physiological mechanisms by which the cerebral cortex stores either memories or knowledge of muscle skills, but what we do know about these will be discussed in Chapter 19.

In most areas, the cerebral cortex is about 6 mm thick, and all together it contains an estimated 50 to 80 billion nerve cell bodies. Also, perhaps a billion nerve fibers lead away from the cortex as well as comparable numbers leading into it, passing to other areas of the cortex, to and from deeper structures of the brain, and some all the way to the spinal cord.

TABLE 2–2 The Cerebrum

	Location and Function
The Cerebral Lobes	
Frontal	Anterior superior
Parietal	Superior midportion
Occipital	Posterior
Temporal	Lateral
Insula	Deep in lateral fissure
Principal Fissures and Sulci	
Longitudinal fissure	Separates the cerebral hemispheres
Central sulcus	Separates frontal and parietal lobes
Lateral fissure	Separates temporal lobe from frontal lobe and part of the parietal lobe
Parieto-occipital sulcus	Separates superior part of parietal lobe from superior part of occipital lobe
Principal Structural Parts	
Cerebral cortex (gray matter)	Thin layer on surface composed mainly of billions of neuronal cell bodies
Deep nuclei (also gray matter) Basal ganglia	Most important: (1) caudate nucleus, (2) putamen, (3) globus pallidus
Some of the limbic structures	
White matter	Composed of billions of nerve fibers, mainly myelinated
Functional Areas	
Motor areas	Located in posterior frontal lobe
Motor cortex	Controls discrete muscle activities
Premotor cortex	Controls patterns of coordinate muscle contractions
Broca's area	Controls speech
Somesthetic cortex	Parietal lobe detects tactile and proprioceptor sensations
Visual area	Occipital lobe detects visual sensations
Auditory area	Superior temporal lobe detects auditory sensations
Wernicke's area	Superior posterior temporal lobe analyzes sensory information from all sources
Short-term memory area	Inferior portions of temporal lobe
Prefrontal area	Anterior half of frontal lobe—"elaboration of thought"

Functional Areas of the Cerebral Cortex

Until World War I we knew the function of only a few areas of the cerebral cortex. But, at that time soldiers with bullet wounds in discrete parts of the brain were studied systematically for brain functional changes. Also, in more recent years neurosurgeons and neurologists have carefully documented changes in brain function caused by tumors or other specific lesions. Figure 2–5 illustrates the principal functional areas of the cerebral cortex that have been determined by these studies. These are as follows:

The Motor Area: Motor Cortex, Premotor Cortex, and Broca's Area. The motor area lies in front of the central sulcus and occupies the posterior half of the frontal lobe. It, in turn, is divided into three subdivisions, the motor cortex, the premotor cortex, and Broca's area, all of which are concerned with the control of muscle activity.

The *motor cortex,* located in a strip about 2 cm wide immediately anterior to the central sulcus, controls the specific muscles throughout the body, especially the muscles that cause fine movements, such as the finger and thumb motions and the lip and mouth motions for talking and eating, and to a much lesser extent the fine motions of the feet and toes.

The *premotor cortex,* located anterior to the motor cortex, elicits coordinate movements that involve either sequences of individual muscle movements or combined movements of a number of different muscles at the same time. It is in this area that much of one's knowledge is stored for controlling learned skilled movements such as the special movements required for playing an athletic game.

Broca's area, located anterior to the motor cortex at the lateral margin of the premotor cortex, controls the coordinate movements of the larynx and mouth to produce the words of speech. This area functions as the person's *speech center* in only one of the two cerebral hemispheres, in the left hemisphere in about 19 out of 20 persons, including all right-handed persons and one half of all left-handed persons.

The Somesthetic Sensory Area. Somesthetic sensations are the sensations from the body such as touch, pressure, temperature, and pain. One can see in Figure 2–5 that the somesthetic sensory area occupies the entire parietal lobe.

Note that this sensory area is divided into a *primary area* and a *secondary area.* This is true also of all the other sensory areas. The primary somesthetic sensory area is the portion of the cortex that receives signals directly from the different sensory receptors located throughout the body. In contrast, signals to the secondary somesthetic sensory area are partly processed in deep brain structures or in the primary somesthetic area before being relayed to the secondary area. The primary area can distinguish the specific types of sensation in discrete regions of the body. The secondary area serves mainly to interpret the sensory signals, not to distinguish among them, such as interpreting that a hand is feeling a chair, a table, a ball, or so forth.

The Visual Area. Figure 2–5 shows that the visual area occupies the entire occipital lobe. Most of the *primary area for vision* is located on the medial surface of the cerebral hemisphere along the course of the calcar-

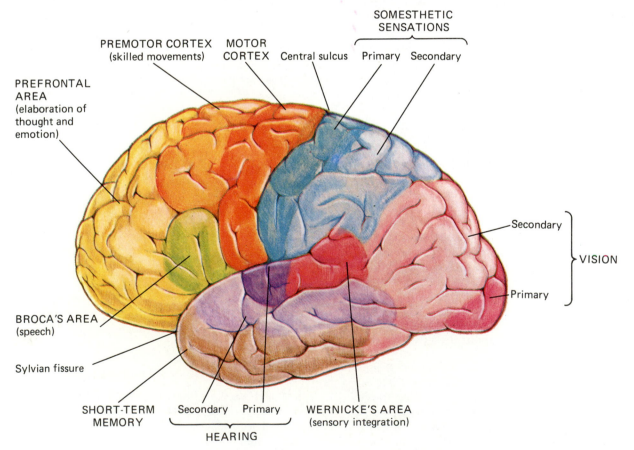

Figure 2–5. The functional areas of the cerebral cortex.

ine sulcus (Figs. 2–3 and 2–4), but a small portion of the primary visual area projects over the outer pole of the occipital lobe as shown in Figure 2–5. This primary area detects specific light and dark spots as well as orientations of lines and borders in the visual scene. The *secondary visual areas* occupy the remainder of the occipital lobe, and their function is to interpret the visual information. For instance, it is in these areas that the meanings of written words are interpreted.

The Area for Hearing (the Auditory Area). The hearing area is located in the upper half of the anterior two thirds of the temporal lobe. The *primary auditory area* is located in the midportion of the superior temporal gyrus. It is here that specific tones, loudness, and other qualities of sound are detected. The *secondary areas* occupy the other parts of the hearing area. It is in these areas that the meanings of spoken words are interpreted; and portions of these areas are also important for music recognition.

Wernicke's Area for Sensory Integration. Wernicke's area lies in the posterior part of the superior temporal lobe at the point where the parietal and occipital lobes both come in contact with the temporal lobe. It is here that sensory signals from all of the three sensory lobes—the temporal, occipital, and parietal lobes—all come together. This area is exceedingly important for interpreting the ultimate meanings of almost all the different types of sensory information such as the meanings of sentences and thoughts, whether they be heard, read, felt, or even generated within the brain itself. Therefore, destruction of this area of the brain causes extreme loss of thinking ability. This area is well developed only in one of the cerebral hemispheres, usually in the left hemisphere. This unilateral development of Wernicke's area prevents confusion of thought processes between the two halves of the brain.

The Short-Term Memory Area of the Temporal Lobe. The lower half of the temporal lobe seems to be mainly of importance for storing short-term memories, memories that last from a few minutes to several weeks.

The Prefrontal Area. The prefrontal area occupies the anterior half of the frontal lobe. Its function is less well defined than that of any other part of the cerebrum. It has been removed in many psychotic patients to bring them out of depressive states. These persons can still function without the prefrontal areas. However, they lose their ability to concentrate for long periods of time and also their abilities to plan for the future or to think through problems. Therefore, this area is said to be important for elaboration of thought.

We shall discuss in further detail the intellectual function of the cerebral cortex in Chapter 19.

THE BASAL GANGLIA

The horizontal sectional view of the brain in Figure 2–4 shows several areas of gray matter, called *nuclei,* located deep inside the brain. A nucleus is a collection of nerve cell bodies congregated into a cohesive area. Two separate groups of nuclei are shown in the figure: (1) the *basal ganglia,* which are part of the cerebrum; and (2) the *thalamus,* which is composed of multiple small nuclei and is part of the diencephalon, to be described later in this chapter.

Figures 2–6 and 2–7 show still other views of the brain with the basal ganglia high-lighted, Figure 2–6 a coronal view and Figure 2–7 a three-dimensional view to show the locations of the basal ganglia in the cerebral masses and also to show their relationships to the thalamus. The three most important of the basal ganglia are the (1) *caudate nucleus,* (2) *putamen,* and (3)

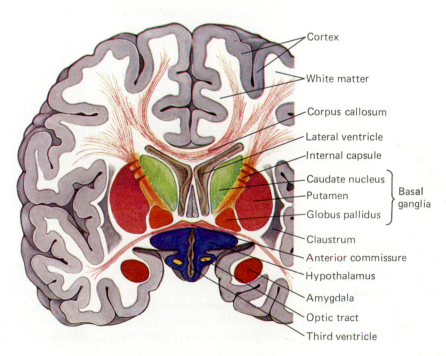

Cortex
White matter
Corpus callosum
Lateral ventricle
Internal capsule
Caudate nucleus
Putamen — } Basal ganglia
Globus pallidus
Claustrum
Anterior commissure
Hypothalamus
Amygdala
Optic tract
Third ventricle

Figure 2–6. A coronal section of the cerebrum in front of the thalamus, showing especially the basal ganglia.

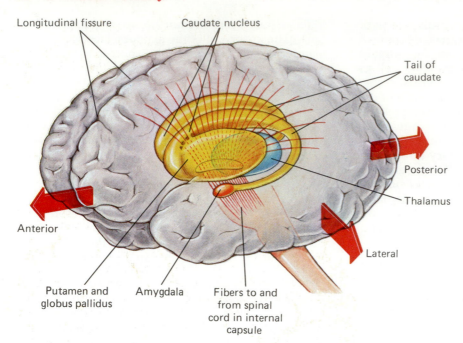

Longitudinal fissure

Caudate nucleus

Tail of caudate

Posterior

Thalamus

Anterior

Lateral

Putamen and globus pallidus

Amygdala

Fibers to and from spinal cord in internal capsule

Figure 2–7. Relationship of the basal ganglia to the thalamus, shown in a three-dimensional view.

globus pallidus. In addition to these three, anatomists also consider the *claustrum* and the *amygdala* to be basal ganglia. However, the function of the claustrum is unknown, and the amygdala functions as part of the limbic system, which we will discuss later in this chapter; it has very little functional relationship to the remainder of the basal ganglia.

On the other hand, the basal ganglia of the cerebrum function in very close association with the *subthalamus* of the diencephalon and the *substantia nigra* and *red nucleus* of the mesencephalon. Therefore, physiologists frequently consider these three bodies also to be part of the basal ganglial system.

The Three-Dimensional Location of the Basal Ganglia in the Cerebrum. Now, let us study in more detail the anatomical locations of the principal basal ganglia. In Figure 2–4 in the horizontal section of the brain, note that the caudate nucleus, putamen, and globus pallidus lie anterior and lateral to the thalamus. Figure 2–6 shows a coronal section of the brain slightly in front of the anterior end of the thalamus, approximately in the motor region of the cerebrum. In this area the three basal ganglia lie astride one of the major fiber pathways of the cerebrum, called the *internal capsule.* This pathway, which will be described later, is the principal communication link between the cortex and the lower regions of the brain and spinal cord. Many of the fibers in the internal capsule originate in the motor cortex and premotor cortex, and branches of them enter the basal ganglia. Portions of the motor signals are then processed and relayed through these ganglia rather than passing directly from the cerebral cortex to the cord.

Note also in Figure 2–6 the relationship of the basal ganglia to the anterior portions of the two *lateral ventricles,* which are fluid-filled cavities in the cerebrum.

The ventricles lie respectively superior and medial to the caudate nucleus in each of the two cerebral hemispheres.

Finally, the three-dimensional diagram in Figure 2–7 shows especially the relationship of the basal ganglia to the thalamus. One can see the central location of the thalamus in the basal portion of the brain and the location of the basal ganglial system mainly anterior and lateral to the thalamus, but note also the long tail of the caudate nucleus that curls posteriorly through the parietal lobe and thence laterally and inferiorly into the temporal lobe. The amygdala lies at the tip of the caudate nucleus tail in the temporal lobe.

Function of the Basal Ganglia. If the cerebral cortex is removed in a cat but without removing the basal ganglia, the cat can still perform most of its normal motor activities, including walking, fighting, arching its back, spitting, and almost any other movement. On the other hand, in the human being a similar loss of the cerebral cortex, but with the basal ganglia intact, leaves the person with only crude motor activities such as gross trunk movements and movements of the limbs with a stiff-legged, uncontrolled walking.

Putting this information together, one can deduce that a major function of the basal ganglia in the human being is to control the background gross body movements, whereas the cerebral cortex is necessary for performance of the more precise movements of the arms, hands, fingers, and feet. When the hand is performing some precise activity that requires a background stance of the body, the basal ganglia provide the body movements while the cerebral cortex provides the precise movements.

To achieve the high degree of coordination that is required among the muscles of the body during most motor functions, a very complex circuitry of nerve fibers interconnects (1) the cerebral cortex and basal ganglia in the cerebrum, (2) the thalamus and subthalamus in the diencephalon, (3) the red nucleus and the substantia nigra in the mesencephalon, and (4) the cerebellum in the hindbrain.

THE WHITE MATTER OF THE CEREBRUM

In almost all areas of the cerebrum besides the cerebral cortex and basal ganglia one finds white matter. The white matter is composed nearly exclusively of nerve fibers that are usually organized into specific bundles of fibers called *fiber tracts*. Three of the principal fiber tracts, each containing millions of fibers, are illustrated in Figure 2–4. These are the following:

1. The *corpus callosum,* which was discussed earlier, connects the respective areas of the cerebral cortex in each cerebral hemisphere with corresponding areas in the opposite hemisphere. The corpus callosum is also seen in sagittal section in Figure 2–3 and in coronal section in Figure 2–6.

2. The *optic radiation,* shown in Figure 2–4, passes from the lateral geniculate body of the thalamus back to the calcarine sulcus area of the occipital lobe. This is the final relay pathway for transmission of visual signals from the eyes to the cerebral cortex.

3. The *internal capsule,* shown in Figure 2–4, is found in the areas between the thalamus, the caudate nucleus, and the putamen. It is through this internal capsule that most signals between the cerebral cortex and the lower brain and spinal cord are transmitted.

Figure 2–8 illustrates even more vividly the gray and white matter of the cerebrum. Shown too is the great mass of nerve fibers that extends upward from the internal capsule to the cerebral cortex through an extensive radiation called the *corona radiata.*

THE DIENCEPHALON

The *diencephalon* is also called the *betweenbrain.* In primitive animals the diencephalon is a nodular structure, distinct from the remainder of the brain, that links the telencephalon (the cerebrum) with the mesencephalon (the midbrain). In the human being, the diencephalon still provides a similar linkage between the cerebrum and the lower parts of the brain, but anatomically it is so tightly fused with the basal portions of the cerebrum that it is difficult to demarcate its boundaries with the cerebrum. However, the diencephalon is defined as those structures that surround the third ventricle (another fluid-filled cavity in the brain, shown in Figure 2–9 and discussed in the following chapter). Table 2–3 summarizes the important structures and functions of the diencephalon. The most important of the structures are the *thalamus* and *hypothalamus.* Both of these are composed of multiple nuclei that perform many different important nervous functions. In addition to these two, smaller nuclear areas of the diencephalon, located posterior and inferior to the thalamus, are the *epithalamus* and *subthalamus.*

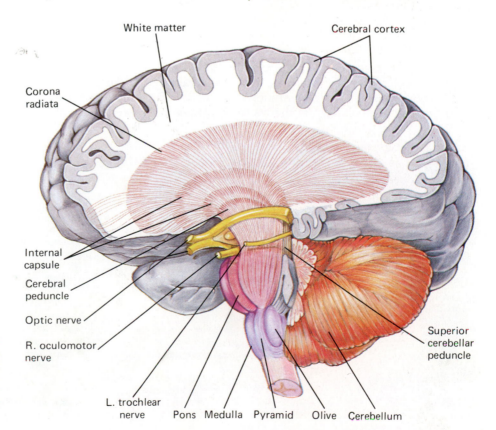

White matter

Cerebral cortex

Corona radiata

Internal capsule

Cerebral peduncle

Optic nerve

R. oculomotor nerve

L. trochlear nerve

Pons   Medulla   Pyramid   Olive   Cerebellum

Superior cerebellar peduncle

Figure 2–8. A deep dissection of the cerebrum showing the radiating nerve fibers, the corona radiata, that conduct signals in both directions between the cerebral cortex and the lower portions of the central nervous system.

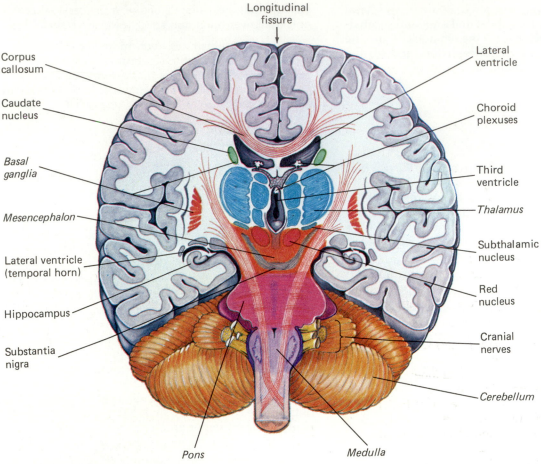

Figure 2—9. A coronal view of the cerebrum looking from anteriorly backward. This section was made immediately anterior to the lower brain stem and through the middle of the thalamus.

The Thalamus

We have already seen in several figures that the thalamus is located in the very center of the brain, enshrouded on all sides except inferiorly by the cerebrum. Figure 2–9 illustrates another view of the thalamus and its location in a coronal section of the brain. Several specific features of the thalamus are described here:

First, the thalamus comprises a number of separate discrete nuclei, as illustrated by the multiple blue-colored areas in the figure.

TABLE 2–3 The Diencephalon and Limbic System

Structure	Location	Function
Diencephalon		
Thalamus	Central base of brain	Relays sensory signals to cortex, sensory analytic functions
Hypothalamus	Inferior to anterior thalamus	Controls internal body functions, stimulates autonomic nervous system
Subthalamus	Inferior to posterior thalamus	Functions with basal ganglia to control subconscious muscle activity
Epithalamus	Posteroinferior to thalamus	Function unknown; includes pineal gland
Limbic System		
Amygdala	Deep inside anterior end of each temporal lobe	Controls behavior for each social occasion
Hippocampus	Medial border of each cerebral hemisphere	Determines which sensory information will be committed to memory
Mammillary body	Posterior to hypothalamus	Perhaps helps to determine mood and degree of wakefulness
Septum pellucidum	Midline of cerebrum anterior and superior to hypothalamus	Perhaps helps to control temper and autonomic nervous system
Limbic cortex: Cingulate gyrus, cingulum, insula, and parahippocampal gyrus	Ring of cerebral cortex in medial part of cerebrum around deeper limbic structures	Conscious components in the control of behavior

Second, the thalamus rests directly on top of the mesencephalon (also called the "midbrain"); almost all signals from the midbrain and other lower regions of the brain, as well as from the spinal cord, are relayed through synapses in the thalamus before proceeding to the cerebral cortex.

Third, the thalamus has numerous two-way connections with all parts of the cerebral cortex, carrying a continual traffic of signals from the thalamus to the cortex and also from the cortex to the thalamus.

Fourth, the thalamus lies in close apposition to the basal ganglia. In fact, the thalamus relays many signals from other lower regions of the brain and spinal cord directly to the basal ganglia. And, in turn, the thalamus also functions as a relay station for signals from some of the basal ganglia to the cortex.

In essence, then, the thalamus is a chief traffic relay station for directing sensory and other signals to appropriate points in both the cerebral cortex and the deeper areas of the cerebrum as well. Some examples of the different types of signals that are relayed through the thalamus include:

1. all the *somesthetic sensory signals* from the body (touch, pressure, pain, temperature, and so forth) to the somesthetic cortex of the parietal lobe;

2. *visual signals* to the calcarine sulcus area of the occipital cortex (the part of the thalamus that relays these signals is sometimes classified as the metathalamus, which is the posterior end of the thalamus);

3. *auditory signals* to the superior temporal gyrus (also relayed by the metathalamus); and

4. *muscle control signals* from the cerebellum, mesencephalon, and other areas of the lower brain stem to the motor cortex and basal ganglia.

Anatomical Relationship of the Thalamus to the Ventricles. Note especially in Figure 2–9 the relationship of the thalamus to several of the ventricles: (1) the two lateral ventricles lie immediately above the two lateral halves of the thalamus and (2) the third ventricle bisects the thalamus into two halves. Each half of the thalamus functions separately with the cerebral hemisphere on the same side, and there is very little direct communication between the two halves of the thalamus.

The Sensory Interpretive Function of the Thalamus. In lower animals, at about the reptile stage, the cerebral cortex is not developed to a great extent, but the thalamus is an established part of the brain. In these animals, the thalamus plays a much greater role in sensory interpretation than it does in man. But, even in man some of its sensory interpretive abilities still persist. This is especially true for pain sensation, for a person can lose most if not all of the somesthetic sensory areas of his cerebral cortex and still retain much if not most of his ability to perceive pain. This is in keeping with the fact that pain is one of the most primitive of our sensations and also that the thalamus is a more primitive portion of the brain than is the cerebrum.

The Thalamocortical Relationship. In addition to the relay pathways through the thalamus to the cerebral cortex, there are also innumerable two-way connections between the thalamus and all areas of the cortex, with nerve fibers going in both directions. Figure 2–10 illustrates by means of a color code the areas of the thalamus that connect with specific areas of the cerebral cortex. For example, the posterior portion of the thalamus (lateral geniculate body and pulvinar) has two-way connections with the occipital lobe of the cortex. The superior medial part of the thalamus (nucleus medialis dorsalis) connects with the prefrontal area of the frontal lobe. And the posterolateral portion of the ventral nucleus (nucleus ventralis posterior lateralis) connects with the primary somesthetic area of the parietal cortex, and so forth.

But what are the purposes of these two-way connections? First, without the thalamus, the cortex is useless. It is the thalamus that drives the cortex to activity, which is another function of the thalamus in addition to relaying signals to the cortex from other areas of the brain and spinal cord. In fact, one can consider most of the cerebral cortex to be mainly an outgrowth of the thalamus; the cortex provides a great memory storehouse to function at the beck and call of the control centers of the thalamus.

The Hypothalamus

Several figures presented earlier in this chapter illustrated a small structure in the middle of the base of the brain called the *hypothalamus*. Study especially Figures 2–2 and 2–3 and locate this structure, but look carefully or otherwise you will miss it. Its small size belies its importance, for it is a major center of the brain for controlling internal body functions.

Figure 2–11 illustrates an enlarged internal view of the hypothalamus, showing in third dimension various nuclei of the hypothalamus. Note that the hypothalamus lies anterior to the red nucleus, which is in the uppermost part of the mesencephalon, and it also lies immediately inferior to the anterior end of the thalamus. There are especially abundant nerve pathways between the hypothalamus and the anterior thalamus and also between the hypothalamus and the mesencephalon.

Some of the Functions of the Nuclei of the Hypothalamus. At many points in this text we shall discuss the importance of one or more of the hypothalamic nuclei for control of some of the internal body functions, and more details of the function of the hypothalamus as a whole will be presented in Chapter 20. However, let us list here some of the important functions of a few of the hypothalamic nuclei.

The *preoptic nucleus*, located anteriorly, is primarily concerned with body temperature control.

The *supraoptic nucleus*, located anteriorly and inferiorly, lying immediately above the optic nerves, controls the secretion of *antidiuretic hormone;* this hormone in turn helps to control the concentration of electrolytes in the body fluids.

The *medial nuclei* of the hypothalamus, when stimulated, gives a person a sense of satiety (that is, he feels satisfied, especially satisfied for food).

Stimulation of the most *lateral regions of the hypo-*

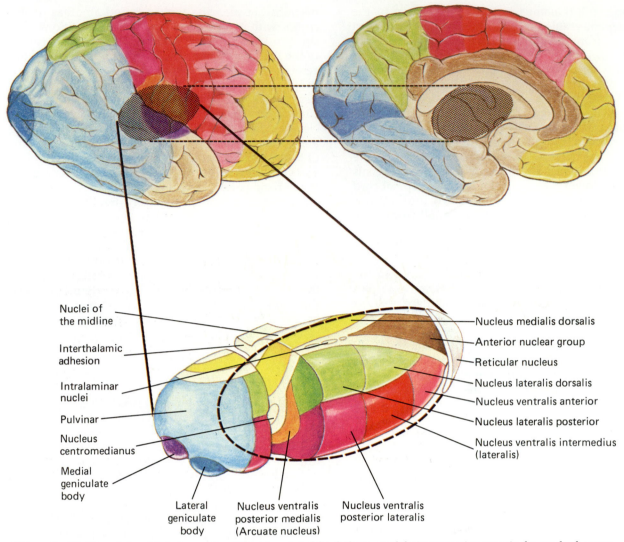

Nuclei of
the midline

Interthalamic
adhesion

Intralaminar
nuclei

Pulvinar

Nucleus
centromedianus

Medial
geniculate
body

Nucleus medialis dorsalis

Anterior nuclear group

Reticular nucleus

Nucleus lateralis dorsalis

Nucleus ventralis anterior

Nucleus lateralis posterior

Nucleus ventralis intermedius
(lateralis)

Lateral
geniculate
body

Nucleus ventralis
posterior medialis
(Arcuate nucleus)

Nucleus ventralis
posterior lateralis

Figure 2–10. The relationship between the different nuclei of the thalamus and their connecting areas in the cerebral cortex. (Reprinted from Warwick and Williams: Gray's Anatomy, 35th British Edition. Philadelphia, W.B. Saunders, 1973.)

thalamus causes a person to become very hungry, and stimulation anteriorly in the lateral hypothalamus causes a person to become very thirsty.

Stimulation of the *posterior hypothalamus* excites the sympathetic nervous system throughout the body, increasing the overall level of activity of many internal parts of the body, especially increasing heart rate and causing blood vessel constriction.

Finally, stimulation of different areas of the hypothalamus causes its neurons to secrete several hormones called *releasing hormones* that are carried in the venous blood directly to the anterior pituitary gland; here they then cause secretion of the anterior pituitary hormones. The pituitary hormones in turn control such varied activities of the body as the metabolism of carbohydrates, metabolism of proteins, metabolism of fats, functions of the sex glands, and several other functions.

Thus, one cannot help but be impressed by the global importance of this small area of the brain, the hypothalamus, and its multiple roles in the control of our bodies. Therefore, we will be discussing its functions at many points in this text.

THE LIMBIC SYSTEM

The word "limbic" means border; the *limbic system,* illustrated in Figure 2–12 and summarized in Table 2–3, comprises the border structures of the cerebrum and the diencephalon that mainly surround the hypothalamus. This limbic system functions especially to control our emotional and behavioral activities. Some of the important parts of the limbic system are as follows:

1. The *amygdala* (also called the *amygdaloid body*) is a small nuclear structure located deep inside each anterior temporal lobe and considered by anatomists to be one of the basal ganglia. However, it functions very closely with the hypothalamus, not with the usual basal ganglia. It is believed that the amygdala helps to control the appropriate behavior of the person for each type of social situation.

2. The *hippocampus,* one on each side, is a primitive portion of the cerebral cortex that lies along the medial-most border of the temporal lobe and folds upward and inward to form the inferior surface of the

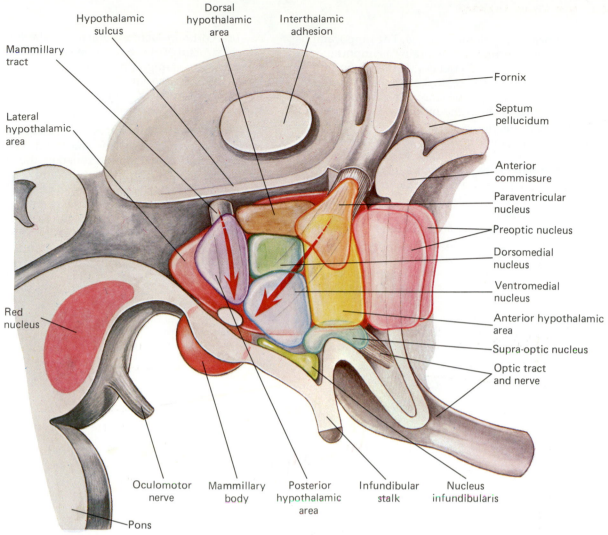

Figure 2–11. A three-dimensional view of one side of the hypothalamus, showing its principal nuclei. (From Nauta and Haymaker: The Hypothalamus. Springfield, Ill., Charles C Thomas, 1969.)

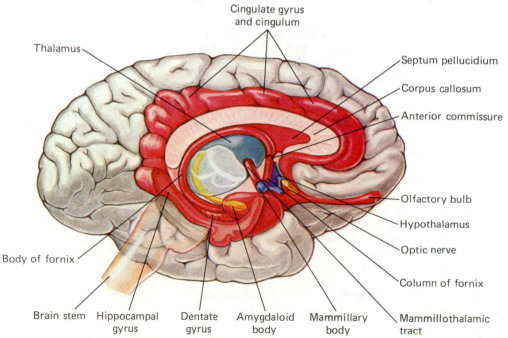

Figure 2–12. The limbic system in the midportion of the cerebrum. (From Warwick and Williams: Gray's Anatomy, 35th British Edition. Philadelphia, W.B. Saunders, 1973.)

inferior horn of the lateral ventricle. The hippocampus is believed to interpret for the brain the importance of most of our sensory experiences. If the hippocampus determines an experience to be important enough, then the experience will be stored as a memory in the cerebral cortex. Without the hippocampus, a person's ability to store memories becomes very deficient.

3. The *mammillary bodies* lie immediately behind the hypothalamus and function in close association with the thalamus, hypothalamus, and brain stem to help control many behavioral functions such as the person's degree of wakefulness and perhaps also his feeling of well-being.

4. The *septum pellucidum* lies anterior to the thalamus, superior to the hypothalamus, and between the basal ganglia in the median plane of the cerebrum. Stimulation in different parts of this septum can cause many different behavioral effects, including the phenomenon of rage.

5. The *cingulate gyrus,* the *cingulum,* the *insula,* and the *parahippocampal gyrus* all together form a ring of cerebral cortex in each cerebral hemisphere around the deeper structures of the limbic system described in the previous few paragraphs. This ring of cortex is believed to allow association between conscious cerebral behavioral functions and subconscious behavioral functions of the deeper limbic system.

Signals from the limbic system leading into the hypothalamus can modify any one or all of the many internal bodily functions controlled by the hypothalamus. And signals feeding from the limbic system into the mesencephalon can control such behavior as wakefulness, sleep, excitement, attentiveness, and even rage or docility. Yet, the precise manner in which the different parts of the limbic system function together to control all these emotional and behavioral functions of the body is still only slightly understood.

REFERENCES

See References, Chapter 4.

Gross Anatomy of the Nervous System:

II. Brain Stem; Cerebellum; Spinal Cord; and Cerebrospinal Fluid System

■ THE BRAIN STEM

The brain stem, illustrated in Figure 3–1, is exactly what its name implies: it is the stem of the brain that connects the forebrain with the spinal cord. Its major divisions are (1) the *mesencephalon,* (2) the *pons,* and (3) the *medulla oblongata* (Table 3–1). Some anatomists also consider the *diencephalon,* which was discussed in the previous chapter, to be part of the brain stem because it too is a connecting link.

Several important *fiber tracts* pass both upward and downward through the brain stem, transmitting sensory signals from the spinal cord mainly to the thalamus and motor signals from the cerebral cortex to the cord. In addition, other fiber tracts either originate or terminate in the brain stem, again mainly for the purpose of carrying sensory and motor signals.

However, the brain stem also contains many very important centers that control such physiological variables as respiration, arterial pressure, equilibrium, and others. In fact, centers in the brain stem even determine the level of activity in the cerebrum and also cause the waking-sleeping cycle of the nervous system.

And, finally, the brain stem serves as the connecting link between the cerebellum and the cerebrum superiorly and between the cerebellum and spinal cord inferiorly.

THE MESENCEPHALON

The surface anatomy of the mesencephalon is shown from the left posterolateral side in Figure 3–1, and a horizontal section across the middle of the mesencephalon is illustrated in Figure 3–2. It is divided from anterior to posterior into two major sections: (1) the two *cerebral peduncles,* which constitute the anterior four fifths of the mesencephalon; and (2) the *tectum,* which comprises the structures near the posterior surface. Passing downward through the posterior part of the mesencephalon near the dividing line between the cerebral peduncles and the tectum is the *cerebral aqueduct,* which is a small tubular canal that connects the third ventricle in the diencephalon with the fourth ventricle in the lower brain stem.

The Cerebral Peduncles. The cerebral peduncles veer anteriorly and laterally at the top of the mesencephalon and then project superiorly into the two lateral halves of the diencephalon. Each cerebral peduncle is divided into three separate areas:

1. A thick anterolateral surface layer of *corticospinal* and *corticopontine fibers* conducts motor signals from the cortex to the spinal cord and to the pons.

2. A deeper layer of darkly pigmented nerve cell bodies called the *substantia nigra* lies behind the fiber layer. The neurons of the substantia nigra function as part of the basal ganglial system to control subconscious muscle activities of the body. Destruction of these neurons causes Parkinson's disease, in which the person develops continuous muscle spasm and a shaking tremor in part or all of the body, sometimes so severe that muscle functions become useless.

3. The *tegmentum* is the major mass of the cerebral peduncles medial and posterior to the substantia nigra.

The tegmentum contains several important fiber tracts and nuclei that provide specific functions as follows:

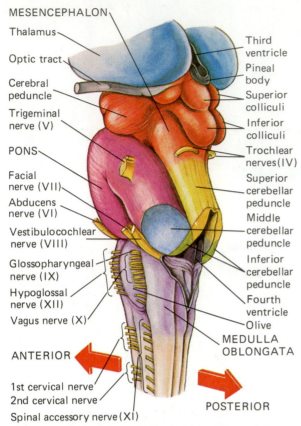

MESENCEPHALON

Thalamus

Optic tract

Cerebral peduncle

Trigeminal nerve (V)

PONS

Facial nerve (VII)

Abducens nerve (VI)

Vestibulocochlear nerve (VIII)

Glossopharyngeal nerve (IX)

Hypoglossal nerve (XII)

Vagus nerve (X)

ANTERIOR

1st cervical nerve
2nd cervical nerve
Spinal accessory nerve (XI)

Third ventricle

Pineal body

Superior colliculi

Inferior colliculi

Trochlear nerves (IV)

Superior cerebellar peduncle

Middle cerebellar peduncle

Inferior cerebellar peduncle

Fourth ventricle

Olive

MEDULLA OBLONGATA

POSTERIOR

Figure 3–1. The brain stem.

1. The *medial lemniscus* is the major fiber tract for transmitting sensory signals from the body to the thalamus.

2. The *medial longitudinal fasciculus* is a tract that connects many of the nuclei of the brain stem with each other and also with the diencephalon.

3. The *red nucleus*, illustrated in Figure 3–2, occupies a major part of each side of the superior mesencephalic tegmentum. This nucleus functions with the basal ganglia and cerebellum to coordinate muscle movements of the body. It also serves as a relay station for signals transmitted from the cerebellum to the thalamus and cerebrum.

4. The *nuclei of the oculomotor and trochlear nerves* are small collections of nerve cells on each side of the mesencephalon that control most of the muscles for eye movements.

5. The *periaqueductal gray* is a collection of diffuse nuclei around the cerebral aqueduct. This area seems to play a major role in the analysis of the reaction to pain.

6. The *reticular formation* is composed of many widely dispersed nuclei in large portions of the tegmentum. The reticular formation is not only present in the mesencephalon but also extends all the way from the superior end of the spinal cord to the diencephalon, passing through the medulla oblongata, the pons, the mesencephalon, and even extending into the middle of the thalamus where it is represented by the thalamic intralaminar nuclei. Various collections of nerve cells within the reticular formation control many of the

stereotyped body movements, such as *turning motions of the trunk, turning and bending motions of the head,* and *postural motions of the limbs.* But, even more important, the reticular formation is a major center of the entire brain for *controlling the brain's overall level of activity.* Generalized stimulation of the mesencephalic and pontine portions of the reticular formation usually causes a high degree of wakefulness in an animal, while also increasing the tone of the muscles throughout the body. Therefore, the reticular formation, though dispersed rather broadly in the brain stem, is functionally one of the most important of all the brain structures, as we shall discuss much more fully in later chapters.

The Tectum. The tectum is the posterior fifth of the mesencephalon, and it consists principally of four small nodular bodies, two superior colliculi and two inferior colliculi, that are arranged in a quadrangle on the posterior surface of the mesencephalon, as illustrated in Figure 3–1.

The two *superior colliculi* are located side by side on the superior portion of the posterior mesencephalon, lying immediately beneath the posterior poles of the thalamus. In lower animals, especially the fish, the superior colliculi are the principal brain terminus for vision. In the human being, the visual functions of these bodies have been lost, but they are still used for *causing eye movements* and even *trunk movements* in response to sudden visual signals such as a flash of light from one side of the field of vision or sudden movement of a person or an animal nearby.

The two *inferior colliculi* are located inferior to the two superior colliculi, also on the posterior surface of the mesencephalon. These serve as way-stations for relaying auditory signals from the ears to the cerebrum. In addition, they play a role in causing a person to turn his or her head or body in response to sounds coming from different directions.

Inferior to the inferior colliculi on the two sides of the mesencephalon are two large bundles of nerve fibers called the *superior cerebellar peduncles* that project inferiorly and posteriorly to connect with the superior portions of the cerebellum. These are one of the major trunk lines between the cerebellum and the remainder of the brain.

THE PONS

The pons, illustrated from the left posterolateral side in Figure 3–1 and in horizontal section in Figure 3–3, has many of the same types of internal structures as the mesencephalon, such as some of the same major fiber pathways that transmit signals both up and down the brain stem and multiple nuclei that perform specific functions. For descriptive purposes, the pons is divided into two parts: the *ventral part* and the *dorsal part,* also called the *tegmentum of the pons,* which is continuous with the tegmentum of the mesencephalon.

The Ventral Part of the Pons. The ventral part of the pons is the large bulbous anterior protrusion illustrated in Figure 3–1 and seen even better in the basal

TABLE 3 – 1 The Brain Stem and Cerebellum

Structure	Function
Mesencephalon	
Cerebral peduncles	
1. Corticospinal and corticopontine tracts	Motor signals to cord and pons
2. Substantia nigra	Part of basal ganglia motor control system
3. Tegmentum	
a. Red nucleus	Relays signals from cerebellum
b. Reticular formation	Excites whole brain, controls muscular tone
c. Nuclei of cranial nerves III and IV	Control eye movements
d. Medial lemnisci	Sensory signals to thalamus
Tectum	
1. Superior colliculi	Help to control eye movements
2. Inferior colliculi	Cause motor reactions to auditory signals
Pons	
Ventral part	
1. Corticospinal tracts	Pass through ventral pons toward cord
2. Pontine nuclei	Terminis of corticopontine tracts
3. Transverse fibers	Fibers from pontine nuclei to opposite cerebellar hemisphere
Tegmentum	
1. Reticular formation	Same as in mesencephalon, also parts of vasomotor and respiratory centers
2. Nuclei of cranial nerves V, VI, VII, and VIII	Eye and facial movements; facial, auditory, and equilibrium sensations
3. Medial lemnisci	Same as in mesencephalon
Medulla Oblongata	
Pyramids and decussation of pyramids	Downward extensions and crossover of corticospinal tracts
Gracile and cuneate nuclei	Origins of fibers in medial lemnisci
Decussation of medial lemnisci	Crossover of the medial lemnisci
Inferior olivary nuclei	Origin of many input fibers to cerebellum
Nuclei of cranial nerves IX, X, XI, and XII	Motor signals to larynx, pharynx, tongue, some neck muscles; sensory signals from viscera; motor signals to parasympathetic nervous system
Reticular formation	
1. Most of vasomotor center	Controls vascular resistance, arterial pressure, heart activity
2. Most of respiratory center	Controls inspiration and expiration
Cerebellum	
Cerebellar peduncles	
1. Inferior peduncle	Extends from medulla; mainly input signals
2. Middle peduncle	Extends from ventral part of pons; entirely input signals
3. Superior peduncle	Extends to mesencephalon; mainly output signals
Vermis	Midline portion of cerebellum; functions with brain stem and spinal cord
Cerebellar hemispheres	Lateral portions of cerebellum; functions mainly with higher motor centers
Cerebellar cortex	Provides delays in motor signals
Deep nuclei	
1. Dentate nuclei	Output nuclei of cerebellar hemispheres
2. Fastigial, globose, and emboliform nuclei	Output nuclei of vernis

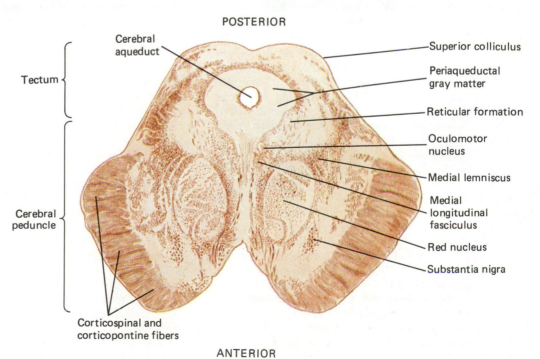

Figure 3 – 2. A horizontal section through the mesencephalon at the level of the superior colliculi.

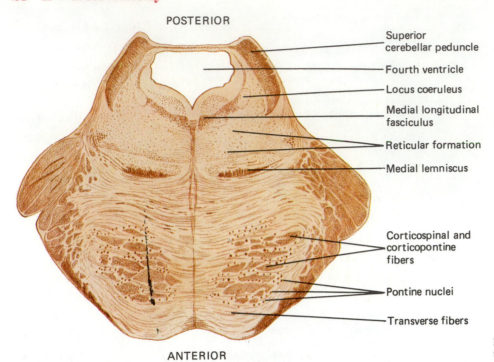

POSTERIOR

Superior
cerebellar peduncle

Fourth ventricle

Locus coeruleus

Medial longitudinal
fasciculus

Reticular formation

Medial lemniscus

Corticospinal and
corticopontine
fibers

Pontine nuclei

Transverse fibers

ANTERIOR

Figure 3–3. A horizontal section
through the pons.

view of the brain in Figure 2–2 of the previous chapter. Its internal structure is illustrated in Figure 3–3.

The same *corticospinal fibers* and *corticopontine fibers* that pass through the cerebral peduncles of the mesencephalon also descend into the ventral part of the pons. The corticospinal fibers then pass on downward through the medulla into the spinal cord. On the other hand, the corticopontine fibers terminate here, synapsing in multiple *pontine nuclei*. From these, transverse fibers cross immediately to the opposite side of the ventral pons and then circle backwards around the two lateral sides of the pons to form the *middle cerebellar peduncles* that extend posteriorly into the two cerebellar hemispheres. In addition, a few fibers also pass directly back into these peduncles on the same side. Because of the crossing of the transverse fibers and because most of the output fiber tracts leaving the cerebellum to reenter the brain stem also cross to the opposite side, the right half of the cerebellum functions mainly with the left half of the cerebrum, and the left half of the cerebellum functions mainly with the right cerebrum.

The Tegmentum of the Pons. Figure 3–3 shows that the pontine tegmentum contains the following three structures that are continuous with those in the mesencephalon: the *medial lemniscus*, the *medial longitudinal fasciculus*, and the *reticular formation*. In addition, it contains the nuclei of several cranial nerves: (1) the *abducens nerve*, which helps to control eye movements; (2) the *facial nerve*, which controls the muscles of expression of the face; (3) the *trigeminal nerve*, which controls the muscles of mastication and also transmits sensory signals from the face, mouth, and scalp; and (4) the *vestibulocochlear nerve*, which transmits sensory signals from the ear and from the vestibular apparatus (the equilibrium apparatus of the inner ear).

THE MEDULLA OBLONGATA

The *medulla oblongata*, usually called simply the *medulla*, is illustrated at the lower end of the brain stem in Figure 3–1, and it is seen in cross-section in Figure 3–4. On the surface are two distinguishing characteristics:

1. On the anterior aspect of the medulla are two protruding longitudinal columns called *pyramids* (seen best in Figure 2–2 of the previous chapter but also illustrated in cross-section in Figure 3–4). These carry the same *corticospinal fibers* that pass from the cerebral cortex through the cerebral peduncles of the mesencephalon and through the ventral pons. The fibers in this trunk pass eventually to all levels of the spinal cord, carrying signals that control muscle contraction. In the inferior portion of the medulla, the fibers of the pyramid cross to the opposite side before passing down the spinal cord, which is called the *decussation of the pyramids*. Therefore, the left cerebral cortex controls muscle contraction in the right half of the body while the right cortex controls the left side.

2. An *olive* protrudes from each anterolateral surface of the medulla lateral to the pyramids (Fig. 3–1). Deep to the external projection of the olive is the *inferior olivary nucleus*, which can be seen in cross-section in Figure 3–4. This nucleus functions to relay signals into the cerebellum, functioning similarly to the pontine nuclei. However, it receives its input signals mainly from the basal ganglia and spinal cord and less from the motor cortex. Its outgoing signals go to the contralateral cerebellum through the *inferior cerebellar peduncle* (Fig. 3–1), which is an upward and posteriorly projecting column of nerve fibers extending from the medulla to the cerebellum.

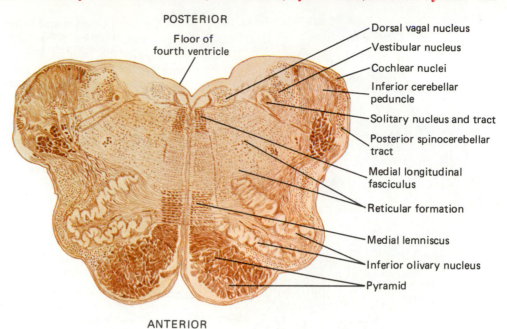

POSTERIOR

Floor of
fourth ventricle

Dorsal vagal nucleus
Vestibular nucleus
Cochlear nuclei
Inferior cerebellar peduncle
Solitary nucleus and tract
Posterior spinocerebellar tract
Medial longitudinal fasciculus
Reticular formation
Medial lemniscus
Inferior olivary nucleus
Pyramid

ANTERIOR

Figure 3–4. A horizontal section through the medulla oblongata.

In addition to these specific surface structures, the medulla contains many of the same components as the mesencephalon and pons. Especially prominent are the bilateral *medial lemnisci,* the large fiber tracts through which sensory signals are conducted from the spinal cord to the cerebrum. The fibers in these tracts originate in large bilateral nuclei in the posterior inferior medulla, the *gracile* and *cuneate nuclei,* which themselves receive sensory signals from sensory fibers in the dorsal column of the spinal cord. After leaving these nuclei, the fibers cross to the opposite side of the medulla in the *decussation of the medial lemnisci* to form the medial lemnisci. Because of this crossing, the left side of the brain is excited by sensory stimuli from the right side of the body and the right brain from the left body.

Also present in the medulla are (1) nuclei for *cranial nerves IX, X, XI, and XII,* which will be discussed more fully in the following chapter, and (2) the *reticular formation* that makes up a large share of the posterior and lateral medulla.

Some Special Functional Areas of the Reticular Formation in the Medulla and Pons. The reticular formation of the medulla and pons contains two especially important control centers:

1. The *vasomotor center* consists of widely dispersed nerve cells in the formation. This center transmits signals to the heart and blood vessels to increase heart pumping activity and to constrict the vessels. These effects acting together can increase the blood pressure greatly.

2. The *respiratory center* is composed of nerve cells also located widely in the reticular formation. This is an automatic, rhythmically active center that causes the rhythmical respiratory muscle contractions necessary for inspiration and expiration.

In addition to these special centers, the medullary and pontine reticular formation also serves as an im-

portant relay station for signals coming from higher brain centers to control many other important internal functions of the body. For instance, signals from the hypothalamus are relayed down the spinal cord to control body temperature, sweating, secretion in the digestive tract, emptying of the urinary bladder, and many other bodily functions. Also, lying in close association with the posterior medullary reticular formation on each side of the midline is the *dorsal motor nucleus of the vagus* (Fig. 3–4). This nucleus relays signals into the vagus nerve to control heart rate, gastric secretion, peristalsis in the gastrointestinal tract, and other internal functions.

■ THE CEREBELLUM

The *cerebellum* is a large structure of the hindbrain (rhombencephalon) located inferior to the occipital lobe of the cerebrum and posterior to the brain stem. Its location in relation to the remainder of the brain is seen best in Figures 2–6 through 2–8 of the previous chapter, and its specific relationship to the brain stem is illustrated in Figure 3–5.

The cerebellum is an important part of the motor control system. Even though it is located far away from both the motor cortex and the basal ganglia, it interconnects with these through special nerve pathways, and it also interconnects with motor areas in both the reticular formation and the spinal cord. Its primary function is to determine the time sequence of contraction of different muscles during complex movements of parts of the body, especially when these movements occur extremely rapidly.

Surface Anatomy of the Cerebellum. Shown in Figure 3–1 are three different cerebellar peduncles that lead from the posterior surface of each side of the brain stem to the cerebellum; these are: (1) the *superior cerebellar peduncle,* which connects with the mesencephalon, as discussed earlier; (2) the *middle cerebellar*

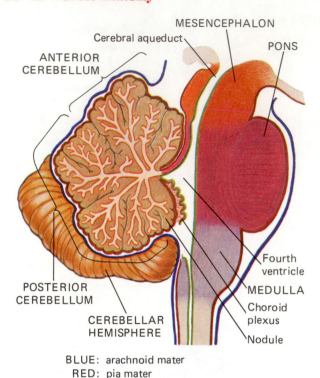

BLUE: arachnoid mater
RED: pia mater
GREEN: ependyma

Figure 3−5. Relationship of the cerebellum to the brain stem.

peduncle, which connects with the pons; and (3) the *inferior cerebellar peduncle,* which connects with the medulla. It is through these that signals are transmitted into and away from the cerebellum.

Now, let us study the structure of the cerebellum itself as shown in Figures 3−5 and 3−6. Its major parts are the *vermis,* which is a midline structure 1 to 2 cm wide extending around the entire cerebellum from anterior to posterior and from superior to inferior, and two *cerebellar hemispheres,* located laterally on the two sides of the cerebellum. Both the vermis and the cerebellar hemispheres can also be divided into the *anterior lobe of the cerebellum,* which is the superior and anterior third of the cerebellum, lying anterior to the primary fissure, and *posterior lobe of the cerebellum,* which

is the posterior and inferior two thirds of the cerebellum, lying posterior to the primary fissure.

In the human being, the cerebellar hemispheres make up by far the greater portion of the cerebellar mass. These hemispheres function in concert with the cerebrum to coordinate voluntary movements of the body. The vermis, on the other hand, functions more for coordinating the stereotyped and subconscious body movements, operating mainly in association with the brain stem and spinal cord.

Internal Structure of the Cerebellum. The internal structure of the cerebellum is illustrated in Figure 3−7. Like the cerebrum, it is composed of three principal structures: (1) the *cerebellar cortex;* (2) the *subcortical white matter* composed almost entirely of nerve fibers; and (3) the *deep nuclei.*

The cerebellar cortex is a 3- to 5-mm thick sheet of nerve cells covering the entire cerebellar surface, containing a total of about 30 billion cells.

The deep nuclei are located in the center of the cerebellar white matter. By far the most prominent of the deep nuclei is the large *dentate nucleus,* which is located in the center of each cerebellar hemisphere. However, three other smaller deep nuclei are located in each side of the vermis: (1) the *fastigial nucleus;* (2) the *emboliform nucleus;* and (3) the *globose nucleus.*

The deep nuclei give rise to the nerve fibers that transmit signals out of the cerebellum to other parts of the nervous system. The cerebellar cortex is a rapidly acting computing area that receives input information from the cerebral cortex, the basal ganglia, the spinal cord, and the peripheral muscles and integrates all of this to help coordinate the muscle movements.

Note also in Figures 3−5, 3−6, and 3−7 the many folds of the cerebellum, called *folia.* If the cerebellar cortex were stretched out to eliminate these folia, it would be represented by a flat sheet about 40 cm in length and 8 cm wide. Each area of the cerebellum has almost exactly the same internal neuronal circuitry as all other areas, suggesting that the cerebellum performs almost exactly the same functions in all its separate parts. What it does is to delay signals for short fractions of a second. Thus, when the motor system

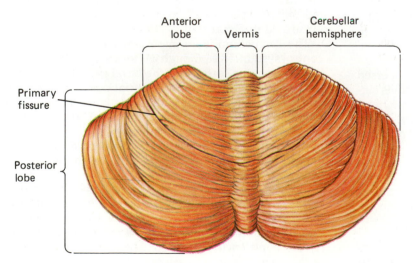

Figure 3−6. Superior view of the cerebellum.

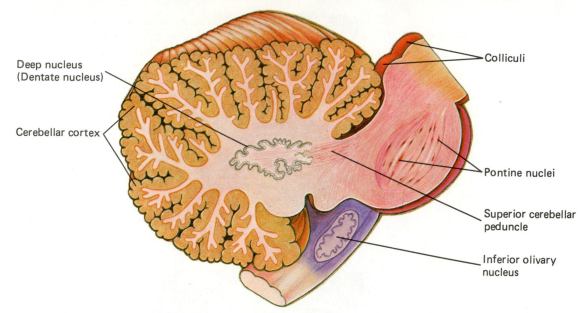

Figure 3–7. The internal anatomy of the cerebellar hemisphere and related structures.

requires contraction of the biceps to start the arm moving and then contraction of the triceps to stop the movement, the cerebellum determines the appropriate delay before turning off the biceps and simultaneously turning on the triceps. When a person is performing rapid movements, such sequential muscle contractions occur in many separate parts of the body one after the other. Without the appropriate turn-on, turn-off sequence of these motor signals, the movements become totally uncoordinated, which is what happens when the cerebellum is destroyed.

■ THE SPINAL CORD

Figure 3–8 illustrates the *spinal cord* and the *spinal nerves* leaving it to be distributed to all parts of the body, and Table 3–2 summarizes the structural parts of the cord and their functions. Note that a spinal nerve leaves the cord on each side through each *intervertebral* foramen between adjacent vertebrae. Some of these nerves are very large because they innervate large areas of the body, such as the spinal nerves of the lower neck that innervate the arms, forearms, and hands and those of the lumbar and sacral regions that innervate the thighs, legs, and feet. In both of these areas the cord itself is also enlarged because of the great number of nerve cell bodies that are required to relay the signals, giving rise to the *cervical enlargement* of the cord in the lower half of the neck and the *lumbosacral enlargement* at the lower end of the spinal cord.

The spinal cord terminates approximately at the lower end of the second lumbar vertebra. The reason for this is that during growth of the fetus and young child the spinal cord does not continue to lengthen as the vertebral column lengthens, so that the cord becomes located progressively more superiorly in the vertebral canal. Yet the lower lumbar and sacral seg-

ments of the spinal cord still exist, and the lumbar and sacral spinal nerves still arise from the cord, but they arise from higher up in the vertebral canal because the levels of the cord segments no longer correspond to the levels of the vertebrae. The nerves then course downward through the lower canal as a large bundle of nerves called the *cauda equina*, and each of them finally emerges through its appropriate lumbar or sacral intervertebral foramen.

The Internal Structure of the Spinal Cord. Like the brain, the spinal cord is composed of areas of *gray matter* and areas of *white matter*, though the white matter is on the surface of the cord, whereas the gray matter is deep. These are illustrated in the cross-sectional view of the cord in Figure 3–9. The *nerve cell bodies* are in the gray matter along with many short nerve fibers as well. But in the white matter, only *fiber tracts* and *glia* exist. Note in Figure 3–9 that the gray matter has the appearance of multiple horns connected by a crossbridge called the *gray commissure* between the two halves of the cord. Many fiber tracts also pass from one side of the cord to the other through *white commissures* that accompany the gray commissure.

The horns of gray matter on each side of the cord are called respectively (1) the *ventral gray horn* (or the *anterior gray horn*), (2) the *dorsal gray horn* (or the *posterior gray horn*), and (3) the *lateral gray horn*.

It is in the ventral horn that the *anterior motor neurons* lie, the nerve cell bodies that send fibers through the spinal nerves to the muscles to cause muscle contraction. In the dorsal gray horn are the nerve cell bodies that receive sensory signals from the spinal nerves. In the lateral gray horn the nerve cells give rise to fibers that lead into the autonomic nervous system, the system that controls many of the internal organs.

Connections of the Spinal Nerves with the Spinal Cord. Note also in Figure 3–9 that each spinal nerve

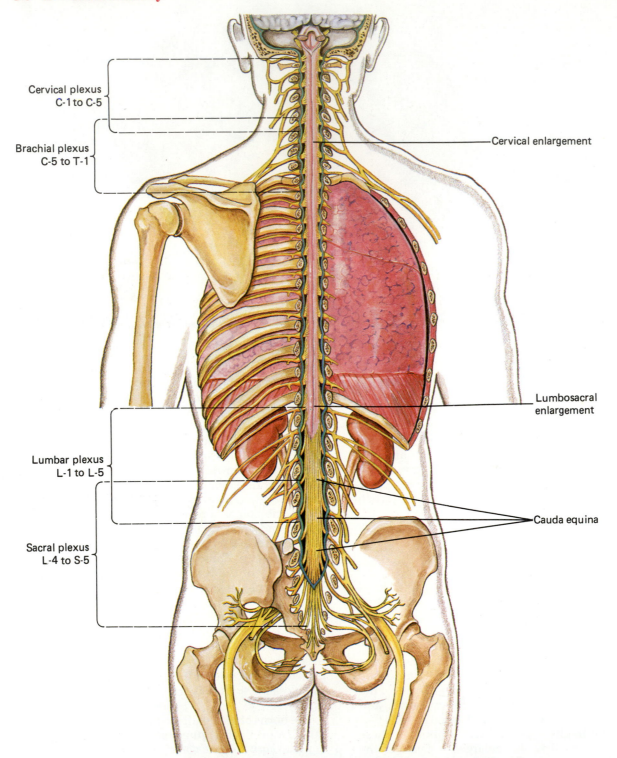

Cervical plexus
C-1 to C-5

Brachial plexus
C-5 to T-1

Cervical enlargement

Lumbosacral
enlargement

Lumbar plexus
L-1 to L-5

Cauda equina

Sacral plexus
L-4 to S-5

Figure 3–8. The spinal cord, its relationship to the peripheral nerves, and the spinal nerve plexuses.

connects with the cord by way of two roots called the *dorsal root* and the *ventral root* (also called *posterior root* and *anterior root*). Each of these roots in turn enters or leaves the cord by way of 7 to 10 small *root filaments*. The dorsal root is also called the *sensory root* because it carries almost entirely sensory fibers; and the anterior root is called the *motor root* because it carries almost entirely motor fibers leading from the cord to the muscles to cause muscle contraction or to the autonomic nervous system to control the activity of the internal organs. The nerve fibers in the ventral root originate from neurons in the ventral and lateral gray horns and then leave the cord along its anterolateral margin in the ventral root filaments. The dorsal root filaments pierce the cord along its posterolateral border, and its nerve fibers then turn either up or down the cord or enter the dorsal gray horn.

On the dorsal root is an enlargement called the *dor-*

TABLE 3–2 The Spinal Cord

Structure	Function
Gray Matter	
Dorsal horns	Loci of sensory input neurons
Lateral horns	Loci of preganglionic autonomic neurons
Ventral horns	Loci of motor neurons for skeletal muscles
White Matter	
Propriospinal tracts	Signals between cord segments
Long motor tracts	
1. Lateral corticospinal	Motor signals from cortex to spinal cord
2. Ventral corticospinal	Same
3. Rubrospinal	Motor signals from brain stem to spinal cord; most are excitatory, a few are
4. Reticulospinal	inhibitory
5. Olivospinal	
6. Vestibulospinal	
7. Tectospinal	
Long sensory tracts	
1. Fasciculus gracillis and fasciculus cuneatus	Discriminatory sensory signals to gracile and cuneate nuclei, thence to thalamus in medial lemnisci
2. Ventral and lateral spinothalamic	Crude touch, pain, and temperature signals to brain stem and thalamus
3. Ventral and dorsal spinocerebellar	Proprioceptor sensory signals to cerebellum
4. Spino-olivary	Cord signals to inferior olivary nuclei, then relayed to cerebellum
Spinal Nerve Roots	
Dorsal	Sensory input
Ventral	Motor output to muscles and preganglionic output to autonomic nervous system

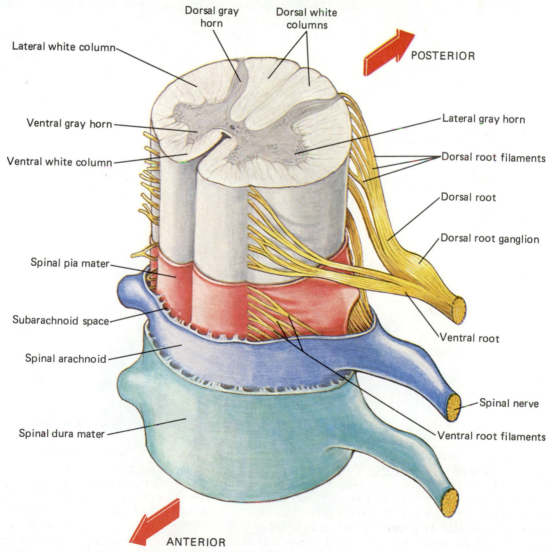

Figure 3–9. Structures of the spinal cord and its connections with the spinal nerves by way of the dorsal and ventral spinal roots. Note also the coverings of the spinal cord, the meninges.

sal root ganglion. This ganglion contains unipolar nerve cells that have no dendrites but do have the usual single axon. However, immediately after the axon leaves the cell body it divides into two branches, a *peripheral branch* and a *central branch*. The peripheral branch passes through the peripheral portions of the spinal nerve to sensory receptors in the body, and the central branch passes into the spinal cord. About two thirds of the sensory fibers entering the cord terminate in the dorsal gray horn near the point of entry. The other third divides immediately into two branches, one branch terminating in the dorsal horn but the other branch passing all the way up the cord in the white matter and then terminating in the gracile and cuneate nuclei in the lower part of the medulla.

Function of the Spinal Cord Gray Matter. The gray matter of the spinal cord serves two functions. First, its synapses *relay signals between the periphery and the brain* in both directions. It is mainly in the dorsal horns that sensory signals are relayed from the sensory roots of the spinal nerves, then pass superiorly in the white matter of the cord to the various sensory areas of the brain. It is mainly in the ventral and lateral horns that motor signals are relayed from the descending nerve tracts from the brain into the motor roots of the spinal nerves.

Second, the gray matter of the cord functions to *integrate some motor activities*. For instance, when the hand is subjected to a painful stimulus, the sensory signals entering the cord cause an immediate reaction in the gray matter of the cord hand region. Within a fraction of a second this leads to motor signals that cause withdrawal of the hand from the painful stimulus. This is called the *withdrawal reflex* (or flexor reflex or pain reflex). This reaction occurs entirely independ-ently of the higher levels of the nervous system. Some other cord reflexes are (1) reflexes that cause tonic contraction of the extensor muscles of the legs when one stands, thus allowing the legs to support the weight of the body; (2) scratch reflexes in lower animals when they are tickled; (3) stretch reflexes that cause muscles to contract when they are stretched (this is the reflex that causes the knee jerk when the patellar tendon is struck); and (4) even walking reflexes.

The Long Fiber Pathways of the Spinal Cord. Figures 3–9 and 3–10 show that the cord white matter is also divided into columns. These are (1) two *dorsal* (or *posterior*) *white columns* lying between the dorsal gray horns; (2) two *lateral white columns*, one lying on each side of the cord lateral to the gray matter; and (3) two *ventral* (or *anterior*) *white columns*, lying between and anterior to the ventral gray horns.

All these columns contain fiber tracts that run lengthwise along the cord. Some of these lie immediately adjacent to the gray matter, as illustrated in Figure 3–10. These, called *propriospinal tracts*, travel for only a few segments of the cord, connecting separate cord segments of gray matter with one another to help in the performance of the cord reflexes. (A "segment" of the cord is that portion of the cord that corresponds to a single pair of spinal nerves.) The remainder of the white matter contains long fiber tracts that carry sensory information to the brain or motor signals from the brain to the cord. To the left in Figure 3–10 are the motor tracts, and to the right are the sensory tracts. These are described here:

Motor Tracts

1. *Lateral corticospinal tract*, from the motor cortex of the brain.

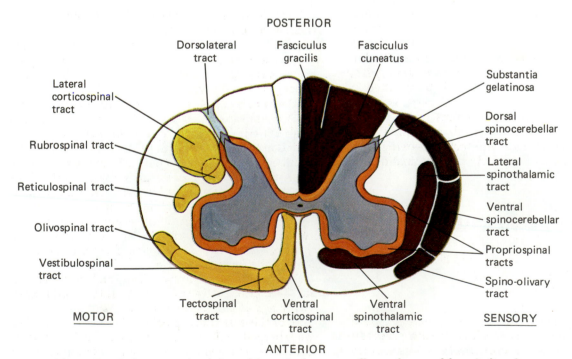

Figure 3–10. A cross-sectional view of the major long nerve fiber pathways of the spinal cord.

2. *Ventral corticospinal tract*, also from the motor cortex of the brain.

3. *Rubrospinal tracts*, from the red nucleus of the mesencephalon.

4. *Reticulospinal tracts*, from the reticular substance of the mesencephalon, pons, and medulla.

5. *Olivospinal tract*, from the inferior olive of the medulla.

6. *Vestibulospinal tract*, from the vestibular nuclei of the medulla and pons.

7. *Tectospinal tract*, from the tectum of the mesencephalon.

Sensory Tracts

1. *Fasciculus gracilis* and *fasciculus cuneatus* (the two of which together make up most of the dorsal white columns), carrying signals directly from the spinal sensory roots all the way to the gracile and cuneate nuclei in the lower end of the medulla.

2. *Ventral* and *lateral spinothalamic tracts*, carrying signals relayed in the posterior gray horn, thence through the anterior white commissure, and finally upward on the opposite side of the cord to the brain stem and thalamus.

3. *Ventral* and *dorsal spinocerebellar tracts*, which relay signals from the posterior gray horns upward to the cerebellum.

4. *Spino-olivary tract*, from the posterior gray horns of the cord to the inferior olive of the medulla.

Stimulation of most motor tracts causes either increased muscle tone or actual muscle contraction, and these tracts are said to be *excitatory tracts*; but stimulation of some tracts can decrease muscle tone, and they are called *inhibitory tracts*.

The sensory signals transmitted in the dorsal column pathways (the fasciculus gracilis and fasciculus cuneatus) are mainly those of fine, discriminatory touch that allow one to recognize the surface locations of sensory stimuli on the skin or the positions of the different parts of the body. The sensory signals transmitted in the spinothalamic tracts are those of crude touch, pain, and temperature. The sensory signals transmitted in the spinocerebellar tracts and also in the spino-olivary tract are mainly signals from the muscles and joints that apprise the cerebellum at all times about the movements and positions of different parts of the body so that the cerebellum can help in coordinating the body's movements.

■ THE CEREBROSPINAL FLUID SYSTEM — A PROTECTIVE FLUID FLOTATION SYSTEM FOR THE BRAIN AND SPINAL CORD

Even though the brain and spinal cord are crucial to the function of our bodies, nevertheless they are extremely delicate structures. For instance, the brain's tissues are so weak that one can push a finger all the way through it with almost no pressure, and the brain can literally be scooped out of the cranial vault with a spoon. Therefore, the brain requires a special protective system. This protection is achieved by the encasement of both the brain and the spinal cord in a rigid, bony vault comprised of the *cranial cavity* in the skull and the *vertebral canal* in the vertebral column. Within this vault, the brain and spinal cord actually "float" in a bath of fluid called the *cerebrospinal fluid*. This flotation system for the brain is illustrated in Figure 3–11. Now, let us describe this system and the mechanisms for maintaining the fluid in the system.

The Ventricular System in the Brain. In several cross-sectional views of the brain presented in the previous chapter and in the first part of this chapter, large fluid cavities called *ventricles* were shown deep inside the cerebrum, diencephalon, and brain stem. These ventricles, four in number, are illustrated in three-dimensonal perspective in Figure 3–12. They are as follows:

1 and 2. The *two lateral ventricles*. Each of these lies near the median plane in each cerebral hemisphere, extending all the way from the center of the frontal lobe anteriorly to the center of the occipital lobe posteriorly. From the parietal region of each of these ventricles an inferior extension turns laterally and anteriorly into the temporal lobe; this is called the *inferior horn of the lateral ventricle*.

3. The *third ventricle*. This lies between the two lateral halves of the thalamus and also extends anteriorly and inferiorly into the midline plane between the two halves of the hypothalamus.

4. The *fourth ventricle*. This lies in the lower brainstem in the space posterior to the pons and medulla but anterior to the cerebellum.

Now, let us study Figure 3–11 to see how the ventricles interconnect with each other. This figure shows the shadow of one of the lateral ventricles deep within a cerebral hemisphere that connects by way of an *interventricular foramen* (also called *foramen of Monro*) with the anterolateral part of the third ventricle. In turn, the third ventricle connects posteriorly and inferiorly with the *cerebral aqueduct (aqueduct of Sylvius)*, which is a small tube passing downward through the mesencephalon to enter the fourth ventricle lying behind the pontine and medullary regions of the brain stem. Finally, three openings occur in the outer wall of the fourth ventricle through which fluid can flow onto the surface of the brain. One of these is the *medial aperture* (also called the *foramen of Magendie*) in the midline inferior to the cerebellum. The other two, the *lateral apertures* (also called the *foramina of Luschka*), are to the sides of the fourth ventricle.

The Fluid Space Surrounding the Brain and Spinal Cord (Subarachnoid Space), and the Meningeal Coverings of the Brain and Cord. Covering all surfaces of the brain and spinal cord is a thin fluid-filled space several millimeters thick called the *subarachnoid space*. This space is bounded by the coverings of the brain and cord, called the *meninges*, which are illus-

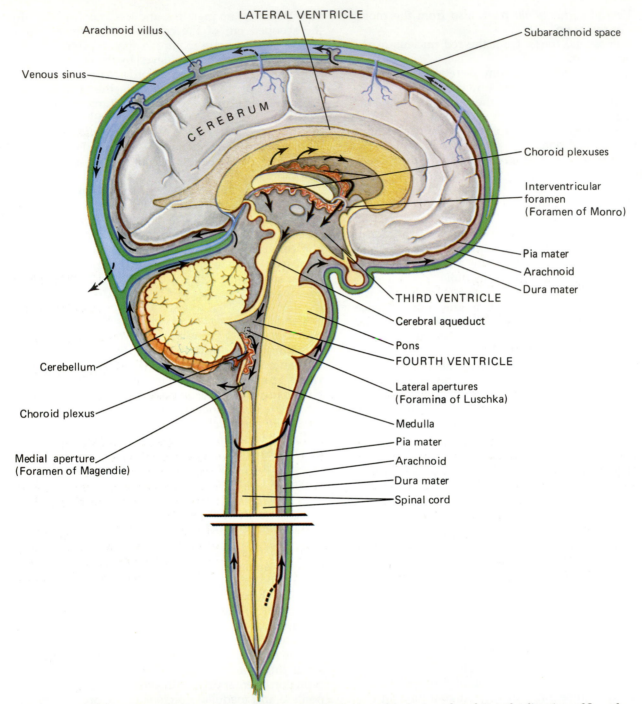

LATERAL VENTRICLE

Arachnoid villus

Venous sinus

CEREBRUM

Subarachnoid space

Choroid plexuses

Interventricular foramen (Foramen of Monro)

Pia mater

Arachnoid

Dura mater

THIRD VENTRICLE

Cerebral aqueduct

Pons

FOURTH VENTRICLE

Lateral apertures (Foramina of Luschka)

Medulla

Pia mater

Arachnoid

Dura mater

Spinal cord

Cerebellum

Choroid plexus

Medial aperture (Foramen of Magendie)

Figure 3–11. The cerebrospinal fluid system and the meningeal coverings of the brain and cord. Note the directions of flow of cerebrospinal fluid indicated by the arrows.

trated for the spinal cord in Figure 3–9 and shown over a section of the brain in Figure 3–13. There are three layers of meninges:

1. The *dura mater* is a strong fibrous covering that surrounds the entire central nervous system; it is bound tightly to the inner surface of the skull but only loosely to the vertebral canal where there is a loose connective tissue space called the *epidural space*.

2. The *arachnoid* is a delicate structure loosely attached to the inner surface of the dura mater. And beneath the arachnoid is the fluid space that surrounds the brain and cord, the *subarachnoid space*. This space is penetrated by large numbers of small *arachnoidal trabeculae* that are part of the arachnoid.

3. The *pia mater* is a thin fibrous and vascular covering of the brain and cord, attached tightly to their surfaces, even dipping into all fissures and sulci.

The blood vessels that serve the brain have special relations to the meninges. First, note in Figure 3–11 the large *venous sinus*, the superior sagittal sinus, that extends along the midline the entire length of the cere-

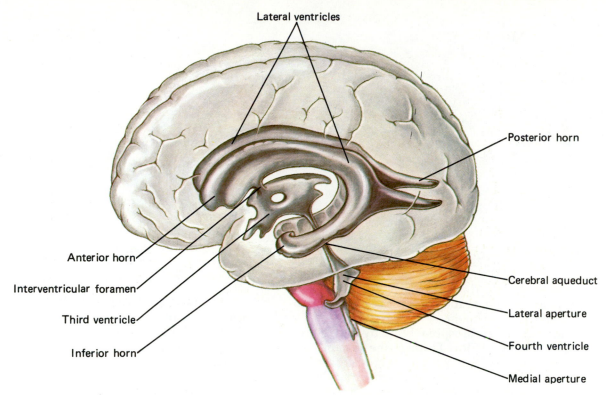

Lateral ventricles

Posterior horn

Anterior horn

Interventricular foramen

Third ventricle

Inferior horn

Cerebral aqueduct

Lateral aperture

Fourth ventricle

Medial aperture

Figure 3–12. A three-dimensional representation of the ventricular system of the brain.

brum from front to back. This large sinus lies within layers of dura mater, and it has a triangular cross-sectional appearance as shown in Figure 3–13. Similar venous sinuses lie over other surfaces of the brain and also in the floor of the cranial cavity. All these sinuses interconnect and eventually give rise to the two internal jugular veins.

Next, note in Figures 3–11 and 3–13 the *arachnoid villi* that protrude into the venous sinuses. These are small penetrations of arachnoidal tissue that have

made small openings in the walls of the sinuses. Cerebrospinal fluid can flow through these openings from the subarachnoid space into the venous blood. However, the villi function like valves to prevent blood from flowing backward from the venous sinuses into the subarachnoid space.

Finally, observe also in Figure 3–13 the large artery lying on the surface of the brain. Though this artery protrudes into the subarachnoid space, it is actually covered by the pia mater. Such arteries on the surface

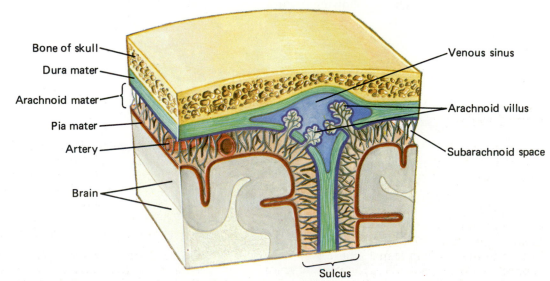

Bone of skull

Dura mater

Arachnoid mater

Pia mater

Artery

Brain

Venous sinus

Arachnoid villus

Subarachnoid space

Sulcus

Figure 3–13. An expanded view of the meninges covering a section of the brain. Note also the venous sinus with arachnoid villi protruding into it.

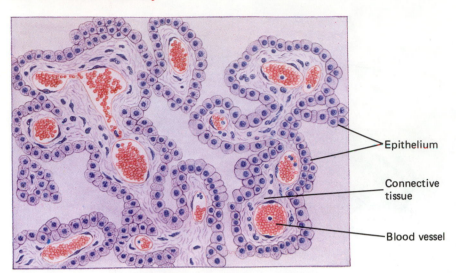

Epithelium

Connective tissue

Blood vessel

Figure 3–14. A microscopic section of a choroid plexus.

of the brain and penetrating branches from them provide nutrition to the brain.

Formation of Cerebrospinal Fluid by the Choroid Plexuses, and Fluid Flow Through the System. Most of cerebrospinal fluid is secreted by special secretory structures called *choroid plexuses* that protrude into each of the four ventricles, as shown in Figure 3–11. The most extensive choroid plexuses lie along the inferior surfaces of the lateral ventricles; therefore, most of the cerebrospinal fluid is formed in the lateral ventricles.

A small section of a choroid plexus is illustrated in Figure 3–14. It has a cauliflower-like growth, with large numbers of small capillary blood vessels embedded in loose connective tissue and covered by a thin layer of cuboidal cells that secrete fluid into the ventricle. The secreted cerebrospinal fluid is a clear, watery fluid that contains nearly the same constituents as the plasma portion of the blood except for the plasma proteins.

Once the fluid has been secreted by the choroid plexuses, it flows through the following pathway:

1. From the two lateral ventricles into the third ventricle by way of the two interventricular foramina.

2. From the third ventricle into the fourth ventricle by way of the cerebral aqueduct.

3. From the fourth ventricle into the subarachnoid space surrounding the brain stem through the medial aperture and the two lateral apertures.

4. Through the subarachnoid spaces upward around the surfaces of the brain to the arachnoid villi.

5. From the subarachnoid spaces into the venous sinuses through the valvelike structures of the arachnoid villi.

The amount of cerebrospinal fluid formed each day is about 800 ml, and the pressure of this fluid in the cerebrospinal fluid system is about 10 mm Hg, only a very low pressure but enough to support the structures of the brain and spinal cord.

REFERENCES

See References, Chapter 4.

4

Gross Anatomy of the Nervous System:

III. The Peripheral Nerves

■ THE CRANIAL NERVES

In several figures of the last two chapters nerves were shown leaving the basal surfaces of the brain. These, called *cranial nerves*, are illustrated in more detail in Figure 4–1, and their connections both in the brain and peripherally are given in Table 4–1.

There are 12 pairs of cranial nerves, numbered in order of their origin on the basal surface of the brain from anterior to posterior, usually using Roman numerals. Each also has its own individual name, which also appears in the figure.

CONNECTIONS OF THE CRANIAL NERVES INSIDE THE BRAIN

Note in Figure 4–1 that the olfactory nerves arise from the cerebrum, the optic nerves from the diencephalon, and all the remaining ten pairs from the brain stem. Because of their special functional importance for smell and vision, we shall discuss the connections of the olfactory nerves with the brain in Chapter 15 in relation to the sense of smell and of the optic nerve in Chapters 12 and 13 in relation to vision. The connecting areas of the other cranial nerves in the brain stem are illustrated in Figure 4–2.

Some of the cranial nerves are entirely *sensory nerves*, some entirely *motor* (that is, they only innervate muscles to cause contraction), and some are combined nerves that have both sensory and motor components. To the left in Figure 4–2 are illustrated the *motor nuclei* to which the different motor and combined motor-sensory nerves connect. To the right are the *sensory nuclei*.

The Motor Nuclei of the Brain Stem. Beginning from the top in Figure 4–2, the important motor nuclei for the cranial nerves are as follows:

The *oculomotor, trochlear,* and *abducens nuclei* (nerves III, IV, and VI) send nerve fibers to the different muscles of the orbit for causing movement of the eye. The upper portion of the oculomotor nucleus is called the *Edinger-Westphal nucleus.* It controls the muscles inside the eye for focusing and for pupillary constriction.

The *trigeminal motor nucleus* (nerve V) controls the muscles of mastication (the muscles for chewing).

The *facial nucleus* (nerve VII) controls the many muscles of expression of the face.

The *dorsal vagal nucleus* (nerve X) is the important nucleus of the parasympathetic nervous system. It controls motor activity in many of the viscera, especially the heart (slowing of the heart), and of the upper digestive tract (increased peristalsis of the stomach and intestines and increased secretion).

The *nucleus ambiguus* sends signals through three different nerves, the glossopharyngeal, the vagus, and the accessory nerves (nerves IX, X, and XI). This nucleus controls such muscles as those for swallowing and the speech muscles of the larynx. The lower end of the nucleus ambiguus is continuous with the anterior horn of the spinal cord from which signals are transmitted through the spinal roots of the accessory nerve to control portions of the trapezius and sternocleidomastoid muscles.

The *hypoglossal nucleus* (nerve XII) controls primarily the movements of the tongue.

THE SENSORY NUCLEI OF THE BRAIN STEM

To the right in Figure 4–2 are the sensory nuclei of the brain stem. From above downward these are the following:

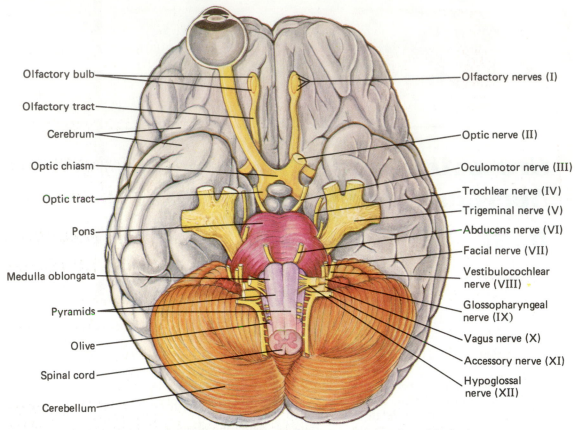

Figure 4–1. Origin of the cranial nerves from the ventral surface of the brain.

TABLE 4–1 The Cranial Nerves

Nerve	Connection with Brain	Function
I. Olfactory nerves and tract	Anterior ventral cerebrum	Sensory: from olfactory epithelium of superior nasal cavity
II. Optic nerve	Lateral geniculate body of the thalamus	Sensory: from retinae of eyes
III. Oculomotor nerve	Mesencephalon	Motor: to four eye-movement muscles and levator palpebrae
		Parasympathetic: smooth muscle in eyeball
IV. Trochlear nerve	Mesencephalon	Motor: to one eye-movement muscle, the superior oblique
V. Trigeminal nerve		
Ophthalmic branch	Pons	Sensory: from forehead, eye, superior nasal cavity
Maxillary branch	Pons	Sensory: from inferior nasal cavity, face, upper teeth, mucosa of superior mouth
Mandibular branch	Pons	Sensory: from surfaces of jaw, lower teeth, mucosa of lower mouth, and anterior tongue
		Motor: to muscles of mastication
VI. Abducens nerve	Pons	Motor: to one eye-movement muscle, the lateral rectus
VII. Facial nerve	Junction pons and medulla	Motor: to facial muscles of expression and cheek muscle, the buccinator
VIII. Vestibulocochlear nerve		
Vestibular branch	Junction pons and medulla	Sensory: from equilibrium sensory organ, the vestibular apparatus
Cochlear branch	Junction pons and medulla	Sensory: from auditory sensory organ, the cochlea
IX. Glossopharyngeal nerve	Medulla	Sensory: from pharynx and posterior tongue, including taste
		Motor: superior pharyngeal muscles
X. Vagus nerve	Medulla	Sensory: much of viscera of thorax and abdomen
		Motor: larynx and middle and inferior pharyngeal muscles
		Parasympathetic: heart, lungs, most of digestive system
XI. Accessory nerve	Medulla and superior spinal segments	Motor: to several neck muscles, sternocleidomastoid and trapezius
XII. Hypoglossal nerve	Medulla	Motor: to intrinsic and extrinsic muscles of tongue

MOTOR   SENSORY

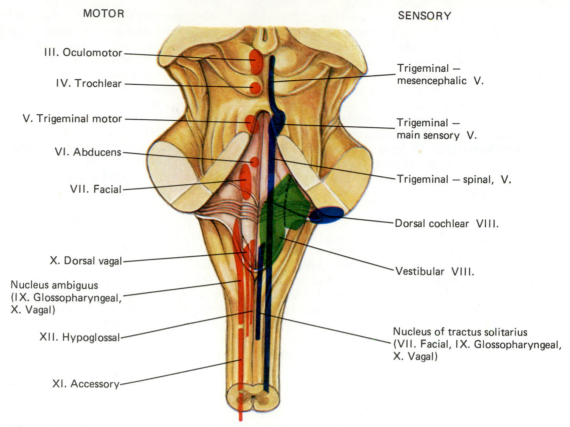

III. Oculomotor

IV. Trochlear

V. Trigeminal motor

VI. Abducens

VII. Facial

X. Dorsal vagal

Nucleus ambiguus
(IX. Glossopharyngeal,
X. Vagal)

XII. Hypoglossal

XI. Accessory

Trigeminal —
mesencephalic V.

Trigeminal —
main sensory V.

Trigeminal — spinal, V.

Dorsal cochlear VIII.

Vestibular VIII.

Nucleus of tractus solitarius
(VII. Facial, IX. Glossopharyngeal,
X. Vagal)

Figure 4–2. The motor and sensory nuclei of the cranial nerves in the brain stem (as seen posteriorly). The motor nuclei are illustrated to the left and the sensory nuclei to the right.

The *trigeminal nuclei* (nerve V) extend all the way from the mesencephalon downward into the upper part of the spinal cord. They have three major divisions: the *main sensory nucleus*, located in the pons, which subserves principally the function of tactile sensation for the face, mouth, and scalp; the *mesencephalic nucleus*, which receives signals mainly from muscles and other deep structures of the head; and the *spinal nucleus*, which is the principal nucleus for receipt of pain signals from the face, mouth, and scalp.

The *cochlear nucleus* (part of nerve VIII) is the receptor area for sound signals from the ear.

The *vestibular nucleus* (the other part of nerve VIII) receives signals from the vestibular apparatus, the sensory organ for equilibrium.

The *nucleus of the tractus solitarius* is the principal nucleus for receipt of visceral sensory signals from such organs as the heart, stomach, special blood pressure receptors (the baroreceptors), and taste buds of the mouth. This nucleus receives signals through the facial, glossopharyngeal, and vagus nerves (nerves VII, IX, and X).

THE EXTERNAL DISTRIBUTIONS OF THE CRANIAL NERVES

The Olfactory Nerves and Olfactory Tract (I). The olfactory nerves and olfactory tract are the sensory pathway for smell. The olfactory nerves are about 20 small nerves, each 1 to 2 cm in length, that arise from the *olfactory epithelium*, the sensory organ for smell located in the superiormost portion of the nasal cavity on the surfaces of the septum and superior concha. These 20 small olfactory nerves pass through an equal number of foramina in the cribriform plate of the ethmoid bone that forms the superior boundary of the nasal cavity and separates this cavity from the anterior fossa of the cranial cavity. Lying on the superior surface of the cribriform plate, between it and the inferior surface of the frontal lobe of the cerebrum, is the *olfactory bulb*, and leading posteriorly from this is the *olfactory tract* that terminates in the *olfactory areas of the cerebrum* located in and between the anterior medial portions of the two temporal lobes. After the olfactory nerves pass through the cribriform plate, some of their fibers terminate at synapses in the olfactory bulb, whereas others continue posteriorly in the olfactory tract, terminating in the cerebral olfactory areas, which will be discussed in Chapter 15.

Optic Nerve (II). Also illustrated in Figure 4–1 is the entire extent of the optic nerve. After leaving the eye, this nerve passes through the posterior recesses of the orbit, then through the optic foramen of the sphenoid bone, finally reaching the basal surface of the brain at the posteromedial limit of the frontal lobes. At this point the lateral half of the optic nerve continues posteriorly along the lateral surface of the hypothalamus, while the medial half abruptly bends medially and crosses to the opposite side through the *optic chiasm* lying anterior to the inferior hypothalamus. The crossed medial fibers then combine with the uncrossed lateral fibers on the opposite sides to form the two *optic tracts*, which course posteriorly along the lateral surfaces of the hypothalamus and terminate in the *lateral geniculate bodies* in the posterior thalamus.

Oculomotor Nerve (III), Trochlear Nerve (IV), and Abducens Nerve (VI). These are the nerves that control the eye movements. Note in Figure 4–1 that the oculomotor nerve leaves the brain stem near the midline of the anterior surface of the mesencephalon. The trochlear nerve arises from the lower posterolateral surface of the mesencephalon and then wraps around its side to its anterior aspect. The abducens nerve arises from the pons at its medullary junction. All these nerves then course through the *superior orbital fissure* into the orbit and innervate the extraocular and intraocular muscles as illustrated in Figure 4–3. The extraocular muscles attach to the eyes to cause the eye movements, and the intraocular muscles control focusing of the eyes and constriction of the pupil.

Trigeminal Nerve (V). As illustrated in Figure 4–1, the trigeminal nerve arises from the anterolateral surface of the midpons. It immediately enlarges for a distance of 1 cm to a diameter about two times its original diameter. This portion of the nerve is called the *trigeminal ganglion*; it contains the cell bodies of the nerve's sensory fibers and therefore is analogous to the dorsal root ganglia of the spinal nerves. Arising from the trigeminal ganglion are three major peripheral branches illustrated in Figure 4–4: (1) the *ophthalmic division*, (2) the *maxillary division*, and (3) the *mandibular division*.

The ophthalmic and maxillary divisions are both entirely sensory. The ophthalmic nerve passes through the upper reaches of the orbit and branches into the skin on the surface of the nose and superiorly over the forehead, supplying sensory nerves to these areas of the face and scalp and to the eye itself. It also sends branches to the nasal cavity and to the air sinuses.

The maxillary nerve leaves the cranial cavity through the *foramen rotundum*; then it passes through the inferior orbit and eventually through a bony canal underneath the eye to distribute over the anterior and lateral sides of the face to provide sensation. This nerve also supplies sensation to the upper teeth, the upper portions of the oral mucosa, and the mucosa of the nasal cavity and nasopharynx.

The mandibular nerve has both a sensory division and a motor division. This nerve passes through the foramen ovale into the space anterior and inferior to the temporal bone and medial to the ramus of the mandible, a space called the *infratemporal fossa*. The sensory division provides sensation to the most lateral portions of the face, the outer surfaces of the lower jaw and chin, the lower teeth, and the lower portions of the oral mucosa, including the anterior two thirds of the tongue. The motor division of the mandibular nerve innervates the muscles of mastication: the *temporalis*, the *masseter*, and the *medial* and *lateral pterygoid muscles*.

Facial Nerve (VII). The facial nerve arises from the brain stem at the posterolateral junction of the pons and the medulla (Fig. 4–1). It immediately passes through the internal auditory meatus and enters the facial canal in the temporal bone (shown in Fig 4–4) and enters the posterior facial region anterior and inferior to the ear, as illustrated in Figure 4–5. It then spreads through the superficial layers of the entire lateral and anterior facial regions to innervate all the muscles of facial expression as well as the buccinator

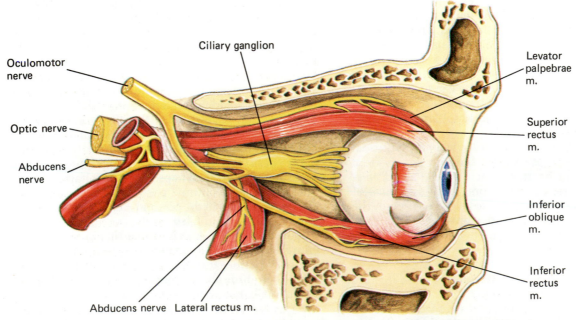

Ciliary ganglion

Oculomotor nerve

Optic nerve

Abducens nerve

Abducens nerve   Lateral rectus m.

Levator palpebrae m.

Superior rectus m.

Inferior oblique m.

Inferior rectus m.

Figure 4–3. The oculomotor nerve (III), trochlear nerve (IV) (not illustrated; in cut away portion), and abducens nerve (VI) innervating the eye muscles and also the internal structures of the eyeball through the ciliary ganglion.

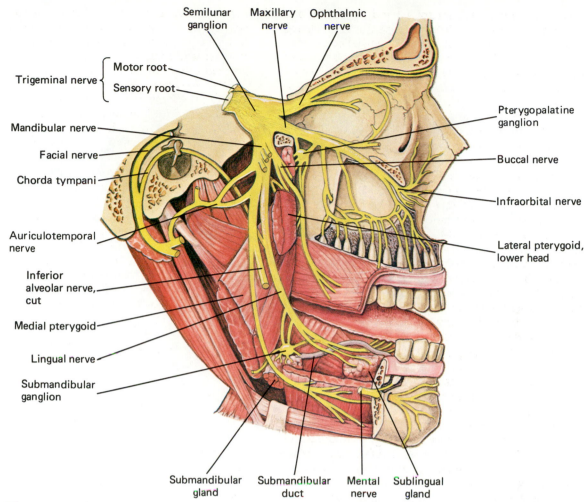

Figure 4—4. The trigeminal nerve (V) and its branches. Note also the chorda tympani that connects the lingual nerve (a branch of the trigeminal) to the facial nerve.

muscle of the cheek. In its early course anterior to the ear it passes through or adjacent to the parotid gland, one of the glands for secretion of saliva. This gland on occasion becomes cancerous, in which case the facial nerve is often destroyed by the cancer or by the surgery required to remove the cancer. The person then loses all capability of emotional expression on that side of the face. He also becomes unable to close his eye completely and cannot keep his lips closed adequately on that side, and his cheek bulges outward with food everytime he eats. This combination is an extremely depressing and even debilitating condition for the patient.

Note also in Figure 4–4 a branch of the facial nerve called the *chorda tympani* that passes through the middle ear and eventually combines with the *lingual nerve,* one of the branches of the mandibular nerve. The fibers from the chorda tympani finally terminate (1) in the *submandibular ganglion,* from which nerves then extend to the submandibular and sublingual glands to control salivary secretion, and (2) in the anterior two thirds of the tongue to provide taste sensation. The sensory ganglion of the facial nerve is called the *geniculate ganglion,* and it is located in the facial canal.

Vestibulocochlear Nerve (VIII). The vestibulocochlear nerve arises from the pons-medullary junction just lateral to the facial nerve (Fig. 4–1). It is a short nerve that immediately enters the *internal auditory meatus* to innervate both the *vestibular apparatus* (the organ of equilibrium) and the *cochlea* (the organ of hearing). Both of these organs are contained within the petrous portion of the temporal bone itself.

Glossopharyngeal Nerve (IX). The glossopharyngeal nerve arises from the upper lateral border of the medulla (Fig. 4–1) and passes immediately from the cranial vault via the jugular foramen into the posterior pharyngeal region. Figure 4–6 illustrates this nerve leaving the cranial vault along with the vagus and accessory nerves. The glossopharyngeal nerve provides sensory innervation for the mucous membrane of the pharynx as well as for the posterior third of the tongue, including both general sensory and taste sensation from this area. A motor branch of the glossopharyngeal nerve also innervates the superior pharyngeal muscles that are important for swallowing.

Vagus Nerve (X). The vagus nerve arises from the lateral border of the medulla (Fig. 4–1) inferior to the glossopharyngeal nerve. Its entry into the cervical re-

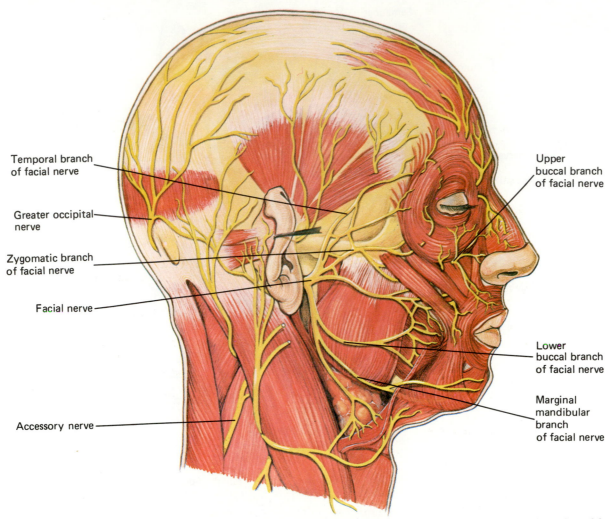

Temporal branch
of facial nerve

Greater occipital
nerve

Zygomatic branch
of facial nerve

Facial nerve

Accessory nerve

Upper
buccal branch
of facial nerve

Lower
buccal branch
of facial nerve

Marginal
mandibular
branch
of facial nerve

Figure 4–5. The facial nerve (VII) and the upper cervical portion of the accessory nerve (XI). Note the many branches of the facial nerve to the muscles of facial expression.

gion via the jugular foramen, along with the glosso-pharyngeal and accessory nerves, is illustrated in Figure 4–6. It then courses inferiorly into the thorax alongside the common carotid artery and internal jugular vein. Branches from the vagus nerves in the neck and upper thorax supply the muscles of the larynx for control of speech. At the superior border of the heart, parasympathetic nerve branches from the vagi, together with sympathetic branches from the thoracic sympathetic chains, form the *cardiac plexus* from which nerves then innervate the heart. The distal portions of the vagus nerves continue inferiorly through the thorax alongside the esophagus and pass through the diaphragm to form *anterior* and *posterior gastric nerves*. These give parasympathetic innervation to the stomach, the entire small intestine, the proximal colon, and other viscera of the abdominal cavity. Thus, the vagus nerve carries the majority of the parasympathetic nerve fibers that help to control the internal organs of the body, such as those controlling heart rate, stomach secretion, intestinal peristalsis, and so forth. The distal colon and pelvic organs receive parasympa-

thetic innervation via sacral spinal nerves, as we shall discuss later.

The vagus nerve also conducts sensory nerve fibers to the medulla from all the same visceral areas that receive vagal sympathetic fibers.

The Accessory Nerve (XI). The accessory nerve arises from the lateral border of the inferior medulla as well as from the anterolateral surface of the upper five segments of the spinal cord (Fig. 4–7). It leaves the cranial vault via the jugular foramen along with the glossopharyngeal and vagus nerves, as illustrated in Figure 4–7. Some of its fibers then join the vagus nerve and innervate the muscles of the larynx and the pharynx, but all of the fibers from the spinal roots of the accessory nerve course downward along the postero-lateral portion of the neck to provide motor control of portions of the sternocleidomastoid and trapezius muscles, as illustrated in Figure 4–6. These muscles also receive fibers from the cervical plexus in the neck region.

The Hypoglossal Nerve (XII). The hypoglossal nerve arises from the lateral border of the lower me-

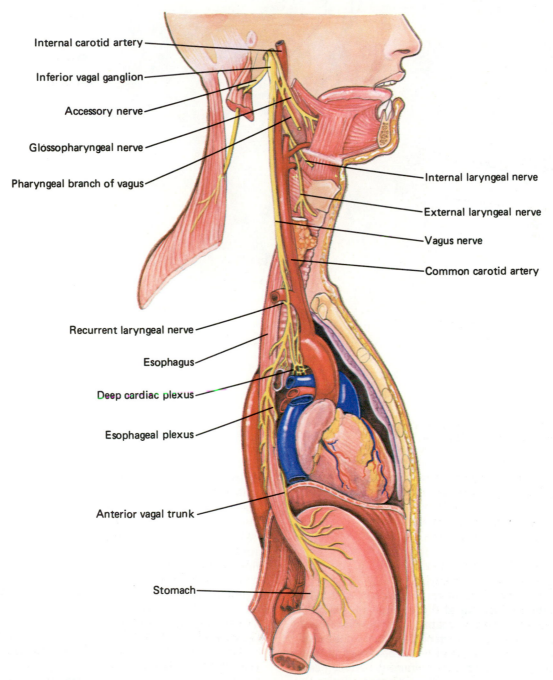

Figure 4 – 6. The glossopharyngeal nerve (IX), the vagus nerve (X), and the accessory nerve (XI).

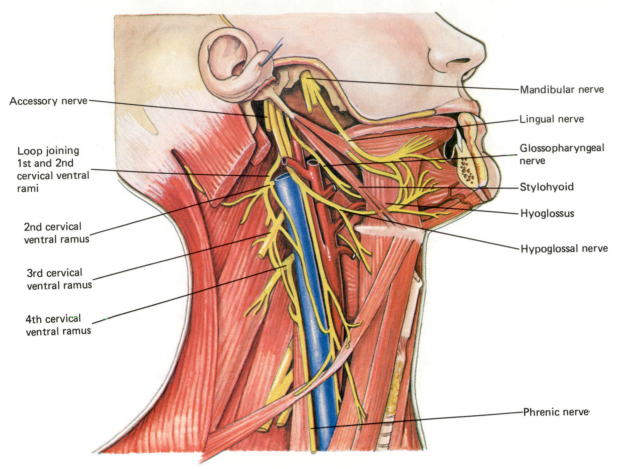

Accessory nerve

Loop joining
1st and 2nd
cervical ventral
rami

2nd cervical
ventral ramus

3rd cervical
ventral ramus

4th cervical
ventral ramus

Mandibular nerve

Lingual nerve

Glossopharyngeal
nerve

Stylohyoid

Hyoglossus

Hypoglossal nerve

Phrenic nerve

Figure 4 – 7. The lingual portion of the trigeminal nerve (V), the glossopharyngeal nerve (IX), the hypoglossal nerve (XII), and branches of the cervical plexus.

dulla (Fig. 4–1) anterior to the origins of the vagus and accessory nerves. It leaves the skull via the hypoglossal foramen. Figure 4–7 illustrates the entry of the hypoglossal nerve into the inframandibular region of the neck and its distribution to all the muscles of the tongue, including the hypoglossus, genioglossus, styloglossus, and intrinsic tongue muscles.

■ THE SPINAL NERVES

The anatomy of the spinal cord was discussed in the previous chapter, as was also the origin of the spinal nerves from the cord. To recapitulate, there is one pair of spinal nerves for each vertebral segment of the cord, and these nerves leave the sides of the vertebral canal through the two *intervertebral foramina* between each two successive vertebrae. The purpose of the remainder of this chapter will be to describe the distributions of the peripheral extensions of the spinal nerves.

Referring again to Figure 3–8 in the previous chapter, showing an overview of the entire spinal cord and its spinal nerves, one can count *8* pairs of cervical spinal nerves, specified as nerves C-1 through C-8; *12* pairs of thoracic spinal nerves, T-1 through T-12; *5* pairs of lumbar spinal nerves, L-1 through L-5; *5* pairs

of sacral spinal nerves, S-1 through S-5; and *1* pair of extremely small coccygeal spinal nerves.

The thoracic spinal nerves (those of the chest region) are relatively small. However, branches from these control the deep back muscles as well as the very large latissimus dorsi, the "climbing" muscle of the arm. Also, the thoracic spinal nerves give rise to the *intercostal nerves* that course around the body inferior to the ribs to supply the intercostal muscles and also to provide cutaneous innervation for the chest and abdomen. Extensions of the lower intercostal nerves also supply most of the muscles of the anterior abdominal wall.

By contrast the spinal nerves of the cervical, lumbar, and sacral regions are very large; it is these nerves that provide motor control and sensation for the neck region, the back of the head, the shoulders, the upper extremities, the lower trunk, and the lower extremities. Note also that all of these nerves, shortly after leaving the vertebral canal, interconnect among themselves to form four major plexuses:

1. The *cervical plexus*, formed by spinal nerves C-1 through C-5, supplies the neck, the back of the head, portions of the shoulder, and the diaphragm.

2. The *brachial plexus*, formed by C-5 through T-1,

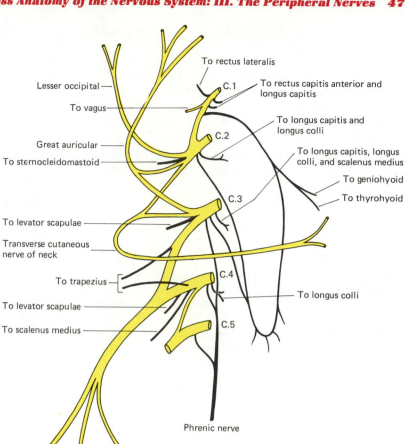

Lesser occipital

To vagus

Great auricular

To sternocleidomastoid

To levator scapulae

Transverse cutaneous nerve of neck

To trapezius

To levator scapulae

To scalenus medius

To rectus lateralis

C.1

To rectus capitis anterior and longus capitis

To longus capitis and longus colli

C.2

To longus capitis, longus colli, and scalenus medius

To geniohyoid

To thyrohyoid

C.3

C.4

To longus colli

C.5

Phrenic nerve

Supraclavicular

Figure 4–8. The cervical plexus and its branches.

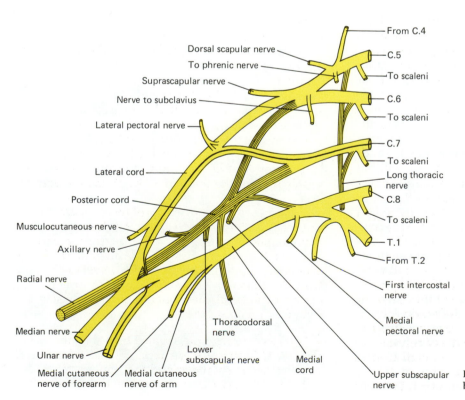

Dorsal scapular nerve

To phrenic nerve

Suprascapular nerve

Nerve to subclavius

Lateral pectoral nerve

Lateral cord

Posterior cord

Musculocutaneous nerve

Axillary nerve

Radial nerve

Median nerve

Ulnar nerve

Medial cutaneous nerve of forearm

Medial cutaneous nerve of arm

Lower subscapular nerve

Thoracodorsal nerve

Medial cord

Upper subscapular nerve

From C.4

C.5

To scaleni

C.6

To scaleni

C.7

To scaleni

Long thoracic nerve

C.8

To scaleni

T.1

From T.2

First intercostal nerve

Medial pectoral nerve

Figure 4–9. The brachial plexus and its branches.

supplies most of the shoulder region, the arm, the forearm, and the hand.

3. The *lumbar plexus*, from L-1 through L-4, supplies some muscles of the lower back, the lower abdomen, and the anterior and medial thigh.

4. The *sacral plexus*, from L-4 through S-5, supplies the gluteal region, the posterior and lateral thigh, the leg, and the foot.

Now, let us look at these plexuses individually to see some of the details of their organizations and distributions.

THE CERVICAL PLEXUS

Figure 4–8 illustrates the cervical plexus that originates mainly between C-1 and C-4 but also receives a small nerve bundle from C-5. Two upper branches of the cervical plexus, the *lesser occipital* and the *greater auricular nerves*, supply sensation to the back of the scalp and the region around the ear. Branching from the lower border of the plexus, several *supraclavicular nerves* supply sensation to the lower neck, and branching anteriorly is the *transverse cutaneous nerve* that provides sensation for the anterior neck. The muscles supplied by the cervical plexus are most of the deep neck muscles, the superficial anterior neck muscles, the levator scapulae, and portions of the trapezius and sternocleidomastoid.

The Phrenic Nerve. Note especially the origin of the *phrenic nerve* from the cervical plexus between C-3 and C-5. This is the principal nerve that controls respiration. The phrenic nerves on the two sides course downward through the neck, then through the thorax on each side of the heart, and finally terminate in the diaphragm to control the breathing movements of this important respiratory muscle. Fractures of the cervical vertebrae, which occur especially in diving accidents, often crush the spinal cord. If the fracture occurs at the fifth through seventh cervical vertebra, as frequently is the case, the connections to the phrenic nerve will remain intact and the person can still breathe. However, all the other spinal nerves below this level will not receive appropriate signals from higher centers, and the entire body except for the neck muscles and this major breathing muscle, the diaphragm, will be paralyzed. This is the condition called *quadriplegia*.

THE BRACHIAL PLEXUS

Figure 4–9 illustrates the brachial plexus, arising from C-5 through T-1. These five spinal nerves are all very large. They unite to form *upper, middle,* and *lower trunks* in the brachial plexus, and each of these in turn splits into an *anterior* and a *posterior division*. All the divisions pass under the clavicle and over the first rib into the axilla (the armpit) where they again fuse into three large bundles called the (1) *lateral cord*, (2) *posterior cord*, and (3) *medial cord*. All along the course of this plexus multiple nerves are given off to provide

both motor and sensory innervation to the shoulder, the superior portion of the anterior and lateral thorax, the arm, the forearm, and the hand. Table 4–2 gives the principal nerve branches of this plexus and the muscles that they innervate. In addition, multiple cutaneous branches conduct sensory signals from skin areas that roughly overlie these muscles.

TABLE 4–2 Major Nerves from the Brachial Plexus, and the Muscles Innervated

Nerve	Spinal Cord Segment	Muscle
Dorsal scapular	C-5	Rhomboideus major
Long thoracic	C-5,6,7	Serratus anterior
Suprascapular	C-5,6	Supraspinatus
		Infraspinatus
Subscapular	C-5,6	Teres major
		Subscapularis
Anterior thoracic	C-5 through T-1	Pectoralis minor
		Pectoralis major
Musculo-cutaneous	C-5,6,7	Biceps brachii
		Coracobrachialis
		Brachialis
Radial	C-5 through T-1	Triceps brachii
		Brachialis
		Brachioradialis
		Supinator
		Extensor carpi radialis longus and brevis
		Extensor carpi ulnaris
		Extensor digitorum
		Extensor pollicis longus
		Extensor pollicis brevis
		Extensor indicis
		Abductor pollicis longus
Median	C-6 through T-1	Pronator teres
		Pronator quadratus
		Palmaris longus
		Flexor carpi radialis
		Flexor digitorum superficialis
		Flexor digitorum profundus (radial half)
		Flexor pollicis longus
		Flexor pollicis brevis (shared with ulnar nerve)
		Abductor pollicis brevis
		Opponens pollicis
		Lumbricals (on radial side of hand)
Ulnar	C-8, T-1	Flexor carpi ulnaris
		Flexor digitorum profundus (ulnar half)
		Flexor pollicis brevis (shaped with median nerve)
		Flexor digiti minimi brevis
		Abductor digiti minimi
		Adductor pollicis
		Opponens digiti minimi
		Lumbricals (on ulnar side of hand)
		Interossei

NOTE: The above nerves are listed in the approximate order that they leave the brachial plexus.

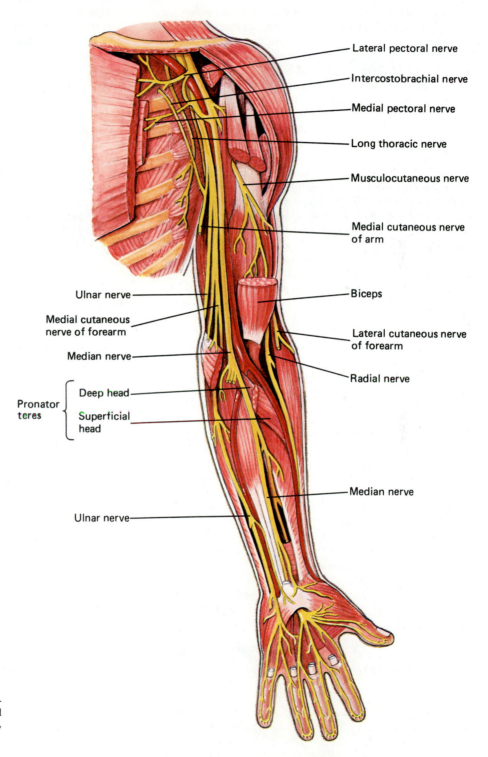

Lateral pectoral nerve

Intercostobrachial nerve

Medial pectoral nerve

Long thoracic nerve

Musculocutaneous nerve

Medial cutaneous nerve of arm

Ulnar nerve

Medial cutaneous nerve of forearm

Median nerve

Pronator teres { Deep head

Superficial head

Biceps

Lateral cutaneous nerve of forearm

Radial nerve

Median nerve

Ulnar nerve

Figure 4–10. The courses of the musculocutaneous, radial, median, and ulnar nerves down the arm, forearm, and hand.

Courses of the Major Nerves in the Upper Extremity. Figure 4–10 illustrates the courses of the four major nerves in the upper extremity. These supply both sensation to the skin and motor signals to the muscles along their courses.

The *musculocutaneous nerve,* upon leaving the plexus, curves laterally through the deep portions of the anterior arm and then continues superficially down the lateral surface of the forearm to provide sensory innervation. As it passes through the arm it innervates the anterior arm muscles listed in Table 4–2, the most important of which is the *biceps brachii,* which causes flexion of the forearm.

The *radial nerve,* after leaving the brachial plexus, curves posteriorly and laterally behind the humerus and enters the forearm over the lateral epicondyle of the humerus. Thereafter it follows mainly the lateral border of the radius and finally continues into the posterior portions of the thumb and first three fingers. The list in Table 4–2 of muscles innervated by this nerve demonstrates its importance in controlling movements of the upper extremity. Careful study will show that they are the muscles in the posterior arm and dorsal and lateral aspects of the forearm and hand. The principal movements that they cause are (1) extension of the elbow, (2) supination of the forearm and hand, (3) extension of the wrist, fingers, and thumb, and (4) abduction of the thumb.

The *median nerve,* after leaving the brachial plexus, passes down the anteromedial portion of the arm, then distally in the anterolateral portions of the forearm, passing next into the hand's lateral palm and into the anterior compartments of the thumb and first two fingers and lateral half of the third finger. Table 4–2 shows that the median nerve innervates approximately the lateral two thirds of the muscles in the anterior compartment of the forearm and lateral third

of the anterior muscles of the hand. The major movements that these muscles cause are (1) pronation of the forearm and hand, (2) flexion of the wrist, fingers, and thumb, (3) abduction of the wrist, (4) abduction of the thumb, and (5) opponens motion of the thumb.

The *ulnar nerve* passes down the posteromedial portion of the arm, then behind the medial epicondyle of the humerus at the elbow joint, and finally alongside the ulna to enter the medial border of the hand, supplying both the anterior and posterior surfaces of the little finger and the medial half of the third finger. Along the course of this nerve, cutaneous branches provide sensation for the anteromedial surface of the forearm and the surface of the hand medial to the midline of the third finger, Also, Table 4–2 shows that the ulnar nerve innervates approximately the medial third of the muscles in the anterior forearm and the medial two thirds of the muscles in the anterior hand. These muscles cause mainly (1) flexion of the wrist and fingers (shared functions with the median nerve), (2) abduction of the fingers, (3) adduction of the fingers and thumb, and (4) opponens motion of the little finger.

Though it is not important at this point to memorize the exact distributions of all these nerves, surgeons find it essential to know these distributions precisely. When a nerve injury occurs, the surgeon can determine which nerve is damaged as well as where it is damaged by studying the areas of sensory loss on the skin and the specific muscles that are paralyzed. One of the most common points for severe damage is in the brachial plexus itself. For instance, if the arm is pulled upward with tearing force, the medial cord of the brachial plexus is especially likely to be damaged. From Figures 4–9 and 4–10 one will see that this can completely sever the fibers to the ulnar nerve and can destroy many of the fibers to the median nerve as well,

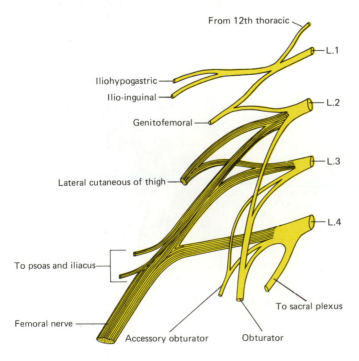

Figure 4–11. The lumbar plexus and its branches, especially the femoral nerve.

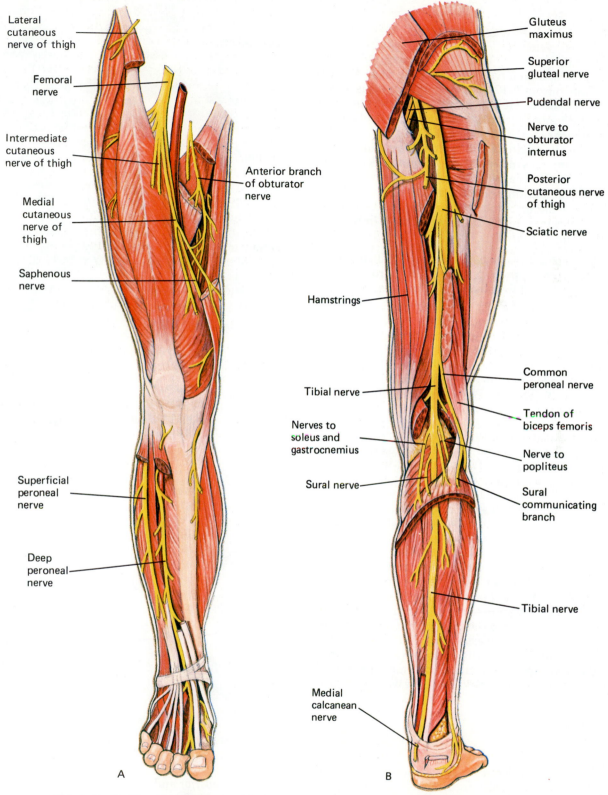

Figure 4–12. The principal nerves of the lower limbs shown in anterior (A) and posterior (B) views.

mainly paralyzing the muscles along the medial side of the anterior forearm and hand and also causing sensory loss along these surface areas.

THE LUMBAR PLEXUS AND THE FEMORAL NERVE

The lumbar spinal nerves between L-1 and L-4 and a small branch from T-12 form the lumbar plexus, illustrated in Figure 4–11. This plexus lies on the posterior wall of the lumbar region of the abdominal cavity, then sends branches inferiorly along the lateral wall of the pelvis. Near its origin, branches from the plexus innervate a few muscles in the abdominal and back regions, including the low back muscles, the psoas major, the quadratus lumborum, and the most inferior portions of the abdominal muscles. However, this plexus mainly sends nerves into the thigh, the three most important of which are as follows:

1. The *lateral femoral cutaneous nerve* (illustrated in Fig. 4–12A) enters the anterolateral thigh from underneath the inguinal ligament. This then passes downward along the lateral side of the thigh to provide sensory innervation to the skin.

2. The *obturator nerve* (Figs. 4–11 and 4–12A) originates from the lower portion of the lumbar plexus and enters the medial side of the thigh. This is mainly a motor nerve, controlling the large array of adductor muscles of the thigh that pull the legs together and are listed in Table 4–3.

3. The *femoral nerve* (Figs. 4–11 and 4–12A) is by far the largest of the nerves from the lumbar plexus. Figure 4–12A shows that this nerve runs for a short distance parallel to the femoral artery in the upper thigh but divides into multiple large branches about 10 cm below the inguinal ligament. Some of these are muscular branches and some cutaneous. As shown in Figure 4–12A, and also in Table 4–3, the *muscular branches* innervate all the muscles of the anterior thigh, the most important of which are the four heads of the very large quadriceps femoris muscle and also the sartorius muscle. These are both the principal flexors of the thigh and the single massive extensor muscle for extending the knee joint. In addition to these muscular branches, there are two principal cutaneous branches: the *anterior femoral cutaneous* nerve that innervates the skin of the anteromedial thigh all the way to the knee and the *saphenous nerve* that innervates the medial surfaces of the leg from the knee all the way to the foot.

THE SACRAL PLEXUS AND THE SCIATIC NERVE

The sacral plexus derives mainly from spinal nerves L-5 through S-3, but also from small branches of L-4 and S-4 to the coccygeal spinal nerves (Co.) as illustrated in Figure 4–13. This plexus lies along the posterior wall of the pelvis. Its principal branches are the following:

TABLE 4–3 Major Nerves from the Lumbar and Sacral Plexuses, and the Muscles Innervated

Nerve	Spinal Cord Segment	Muscle
Lumbar Plexus		
Obturator	L-2,3,4	Pectineus (shared with femoral nerve)
		Adductor longus
		Adductor magnus (shared with sciatic)
		Adductor brevis
		Gracillis
Femoral	L-2,3,4	Sartorius
		Iliacus
		Pectineus (shared with obturator nerve)
		Quadriceps femoris
		1. Rectus femoris
		2. Vastus medialis
		3. Vastus lateralis
		4. Vastus intermedius
Other muscular branches	L-2,3	Psoas major
		Quadratus lumborum
Sacral Plexus		
Superior gluteal	L-4,5,S-1	Gluteus medius
		Gluteus minimus
		Tensor fasciae latae
Inferior gluteal	L-5,S-1,2	Gluteus maximus
Sciatic	L-4 through S-3	Adductor magnus (shared with obturator nerve)
		Obturator internus
		Superior gemellus
		Inferior gemellus
		Quadratus femoris
Tibial portion of the sciatic	L-4 through S-3	Biceps femoris (shared with peroneal nerve)
		Semitendinosus
		Semimembranosus
		Gastrocnemius
		Soleus
		Popliteus
		Tibialis posterior
		Flexor digitorum longus
		Flexor hallucis longus
		Plantar and medial foot muscles
Peroneal portion of the sciatic	L-4 through S-2	Biceps femoris (shared with tibial nerve)
		Tibialis anterior
		Peroneus longus
		Peroneus brevis
		Extensor digitorum longus
		Extensor hallucis longus
		Dorsal and lateral foot muscles
Pudendal	S-2,3,4	Muscles of urogenital triangle
Other muscular branches	S-3,4	Levator ani
		Coccygeus
		External anal sphincter

1. The *superior* and *inferior gluteal nerves* exit laterally from the pelvis to control mainly the gluteal muscles of the buttock and lateral hip, which cause both backward extension and lateral abduction at the hip joint.

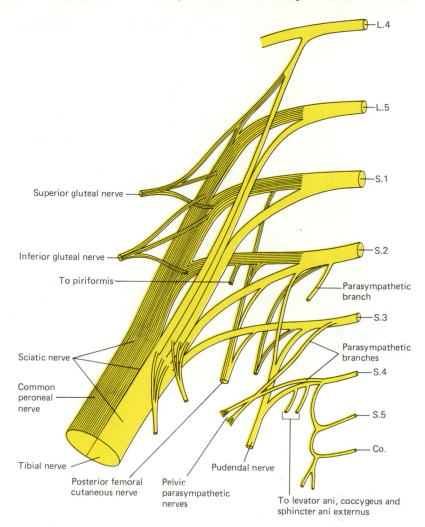

Figure 4 – 13. The sacral plexus and its branches, especially the sciatic nerve.

2. The *posterior femoral cutaneous nerve* passes down the back of the thigh and upper portion of the leg to provide sensation.

3. The *pudendal nerve* passes to the perineum and to the external genital organs, including the penis and scrotum in the male and the vagina in the female, to subserve sexual functions and sensations.

4. *Pelvic parasympathetic nerve branches* derived from sacral spinal nerves S-2 through S-4 pass to the pelvic organs to initiate such functions as defecation (emptying the rectum) and micturition (emptying the bladder), as well as playing roles in the acts of erection, orgasm, and ejaculation during sexual intercourse.

5. Several small nerves derived from spinal nerves S-3 and S-4 control the voluntary muscle sphincters around the anus and around the external urethra. These allow the person to prevent defecation or micturition when these are inconvenient.

6. The very large *sciatic nerve* is so important that it deserves special consideration as follows.

The Sciatic Nerve. The sciatic nerve, illustrated in Figures 4 – 12B and 4 – 13, is by far the largest nerve of the body. It originates in the sacral plexus mainly from spinal segments L-5 to S-2, then leaves the posterior

pelvis medial to the ischial tuberosity and courses distally in the posterior compartment of the thigh embedded between the hamstring muscles. Along this course, it supplies muscular branches to all the deep muscles posterior to the hip joint and also those in the posterior thigh, all listed in Table 4 – 3. These muscles cause extension of the thigh, and the hamstrings in the posterior thigh are the strong flexors of the knee joint.

At the lower end of the thigh, immediately above the knee joint, the sciatic nerve divides into two major branches, the *tibial nerve* and the *common peroneal nerve*. The tibial nerve continues distally in the posterior compartment of the leg, lying in the interval between the tibia and the fibula. It finally enters the medial side of the foot behind the medial malleolus. In this course it supplies sensory branches to the skin as well as branches to all the muscles of the back of the leg, especially to the soleus, the gastrocnemius, the tibialis posterior, and the flexors of the toes. The principal functions of these muscles are to flex the foot and toes downward and also to invert the foot.

The common peroneal nerve wraps around the lateral side of the fibula, where it divides into the *superficial* and *deep peroneal nerves*. The superficial peroneal nerve descends in the lateral leg to provide motor in-

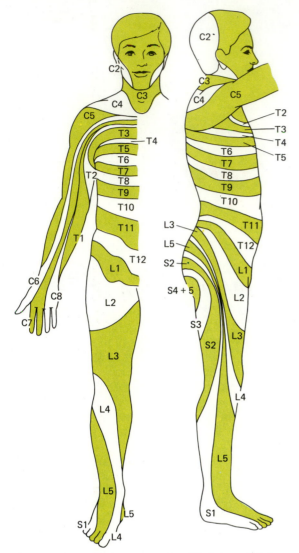

Figure 4–14. The dermatomes. (Modified from Grinker and Sahs: Neurology. Springfield, Ill., Charles C Thomas, 1966.)

nervation to the peroneus muscles and cutaneous innervation to the dorsum of the foot. These muscles are the everters of the foot. The deep peroneal nerve descends in the anterior compartment of the leg in relation to the anterolateral muscles (the tibialis anterior and the extensor muscles of the toes) and controls them. Their principal function is upward flexion of the foot.

THE DERMATOMES

Each spinal nerve provides sensory innervation to a "segmental field" of the skin called a *dermatome*. This

is true even though most of the spinal nerves appear to become mixed up with other spinal nerves as they pass through the plexuses. The different dermatomes for the separate spinal nerves are illustrated in Figure 4–14. However, in this figure, the dermatomes are shown as if there were distinct borders between the adjacent dermatomes. This is only partly true because the distal branches of the nerves invade one another's territory. For this reason, an entire single spinal nerve can often be destroyed without significant loss of sensation in the skin, but when several adjacent spinal nerves are destroyed, one can easily establish the extent of sensory loss and from this determine the segmental level of the nerve injury.

Figure 4–14 shows that the anal region of the body lies in the dermatome of the most distal cord segments, S-4 and S-5. In the embryo, this is the tail region and the most distal portion of the body. The lower limbs develop from the lumbar and upper sacral levels of the embryo rather than from the distal sacral segments, which is also evident from the dermatomal map, for the dermatomes of this limb are L-2 through S-2.

Note also in Figure 4–14 that the face and anterior half of the head are not designated by spinal nerve dermatomes. But remember that sensation in these areas is served by the three branches of the fifth cranial nerve (the trigeminal nerve).

REFERENCES (CHAPTERS 2 TO 4)

Anderson, J. E.: The cranial nerves. *In* Grant's Atlas of Anatomy. Baltimore, Williams & Wilkins, 1978, pp. 8–1 to 8–12.
Bloom, W., and Fawcett, D. W.: The nervous tissue. *In* A Textbook of Histology, 11th Ed. Philadelphia, W. B. Saunders, 1986.
Carpenter, M. D.: Human Neuroanatomy, 18th Ed. Baltimore, Williams & Wilkins, 1983.
Copenhaver, W. M., Kelly, D. E., and Wood, R. L.: Nervous system. *In* Bailey's Textbook of Histology, 17th Ed. Baltimore, Williams & Wilkins, 1978, pp. 290–357.
Figge, F. H. J.: The central nervous system. *In* Sobotta/Figge Atlas of Human Anatomy. Vol. III. Baltimore, Urban & Schwarzenberg, 1977, pp. 1–131.
Fujita, T., Tanaka, K., and Tokunaga, J.: Muscles, nerves, and brain. *In* SEM Atlas of Cells and Tissues. New York, Igaku-Shoin, 1981, pp. 312–328.
Ham, A. W., and Cromack, D. H.: Nervous tissue. *In* Histology, 8th Ed. Philadelphia, J. B. Lippincott, 1979, pp. 483–539.
Hammersen, F.: Nervous system. *In* Sobotta/Hammersen Histology. Baltimore, Urban & Schwarzenberg, 1980, pp. 203–216.
Hammersen, F.: Nervous tissue and neuroglia. *In* Sobotta/Hammersen Histology. Baltimore, Urban & Schwarzenberg, 1980, pp. 80–93.
Kandel, E. R., and Schwartz, J. H.: Principles of Neural Science, 2nd Ed. New York, Elsevier, 1985.
Langman, J., and Woerdeman, M. W.: Head and neck. *In* Atlas of Medical Anatomy. Philadelphia, W. B. Saunders, 1978, pp. 351–472.
Leeson, T. S., and Leeson, C. R.: Nervous tissue. *In* A Brief Atlas of Histology. Philadelphia, W. B. Saunders, 1979, pp. 89–104.
Leeson, T. S., and Leeson, C. R.: Nervous tissue. *In* Histology, 4th Ed. Philadelphia, W. B. Saunders, 1981, pp. 216–256.
Netter, F. H.: Nervous system. *In* The CIBA Collection of Medical Illustrations. Vol. 1, Summit, N. J., CIBA Medical Education Division, 1972.
Pernkopf, E.: Brain and meninges. *In* Atlas of Topographical and Applied Human Anatomy. Vol. I. Philadelphia, W. B. Saunders, 1980, pp. 29–135.
Snell, R. S.: Clinical Neuroanatomy. Boston, Little, Brown, 1980.
Williams, P. L., and Warwick, R.: The nervous system. *In* Gray's Anatomy, 36th British Edition. Philadelphia, W. B. Saunders, 1980, pp. 802–1215.

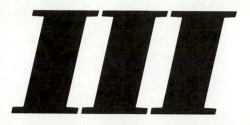

MEMBRANE BIOPHYSICS AND NERVE SIGNALS

5

Transport of Ions Through the Cell Membrane

The transmission of nerve signals is the basis of function in the nervous system. However, to understand nerve transmission, one must first be familiar with the biophysics of the nerve cell membrane, especially with the transport of ions through this membrane and the development of electrical potentials across it. It is the purpose of this chapter to discuss the basic principles of these phenomena; the following chapter will utilize these basic principles to explain nerve transmission itself.

■ CONCENTRATIONS OF IONS AND OTHER SUBSTANCES ON THE OUTSIDE AND INSIDE OF THE CELL MEMBRANE

Figure 5–1 gives the approximate compositions of the *extracellular fluid*, which lies outside the cell membranes, and the *intracellular fluid,* inside the cells. Note that the extracellular fluid contains large quantities of *sodium* but only small quantities of *potassium*. Exactly the opposite is true of the intracellular fluid. Also, the extracellular fluid contains large quantities of chloride, whereas the intracellular fluid contains very little. But the concentrations of phosphates, essentially all of which are organic metabolic intermediates, and proteins in the intracellular fluid are considerably greater than in the extracellular fluid. These many differences are extremely important to the life of the cell and especially to nerve transmission of signals.

THE LIPID BARRIER AND THE TRANSPORT PROTEINS OF THE CELL MEMBRANE

The cell membrane consists almost entirely of a *lipid bilayer* with large numbers of protein molecules floating in the lipid, many penetrating all the way through, as illustrated in Figure 5–2.

The lipid bilayer is not miscible with either the extracellular fluid or the intracellular fluid. Therefore, it constitutes a barrier for the movement of most water molecules and water-soluble substances between the extracellular and intracellular fluid compartments. However, as illustrated by the left-hand arrow of Figure 5–2, a few substances can penetrate this bilayer and can either enter the cell or leave it, passing directly through the lipid substance itself.

The protein molecules, on the other hand, have entirely different transport properties. Their molecular structures interrupt the continuity of the lipid bilayer and therefore constitute an alternate pathway through the cell membrane. Most of these penetrating proteins, therefore, are *transport proteins*. Different proteins function differently. Some have watery spaces all the way through the molecule and allow free movement of certain ions or molecules; these are called *channel proteins*. Others, called *carrier proteins,* bind with substances that are to be transported, and conformational changes in the protein molecules then move the substances through the interstices of the molecules to the other side of the membrane. Both the channel proteins and the carrier proteins are highly selective in the type or types of molecules or ions that are allowed to cross the membrane.

Diffusion Versus Active Transport. Transport through the cell membrane, either directly through the lipid bilayer or through the proteins, occurs by one of two basic processes, *diffusion* (which is also called "passive transport") or *active transport*. Although there are many different variations of these two basic mechanisms, as we see later in this chapter, diffusion means random molecular movement of substances molecule by molecule either through intermolecular spaces in the membrane or in combination with a carrier protein. The energy that causes diffusion is the energy of the normal kinetic motion of matter. By contrast, active transport means movement of ions or

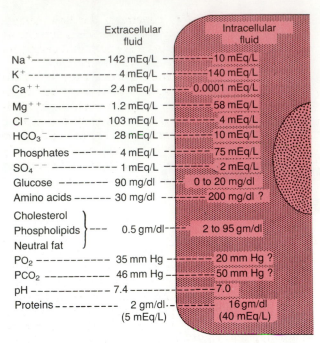

Extracellular fluid
Intracellular fluid
Na⁺ 142 mEq/L 10 mEq/L
K⁺ 4 mEq/L 140 mEq/L
Ca⁺⁺ 2.4 mEq/L 0.0001 mEq/L
Mg⁺⁺ 1.2 mEq/L 58 mEq/L
Cl⁻ 103 mEq/L 4 mEq/L
HCO₃⁻ 28 mEq/L 10 mEq/L
Phosphates 4 mEq/L 75 mEq/L
SO₄⁻⁻ 1 mEq/L 2 mEq/L
Glucose 90 mg/dl 0 to 20 mg/dl
Amino acids 30 mg/dl 200 mg/dl ?
Cholesterol
Phospholipids 0.5 gm/dl 2 to 95 gm/dl
Neutral fat
PO₂ 35 mm Hg 20 mm Hg ?
PCO₂ 46 mm Hg 50 mm Hg ?
pH 7.4 7.0
Proteins 2 gm/dl (5 mEq/L) 16 gm/dl (40 mEq/L)

Figure 5–1. Chemical compositions of extracellular and intracellular fluids.

other substances across the membrane in combination with a carrier protein but additionally *against an energy gradient,* such as from a low concentration state to a high concentration state, a process that requires an additional source of energy besides kinetic energy to cause the movement. Let us explain in more detail the basic physics and physical chemistry of these two separate processes.

DIFFUSION

All molecules and ions in the body fluids, including both water molecules and dissolved substances, are in constant motion, each particle moving its own separate way. Motion of these particles is what physicists call heat—the greater the motion, the higher the temperature—and motion never ceases under any conditions except at absolute zero temperature. When a moving molecule, A, approaches a stationary molecule, B, the electrostatic and internuclear forces of molecule A repel molecule B, transferring some of the energy of motion to molecule B. Consequently, molecule B gains kinetic energy of motion, while molecule A slows down, losing some of its kinetic energy. Thus, as shown in Figure 5–3, a single molecule in solution bounces among the other molecules first in one direction, then another, then another, and so forth, bouncing randomly billions of times each second.

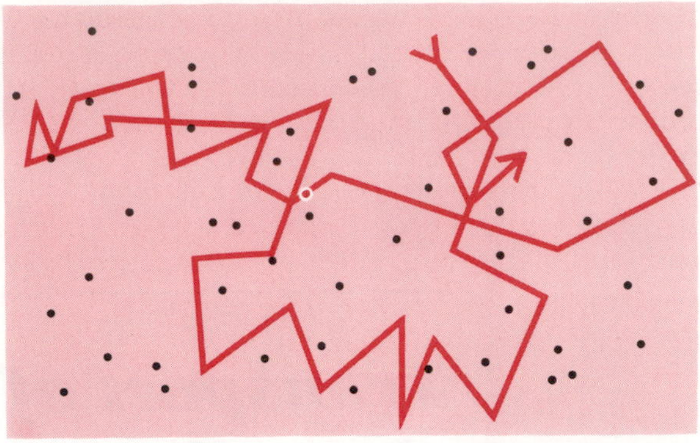

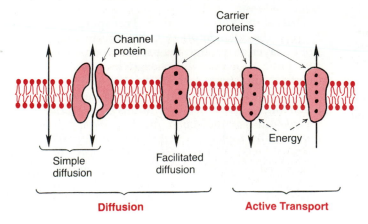

Figure 5–3. Diffusion of a fluid molecule during a billionth of a second.

This continual movement of molecules among each other in liquids, or in gases, is called *diffusion.* Ions diffuse in exactly the same manner as whole molecules, and even suspended colloid particles diffuse in a similar manner except that they diffuse far less rapidly than molecular substances because of their very large sizes.

DIFFUSION THROUGH THE CELL MEMBRANE

Diffusion through the cell membrane is divided into two separate subtypes called *simple diffusion* and *facilitated diffusion.* Simple diffusion means the molecular kinetic movement of molecules or ions through a membrane opening or intermolecular spaces without the necessity of binding with carrier proteins in the membrane. The rate of diffusion is determined by the amount of substance available, by the velocity of kinetic motion, and by the number of openings in the cell membrane through which the molecules or ions can

Carrier proteins
Channel protein
Energy
Simple diffusion
Facilitated diffusion
Diffusion
Active Transport

Figure 5–2. Transport pathways through the cell membrane and the basic mechanisms of transport.

move. On the other hand, facilitated diffusion requires the interaction of the molecules or ions with a carrier protein that aids its passage through the membrane, probably by binding chemically with it and shuttling it through the membrane in this form.

Simple diffusion can occur through the cell membrane by two pathways: through the interstices of the lipid bilayer and through watery channels in some of the transport proteins, as illustrated to the left in Figure 5–2.

Simple Diffusion Through the Lipid Bilayer

Diffusion of Lipid-Soluble Substances. In experimental studies, the lipids of cells have been separated from the proteins and then reconstituted as artificial membranes consisting of a lipid bilayer but without any transport proteins. Using such an artificial membrane, the transport properties of the lipid bilayer by itself have been determined.

One of the most important factors that determines how rapidly a substance will move through the lipid bilayer is the *lipid solubility* of the substance. For instance, the lipid solubilities of oxygen, nitrogen, carbon dioxide, and alcohols are very high, so that all these can dissolve directly in the lipid bilayer and diffuse through the cell membrane in exactly the same manner that diffusion occurs in a watery solution. For obvious reasons, the rate of diffusion of these substances through the membrane is directly proportional to their lipid solubility. Especially large quantities of oxygen can be transported in this way; therefore, oxygen is delivered to the interior of the cell almost as though the cell membrane did not exist.

Transport of Water and Other Lipid-Insoluble Molecules. Even though water is highly insoluble in the membrane lipids, nevertheless it penetrates the cell membrane very readily, much of it passing directly through the lipid bilayer and still more passing through protein channels. The rapidity with which water molecules can penetrate the cell membrane is astounding. As an example, the total amount of water that diffuses in each direction through the red cell membrane during each second is approximately 100 times as great as the volume of the red cell itself.

The reason for the large amount of diffusion of water through the lipid bilayer is still not certain, but it is believed that water molecules are small enough and their kinetic energy great enough that they can simply penetrate like bullets through the lipid portion of the membrane before the "hydrophobic" character of the lipids can stop them.

Other lipid-insoluble molecules also can pass through the lipid bilayer in the same way as water molecules if they are small enough. However, as they become larger, their penetration falls off extremely rapidly. For instance, the diameter of the urea molecule is only 20 per cent greater than that of water. Yet its penetration through the cell membrane is about a thousand times less than that of water. The glucose molecule, which has a diameter only three times that

of the water molecule, penetrates the lipid bilayer 100,000 times less rapidly than water, thus illustrating that the only lipid-insoluble molecules that can penetrate the lipid bilayer are the very small ones.

Failure of Ions to Diffuse Through the Lipid Bilayer. Even though water and other very small uncharged molecules diffuse easily through the lipid bilayer, ions — even small ones, such as hydrogen ions, sodium ions, potassium ions, and so forth — penetrate the lipid bilayer about one million times less rapidly than does water. Therefore, any significant transport of these through the cell membrane must occur through channels in the proteins, as we will discuss shortly.

The reason for the impenetrability of the lipid bilayer to ions is the electrical charge of the ions; this impedes ionic movement in two separate ways: (1) The electrical charge of these ions causes multiple molecules of water to become bonded to the ions, forming so-called *hydrated ions.* This greatly increases the sizes of ions, which alone impedes penetration of the lipid bilayer. (2) Even more important, the electrical charge of the ion also interacts with the charges of the lipid bilayer in the following way. It will be recalled that each half of the bilayer is composed of "polar" lipids that have an excess of negative charge facing toward the surfaces of the membrane. Therefore, when a charged ion tries to penetrate either the negative or the positive electrical barrier, it is instantaneously repulsed.

To summarize, Table 5–1 gives the relative permeabilities of the lipid bilayer to a number of different molecules or ions of different diameters. Note especially the *extremely poor permeance of the ions* because of their electrical charges and the *poor permeance of glucose* because of its molecular diameter. Note also that glycerol penetrates the membrane almost as easily as urea even though its diameter is almost twice as great. The reason for this is a slight degree of lipid solubility.

Simple Diffusion Through Protein Channels and "Gating" of These Channels

The protein channels are believed to be watery pathways through the interstices of the protein molecules. In fact, computerized three-dimensional reconstructions of some of these proteins have demonstrated

TABLE 5–1 Relationship of Effective Diameters of Different Substances to Their Lipid Bilayer Permeabilities

Substance	Diameter (nm)	Relative Permeability
Water molecule	0.3	1.0
Urea molecule	0.36	0.0006
Hydrated chloride ion	0.386	0.00000001
Hydrated potassium ion	0.396	0.0000000006
Hydrated sodium ion	0.512	0.0000000002
Glycerol	0.62	0.0006
Glucose	0.86	0.000009

tube-shaped channels from the extracellular to the intracellular ends. Therefore, substances can diffuse directly through these channels from one side of the membrane to the other. However, the protein channels are distinguished by two important characteristics: (1) They are often selectively permeable to certain substances. (2) Many of the channels can be opened or closed by *gates.*

Selective Permeability of Different Protein Channels. Most, but not all, protein channels are highly selective for the transport of one or more specific ions or molecules. This results from the characteristics of the channel itself, such as its diameter, its shape, and the nature of the electrical charges along its inside surfaces. To give an example, one of the most important of the protein channels, the so-called *sodium channels,* calculate to be only 0.3 by 0.5 nm in size, but more importantly the inner surfaces of these channels are *strongly negatively charged,* as illustrated by the negative signs inside the channel proteins in the top panel of Figure 5–4. These strong negative charges are postulated to pull sodium ions more than they pull other physiologically important ions into the channels because of the smaller ionic diameter of the dehydrated sodium ions than for the others. Once in the channel, the sodium ions then diffuse in either direction according to the usual laws of diffusion. Thus, the sodium channel is specifically selective for the passage of sodium ions.

On the other hand, another set of protein channels is selective for potassium transport, illustrated in the lower panel of Figure 5–4. These channels calculate to be slightly smaller than the sodium channels, only 0.3 by 0.3 nm, but *they are not negatively charged.* Therefore, no strong attractive force is pulling ions into the

channels, and the ions are not pulled away from the water molecules that hydrate them. The hydrated form of the potassium ion is considerably smaller than the hydrated form of sodium because the sodium ion has one whole orbital set of electrons less than the potassium ion, which allows the sodium nucleus to attract far more water molecules than can the potassium. Therefore, the smaller hydrated potassium ions can pass easily through this smaller channel, whereas sodium ions are mainly rejected, thus once again providing selective permeability for a specific ion.

Gating of Protein Channels. Gating of protein channels provides a means for controlling the permeability of the channels. This is illustrated in both the upper and the lower panels of Figure 5–4 for the sodium and the potassium ion. It is believed that the gates are actual gatelike extensions of the transport protein molecule, which can close over the opening of the channel or can be lifted away from the opening by a conformational change in the shape of the protein molecule itself. In the case of the sodium channels, this gate opens and closes on the outer surface of the cell membrane, whereas for the potassium channels it opens and closes on the inner surface.

The opening and closing of gates are controlled in two principal ways:

1. *Voltage gating.* In this instance, the molecular conformation of the gate responds to the electrical potential across the cell membrane. For instance, when there is a strong negative charge on the inside of the cell membrane, the sodium gates remain tightly closed; on the other hand, when the inside of the membrane loses its negative charge, these gates open suddenly and allow tremendous quantities of sodium to pass inward through the sodium pores (until still another set of gates at the cytoplasmic ends of the channels close, as explained in Chapter 6). This is the basic cause of action potentials in nerves that are responsible for nerve signals. The potassium gates also open when the inside of the cell membrane becomes positively charged, but this response is much slower than that for the sodium gates. These events are discussed in the following chapter.

2. *Ligand gating.* Some protein channel gates are opened by the binding of another molecule with the protein; this causes a conformational change in the protein molecule that opens or closes the gate. This is called *ligand gating,* and the substance that binds is the *ligand.* One of the most important instances of ligand gating is the effect of acetylcholine on the so-called *acetylcholine channel.* This opens the gate of this channel, providing a pore about 0.65 nm in diameter that allows all molecules and positive ions smaller than this diameter to pass through. This gate is exceedingly important in the transmission of signals from one nerve cell to another (Chapter 47) and from nerve cells to muscle cells (Chapter 25).

The Open-State, Closed-State of Gated Channels. Figure 5–5 illustrates an especially interesting characteristic of voltage-gated channels. This figure shows two recordings of electrical current flowing through a

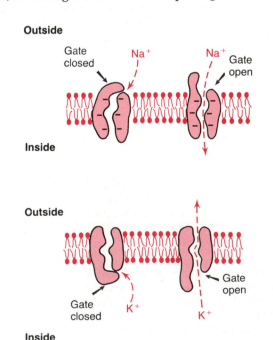

Figure 5–4. Transport of sodium and potassium ions through protein channels. Also shown are conformational changes of the channel protein molecules that open or close the "gates" guarding the channels.

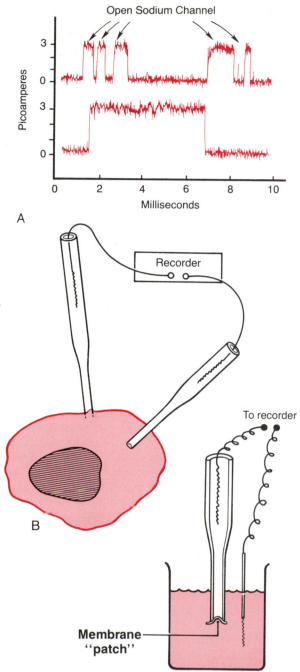

A

B

Figure 5–5. *A*, Record of current flow through a single voltage-gated sodium channel, demonstrating the all-or-none principle for opening of the channel. *B*, The "patch-clamp" method for recording current flow through a single protein channel. To the left, recording is performed from a "patch" of a living cell membrane. To the right, recording is from a membrane patch that has been torn away from the cell.

time or almost all the time, whereas at another voltage level it may remain open either all or most of the time. However, at in-between voltages, the gates tend to snap open and closed intermittently, as illustrated in the upper recording, giving an average current flow somewhere between the minimum and the maximum.

THE PATCH-CLAMP METHOD FOR RECORDING ION CURRENT FLOW THROUGH SINGLE CHANNELS. One might wonder how it is technically possible to record ion current flow through single channels as shown in Figure 5–5A. This has been achieved by using the "patch-clamp" method illustrated in Figure 5–5B. Very simply, a micropipette, having a tip diameter of only 1 or 2 μm, is abutted against the outside of a cell membrane. Then suction is applied inside the pipette to pull the membrane slightly into the tip of the pipette. This creates a seal where the edges of the pipette touch the cell membrane. The result is a minute "patch" at the tip of the pipette through which current flow can be recorded.

Alternatively, as shown to the right in Figure 5–5B, the small cell membrane patch at the end of the pipette can be torn away from the cell. The pipette with its sealed patch is then inserted into a free solution. This allows the concentration of the ions both inside the micropipette and in the outside solution to be altered as desired. Also, the voltage between the two sides of the membrane can be set at will — that is, "clamped" to a given voltage.

Fortunately, it has been possible to make such patches small enough that one often finds only a single channel protein in the membrane patch that is being studied. By varying the concentrations of different ions and the voltage across the membrane, one can determine the transport characteristics of the channel as well as its gating properties.

Facilitated Diffusion

Facilitated diffusion is also called *carrier-mediated diffusion* because a substance transported in this manner usually cannot pass through the membrane without a specific carrier protein helping it. That is, the carrier *facilitates* the diffusion of the substance to the other side.

Facilitated diffusion differs from simple diffusion through an open channel in the following very important way: Although the rate of diffusion through an open channel increases proportionately with the concentration of the diffusing substance, in facilitated diffusion the rate of diffusion approaches a maximum, called V_{max}, as the concentration of the substance increases.

What is it that limits the rate of facilitated diffusion? A probable answer is that the molecule to be transported enters the protein channel and becomes bound to a "receptor" in the carrier protein molecule. Then in a fraction of a second a conformational change occurs in the carrier protein, so that the channel now opens to the opposite side of the membrane. Because the binding force of the receptor is weak, the thermal motion of the attached molecule causes it to break away and to be released on the opposite side. Obviously, the rate at which molecules can be transported by this mechanism can never be greater than the rate at which the carrier protein molecule can undergo conformational change back and forth between its two states. Note specifically that this mechanism allows the trans-

single sodium channel when there was an approximate 25 mV potential gradient across the membrane. Note that the channel conducts current either all or none. That is, the gate of the channel snaps open and then snaps closed, each snapping event occurring within a few millionths of a second. This illustrates the rapidity with which conformational changes can occur in the shape of the protein molecular gates. At one voltage potential the channel may remain closed all the ported molecule to "diffuse" in either direction through the membrane.

Among the most important substances that cross cell membranes by facilitated diffusion are *glucose* and most of the *amino acids*.

FACTORS THAT AFFECT NET RATE OF DIFFUSION

By now it is evident that many different substances can diffuse either through the lipid bilayer of the cell membrane or through protein channels. However, please understand clearly that substances that diffuse in one direction can also diffuse in the opposite direction. Usually, what is important to the cell is not the total substance diffusing in both directions but the difference between these two, which is the *net rate of diffusion* in one direction. The factors that affect this are (1) the permeability of the membrane, (2) the difference in concentration of the diffusing substance between the two sides of the membrane, (3) the pressure difference across the membrane, and (4) in the case of ions, the electrical potential difference between the two sides of the membrane.

Effect of a Concentration Difference. Figure 5–6A illustrates a cell membrane with a substance in high concentration on the outside and low concentration on the inside. The rate at which the substance diffuses *inward* is proportional to the concentration of molecules on the *outside*, for this concentration determines how many molecules strike the outside of the channels each second. On the other hand, the rate at which molecules diffuse *outward* is proportional to their concentration *inside* the membrane. Obviously, therefore, the rate of net diffusion into the cell is proportional to the concentration on the outside *minus* the concentration on the inside or

$$\text{Net diffusion} \propto P(C_o - C_i)$$

in which C_o is the concentration on the outside, C_i is the concentration on the inside, and P is the permeability of the membrane for the substance.

Effect of an Electrical Potential on the Diffusion of Ions. If an electrical potential is applied across the membrane as shown in Figure 5–6B, because of their electrical charges, ions will move through the membrane even though no concentration difference exists to cause their movement. Thus, in the left panel of the figure, the concentrations of negative ions are exactly the same on both sides of the membrane, but a positive charge has been applied to the right side of the membrane and a negative charge to the left, creating an electrical gradient across the membrane. The positive charge attracts the negative ions, while the negative charge repels them. Therefore, net diffusion occurs from left to right. After much time, large quantities of negative ions will have moved to the right (if we neglect, for the time being, the disturbing effects of the positive ions of the solution), creating the condition illustrated in the right panel of Figure 5–6B, in which a

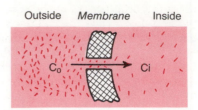

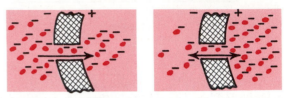

Figure 5–6. Effect of (A) concentration difference and (B) electrical difference on net diffusion of molecules and ions through a cell membrane.

concentration difference of the same ions has developed in the direction opposite to the electrical potential difference. Obviously, the concentration difference is now tending to move the ions to the left, while the electrical difference is tending to move them to the right. When the concentration difference rises high enough, the two effects exactly balance each other. At normal body temperature (37°C), the electrical difference that will exactly balance a given concentration difference of *univalent* ions, such as sodium (Na^+), potassium (K^+), or chloride (Cl^-), can be determined from the following formula called the *Nernst equation*:

$$\text{EMF (in millivolts)} = \pm 61 \log \frac{C_1}{C_2}$$

in which EMF is the electromotive force (voltage) between side 1 and side 2 of the membrane, C_1 is the concentration on side 1, and C_2 is the concentration on side 2. The polarity of the voltage on side 1 in the above equation is $+$ for negative ions and $-$ for positive ions. This relationship is extremely important in understanding the transmission of nerve impulses, for which reason it is discussed in even greater detail in Chapter 6.

■ ACTIVE TRANSPORT

From the discussion thus far, it is evident that *no substances can diffuse against an "electrochemical gradient,"* which is the sum of all the diffusion forces acting at the membrane — the forces caused by concentration difference, electrical difference, and pressure difference. That is, it is often said that substances cannot diffuse "uphill."

Yet, at times a large concentration of a substance is required in the intracellular fluid even though the extracellular fluid contains only a minute concentration.

This is true, for instance, for potassium ions. Conversely, it is important to keep the concentrations of other ions very low inside the cell even though their concentrations in the extracellular fluid are very great. This is especially true for sodium ions. Obviously, neither of these two effects could occur by the process of simple diffusion, for simple diffusion tends always to equilibrate the concentrations on the two sides of the membrane. Instead, some energy source must cause movement of potassium ions "uphill" to the inside of cells and cause movement of sodium ions also "uphill" but in this instance to the outside of the cell. When a cell membrane moves molecules or ions uphill against a concentration gradient (or uphill against an electrical or pressure gradient), the process is called *active transport.*

Among the different substances that are actively transported through cell membranes are sodium ions, potassium ions, calcium ions, iron ions, hydrogen ions, chloride ions, iodide ions, urate ions, several different sugars, and most of the amino acids.

Primary Active Transport and Secondary Active Transport. Active transport is divided into two types according to the source of the energy used to cause the transport. These are called *primary active transport* and *secondary active transport.* In primary active transport, the energy is derived directly from the breakdown of adenosine triphosphate (ATP) or some other high-energy phosphate compound. In secondary active transport, the energy is derived secondarily from ionic concentration gradients that have been created in the first place by primary active transport. In both instances, transport depends on *carrier proteins* that penetrate through the membrane, the same as is true for facilitated diffusion. However, in active transport, the carrier protein functions differently from the carrier in facilitated diffusion, for it is capable of imparting energy to the transported substance to move it against an electrochemical gradient. Let us give some examples of primary active transport and secondary active transport and explain their principles of function more fully.

PRIMARY ACTIVE TRANSPORT — THE SODIUM-POTASSIUM "PUMP"

Among the substances that are transported by primary active transport are sodium, potassium, calcium, hydrogen, chloride, and a few other ions. However, not all of these substances are transported by the membranes of all cells. Furthermore, some of the pumps function at intracellular membranes rather than (or in addition to) the surface membrane of the cell, such as at the membrane of the muscle sarcoplasmic reticulum or at one of the two membranes of the mitochondria. Nevertheless, they all operate by essentially the same basic mechanism.

The active transport mechanism that has been studied in greatest detail is the *sodium-potassium pump,* a transport process that pumps sodium ions outward through the cell membrane and at the same time

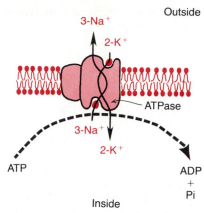

Figure 5 – 7. The postulated mechanism of the Na$^+$-K$^+$ pump.

pumps potassium ions from the outside to the inside. This pump is present in all cells of the body, and it is responsible for maintaining the sodium and potassium concentration differences across the cell membrane as well as for establishing a negative electrical potential inside the cells. Indeed, we see in the next chapter that this pump is the basis of nerve function to transmit nerve signals throughout the nervous system.

Figure 5 – 7 illustrates the basic components of the Na$^+$-K$^+$ pump. The *carrier protein* is a complex of two separate globular proteins, a larger one with a molecular weight of about 100,000 and a smaller one with a molecular weight of 55,000. Although the function of the smaller protein is not known, the larger protein has three specific features that are important for function of the pump:

1. It has three *receptor sites for binding sodium ions* on the portion of the protein that protrudes to the interior of the cell.

2. It has two *receptor sites for potassium ions* on the outside.

3. The inside portion of this protein adjacent to or near to the sodium binding sites has ATPase activity.

Now to put the pump into perspective: When three sodium ions bind on the inside of the carrier protein and two potassium ions on the outside, the ATPase function of the protein becomes activated. This then cleaves one molecule of ATP, splitting it to adenosine diphosphate (ADP) and liberating a high-energy phosphate bond of energy. This energy is then believed to cause a conformational change in the protein carrier molecule, extruding the sodium ions to the outside and the potassium ions to the inside. Unfortunately, the precise mechanism of the conformational change of the carrier is still to be discovered.

The Electrogenic Nature of the Na$^+$-K$^+$ Pump. The fact that the Na$^+$-K$^+$ pump moves three sodium ions to the exterior for every two potassium ions to the interior means that a net of one positive charge is moved from the interior of the cell to the exterior for each revolution of the pump. This obviously creates positivity outside the cell but leaves a deficit of positive ions inside the cell; that is, it causes negativity on the inside. Therefore, the Na$^+$-K$^+$ pump is said to be *electrogenic* because it creates an electrical potential across the cell membrane as it pumps.

The Calcium Pump

Another very important primary active transport mechanism is the calcium pump. Calcium ions are normally maintained at an extremely low concentration in the intracellular cytosol, at a concentration about 10,000 times less than that in the extracellular fluid. This is achieved by two calcium pumps. One is in the cell membrane and pumps calcium to the outside of the cell. The other pumps calcium ions into one or more of the internal vesicular organelles of the cell, such as into the sarcoplasmic reticulum of muscle cells and into the mitochondria in all cells. In both instances, the carrier protein penetrates the membrane from side to side and also serves as an ATPase having the same capability to cleave ATP as the ATPase sodium carrier protein. The difference is that this protein has a binding site for calcium instead of sodium.

SECONDARY ACTIVE TRANSPORT— CO-TRANSPORT AND COUNTER- TRANSPORT

When sodium ions are transported out of cells by primary active transport, a very large concentration gradient of sodium usually develops — very high concentration outside the cell and very low concentration inside. This gradient represents a storehouse of energy because the excess sodium outside the cell membrane is always attempting to diffuse to the interior. Under the appropriate conditions, this diffusion energy of sodium can literally pull other substances along with the sodium through the cell membrane. This phenomenon is called co-transport; it is one form of secondary active transport.

For sodium to pull another substance along with it, a coupling mechanism is required. This is achieved by means of still another carrier protein in the cell membrane. The carrier in this instance serves as an attachment point for both the sodium ion and the substance to be co-transported. Once they both are attached, a conformational change occurs in the carrier protein, and the energy gradient of the sodium ion causes both the sodium ion and the other substance to be transported together to the interior of the cell.

In counter-transport, sodium ions again attempt to diffuse to the interior of the cell because of their large concentration gradient. However, this time, the substance to be transported is on the inside of the cell and must be transported to the outside. Therefore, the sodium ion binds to the carrier protein where it projects through the exterior surface of the membrane, while the substance to be counter-transported binds to the interior projection of the carrier protein. Once both have bound, a conformational change occurs again, with the sodium ion moving to the interior and thus causing the other substance to move to the exterior.

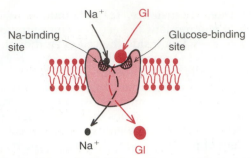

Figure 5–8. A postulated mechanism for sodium co-transport of glucose.

Sodium Co-transport of Glucose, Amino Acids, and Chloride Ions. Glucose and many amino acids are transported into most cells against very large concentration gradients; the mechanism of this is entirely by the co-transport mechanism illustrated in Figure 5–8. Note that the transport carrier protein has two binding sites on its exterior side, one for sodium and one for glucose. Also, the concentration of sodium ions is very high on the outside and very low inside, which provides the energy for the transport. A special property of the transport protein is that the conformational change to allow sodium movement to the interior will not occur until a glucose molecule also attaches. But when they are both attached, the conformational change takes place automatically, and both the sodium and the glucose are transported to the inside of the cell at the same time. Hence, this is a sodium-glucose co-transport mechanism.

Sodium co-transport of the amino acids occurs in the same manner as for glucose except that it uses a different set of transport proteins.

Two other important co-transport mechanisms are (1) a sodium-potassium-two chloride co-transporter that allows two chloride ions to be carried into cells along with one sodium and one potassium ion, all moving in the same direction, and (2) a potassium-chloride co-transporter that allows potassium and chloride ions to be transported from inside cells to the exterior together. Still other co-transport mechanisms into at least some cells include co-transport of iodine ions, iron ions, and urate ions.

REFERENCES

Agnew, W. S.: Voltage-regulated sodium channel molecules. Annu. Rev. Physiol., 46:517, 1984.
Andreoli, T. E., et al. (eds.): Physiology of Membrane Disorders. 2nd Ed. New York, Plenum Publishing Corp., 1986.
Auerbach, A., and Sachs, F.: Patch clamp studies of single ionic channels. Annu. Rev. Biophys. Bioeng., 13:269, 1984.
Biggio, G., and Costa, E. (eds.): Chloride Channels and Their Modulation by Neurotransmitters and Drugs. New York, Raven Press, 1988.
Bretag, A. H.: Muscle chloride channels. Physiol. Rev., 67:618, 1987.
Dinno, M. A., and Armstrong, W. M. (eds.): Membrane Biophysics III: Biological Transport. New York, Alan R. Liss, Inc., 1988.
DiPolo, R., and Beauge, L.: The calcium pump and sodium-calcium exchange in squid axons. Annu. Rev. Physiol., 45:313, 1983.
Ellis, D.: Na-Ca exchange in cardiac tissues. Adv. Myocardiol., 5:295, 1985.
Forgac, M.: Structure and function of vacuolar class of ATP-driven proton pumps. Physiol. Rev., 69:765, 1989.
Gadsby, D. C.: The Na/K pump of cardiac cells. Annu. Rev. Biophys. Bioeng., 13:373, 1984.

Haas, M.: Properties and diversity of Na-K-Cl cotransporters. Annu. Rev. Physiol., 51:443, 1989.

Haynes, D. H., and Mandveno, A.: Computer modeling of Ca^{2+} pump function of Ca^{2+}-Mg^{2+}-ATPase of sarcoplasmic reticulum. Physiol. Rev., 67:244, 1987.

Hidalgo, C. (ed.): Physical Properties of Biological Membranes and Their Functional Implications. New York, Plenum Publishing Corp., 1988.

Jacobson, K., et al.: Lateral diffusion of proteins in membranes. Annu. Rev. Physiol., 49:163, 1987.

Kaplan, J. H.: Ion movements through the sodium pump. Annu. Rev. Physiol., 47:535, 1985.

Latorre, R., et al.: Varieties of calcium-activated potassium channels. Annu. Rev. Physiol., 51:385, 1989.

Lauger, P.: Dynamics of ion transport systems in membranes. Physiol. Rev., 67:1296, 1987.

Malhotra, S. K.: The Plasma Membrane. New York, John Wiley & Sons, 1983.

Narahashi, T.: Ion Channels. New York, Plenum Publishing Corp., 1988.

Petersen, O. H.: Potassium channels and fluid secretion. News Physiol. Sci., 1:92, 1986.

Petersen, O. H., and Petersen, C. C. H.: The patch-clamp technique: Recording ionic currents through single pores in the cell membrane. News Physiol. Sci., 1:5, 1986.

Reuter, H.: Ion channels in cardiac cell membranes. Annu. Rev. Physiol., 46:473, 1984.

Reuter, H.: Modulation of ion channels by phosphorylation and second messengers. News Physiol. Sci., 2:168, 1987.

Sakmann, B., and Neher, E.: Patch clamp techniques for studying ionic channels in excitable membranes. Annu. Rev. Physiol., 46:455, 1984.

Schatzmann, H. J.: The calcium pump of the surface membrane and of the sarcoplasmic reticulum. Annu. Rev. Physiol., 51:473, 1989.

Stein, W. D. (ed.): The Ion Pumps: Structure, Function, and Regulation. New York, Alan R. Liss, Inc., 1988.

Trimmer, J. S., and Agnew, W. S.: Molecular diversity of voltage-sensitive Na channels. Annu. Rev. Physiol., 51:401, 1989.

Verner, K., and Schatz, G.: Protein translocation across membranes. Science, 241:1307, 1988.

Wright, E. M.: Electrophysiology of plasma membrane vesicles. Am. J. Physiol., 246:F363, 1984.

6

Membrane Potentials and Action Potentials

Electrical potentials exist across the membranes of essentially all cells of the body, and some cells, such as nerve and muscle cells, are "excitable" — that is, capable of self-generation of electrochemical impulses at their membranes and, in most instances, employment of these impulses to transmit signals along the membranes. In still other types of cells, such as glandular cells, macrophages, and ciliated cells, other types of changes in membrane potentials probably play significant roles in controlling many of the cell's functions. However, the present discussion is concerned with membrane potentials generated both at rest and during action by nerve and muscle cells.

■ BASIC PHYSICS OF MEMBRANE POTENTIALS

MEMBRANE POTENTIALS CAUSED BY DIFFUSION

Figure 6–1A and B illustrates a nerve fiber when there is no active transport of either sodium or potassium ions. In Figure 6–1A, the potassium concentration is very great inside the membrane, whereas that outside is very low. Let us also assume that the membrane in this instance is very permeable to the potassium ions but not to any other ions. Because of the large potassium concentration gradient from the inside toward the outside, there is a strong tendency for potassium ions to diffuse outward. As they do so, they carry positive charges to the outside, thus creating a state of electropositivity outside the membrane and electronegativity on the inside because of the negative anions that remain behind, which do not diffuse outward along with the potassium. This new potential difference repels the positively charged potassium ions in a backward direction from the outside to the inside. Within a millisecond or so, the potential change be-

comes great enough to block further net diffusion of potassium ions to the exterior despite the high potassium ion concentration gradient. In the normal large mammalian nerve fiber, the potential difference required is about 94 mV, with negativity inside the fiber membrane.

Figure 6–1B illustrates the same phenomenon as that in Figure 6–1A but this time with a high concentration of sodium ions outside the membrane and a low sodium concentration inside. These ions are also positively charged, and this time the membrane is highly permeable to the sodium ions but impermeable to all other ions. Diffusion of the sodium ions to the inside creates a membrane potential now of opposite polarity, with negativity outside and positivity inside. Again, the membrane potential rises high enough within milliseconds to block further net diffusion of the sodium ions to the inside; however, this time, for the large mammalian nerve fiber, the potential is about 61 mV and with positivity inside the fiber.

Thus, in both parts of Figure 6–1 we see that a concentration difference of ions across a selectively permeable membrane can, under appropriate conditions, cause the creation of a membrane potential. In later sections of this chapter, we see that many of the rapid changes in membrane potentials observed during the course of nerve and muscle impulse transmission result from the occurrence of rapidly changing diffusion potentials of this nature.

Relationship of the Diffusion Potential to the Concentration Difference — The Nernst Equation. The potential level across the membrane that will exactly prevent net diffusion of an ion in either direction through the membrane is called the *Nernst potential* for that ion. The magnitude of this potential is determined by the *ratio* of the ion concentrations on the two sides of the membrane — the greater this ratio, the greater the tendency for the ions to diffuse in one direction, and therefore the greater is the Nernst potential. The following equation, called the *Nernst equation*, can be

Diffusion Potentials

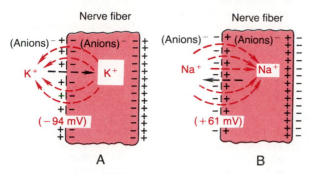

Figure 6–1. *A*, Establishment of a diffusion potential across a cell membrane, caused by potassium ions diffusing from inside the cell to the outside through a membrane that is selectively permeable only to potassium. *B*, Establishment of a diffusion potential when the membrane is permeable only to sodium ions. Note that the internal membrane potential is negative when potassium ions diffuse and positive when sodium ions diffuse because of opposite concentration gradients of these two ions.

used to calculate the Nernst potential for any univalent ion at normal body temperature of 37°C:

$$\text{EMF (millivolts)} = \pm 61 \log \frac{\text{Conc. inside}}{\text{Conc. outside}}$$

When using this formula, it is assumed that the potential outside the membrane always remains at exactly zero potential, and the Nernst potential that is calculated is the potential inside the membrane. Also, the sign of the potential is positive (+) when the ion under consideration is a negative ion and negative (−) when it is a positive ion.

Thus, when the concentration of a positive ion (potassium ions, for instance) on the inside is 10 times that on the outside, the log of 10 is 1, so that the Nernst potential calculates to be −61 mV inside the membrane.

Calculation of the Diffusion Potential When the Membrane is Permeable to Several Different Ions

When a membrane is permeable to several different ions, the diffusion potential that develops depends on three factors: (1) the polarity of the electrical charge of each ion, (2) the permeability of the membrane (*P*) to each ion, and (3) the concentrations (*C*) of the respective ions on the inside (*i*) and outside (*o*) of the membrane. Thus, the following formula, called the *Goldman equation*, or the *Goldman-Hodgkin-Katz equation*, gives the calculated membrane potential on the *inside* of the membrane when two univalent positive ions, sodium (Na⁺) and potassium (K⁺), and one univalent negative ion, chloride (Cl⁻), are involved.

$$\text{EMF (millivolts)} =$$
$$-61 \cdot \log \frac{C_{Na^+_i} P_{Na^+} + C_{K^+_i} P_{K^+} + C_{Cl^-_o} P_{Cl^-}}{C_{Na^+_o} P_{Na^+} + C_{K^+_o} P_{K^+} + C_{Cl^-_i} P_{Cl^-}}$$

Now, let us study the importance and the meaning of this equation. First, sodium, potassium, and chloride ions are the ions most importantly involved in the development of membrane potentials in nerve and muscle fibers as well as in the neuronal cells in the central nervous system. The concentration gradient of each of these ions across the membrane helps determine the voltage of the membrane potential.

Second, the degree of importance of each of the ions in determining the voltage is proportional to the membrane permeability for that particular ion. Thus, if the membrane is impermeable to both potassium and chloride ions, the membrane potential becomes entirely dominated by the concentration gradient of sodium ions alone, and the resulting potential will be exactly equal to the Nernst potential for sodium. The same principle holds for each of the other two ions if the membrane should become selectively permeable for either one of them alone.

Third, a positive ion concentration gradient from *inside* the membrane *to the outside* causes electronegativity inside the membrane. The reason for this is that positive ions diffuse to the outside when their concentration is higher inside than outside. This carries positive charges to the outside but leaves the nondiffusible negative anions on the inside. Exactly the opposite effect occurs when there is a negative ion gradient. That is, a chloride ion gradient from the *outside to the inside* causes negativity inside the cell because negatively charged chloride ions then diffuse to the inside, while leaving the positive ions on the outside.

Fourth, we see later that the permeabilities of the sodium and potassium channels undergo very rapid changes during conduction of the nerve impulse, whereas the permeability of the chloride channels does not change greatly during this process. Therefore, the changes in the sodium and potassium permeabilities are primarily responsible for signal transmission in the nerves, which is the subject of most of the remainder of this chapter.

Measuring the Membrane Potential

The method for measuring the membrane potential is simple in theory but often very difficult in practice because of the small sizes of many of the fibers. Figure 6–2 illustrates a small pipette, filled with a very strong electrolyte solution

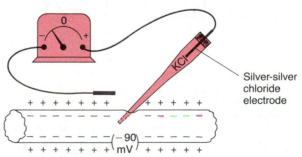

Figure 6–2. Measurement of the membrane potential of the nerve fiber using a microelectrode.

(KCl), that is impaled through the cell membrane to the interior of the fiber. Then another electrode, called the "indifferent electrode," is placed in the interstitial fluids, and the potential difference between the inside and outside of the fiber is measured using an appropriate voltmeter. This is a highly sophisticated electronic apparatus that is capable of measuring very small voltages despite extremely high resistance to electrical flow through the tip of the micropipette, which has a diameter usually less than 1 μm and a resistance often as great as a billion ohms. For recording rapid *changes* in the membrane potential during the transmission of nerve impulses, the microelectrode is connected to an oscilloscope, as explained later in the chapter.

The Cell Membrane as an Electrical Capacitor

In each of the figures shown thus far, the negative and positive ionic charges that cause the membrane potential have been shown to be lined up against the membrane, and we have not spoken of the arrangement of the charges elsewhere in the fluids, either inside the nerve fiber or on the outside in the interstitial fluid. However, Figure 6–3 illustrates this, showing that everywhere except adjacent to the surfaces of the cell membrane itself, the negative and positive charges are exactly equal. This is called the principle of *electrical neutrality*; that is, for every positive ion there is a negative ion nearby to neutralize it, or otherwise electrical potentials of billions of volts would appear within the fluids.

When positive charges are pumped to the outside of the membrane, these positive charges line up along the outside of the membrane, and on the inside the anions line up that have been left behind. This creates a *dipole layer* of positive and negative charges between the outside and inside of the membrane, but it still leaves equal numbers of negative and positive charges everywhere else within the fluids. This is the same effect that occurs when the plates of an electrical capacitor become electrically charged—that is, lining up of negative and positive charges on the opposite sides of the dielectric membrane of the capacitor. Therefore, the lipid bilayer of the cell membrane actually functions as a *dielectric*

of a cell membrane capacitor, much as mica, paper, and Mylar function as dielectrics in electrical capacitors.

Because of the extreme thinness of the cell membrane (only 7 to 10 nm), its capacitance is tremendous for its area—about $1 \mu f/cm^2$.

The lower part of Figure 6–3 illustrates the electrical potential that will be recorded at each point in or near the nerve fiber membrane, beginning at the left side of the figure and passing to the right. As long as the electrode is outside the nerve membrane, the potential that is recorded is zero, which is the potential of the extracellular fluid. Then, as the recording electrode passes through the electrical dipole layer at the cell membrane, the potential decreases immediately to −90 mV. Again, the electrical potential remains at a steady level as the electrode passes across the interior of the fiber but reverses back to zero the instant it passes through the opposite side of the membrane.

The fact that the nerve membrane functions as a capacitor has one especially important point of significance: To create a negative potential inside the membrane, only enough positive ions must be transported outward to develop the electrical dipole layer at the membrane itself. All the remaining ions inside the nerve fiber can still be both positive and negative ions. Therefore, an incredibly small number of ions needs to be transferred through the membrane to establish the normal potential of −90 mV inside the nerve fiber— only about $1/5,000,000$ to $1/100,000,000$ of the total positive charges inside the fiber need be so transferred. Also, an equally small number of positive ions moving from outside to the inside of the fiber can reverse the potential from −90 mV to as much as +35 mV within as little as $1/10,000$ of a second. This rapid shifting of ions in this manner causes the nerve signals that we discuss in the subsequent sections of this chapter.

■ THE RESTING MEMBRANE POTENTIAL OF NERVES

The membrane potential of large nerve fibers when they are not transmitting nerve signals is about −90 mV. That is, the potential *inside the fiber* is 90 mV more negative than the potential in the interstitial fluid on the outside of the fiber. In the next few paragraphs, we explain all the factors that determine the level of this potential, but before doing so we must describe the transport properties of the resting nerve membrane for sodium and potassium.

Active Transport of Sodium and Potassium Ions Through the Membrane — The Sodium-Potassium Pump. First, let us recall from the discussions of the previous chapter that all cell membranes of the body have a powerful sodium-potassium pump and that this continually pumps sodium to the outside of the fiber and potassium to the inside. Further, let us remember that this is an *electrogenic pump* because more positive charges are pumped to the outside than to the inside (three Na^+ ions to the outside for each two K^+ ions to the inside), leaving a net deficit of positive ions on the inside; this is the same as causing a negative charge inside the cell membrane.

This sodium-potassium pump also causes the tremendous concentration gradients for sodium and potassium across the resting nerve membrane. These gradients are the following:

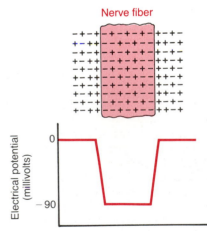

Figure 6–3. Distribution of positively and negatively charged ions in the interstitial fluid surrounding a nerve fiber and in the fluid inside the fiber; note the dipolar alignment of negative charges along the inside surface of the membrane and positive charges along the outside surface. In the lower panel are illustrated the abrupt changes in membrane potential that occur at the membranes on the two sides of the fiber.

Na⁺ (outside):	142 mEq/L
Na⁺ (inside):	14 mEq/L
K⁺ (outside):	4 mEq/L
K⁺ (inside):	140 mEq/L

The ratios of these two respective ions from the inside to the outside are

$$Na^+_{inside}/Na^+_{outside} = 0.1$$
$$K^+_{inside}/K^+_{outside} = 35.0$$

Leakage of Potassium and Sodium Through the Nerve Membrane. To the right in Figure 6–4 is illustrated a channel protein in the cell membrane, through which potassium and sodium ions can leak, called a *potassium-sodium "leak" channel*. There are actually multiple different proteins of this type with different leak characteristics. However, the emphasis is on potassium leakage because on the average the channels are far more permeable to potassium than to sodium, normally about 100 times as permeable. We see later that this differential in permeability is exceedingly important in determining the level of the normal resting membrane potential.

ORIGIN OF THE NORMAL RESTING MEMBRANE POTENTIAL

Figure 6–5 illustrates the important factors in the establishment of the normal resting membrane potential of −90 mV. These are as follows:

Contribution of the Potassium Diffusion Potential. In Figure 6–5A, we make the assumption that the only movement of ions through the membrane is the diffusion of potassium ions, as illustrated by the open channels between the potassium inside the membrane and the outside. Because of the high ratio of potassium ions inside to outside, 35 to 1, the Nernst potential corresponding to this ratio is −94 mV, for the logarithm of 35 is 1.54, and this times −61 mV is −94 mV. Therefore, if potassium ions were the only factor causing the resting potential, this resting potential would also be equal to −94 mV, as illustrated in the figure.

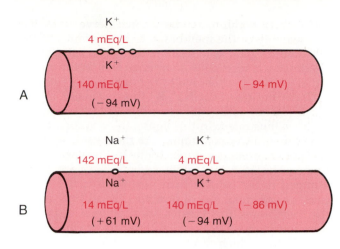

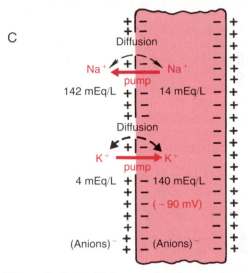

Figure 6–5. Establishment of resting membrane potentials in nerve fibers under three separate conditions: *A*, when the membrane potential is caused entirely by potassium diffusion alone; *B*, when the membrane potential is caused by diffusion of both sodium and potassium ions; and *C*, when the membrane potential is caused by diffusion of both sodium and potassium ions plus pumping of both these ions by the Na⁺-K⁺ pump.

Contribution of Sodium Diffusion Through the Nerve Membrane. Figure 6–5B illustrates the addition of very slight permeability of the nerve membrane to sodium ions, caused by the minute diffusion of the sodium ions through the K⁺-Na⁺ leak channels. The ratio of sodium ions from inside to outside the membrane is 0.1, and this gives a calculated Nernst potential for the inside of the membrane of +61 mV. But also shown in Figure 6–5B is the Nernst potential for potassium diffusion of −94 mV. How do these interact with each other, and what will be the summated potential? This can be answered by using the Goldman equation described earlier. However, intuitively, one can see that if the membrane is highly permeable to potassium but only very slightly permeable to sodium, it is logical that the diffusion of potassium will contribute far more to the membrane potential than will the

Outside

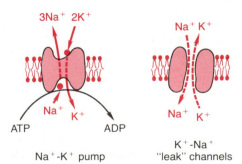

Figure 6–4. The functional characteristics of the Na⁺-K⁺ pump and also of the potassium-sodium "leak" channels.

diffusion of sodium. In the normal nerve fiber, the permeability of the membrane to potassium is about 100 times as great as to sodium. Using this value in the Goldman equation gives an internal membrane potential of −86 mV, as shown to the right in the figure.

Contribution of the Na⁺-K⁺ Pump. Finally, in Figure 6−5C an additional contribution of the Na⁺-K⁺ pump is illustrated. In this figure, there is continuous pumping of three sodium ions to the outside for each two potassium ions pumped to the inside of the membrane. The fact that more sodium ions are being pumped to the outside than potassium to the inside causes a continual loss of positive charges from inside the membrane; this creates an additional degree of negativity (about −4 mV additional) on the inside beyond that which can be accounted for by diffusion alone. Therefore, as illustrated in Figure 6−5C, the net membrane potential with all these factors operative at the same time is −90 mV.

In summary, the diffusion potentials alone caused by potassium and sodium diffusion would give a membrane potential of approximately −86 mV, almost all of this being determined by potassium diffusion. Then, an additional −4 mV is contributed to the membrane potential by the electrogenic Na⁺-K⁺ pump, giving a net resting membrane potential of −90 mV.

The resting membrane potential in large skeletal muscle fibers is approximately the same as that in large nerve fibers, also −90 mV. However, in both small nerve fibers and small muscle fibers — smooth muscle, for instance — as well as in many of the neurons of the central nervous system, the membrane potential is often as little as −40 to −60 mV instead of −90 mV.

■ THE NERVE ACTION POTENTIAL

Nerve signals are transmitted by *action potentials*, which are rapid changes in the membrane potential. Each action potential begins with a sudden change from the normal resting negative potential to a positive membrane potential and then ends with an almost equally rapid change back again to the negative potential. To conduct a nerve signal, the action potential moves along the nerve fiber until it comes to the fiber's end. The upper panel of Figure 6−6 shows the disturbances that occur at the membrane during the action potential, with transfer of positive charges to the interior of the fiber at its onset and return of positive charges to the exterior at its end. The lower panel illustrates graphically the successive changes in the membrane potential over a few 10,000ths of a second, illustrating the explosive onset of the action potential and the almost equally as rapid recovery.

The successive stages of the action potential are as follows:

Resting Stage. This is the resting membrane potential before the action potential occurs. The membrane is said to be "polarized" during this stage because of

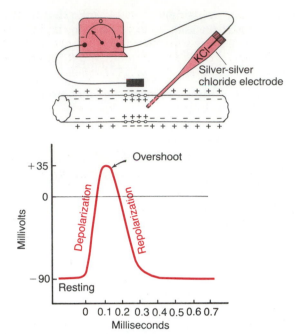

Figure 6−6. A typical action potential recorded by the method illustrated in the upper panel of the figure.

the very large negative membrane potential that is present.

Depolarization Stage. At this time, the membrane suddenly becomes very permeable to sodium ions, allowing tremendous numbers of sodium ions to flow to the interior of the axon. The normal "polarized" state of −90 mV is lost, with the potential rising rapidly in the positive direction. This is called *depolarization*. In large nerve fibers, the membrane potential actually "overshoots" beyond the zero level and becomes somewhat positive, but in some smaller fibers as well as many central nervous system neurons, the potential merely approaches the zero level and does not overshoot to the positive state.

Repolarization Stage. Within a few 10,000ths of a second after the membrane becomes highly permeable to sodium ions, the sodium channels begin to close, and the potassium channels open more than normally. Then, rapid diffusion of potassium ions to the exterior re-establishes the normal negative resting membrane potential. This is called *repolarization* of the membrane.

To explain more fully the factors that cause both the depolarization and the repolarization processes, we need now to describe the special characteristics of yet two other types of transport channels through the nerve membrane: the voltage-gated sodium and potassium channels.

THE VOLTAGE-GATED SODIUM AND POTASSIUM CHANNELS

The necessary factor causing both depolarization and repolarization of the nerve membrane during the action potential is the *voltage-gated sodium channel.*

However, the *voltage-gated potassium channel* also plays an important role in increasing the rapidity of repolarization of the membrane. *These two voltage-gated channels are in addition to the Na⁺-K⁺ pump and also in addition to the Na⁺-K⁺ leak channels.*

The Voltage-Gated Sodium Channel — "Activation" and "Inactivation" of the Channel

The upper panel of Figure 6–7 illustrates the voltage-gated sodium channel in three separate states. This channel has two *gates*, one near the outside of the channel called the *activation gate* and another near the inside called the *inactivation gate*. To the left is shown the state of these two gates in the normal resting membrane when the membrane potential is −90 mV. In this state, the activation gate is closed, which prevents any entry of sodium ions to the interior of the fiber through these sodium channels. On the other hand, the inactivation gate is open and does not at this time constitute any barrier to the movement of the sodium ions.

Activation of the Sodium Channel. When the membrane potential becomes less negative than during the resting state, rising from −90 mV toward zero, it finally reaches a voltage, usually somewhere between −70 and −50 mV, that causes a sudden conformational change in the activation gate, flipping it to the open position. This is called the *activated state*; during this state, sodium ions can literally pour inward through the channel, increasing the sodium permeability of the membrane as much as 500-fold to 5000-fold.

Inactivation of the Sodium Channel. To the far right in the upper panel of Figure 6–7 is illustrated a third state of the sodium channel. The same increase in voltage that opens the activation gate also closes the inactivation gate. However, closure of the inactivation gate occurs a few 10,000ths of a second after the activation gate opens. That is, the conformational change that flips the inactivation gate to the closed state is a slower process, whereas the conformational change that opens the activation gate is a very rapid process. Therefore, after the sodium channel has remained open for a few 10,000ths of a second, it closes, and sodium ions can no longer pour to the inside of the membrane. At this point the membrane potential begins to recover back toward the resting membrane state, which is the repolarization process.

A very important characteristic of the sodium channel inactivation process is that *the inactivation gate will not reopen again until the membrane potential returns either to or nearly to the original resting membrane potential level.* Therefore, it is not possible for the sodium channels to open again without the nerve fiber first repolarizing.

The Voltage-Gated Potassium Channels and Their Activation

The lower panel of Figure 6–7 illustrates the voltage-gated potassium channel in two separate states: during the resting state and toward the end of the action potential. During the resting state, the gate of the potassium channel is closed, as illustrated to the left in the figure, and potassium ions are prevented from passing through this channel to the exterior. When the membrane potential rises from −90 mV toward zero, this voltage change causes a slow conformational opening of the gate and allows increased potassium diffusion outward through the channel. However, because of the slowness of opening of these potassium channels, they mainly open just at the same time that the sodium channels are beginning to become inactivated and therefore are closing. Thus, the decrease in sodium entry to the cell and simultaneous increase in potassium exit from the cell greatly speeds the repolarization process, leading within a few 10,000ths of a second to full recovery of the resting membrane potential.

The Research Method for Measuring the Effect of Voltage on Opening and Closing of the Voltage-Gated Channels — The "Voltage Clamp." The original research that led to our quantitative understanding of the sodium and potassium channels was so ingenious that it led to Nobel prizes for the scientists responsible, Hodgkin and Huxley. The essence of these studies is illustrated in Figures 6–8 and 6–9.

Figure 6–8 illustrates the experimental apparatus called the *voltage clamp*, which is used to measure the flow of ions through the different channels. In using this apparatus, two electrodes are inserted into the nerve fiber. One of these is for the purpose of measuring the voltage of the membrane potential. The other is to conduct electrical current either into or out of the nerve fiber. This apparatus is used in the following

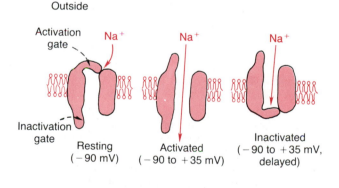

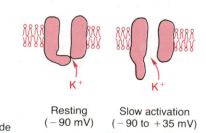

Figure 6–7. Characteristics of the voltage-gated sodium and potassium channels, showing both activation and inactivation of the sodium channels but activation of the potassium channels only when the membrane potential is changed from the normal resting negative value to a positive value.

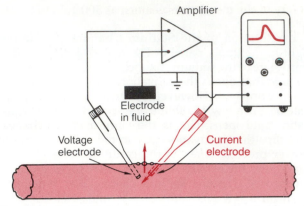

Figure 6–8. The "voltage clamp" method for studying flow of ions through specific channels.

way: The investigator decides what voltage he or she wishes to establish inside the nerve fiber. He or she then adjusts the electronic portion of the apparatus to the desired voltage, and this automatically injects either positive or negative electricity through the current electrode at whatever rate is required to hold the voltage, as measured by the voltage electrode, at the level set by the operator. For instance, when the membrane potential is suddenly increased by this voltage clamp from −90 mV to zero, the voltage-gated sodium and potassium channels open, and sodium and potassium ions begin to pour through the channels. To counterbalance the effect of these ion movements on the set potential, electrical current is injected automatically through the current electrode of the voltage clamp to maintain the intracellular voltage at the zero level. To achieve this, the current injected must be exactly equal to but of opposite polarity to the net current flow through the membrane channels. To measure how much current flow is occurring at each instant, the current electrode is connected to an oscilloscope that records the current flow, as illustrated on the screen of the oscilloscope in the figure. Finally, the investigator adjusts the concentrations of the ions to desired levels both inside and outside the nerve fiber and repeats the study. This can be done very easily when using very large nerve fibers removed from some crustaceans, especially the giant squid axon that is sometimes as large as 1 mm in diameter. When sodium is the only permeant ion in the solutions inside and outside the squid axon, the voltage clamp measures current flow only

through the sodium channels. When potassium is the only permeant ion, current flow only through the potassium channels is measured.

Another means for studying the flow of ions through individual channels is to block one type of channel at a time. For instance, the sodium channels can be blocked by a toxin called *tetrodotoxin* by applying this toxin to the outside of the cell membrane where the sodium activation gates are located. Conversely, *tetraethylammonium ion* blocks the potassium pores when it is applied to the interior of the nerve fiber.

Figure 6–9 illustrates the typical changes in conductance of the voltage-gated sodium and potassium channels when the membrane potential is suddenly changed by use of the voltage clamp from −90 mV to +10 mV and 2 msec later back again to −90 mV. Note the sudden opening of the sodium channels (the activation stage) within a very small fraction of a millisecond after the membrane potential is increased to the positive value. However, during approximately the next millisecond or so, the sodium channels automatically close (the inactivation stage).

Now, note the opening (activation) of the potassium channels. These open slowly and reach the full open state only after the sodium channels have already become almost completely closed. Furthermore, once the potassium channels open, they remain open for the entire duration of the positive membrane potential and do not close again until after the membrane potential is decreased back to a very negative value.

Finally, let us recall that the voltage-gated channels normally flip to the open state or the closed state very suddenly, which was illustrated in Figure 5–5 in the previous chapter. Therefore, how is it that the curves in Figure 6–9 are so smooth? The answer is that these curves represent the flow of sodium and potassium ions through literally thousands of channels at the same time. Some open at one voltage level, others at another voltage, and so forth. Likewise, some become inactivated at different points in the cycle from the others. Thus, the illustrated curves represent summated current flows through the many channels.

SUMMARY OF THE EVENTS THAT CAUSE THE ACTION POTENTIAL

Figure 6–10 illustrates in summary form the sequential events that occur during and shortly after the action potential. These are as follows:

At the bottom of the figure are shown the changes in membrane conductances for sodium and potassium ions. During the resting state, before the action potential begins, the conductance for potassium ions is shown to be 50 to 100 times as great as the conductance for sodium ions. This is caused by much greater leakage of potassium ions than sodium ions through the leak channels. However, at the onset of the action potential, the sodium channels instantaneously become activated and allow an up to 5000-fold increase in sodium conductance. Then the inactivation process closes the sodium channels within another few fractions of a millisecond. The onset of the action potential also causes voltage gating of the potassium channels, causing them to begin opening a fraction of a millisecond after the sodium channels open. And at the end of the action potential, the return of the membrane po-

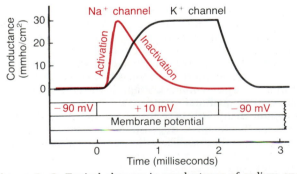

Figure 6–9. Typical changes in conductance of sodium and potassium ion channels when the membrane potential is suddenly increased from the normal resting value of −90 mV to a positive value of +10 mV for 2 msec. This figure illustrates that the sodium channels open (activate) and then close (inactivate) before the end of the 2 msec, whereas the potassium channels only open (activate).

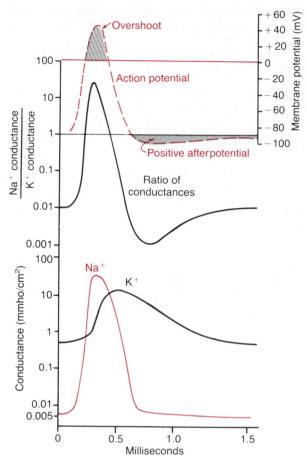

Figure 6–10. Changes in sodium and potassium conductances during the course of the action potential. Note that sodium conductance increases several thousandfold during the early stages of the action potential, whereas potassium conductance increases only about 30-fold during the latter stages of the action potential and for a short period thereafter. (Curves constructed from data in Hodgkin and Huxley papers but transposed from squid axon to apply to the membrane potentials of large mammalian nerve fibers.)

tential to the negative state causes the potassium channels to close back to their original status, but again only after a short delay.

In the middle portion of Figure 6–10 is shown the ratio of sodium conductance to potassium conductance at each instant during the action potential, and above this is shown the action potential itself. During the early portion of the action potential, this ratio increases more than a thousandfold. Therefore, far more sodium ions now flow to the interior of the fiber than do potassium ions to the exterior. This is what causes the membrane potential to become positive. Then the sodium channels begin to become inactivated, and at the same time the potassium channels open, so that the ratio of conductance now shifts far in favor of high potassium conductance but low sodium conductance. This allows extremely rapid loss of potassium ions to the exterior, while essentially no sodium ions flow to the interior. Consequently, the action potential quickly returns to its baseline level.

The "Positive" Afterpotential

Note also in Figure 6–10 that the membrane potential becomes even more negative than the original resting membrane potential for a few milliseconds after the action potential is over. Strangely enough, this is called the *"positive" afterpotential,* which is a misnomer because the positive afterpotential is actually even more negative than the resting potential. The reason for calling it "positive" is that, historically, the first potential measurements were made on the outside of the nerve fiber membrane rather than inside, and when measured on the outside this potential causes a positive record on the recording meter rather than a negative one.

The cause of the positive afterpotential is mainly that many potassium channels remain open for several milliseconds after the repolarization process of the membrane is complete. This allows excess potassium ions to diffuse out of the nerve fiber, leaving an extra deficit of positive ions on the inside, which means more negativity.

ROLES OF OTHER IONS DURING THE ACTION POTENTIAL

Thus far, we have considered only the roles of sodium and potassium ions in the generation of the action potential. However, at least three other types of ions must be considered. These are as follows:

The Impermeant Negatively Charged Ions (Anions) Inside the Axon. Inside the axon are many negatively charged ions that cannot go through the membrane channels. These include protein molecules, many organic phosphate compounds, sulfate compounds, and so forth. Because these cannot leave the interior of the axon, any deficit of positive ions inside the membrane leaves an excess of the impermeant negative ions. Therefore, these impermeant negative ions are responsible for the negative charge inside the fiber when there is a deficit of the positively charged potassium ions and other positive ions.

Calcium Ions. The cell membranes of almost all, if not all, cells of the body have a calcium pump similar to the sodium pump. Like the sodium pump, this device pumps calcium ions from the interior to the exterior of the cell membrane (or into the endoplasmic reticulum), creating a calcium ion gradient of about 10,000-fold, leaving an internal concentration of calcium ions of about 10^{-7} M in contrast to an external concentration of about 10^{-3} M.

In addition, there are also voltage-gated calcium channels. These channels are slightly permeable to sodium ions as well as to calcium ions; when they open, both calcium and sodium ions flow to the interior of the fiber. Therefore, these channels are sometimes also called Ca^{++}-Na^+ channels. The calcium channels are very slow to become activated, requiring 10 to 20 times as long for activation as the sodium channels. Therefore, they are also frequently called *slow channels,* in contrast to the sodium channels that are called *fast channels.*

Calcium channels are very numerous in both cardiac muscle and smooth muscle. In fact, in some types of smooth muscle, the fast sodium channels are hardly present at all, so the action potentials then are caused almost entirely by activation of the slow calcium channels.

Increased Permeability of the Sodium Channels When There Is a Deficit of Calcium Ions. The concentration of calcium ions in the interstitial fluid also has a profound effect on the voltage level at which the sodium channels become activated. When there is a deficit of calcium ions, the sodium channels are activated (opened) by very little increase of the membrane potential above the normal resting level. Therefore, the nerve fiber becomes highly excitable, sometimes discharging repetitively without any provocation rather than remaining in the resting state. In fact, the calcium ion concentration needs to fall only 30 to 50 per cent below normal before spontaneous discharge occurs in many peripheral nerves, often causing muscle "tetany" that can actually be lethal because of tetanic contraction of the respiratory muscles.

The probable way in which calcium ions affect the sodium channels is the following: These ions appear to bind to the exterior surfaces of the sodium channel protein molecule. The positive charges of these calcium ions, in turn, alter the electrical state of the channel protein itself, in this way increasing the voltage level required to open the gate.

Chloride Ions. Chloride ions leak through the resting membrane in the same way that small quantities of potassium and sodium ions leak through. In the usual nerve fiber, the rate of chloride diffusion through the membrane is about one half as great as the diffusion of potassium ions. Therefore, the question must be asked: Why have we not considered the chloride ions in our explanation of the action potential? The answer is that the chloride ions function passively in this process. Also, the permeability of the chloride leak channels does not change significantly during the action potential.

In the normal resting state of the nerve fiber, the −90 mV inside the fiber repels most of the chloride ions from the fiber. Therefore, the concentration of chloride ions inside the fiber is only 3 to 4 mEq/L, whereas the concentration outside the fiber is about 103 mEq/L. The Nernst potential for this ratio of chloride ions is exactly equal to the −90 mV membrane potential, which is what one would expect for an ion that is not actively pumped.

During the action potential, small quantities of chloride ions do diffuse into the nerve fiber because of the temporary loss of the internal negativity. This movement of chloride ions serves to alter slightly the timing of the successive voltage changes during the action potential, but it does not alter the fundamental process.

INITIATION OF THE ACTION POTENTIAL

Up to this point, we have explained the changing sodium and potassium permeabilities of the membrane as well as the development of the action potential itself, but we have not explained what initiates the action potential. The answer to this, as follows, is really quite simple:

A Positive Feedback Opens the Sodium Channels. First, as long as the membrane of the nerve fiber remains totally undisturbed, no action potential occurs in the normal nerve. However, if any event at all causes enough initial rise in the membrane potential from −90 mV up toward the zero level, the rising voltage itself will cause many voltage-gated sodium channels to begin opening. This allows rapid inflow of sodium ions, which causes still further rise of the membrane potential, thus opening still more voltage-gated sodium channels and more streaming of sodium ions to the interior of the fiber. Obviously, this process is a positive-feedback vicious circle that, once the feedback is strong enough, will continue until all the

voltage-gated sodium channels have become totally activated (opened). Then, within another fraction of a millisecond, the rising membrane potential causes beginning inactivation of the sodium channels as well as opening of potassium channels, and the action potential soon terminates.

Threshold for Initiation of the Action Potential. An action potential will not occur until the initial rise in membrane potential is great enough to create the vicious circle described in the last paragraph. Usually, a sudden rise in membrane potential of 15 to 30 mV is required. Therefore, a sudden increase in the membrane potential in a large nerve fiber of from −90 mV up to about −65 mV will usually cause the explosive development of the action potential. This level of −65 mV, therefore, is said to be the *threshold* for stimulation.

Accommodation of the Membrane — Failure to Fire Despite Rising Voltage. If the membrane potential rises very slowly — over many milliseconds instead of a fraction of a millisecond — the slow, inactivating gates of the sodium channels will have time to close at the same time that the activating gates are opening. Consequently, the opening of the activating gates will not be as effective in increasing the flow of sodium ions as normally. Therefore, a slow increase in the internal potential of a nerve fiber either requires a higher threshold voltage than normal to cause firing or prevents firing entirely at times, even with a voltage rise all the way to zero or even to positive voltage. This phenomenon is called *accommodation* of the membrane to the stimulus.

■ PROPAGATION OF THE ACTION POTENTIAL

In the preceding paragraphs we have discussed the action potential as it occurs at one spot on the membrane. However, an action potential elicited at any one point on an excitable membrane usually excites adjacent portions of the membrane, resulting in propagation of the action potential. The mechanism of this is illustrated in Figure 6–11. Figure 6–11A shows a normal resting nerve fiber, and Figure 6–11B shows a nerve fiber that has been excited in its midportion — that is, the midportion has suddenly developed increased permeability to sodium. The arrows illustrate a "local circuit" of current flow between the depolarized areas of the membrane and the adjacent resting membrane areas; positive electrical charges carried by the inward diffusing sodium ions flow inward through the depolarized membrane and then for several millimeters along the core of the axon. These positive charges increase the voltage for a distance of 1 to 3 mm inside large fibers to above the threshold voltage value for initiating an action potential. Therefore, the sodium channels in these new areas immediately activate, and, as illustrated in Figure 6–11C and D, the explosive action potential spreads. Then these newly depolarized areas cause local circuits of current flow still further along the membrane, causing progressively more and more depolarization. Thus, the depolarization process travels along the entire extent of the fiber. The transmission of the depolarization process

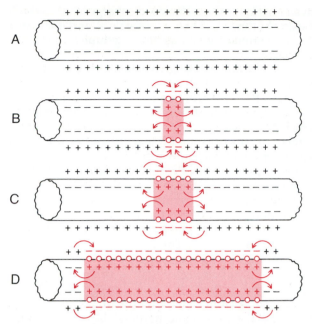

Figure 6–11. Propagation of action potentials in both directions along a conductive fiber.

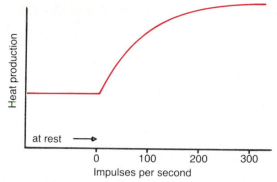

Figure 6–12. Heat production in a nerve fiber at rest and at progressively increasing rates of stimulation.

along a nerve or muscle fiber is called a *nerve* or *muscle impulse.*

Direction of Propagation. It is obvious, as illustrated in Figure 6–11, that an excitable membrane has no single direction of propagation, but that the action potential can travel in both directions away from the stimulus—and even along all branches of a nerve fiber—until the entire membrane has become depolarized.

The All-or-Nothing Principle. It is equally obvious that, once an action potential has been elicited at any point on the membrane of a normal fiber, the depolarization process will travel over the entire membrane if conditions are right, or it might not travel at all if conditions are not right. This is called the all-or-nothing principle, and it applies to all normal excitable tissues. Occasionally, though, the action potential will reach a point on the membrane at which it does not generate sufficient voltage to stimulate the next area of the membrane. When this occurs, the spread of depolarization stops. Therefore, for continued propagation of an impulse to occur, the ratio of action potential to threshold for excitation must at all times be greater than 1. This is called the *safety factor* for propagation.

■ RE-ESTABLISHING SODIUM AND POTASSIUM IONIC GRADIENTS AFTER ACTION POTENTIALS— IMPORTANCE OF ENERGY METABOLISM

Transmission of each impulse along the nerve fiber reduces infinitesimally the concentration differences of sodium and potassium between the inside and out-

side of the membrane because of diffusion of sodium ions to the inside during depolarization and diffusion of potassium ions to the outside during repolarization. For a single action potential, this effect is so minute that it cannot be measured. Indeed, 100,000 to 50,000,000 impulses can be transmitted by nerve fibers, the number depending on the size of the fiber and several other factors, before the concentration differences have run down to the point that action potential conduction ceases. Yet, even so, with time it becomes necessary to re-establish the sodium and potassium membrane concentration differences. This is achieved by the action of the Na^+-K^+ pump in exactly the same way as that described earlier in the chapter for original establishment of the resting potential. That is, the sodium ions that have diffused to the interior of the cell during the action potentials and the potassium ions that have diffused to the exterior are returned to their original state by the Na^+-K^+ pump. Since this pump requires energy for operation, this process of "recharging" the nerve fiber is an active metabolic one, using energy derived from the adenosine triphosphate energy system of the cell. Figure 6–12 shows that the nerve fiber produces excess heat, which is a measure of its energy expenditure, when the impulse frequency increases.

A special feature of the sodium-potassium ATPase pump is that its degree of activity is strongly stimulated when excess sodium ions accumulate inside the cell membrane. In fact, the pumping activity increases approximately in proportion to the third power of the sodium concentration. That is, as the internal sodium concentration rises from 10 to 20 mEq/L, the activity of the pump does not merely double but instead increases approximately eightfold. Therefore, it can easily be understood how the "recharging" process of the nerve fiber can rapidly be set into motion whenever the concentration differences of sodium and potassium ions across the membrane begin to "run down."

■ PLATEAU IN SOME ACTION POTENTIALS

In some instances the excitable membrane does not repolarize immediately after depolarization, but, in-

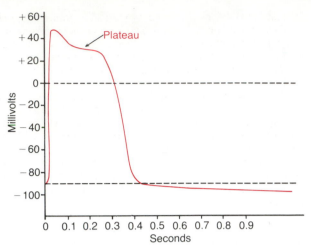

Figure 6–13. An action potential from a Purkinje fiber of the heart, showing a "plateau."

stead, the potential remains on a plateau near the peak of the spike for many milliseconds before repolarization begins. Such a plateau is illustrated in Figure 6–13; one can readily see that the plateau greatly prolongs the period of depolarization. This type of action potential occurs in heart muscle fibers, where the plateau lasts for as long as two to three tenths of a second and causes contraction of the heart muscle also to last for this same long period of time.

The cause of the plateau is a combination of several different factors. First, in heart muscle, two separate types of channels enter into the depolarization process: (1) the usual voltage-activated sodium channels, called *fast channels*, and (2) voltage-activated calcium channels, which are slow to be activated and therefore are called *slow channels*—these channels allow diffusion mainly of calcium ions but of some sodium ions as well. Activation of the fast channels causes the spike portion of the action potential, whereas the slow but prolonged activation of the slow channels is mainly responsible for the plateau portion of the action potential.

A second factor partly responsible for the plateau is that the voltage-gated potassium channels are slow to be activated in some instances, often not opening until the very end of the plateau. This delays the return of the membrane potential toward the resting value. But, then, this opening of the potassium channels at the same time that the slow channels begin to close causes rapid return of the action potential from its plateau level back to the negative resting level, accounting for the rapid downslope at the end of the action potential.

■ RHYTHMICITY OF CERTAIN EXCITABLE TISSUES — REPETITIVE DISCHARGE

Repetitive self-induced discharges, or *rhythmicity*, occur normally in the heart, in most smooth muscle, and in many of the neurons of the central nervous

system. It is these rhythmical discharges that cause the heart rhythm, that cause peristalsis, and that cause such neuronal events as the rhythmical control of breathing.

Also, all other excitable tissues can discharge repetitively if the threshold for stimulation is reduced low enough. For instance, even nerve fibers and skeletal muscle fibers, which normally are highly stable, discharge repetitively when they are placed in a solution containing the drug veratrine or when the calcium ion concentration falls below a critical value.

The Re-excitation Process Necessary for Rhythmicity. For rhythmicity to occur, the membrane, even in its natural state, must already be permeable enough to sodium ions (or to calcium and sodium ions through the slow calcium channels) to allow automatic membrane depolarization. Thus, Figure 6–14 shows that the "resting" membrane potential is only −60 to −70 mV. This is not enough negative voltage to keep the sodium and calcium channels closed. That is, (1) sodium and calcium ions flow inward, (2) this further increases the membrane permeability, (3) still more ions flow inward, (4) the permeability increases more, and so forth, thus eliciting the regenerative process of sodium and calcium channel openings until an action potential is generated. Then, at the end of the action potential, the membrane repolarizes. But shortly thereafter, the depolarization process begins again, and a new action potential occurs spontaneously. This cycle continues again and again and causes self-induced rhythmical excitation of the excitable tissue.

Yet, why does the membrane not depolarize immediately after it has become repolarized rather than delaying for nearly a second before the onset of the next action potential? The answer to this can be found by referring back to Figure 6–10, which shows that toward the end of all action potentials, and continuing for a short period thereafter, the membrane becomes excessively permeable to potassium. The excessive outflow of potassium ions carries tremendous numbers of positive charges to the outside of the membrane, creating inside the fiber considerably more negativity than would otherwise occur for a short period after the preceding action potential is over, thus drawing the membrane potential nearer to the potassium

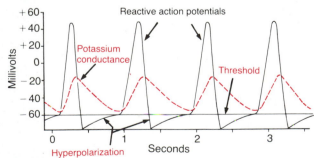

Figure 6–14. Rhythmic action potentials similar to those recorded in the rhythmical control center of the heart. Note their relationship to potassium conductance and to the state of hyperpolarization.

Nernst potential. This is a state called *hyperpolarization,* which is illustrated in Figure 6–14. As long as this state exists, re-excitation will not occur; but gradually the excess potassium conductance (and the state of hyperpolarization) disappears, as shown in the figure, thereby allowing the membrane potential to increase until it reaches the *threshold* for excitation; then suddenly a new action potential results, the process occurring again and again.

■ SPECIAL ASPECTS OF SIGNAL TRANSMISSION IN NERVE TRUNKS

Myelinated and Unmyelinated Nerve Fibers. Figure 6–15 illustrates a cross-section of a typical small nerve trunk, showing many large nerve fibers that compose most of the cross-sectional area. However, look very carefully and you will see many more small fibers lying between the large ones. The large fibers are *myelinated,* and the small ones are *unmyelinated.* The average nerve trunk contains about twice as many unmyelinated fibers as myelinated fibers.

Figure 6–16 illustrates a typical myelinated fiber. The central core of the fiber is the *axon,* and the membrane of the axon is the actual *conductive membrane.* The axon is filled in its center with *axoplasm,* which is a viscid intracellular fluid. Surrounding the axon is a *myelin sheath* that is often thicker than the axon itself, and about once every 1 to 3 mm along the length of the axon the myelin sheath is interrupted by a *node of Ranvier.*

The myelin sheath is deposited around the axon by Schwann cells in the following manner: The membrane of a Schwann cell first envelops the axon. Then the cell rotates around the axon many times, laying down multiple layers of cellular membrane containing the lipid substance *sphingomyelin.* This substance is an excellent insulator that decreases the ion flow through the membrane approximately 5000-

fold and also decreases the membrane capacitance as much as 50-fold. However, at the juncture between each two successive Schwann cells along the axon, a small, uninsulated area remains, only 2 to 3 μm in length, where ions can still flow with ease between the extracellular fluid and the axon. This area is the node of Ranvier.

"Saltatory" Conduction in Myelinated Fibers from Node to Node. Even though ions cannot flow significantly through the thick myelin sheaths of myelinated nerves, they can flow with considerable ease through the nodes of Ranvier. Therefore, action potentials can occur *only at the nodes.* Yet, the action potentials are conducted from node to node, as illustrated in Figure 6–17; this is called *saltatory conduction.* That is, electrical current flows through the surrounding extracellular fluids and also through the axoplasm from node to node, exciting successive nodes one after another. Thus, the nerve impulse jumps down the fiber, which is the origin of the term "saltatory."

Saltatory conduction is of value for two reasons. First, by causing the depolarization process to jump long intervals along the axis of the nerve fiber, this mechanism increases the velocity of nerve transmission in myelinated fibers as much as 5-fold to 50-fold. Second, saltatory conduction conserves energy for the axon, for only the nodes depolarize, allowing perhaps a hundred times smaller loss of ions than would otherwise be necessary and therefore requiring little extra metabolism for re-establishing the sodium and potassium concentration differences across the membrane after a series of nerve impulses.

Still another feature of saltatory conduction in large myelinated fibers is the following: The excellent insulation afforded by the myelin membrane and the 50-fold decrease in membrane capacitance allows the repolarization process to occur with very little transfer of ions. Therefore, at the end of the action potential when the sodium channels begin to close, repolarization occurs so rapidly that the potassium channels usually have not yet begun to open significantly. Therefore, conduction of the nerve impulse in the myeli-

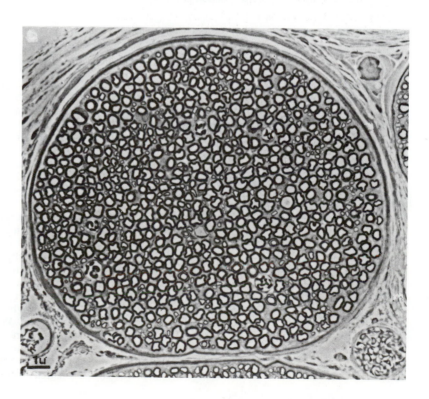

Figure 6–15. Cross-section of a small nerve trunk containing myelinated and unmyelinated fibers.

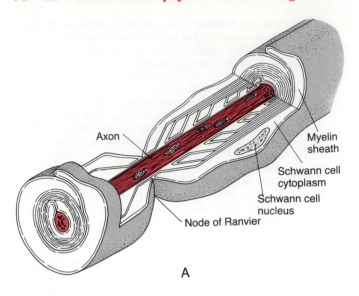

A

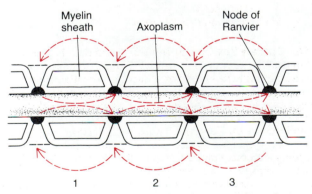

B

Figure 6–16. Function of the Schwann cell to insulate nerve fibers. *A,* The wrapping of a Schwann cell membrane around a large axon to form the myelin sheath of the myelinated nerve fiber. (Modified from Leeson and Leeson: Histology. Philadelphia, W. B. Saunders Company, 1979.) *B,* Evagination of the membrane and cytoplasm of a Schwann cell around multiple unmyelinated nerve fibers.

nated nerve fiber is accomplished almost entirely by the sequential changes in the voltage-gated sodium channels, with very little contribution by the potassium channels.

VELOCITY OF CONDUCTION IN NERVE FIBERS

The velocity of conduction in nerve fibers varies from as little as 0.5 m/sec in very small unmyelinated fibers to as high as 100 m/sec (the length of a football field in 1 sec) in very large myelinated fibers. The velocity increases approximately with the fiber diameter in myelinated nerve fibers and approxi-

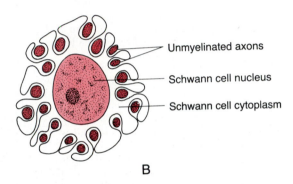

Figure 6–17. Saltatory conduction along a myelinated axon.

mately with the square root of fiber diameter in unmyelinated fibers.

■ EXCITATION—THE PROCESS OF ELICITING THE ACTION POTENTIAL

Basically, any factor that causes sodium ions to begin to diffuse inward through the membrane in sufficient numbers will set off the automatic, regenerative opening of the sodium channels. This can result from simple *mechanical* disturbance of the membrane, *chemical* effects on the membrane, or passage of *electricity* through the membrane. All these are used at different points in the body to elicit nerve or muscle action potentials: mechanical pressure to excite sensory nerve endings in the skin, chemical neurotransmitters to transmit signals from one neuron to the next in the brain, and electrical current to transmit signals between muscle cells in the heart and intestine. For the purpose of understanding the excitation process, let us begin by discussing the principles of electrical stimulation.

Excitation of a Nerve Fiber by a Negatively Charged Metal Electrode. The usual means for exciting a nerve or muscle in the experimental laboratory is to apply electricity at the nerve or muscle surface through two small electrodes, one of which is negatively charged and the other positively charged. When this is done, one finds that the excitable membrane becomes stimulated at the negative electrode.

The cause of these effects is the following: Remember that the action potential is initiated by the opening of voltage-gated sodium channels. Furthermore, these channels are opened by a decrease in the electrical voltage across the membrane. The negative current from the negative electrode reduces the voltage immediately outside the membrane, drawing this voltage nearer to the voltage of the negative membrane potential inside the fiber. This decreases the electrical voltage across the membrane and allows activation of the sodium channels, thus resulting in an action potential. Conversely, at the anode, the injection of positive charges on the outside of the nerve membrane heightens the voltage difference across the membrane rather than lessening it. And this causes a state of "hyperpolarization," which decreases the excitability of the fiber.

Threshold for Excitation and "Acute Local Potentials." A weak electrical stimulus may not be able to excite a fiber. However, when the stimulus is progressively increased, there comes a point at which excitation does take place. Figure 6–18 illustrates the effects of successively applied stimuli of progressing strength. A very weak stimulus at point A causes the membrane potential to change from −90 to −85 mV, but this is not sufficient change for the automatic regenerative processes of the action potential to develop. At point B, the stimulus is greater, but, here again, the intensity still is not enough. Nevertheless, the stimulus does disturb the membrane potential locally for as long as a millisecond or more after both of these weak stimuli. These local potential changes are called *acute local potentials,* and when they fail to elicit an action potential, they are called *acute subthreshold potentials.*

At point C in Figure 6–18, the stimulus is even stronger. Now the local potential has reached barely the level to elicit an action potential, called the threshold level, but this occurs only after a short "latent period." At point D, the stimulus is still stronger, the acute local potential is also stronger, and the action potential occurs after less of a latent period.

Thus, this figure shows that even a very weak stimulus always causes a local potential change at the membrane, but the intensity of the local potential must rise to a *threshold level* before the action potential will be set off.

The "Refractory Period" During Which New Stimuli Cannot Be Elicited

A new action potential cannot occur in an excitable fiber as long as the membrane is still depolarized from the preceding action potential. The reason for this is that shortly after the action potential is initiated the sodium channels (or calcium channels, or both) become inactivated, and any amount of excitatory signal applied to these channels at this point will not open the inactivation gates. The only condition that will

reopen them is for the membrane potential to return either to or almost to the original resting membrane potential level. Then, within another small fraction of a second, the inactivation gates of the channels open, and a new action potential can then be initiated.

The period of time during which a second action potential cannot be elicited, even with a very strong stimulus, is called the *absolute refractory period.* This period for large myelinated nerve fibers is about 1/2500 sec. Therefore, one can readily calculate that such a fiber can carry a maximum of about 2500 impulses/sec.

Following the absolute refractory period is a *relative refractory period,* lasting about one quarter to one half as long as the absolute period. During this time, stronger than normal stimuli can excite the fiber. The cause of this relative refractoriness is twofold: (1) During this time some of the sodium channels still have not been reversed from their inactivation state, and (2) the potassium channels are usually wide open at this time, causing a state of hyperpolarization that makes it more difficult to stimulate the fiber.

INHIBITION OF EXCITABILITY— "STABILIZERS" AND LOCAL ANESTHETICS

In contrast to the factors that increase nerve excitability, still others, called *membrane-stabilizing factors, can decrease excitability. For instance, a high extracellular fluid calcium ion concentration* decreases the membrane permeability and simultaneously reduces its excitability. Therefore, calcium ions are said to be a "stabilizer." Also, *low potassium ion concentration* in the extracellular fluids, because it has a direct effect of decreasing the permeability of the potassium channels, likewise acts as a stabilizer and reduces membrane excitability. Indeed, in a hereditary disease known as *familial periodic paralysis,* the extracellular potassium ion concentration is often so much reduced that the person actually becomes paralyzed but reverts to normal instantly after intravenous administration of potassium.

Local Anesthetics. Among the most important stabilizers are the many substances used clinically as local anesthetics, including *procaine, tetracaine,* and many other drugs. Most of these act directly on the activation gates of the sodium channels, making it much more difficult for these gates to open and thereby reducing the membrane excitability. When the excitability has been reduced so low that the ratio of *action potential strength to excitability threshold* (called the "safety factor") is reduced below 1.0, a nerve impulse fails to pass through the anesthetized area.

■ RECORDING MEMBRANE POTENTIALS AND ACTION POTENTIALS

The Cathode Ray Oscilloscope. Earlier in this chapter we noted that the membrane potential changes occur very rapidly throughout the course of an action potential. Indeed, most of the action potential complex of large nerve fibers takes place in less than 1/1000 sec. In some figures of this chapter an electrical meter has been shown recording these potential changes. However, it must be understood that any meter capable of recording them must be capable of responding extremely rapidly. For practical purposes, the only common type of meter that is capable of responding accurately to the very rapid membrane potential changes is the cathode ray oscilloscope.

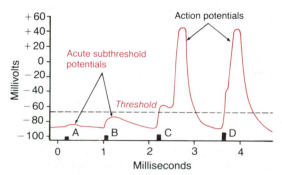

Figure 6–18. Effect of stimuli on the potential of the excitable membrane, showing the development of "acute subthreshold potentials" when the stimuli are below the threshold value required for eliciting an action potential.

Figure 6–19 illustrates the basic components of a cathode ray oscilloscope. The cathode ray tube itself is composed basically of an *electron gun* and a *fluorescent surface* against which electrons are fired. Where the electrons hit the surface, the fluorescent material glows. If the electron beam is moved across the surface, the spot of glowing light also moves and draws a fluorescent line on the screen.

In addition to the electron gun and fluorescent surface, the cathode ray tube is provided with two sets of electrically charged plates, one set positioned on either side of the electron beam and the other set positioned above and below. Appropriate electronic control circuits change the voltage on these plates so that the electron beam can be bent up or down in response to electrical signals coming from recording electrodes on nerves. The beam of electrons is also swept horizontally across the screen at a constant rate. This gives the record illustrated on the face of the cathode ray tube, giving a time base horizontally and the voltage changes at the nerve electrodes vertically. Note at the left end of the record a small *stimulus artifact* caused by the electrical stimulus used to elicit the action potential. Then further to the right is the recorded action potential itself.

Recording the Monophasic Action Potential. Throughout this chapter "monophasic" action potentials have been shown in the different diagrams. To record these, a micropipette electrode, illustrated earlier in the chapter in Figure 6–2, is inserted into the interior of the fiber. Then, as the action potential spreads down the fiber, the changes in the potential inside the fiber are recorded, as illustrated earlier in the chapter in Figures 6–6, 6–10, and 6–13.

Recording a Biphasic Action Potential. When one wishes to record impulses from a whole nerve trunk, it is not feasible to place electrodes inside the nerve fibers. Therefore, the usual method of recording is to place two electrodes on the outside of fibers. However, the record that is obtained is then biphasic for the following reasons: When an action potential moving down the nerve fiber reaches the first electrode, it becomes charged negatively, while the second electrode is still unaffected. This causes the oscilloscope to record in the negative direction. Then as the action potential proceeds still farther down the nerve, there comes a point when the membrane beneath the first electrode becomes repolarized, while the second electrode is negative, and the oscilloscope records in the opposite direction. Thus, a graphic

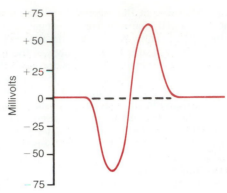

Figure 6–20. Recording of a biphasic action potential.

record such as that illustrated in Figure 6–20 is recorded by the oscilloscope, showing a potential change first in one direction and then in the opposite direction.

REFERENCES

Agnew, W. S.: Voltage-regulated sodium channel molecules. Annu. Rev. Physiol., 45:517, 1984.

Armstrong, C. M.: Sodium channels and gating currents. Physiol. Rev., 61:644, 1981.

Auerbach, A., and Sachs, F.: Patch clamp studies of single ionic channels. Annu. Rev. Biophys. Bioeng., 13:269, 1984.

Biggio, G., and Costa, E.: Chloride Channels and Their Modulation by Neurotransmitters and Drugs. New York, Raven Press, 1988.

Bretag, A. H.: Muscle chloride channels. Physiol. Rev., 67:618, 1987.

Byrne, J. H., and Schultz, S. G.: An Introduction to Membrane Transport and Bioelectricity. New York, Raven Press, 1988.

Clausen, T.: Regulation of active Na+-K+ transport in skeletal muscle. Physiol. Rev., 66:542, 1986.

Cole, K. S.: Electrodiffusion models for the membrane of squid giant axon. Physiol. Rev., 45:340, 1965.

Cooper, S. A.: New peripherally-acting oral analgesic agents. Annu. Rev. Pharmacol. Toxicol., 23:617, 1983.

DeWeer, P., et al.: Voltage dependence of the Na-K pump. Annu. Rev. Physiol., 50:225, 1988.

DiFrancesco, D., and Noble, D.: A model of cardiac electrical activity incorporating ionic pumps and concentration changes. Phil. Trans. R. Soc. Lond. (Biol.), 307:353, 1985.

DiPolo, R., and Beauge, L.: The calcium pump and sodium-calcium exchange in squid axons. Annu. Rev. Physiol., 45:313, 1983.

French, R. J., and Horn, R.: Sodium channel gating: Models, mimics, and modifiers. Annu. Rev. Biophys. Bioeng., 12:319, 1983.

Garty, H., and Benos, D. J.: Characteristics and regulatory mechanisms of the amiloride-blockable Na+ channel. Physiol. Rev., 68:309, 1988.

Grinnell, A. D., et al. (eds.): Calcium and Ion Channel Modulation. New York, Plenum Publishing Corp., 1988.

Hille, B.: Gating in sodium channels of nerve. Annu. Rev. Physiol., 38:139, 1976.

Hodgkin, A. L.: The Conduction of the Nervous Impulse. Springfield, Ill., Charles C Thomas, 1963.

Hodgkin, A. L., and Horowicz, P.: The effect of sudden changes in ionic concentrations on the membrane potential of single muscle fibers. J. Physiol. (Lond.), 153:370, 1960.

Hodgkin, A. L., and Huxley, A. F.: Movement of sodium and potassium ions during nervous activity. Cold Spr. Harb. Symp. Quant. Biol., 17:43, 1952.

Hodgkin, A. L., and Huxley, A. F.: Quantitative description of membrane current and its application to conduction and excitation in nerve. J. Physiol. (Lond.), 117:500, 1952.

Kaplan, J. H.: Ion movements through the sodium pump. Annu. Rev. Physiol., 47:535, 1985.

Katz, B.: Nerve, Muscle, and Synapse. New York, McGraw-Hill, 1968.

Keynes, R. D.: Ion channels in the nerve-cell membrane. Sci. Am., 240:126, 1979.

Kostyuk, P. G.: Intracellular perfusion of nerve cells and its effects on membrane currents. Physiol. Rev., 64:435, 1984.

Krueger, B. K.: Toward an understanding of structure and function of ion channels. FASEB J., 3:1906, 1989.

Latorre, R., and Alvarez, O.: Voltage-dependent channels in lipid bilayer membranes. Physiol. Rev., 61:77, 1981.

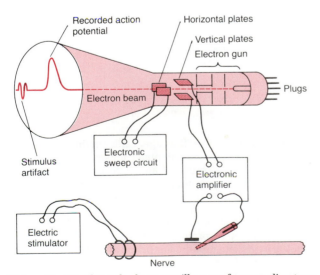

Figure 6–19. The cathode ray oscilloscope for recording transient action potentials.

Latorre, R., et al.: K$^+$ channels gated by voltage and ions. Annu. Rev. Physiol., 46:485, 1984.

Levitan, I. B.: Modulation of ion channels in neutrons and other cells. Annu. Rev. Neurosci., 11:119, 1988.

Malhotra, S. K.: The Plasma Membrane. New York, John Wiley & Sons, 1983.

Miller, R. J.: Multiple calcium channels and neuronal function. Science, 235:46, 1987.

Moody, W., Jr.: Effects of intracellular H$^+$ on the electrical properties of excitable cells. Annu. Rev. Neurosci., 7:257, 1984.

Naftalin, R. J.: The thermostatics and thermodynamics of cotransport. Biochem. Biophys. Acta., 778:155, 1984.

Narahashi, T.: Ion Channels. New York, Plenum Publishing Corp., 1988.

Requena, J.: Calcium transport and regulation in nerve fibers. Annu. Rev. Biophys. Bioeng., 12:237, 1983.

Reuter, H.: Modulation of ion channels by phosphorylation and second messengers. News Physiol. Sci., 2:168, 1987.

Rogart, R.: Sodium channels in nerve and muscle membrane. Annu. Rev. Physiol., 43:711, 1981.

Ross, W. N.: Changes in intracellular calcium during neuron activity. Annu. Rev. Physiol., 51:491, 1989.

Sakmann, B., and Neher, E.: Patch clamp techniques for studying ionic channels in excitable membranes. Annu. Rev. Physiol., 46:455, 1984.

Schubert, D.: Developmental Biology of Cultured Nerve, Muscle and Glia. New York, John Wiley & Sons, 1984.

Schwartz, W., and Passow, H.: Ca^{2+}-activated K$^+$ channels in erythrocytes and excitable cells. Annu. Rev. Physiol., 45:359, 1983.

Shepherd, G. M.: Neurobiology. New York, Oxford University Press, 1987.

Sjodi, R. A.: Ion Transport in Skeletal Muscle. New York, John Wiley & Sons, 1982.

Skene, J. H. P.: Axonal growth-associated proteins. Annu. Rev. Neurosci., 12:127, 1989.

Snell, R. M. (ed.): Transcellular Membrane Potentials and Ionic Fluxes. New York, Gordon Press Pubs., 1984.

Sperelakis, N.: Hormonal and neurotransmitter regulation of Ca^{++} influx through voltage-dependent slow channels in cardiac muscle membrane. Membr. Biochem., 5:131, 1984.

Stefani, E., and Chiarandini, D. J.: Ionic channels in skeletal muscle. Annu. Rev. Physiol., 44:357, 1982.

Swadlow, H. A., et al.: Modulation of impulse conduction along the axonal tree. Annu. Rev. Biophys. Bioeng., 9:143, 1980.

Trimmer, J. A., and Agnew, W. S.: Molecular diversity of voltage-sensitive Na channels. Annu. Rev. Physiol., 51:401, 1989.

Tsien, R. W.: Calcium channels in excitable cell membranes. Annu. Rev. Physiol., 45:341, 1983.

Ulbricht, W.: Kinetics of drug action and equilibrium results at the node of Ranvier. Physiol. Rev., 61:785, 1981.

Vinores, S., and Guroff, G.: Nerve growth factor: Mechanisms of action. Annu. Rev. Biophys. Bioeng., 9:223, 1980.

Weiss, D. C. (ed.): Axioplasmic Transport in Physiology and Pathology. New York, Springer-Verlag, 1982.

Windhager, E. E., and Taylor, A.: Regulatory role of intracellular calcium ions in epithelial Na transport. Annu. Rev. Physiol., 45:519, 1983.

Wright, E. M.: Electrophysiology of plasma membrane vesicles. Am. J. Physiol., 246:F363, 1984.

Zigmond, R. E., and Bowers, C. W.: Influence of nerve activity on the macromolecular content of neurons and their effector organs. Annu. Rev. Physiol., 43:673, 1981.

THE CENTRAL NERVOUS SYSTEM:

A. General Principles and Sensory Physiology

7

Organization of the Central Nervous System; Basic Functions of Synapses and Transmitter Substances

The nervous system, along with the endocrine system, provides most of the control functions for the body. In general, the nervous system controls the rapid activities of the body, such as muscular contractions, rapidly changing visceral events, and even the rates of secretion of some endocrine glands. The endocrine system, by contrast, regulates principally the metabolic functions of the body.

The nervous system is unique in the vast complexity of the control actions that it can perform. It receives literally millions of bits of information from the different sensory organs and then integrates all these to determine the response to be made by the body. The purpose of this chapter is to present, first, a general outline of the overall mechanisms by which the nervous system performs such functions. Then we will discuss the function of central nervous system synapses, the basic structures that control the passage of signals into, through, and out of the nervous system. In succeeding chapters we will analyze in detail the functions of the individual parts of the nervous system. Before beginning this discussion, however, the reader should refer to Chapters 5, 6, and 25, which present, respectively, the principles of membrane potentials and transmission of signals in nerves and through neuromuscular junctions.

■ GENERAL DESIGN OF THE CENTRAL NERVOUS SYSTEM

THE CENTRAL NERVOUS SYSTEM NEURON — THE BASIC FUNCTIONAL UNIT

The central nervous system is composed of more than 100 billion neurons. Figure 7–1 illustrates a typical neuron of the type found in the cerebral motor cortex. The incoming information enters the cell almost entirely through synapses on the neuronal dendrites or cell body; there may be only a few hundred or as many as 200,000 such synaptic connections from the input fibers. On the other hand, the output signal travels by way of a single axon, but this axon gives off many separate branches to other parts of the brain, the spinal cord, or the peripheral body. These terminals then provide synapses with the next order of neurons or with muscle cells or secretory cells.

A special feature of most synapses is that the signal normally passes only in the forward direction except under rare conditions. This allows the signals to be conducted in the required directions for performing necessary nervous functions. We shall also see that the neurons are organized into a great multitude of neural networks that determine the functions of the nervous system.

THE SENSORY DIVISION OF THE CENTRAL NERVOUS SYSTEM — SENSORY RECEPTORS

Most activities of the nervous system are initiated by sensory experience emanating from *sensory receptors*, whether visual receptors, auditory receptors, tactile receptors on the surface of the body, or other kinds of receptors. This sensory experience can cause an immediate reaction, or its memory can be stored in the brain for minutes, weeks, or years and then can help determine the bodily reactions at some future date.

Figure 7–2 illustrates a portion of the sensory system, the *somatic* portion, which transmits sensory information from the receptors of the entire surface of the body and some deep structures. This information

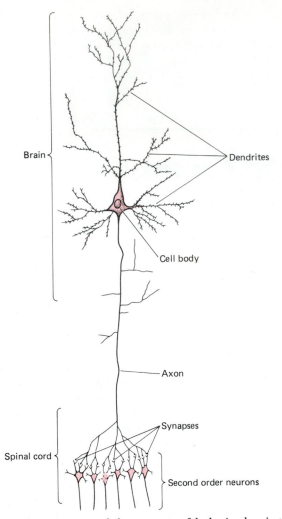

Figure 7–1. Structure of a large neuron of the brain, showing its important functional parts.

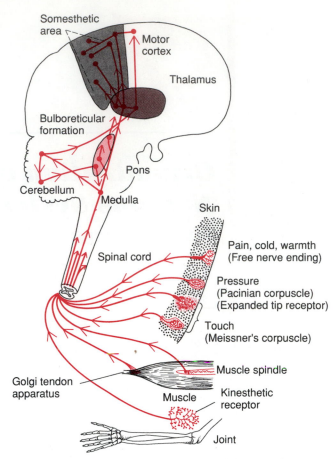

Figure 7–2. The somatic sensory axis of the central nervous system.

enters the central nervous system through the spinal nerves and is conducted to multiple "primary" sensory areas in (1) the spinal cord at all levels, (2) the reticular substance of the medulla, pons, and mesencephalon, (3) the cerebellum, (4) the thalamus, and (5) the somesthetic areas of the cerebral cortex. But in addition to these primary sensory areas, signals are then relayed to essentially all other parts of the nervous system as well.

THE MOTOR DIVISION— THE EFFECTORS

The most important ultimate role of the nervous system is to control the various bodily activities. This is achieved by controlling (1) contraction of skeletal muscles throughout the body, (2) contraction of smooth muscle in the internal organs, and (3) secretion by both exocrine and endocrine glands in many parts of the body. These activities are collectively called *motor functions* of the nervous system, and the muscles and glands are called *effectors,* because they perform the functions dictated by the nerve signals.

Figure 7–3 illustrates the *motor axis* of the nervous system for controlling skeletal muscle contraction. Operating parallel to this axis is another similar system for control of the smooth muscles and glands called the *autonomic nervous system,* which is presented in Chapter 22. Note in Figure 7–3 that the skeletal muscles can be controlled from many different levels of the central nervous system, including (1) the spinal cord, (2) the reticular substance of the medulla, pons, and mesencephalon, (3) the basal ganglia, (4) the cerebellum, and (5) the motor cortex. Each of these different areas plays its own specific role in the control of body movements, the lower regions being concerned primarily with automatic, instantaneous responses of the body to sensory stimuli; and the higher regions, with deliberate movements controlled by the thought process of the cerebrum.

PROCESSING OF INFORMATION— "INTEGRATIVE" FUNCTION OF THE CENTRAL NERVOUS SYSTEM

The major function of the nervous system is to process incoming information in such a way that *appropriate* motor responses occur. More than 99 per cent of all sensory information is discarded by the brain as irrele-

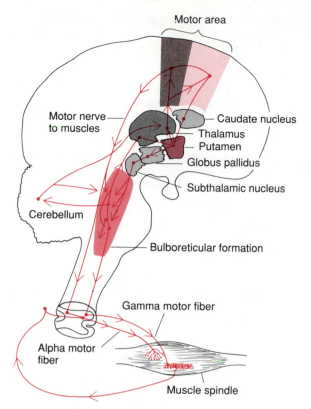

Figure 7–3. The motor axis of the central nervous system.

vant and unimportant. For instance, one is ordinarily totally unaware of the parts of the body that are in contact with clothing and is also unaware of the seat pressure when sitting. Likewise, attention is drawn only to an occasional object in one's field of vision, and even the perpetual noise of our surroundings is usually relegated to the background.

After the important sensory information has been selected, it is then channeled into proper motor regions of the brain to cause the desired responses. This channeling of information is called the *integrative function* of the nervous system. Thus, if a person places a hand on a hot stove, the desired response is to lift the hand. There are other associated responses, too, such as moving the entire body away from the stove and perhaps even shouting with pain. Yet even these responses represent activity by only a small fraction of the total motor system of the body.

Role of Synapses in Processing Information. The synapse is the junction point from one neuron to the next and, therefore, is an advantageous site for control of signal transmission. Later in this chapter we discuss the details of synaptic function. However, it is important to point out here that the synapses determine the directions that the nervous signals spread in the nervous system. Some synapses transmit signals from one neuron to the next with ease, and others transmit signals only with difficulty. Also, facilitatory and inhibitory signals from other areas in the nervous system can control synaptic activity, sometimes opening the synapses for transmission and at other times closing them. In addition, some postsynaptic neurons respond with large numbers of impulses, and others respond with

only a few. Thus, the synapses perform a selective action, often blocking the weak signals while allowing the strong signals to pass, often selecting and amplifying certain weak signals, and often channeling the signals in many different directions, rather than simply in one direction.

STORAGE OF INFORMATION — MEMORY

Only a small fraction of the important sensory information causes an immediate motor response. Much of the remainder is stored for future control of motor activities and for use in the thinking processes. Most of this storage occurs in the *cerebral cortex*, but not all, for even the basal regions of the brain and perhaps even the spinal cord can store small amounts of information.

The storage of information is the process we call *memory*, and this, too, is a function of the synapses. That is, each time certain types of sensory signals pass through sequences of synapses, these synapses become more capable of transmitting the same signals the next time, which process is called *facilitation*. After the sensory signals have passed through the synapses a large number of times, the synapses become so facilitated that signals generated within the brain itself can also cause transmission of impulses through the same sequences of synapses even though the sensory input has not been excited. This gives the person a perception of experiencing the original sensations, although in effect they are only memories of the sensations.

Unfortunately, we do not know the precise mechanism by which facilitation of synapses occurs in the memory process, but what is known about this and other details of the memory process will be discussed in Chapter 19.

Once memories have been stored in the nervous system, they become part of the processing mechanism. The thought processes of the brain compare new sensory experiences with the stored memories; the memories help to select the important new sensory information and to channel this into appropriate storage areas for future use or into motor areas to cause bodily responses.

■ THE THREE MAJOR LEVELS OF CENTRAL NERVOUS SYSTEM FUNCTION

The human nervous system has inherited specific characteristics from each stage of evolutionary development. From this heritage, three major levels of the central nervous system have specific functional attributes: (1) the *spinal cord level*; (2) the *lower brain level*; and (3) the *higher brain* or *cortical level*.

THE SPINAL CORD LEVEL

We often think of the spinal cord as being only a conduit for signals from the periphery of the body to the brain or in the opposite direction from the brain back to the body. However, this is far from the truth. Even after the spinal cord has been cut in the high neck region, many spinal cord functions still occur. For instance, neuronal circuits in the cord can cause (1) walking movements, (2) reflexes that withdraw portions of the body from objects, (3) reflexes that stiffen the legs to support the body against gravity, and (4) reflexes that control local blood vessels, gastrointestinal movements, and so forth, in addition to many other functions.

In fact, the upper levels of the nervous system often operate not by sending signals directly to the periphery of the body but instead by sending signals to the control centers of the cord, simply "commanding" the cord centers to perform their functions.

THE LOWER BRAIN LEVEL

Many if not most of what we call subconscious activities of the body are controlled in the lower areas of the brain—in the medulla, pons, mesencephalon, hypothalamus, thalamus, cerebellum, and basal ganglia. Subconscious control of arterial pressure and respiration is achieved mainly in the medulla and pons. Control of equilibrium is a combined function of the older portions of the cerebellum and the reticular substance of the medulla, pons, and mesencephalon. Feeding reflexes, such as salivation in response to the taste of food and the licking of the lips, are controlled by areas in the medulla, pons, mesencephalon, amygdala, and hypothalamus; and many emotional patterns, such as anger, excitement, sexual activities, reaction to pain, or reaction of pleasure, can occur in animals without a cerebral cortex.

THE HIGHER BRAIN OR CORTICAL LEVEL

After recounting all the nervous system functions that can occur at the cord and lower brain levels, what is left for the cerebral cortex to do? The answer to this is a complex one, but it begins with the fact that the cerebral cortex is an extremely large memory storehouse. The cortex never functions alone, but always in association with the lower centers of the nervous system.

Without the cerebral cortex, the functions of the lower brain centers are often very imprecise. The vast storehouse of cortical information usually converts these functions to very determinative and precise operations.

Finally, the cerebral cortex is essential for most of our thought processes even though it also cannot function alone in this. In fact, it is the lower centers that cause _wakefulness_ in the cerebral cortex, thus opening its bank of memories to the thinking machinery of the brain.

Thus, each portion of the nervous system performs specific functions. Many integrative functions are well developed in the spinal cord, and many of the subconscious functions originate and are executed entirely in the lower regions of the brain. But it is the cortex that opens the world up for one's mind.

■ COMPARISON OF THE CENTRAL NERVOUS SYSTEM WITH AN ELECTRONIC COMPUTER

When electronic computers were first developed in many different laboratories of the world by as many different scientists, it soon became apparent that all these machines have many features in common with the nervous system. First, they all have input circuits that are comparable to the sensory portion of the nervous system and output circuits that are comparable to the motor portion of the nervous system. In the conducting pathway between the inputs and the outputs are the mechanisms for performing the different types of computations.

In simple computers, the output signals are controlled directly by the input signals, operating in a manner similar to that of the simple reflexes of the spinal cord. However, in the more complex computers, the output is determined both by input signals and by information that has already been stored in memory in the computer, which is analogous to the more complex reflex and processing mechanisms of our higher nervous system. Furthermore, as the computers become even more complex, it is necessary to add still another unit, called the _central programming unit_, which determines the sequence of all operations. This unit is analogous to the mechanism in our brain that allows us to direct our attention first to one thought or sensation or motor activity, then to another, and so forth, until complex sequences of thought or action take place.

Figure 7–4 is a simple block diagram of a modern computer. Even a rapid study of this diagram will demonstrate its similarity to the nervous system. The fact that the basic components of the general purpose computer are analogous to those of the human nervous system demonstrates that the brain is basically a computer that continuously collects sensory information and uses this along with stored information to compute the daily course of bodily activity.

■ THE CENTRAL NERVOUS SYSTEM SYNAPSES

Every medical student is aware that information is transmitted in the central nervous system mainly in the form of nerve impulses through a succession of neu-

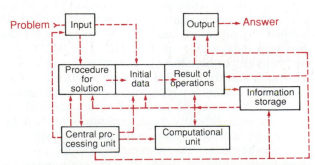

Figure 7–4. Block diagram of a general purpose electronic computer, showing the basic components and their interrelationships.

rons, one after another. However, it is not immediately apparent that each impulse (1) may be blocked in its transmission from one neuron to the next, (2) may be changed from a single impulse into repetitive impulses, or (3) may be integrated with impulses from other neurons to cause highly intricate patterns of impulses in successive neurons. All these functions can be classified as *synaptic functions of neurons.*

TYPES OF SYNAPSES — CHEMICAL AND ELECTRICAL

Nerve signals are transmitted from one neuron to the next through interneuronal junctions called *synapses.* In the animal world there are basically two different types of synapses: (1) the *chemical synapse* and (2) the *electrical synapse.*

Almost all the synapses utilized for signal transmission in the central nervous system of the human being are *chemical synapses.* In these, the first neuron secretes a chemical substance called a *neurotransmitter* at the synapse, and this transmitter in turn acts on receptor proteins in the membrane of the next neuron to excite the neuron, to inhibit it, or to modify its sensitivity in some other way. Over 40 different transmitter substances have been discovered thus far. Some of the best known are acetylcholine, norepinephrine, histamine, gamma-aminobutyric acid (GABA), and glutamate.

Electrical synapses, on the other hand, are characterized by direct channels that conduct electricity from one cell to the next. Most of these consist of small protein tubular structures called *gap junctions* that allow free movements of ions from the interior of one cell to the next. Only a few gap junctions have been found in the central nervous system, and their significance in general is not known. On the other hand, it is by way of gap junctions and other similar junctions that action potentials are transmitted from one smooth muscle fiber to the next in visceral smooth muscle (Chapter 25) and also from one cardiac muscle cell to the next in cardiac muscle (Chapter 26).

One-Way Conduction Through Chemical Synapses. Chemical synapses have one exceedingly important characteristic that makes them highly desirable as the form of transmission of nervous system signals: they always transmit the signals in one direction — that is, from the neuron that secretes the transmitter, called the *presynaptic neuron,* to the neuron on which the transmitter acts, called the *postsynaptic neuron.* This is the principle of *one-way conduction* through chemical synapses, and it is quite different from conduction through electrical synapses that can transmit signals in either direction.

Think for a moment about the extreme importance of the one-way conduction mechanism. It allows signals to be directed toward specific goals. Indeed, it is this specific transmission of signals to discrete and highly focused areas in the nervous system that allows the nervous system to perform its myriad functions of sensation, motor control, memory, and many others.

PHYSIOLOGIC ANATOMY OF THE SYNAPSE

Figure 7–5 illustrates a typical *motor neuron* in the anterior horn of the spinal cord. It is composed of three major parts: the *soma,* which is the main body of the neuron; a single *axon,* which extends from the soma into the peripheral nerve; and the *dendrites,* which are thin projections of the soma that extend up to 1 mm into the surrounding areas of the cord.

Up to as many as 100,000 small knobs called *presynaptic terminals* lie on the surfaces of the dendrites and soma of the motor neuron, approximately 80 to 95 per cent of them on the dendrites and only 5 to 20 per cent on the soma. These terminals are the ends of nerve fibrils that originate in many other neurons; usually not more than a few of the terminals are derived from any single previous neuron. Later it will become evident that many of these presynaptic terminals are *excitatory* and secrete a substance that excites the postsynaptic neuron, but many others are *inhibitory* and secrete a substance that inhibits the neuron.

Neurons in other parts of the cord and brain differ markedly from the motor neuron in (1) the size of the cell body; (2) the length, size, and number of dendrites, ranging in length to as long as many centimeters; (3) the length and size of the axon; and (4) the number of presynaptic terminals, which may range from only a few to several hundred thousand. These differences make neurons in different parts of the nervous system react differently to incoming signals and therefore perform different functions.

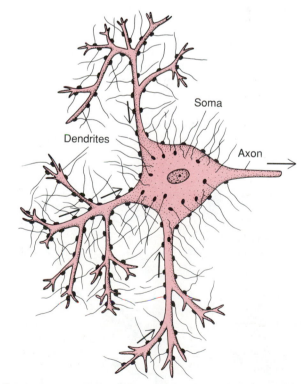

Figure 7–5. A typical motor neuron, showing presynaptic terminals on the neuronal soma and dendrites. Note also the single axon.

The Presynaptic Terminals. Electron microscopic studies of the presynaptic terminals show that these have varied anatomical forms, but most resemble small round or oval knobs and therefore are frequently called *terminal knobs, boutons, end-feet,* or *synaptic knobs.*

Figure 7–6 illustrates the basic structure of the presynaptic terminal. It is separated from the neuronal soma by a *synaptic cleft* having a width usually of 200 to 300 Å. The terminal has two internal structures important to the excitatory or inhibitory functions of the synapse: the *synaptic vesicles* and the *mitochondria.* The synaptic vesicles contain *transmitter substances* that, when released into the synaptic cleft, either *excite* or *inhibit* the postsynaptic neuron—excite if the neuronal membrane contains *excitatory receptors,* inhibit if it contains *inhibitory receptors.* The mitochondria provide adenosine triphosphate (ATP), which is required to synthesize new transmitter substance.

When an action potential spreads over a presynaptic terminal, the membrane depolarization causes emptying of a small number of vesicles into the cleft; and the released transmitter in turn causes an immediate change in the permeability characteristics of the postsynaptic neuronal membrane, which leads to excitation or inhibition of the neuron, depending on its receptor characteristics.

Mechanism by Which Action Potentials Cause Transmitter Release at the Presynaptic Terminals — Role of Calcium Ions

The synaptic membrane of the presynaptic terminals contains large numbers of *voltage-gated calcium channels.* This is quite different from the other areas of the nerve fiber, which contain very few of these channels. When the action potential depolarizes the terminal, large numbers of calcium ions, along with the sodium ions that cause most of the action potential, flow into the terminal. The quantity of transmitter substance that is released into the synaptic cleft is directly related to the number of calcium ions that enter the terminal.

The precise mechanism by which the calcium ions cause this release is not known but is believed to be the following:

When the calcium ions enter the synaptic terminal, it is believed that they bind with protein molecules on the inner surfaces of the synaptic membrane, called *release sites.* This in turn causes the transmitter vesicles in the local vicinity to bind also with the membrane and actually to fuse with it, and finally to open to the exterior by the process called *exocytosis.* Usually, a few vesicles release their transmitter into the cleft following each single action potential. For the vesicles that store the neurotransmitter acetylcholine, between 2000 and 10,000 molecules of acetylcholine are present in each vesicle, and there are enough vesicles in the presynaptic terminal to transmit from a few hundred to more than 10,000 action potentials.

Action of the Transmitter Substance on the Postsynaptic Neuron — The Function of Receptors

At the synapse, the membrane of the postsynaptic neuron contains large numbers of *receptor proteins,* also illustrated in Figure 7–6. These receptors have two important components: (1) a *binding component* that protrudes outward from the membrane into the synaptic cleft—it binds with the neurotransmitter from the presynaptic terminal—and (2) an *ionophore component* that passes all the way through the membrane to the interior of the postsynaptic neuron. The ionophore in turn is one of two types: (1) a *chemically activated ion channel* or (2) an *enzyme that activates an internal metabolic change inside the cell.*

The Ion Channels. The chemically activated ion channels (also called ligand-activated channels) are usually of three types: (1) *sodium channels* that allow mainly sodium ions (but some potassium ions as well) to pass through; (2) *potassium channels* that allow mainly potassium ions to pass; and (3) *chloride channels* that allow chloride and a few other anions to pass. We learn later that opening the sodium channels excites the postsynaptic neuron. Therefore, a transmitter substance that opens the sodium channels is called an *excitatory transmitter.* On the other hand, opening of potassium and chloride channels inhibits the neuron, and transmitters that open either or both of these are called *inhibitory transmitters.*

The Enzyme Receptors. Activation of an enzymatic type of receptor causes other effects on the postsynaptic neuron. One effect is to *activate the metabolic machinery of the cell,* such as causing formation of cyclic adenosine monophosphate (AMP), which in turn excites many other intracellular activities. Another is to *activate cellular genes,* which in turn manufacture additional receptors for the postsynaptic membrane. Still a third effect is to *activate protein kinases,* which decrease the numbers of receptors. Changes such as these can alter the reactivity of the synapse for minutes, days, months, or even years. Therefore, transmitter substances that cause such effects are sometimes called synaptic *modulators.* Recent experiments have

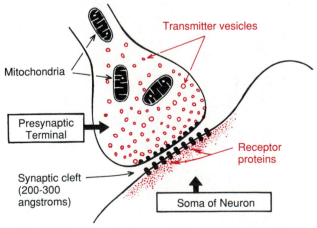

Figure 7–6. Physiologic anatomy of the synapse.

demonstrated that such modulators are important in at least some of the memory processes, which we discuss in Chapter 19.

Excitatory and Inhibitory Receptors

Some postsynaptic receptors, when activated, cause excitation of the postsynaptic neuron and others cause inhibition. The importance of having inhibitory as well as excitatory types of receptors is that this gives an additional dimension to nervous function, allowing restraint of nervous action as well as excitation.

The different molecular and membrane mechanisms employed by the different receptors to cause excitation or inhibition include the following:

Excitation

1. Opening of sodium channels to allow large numbers of positive electrical charges to flow to the interior of the postsynaptic cell. This raises the membrane potential in the positive direction up toward the threshold level for excitation. It is by far the most widely used means of causing excitation.

2. Depressed conduction through potassium or chloride channels, or both. This decreases the diffusion of positively charged potassium ions out of the postsynaptic neuron or decreases the diffusion of negatively charged chloride ions to the inside. In either instance, the effect is to make the internal membrane potential more positive than normally, which is excitatory.

3. Various changes in the internal metabolism of the cell to excite cell activity, or in some instances increase in the number of excitatory membrane receptors or decrease in the number of inhibitory membrane receptors.

Inhibition

1. Opening of potassium ion channels through the receptor molecule. This allows rapid diffusion of positively charged potassium ions from inside the postsynaptic neuron to the outside, thereby carrying positive charges outward and increasing the negativity inside, which is inhibitory.

2. Increase in the conductance of chloride ions through the receptor. This allows negative chloride ions to diffuse to the interior, which is also inhibitory.

3. Activation of receptor enzymes that inhibit cellular metabolic functions or that increase the number of inhibitory synaptic receptors or decrease the number of excitatory receptors.

CHEMICAL SUBSTANCES THAT FUNCTION AS SYNAPTIC TRANSMITTERS

More than 40 different chemical substances have been proved or postulated to function as synaptic transmitters. Most of these are listed in Tables 7–1 and 7–2, which give two different groups of synaptic transmitters. One is composed of small-molecule, rapidly act-

TABLE 7–1 Small-Molecule, Rapidly Acting Transmitters

Class I
  Acetylcholine
Class II: *The Amines*
  Norepinephrine
  Epinephrine
  Dopamine
  Serotonin
  Histamine
Class III: *Amino Acids*
  Gamma-aminobutyric acid (GABA)
  Glycine
  Glutamate
  Aspartate

ing transmitters. The other comprises a large number of neuropeptides of much larger molecular size and much more slowly acting.

The small-molecule, rapidly acting transmitters are the ones that cause most of the acute responses of the nervous system, such as transmission of sensory signals to and inside the brain and motor signals back to the muscles. The neuropeptides, on the other hand, usually cause more prolonged actions, such as long-term changes in numbers of receptors, long-term closure of certain ion channels, and possibly even long-term changes in numbers of synapses.

The Small-Molecule, Rapidly Acting Transmitters

Almost without exception, the small-molecule types of transmitters are synthesized in the cytosol of the pre-

TABLE 7–2 Neuropeptide, Slowly Acting Transmitters

A. *Hypothalamic-releasing hormones*
   Thyrotropin-releasing hormone
   Luteinizing hormone-releasing hormone
   Somatostatin (growth hormone-inhibitory factor)
B. *Pituitary peptides*
   ACTH
   β-Endorphin
   α-Melanocyte-stimulating hormone
   Prolactin
   Luteinizing hormone
   Thyrotropin
   Growth hormone
   Vasopressin
   Oxytocin
C. *Peptides that act on gut and brain*
   Leucine enkephalin
   Methionine enkephalin
   Substance P
   Gastrin
   Cholecystokinin
   Vasoactive intestinal polypeptide (VIP)
   Neurotensin
   Insulin
   Glucagon
D. *From other tissues*
   Angiotensin II
   Bradykinin
   Carnosine
   Sleep peptides
   Calcitonin

synaptic terminal and then are absorbed by active transport into the many transmitter vesicles in the terminal. Then, each time an action potential reaches the presynaptic terminal, a few vesicles at a time release their transmitter into the synaptic cleft, usually within a millisecond or less, by the mechanism described earlier. The subsequent action of the small-molecule type of transmitter on the postsynaptic membrane receptors usually also occurs within another millisecond or less. Most often the effect is to increase or decrease conductance through ion channels; an example is to increase sodium conductance, which causes excitation, or to increase potassium or chloride conductance, which causes inhibition. However, occasionally the small-molecule types of transmitters can stimulate receptor-activated enzymes instead of opening ion channels, thus changing the internal metabolic machinery of the cell.

Recycling of the Small-Molecule Types of Vesicles. The vesicles that store and release small-molecule transmitters are continually recycled, that is, used over and over again. After they fuse with the synaptic membrane and open to release their transmitters, the vesicle membrane at first simply becomes part of the synaptic membrane. However, within seconds to minutes, the vesicle portion of the membrane invaginates back to the inside of the presynaptic terminal and pinches off to form a new vesicle. It still contains the appropriate transport proteins required for concentrating new transmitter substance inside the vesicle.

Acetylcholine is a typical small-molecule transmitter that obeys the above principles of synthesis and release. It is synthesized in the presynaptic terminal from acetyl coenzyme A (acetyl-CoA) and choline in the presence of the enzyme *choline acetyltransferase.* Then it is transported into its specific vesicles. When the vesicles later release the acetylcholine into the synaptic cleft, the acetylcholine is rapidly split again to acetate and choline by the enzyme *cholinesterase,* which is bound to the proteoglycan reticulum that fills the space of the synaptic cleft. Then the vesicles are recycled, and choline also is actively transported back into the terminal to be used again for synthesis of new acetylcholine.

Characteristics of Some of the More Important Small-Molecule Transmitters. The most important of the small-molecule transmitters are the following:

Acetylcholine is secreted by neurons in many areas of the brain, but specifically by the large pyramidal cells of the motor cortex, by many different neurons in the basal ganglia, by the motor neurons that innervate the skeletal muscles, by the preganglionic neurons of the autonomic nervous system, by the postganglionic neurons of the parasympathetic nervous system, and by some of the postganglionic neurons of the sympathetic nervous system. In most instances acetylcholine has an excitatory effect; however, it is known to have inhibitory effects at some of the peripheral parasympathetic nerve endings, such as inhibition of the heart by the vagus nerves.

Norepinephrine is secreted by many neurons whose cell bodies are located in the brain stem and hypothala-

mus. Specifically, norepinephrine-secreting neurons located in the *locus ceruleus* in the pons send nerve fibers to widespread areas of the brain and help control the overall activity and mood of the mind. In most of these areas it probably activates excitatory receptors, but in a few areas, inhibitory receptors instead. Norepinephrine is also secreted by most of the postganglionic neurons of the sympathetic nervous system, where it excites some organs but inhibits others.

Dopamine is secreted by neurons that originate in the substantia nigra. The termination of these neurons are mainly in the striatal region of the basal ganglia. The effect of dopamine is usually inhibition.

Glycine is secreted mainly at synapses in the spinal cord. It probably always acts as an inhibitory transmitter.

Gamma-aminobutyric acid (GABA) is secreted by nerve terminals in the spinal cord, the cerebellum, the basal ganglia, and many areas of the cortex. It is believed always to cause inhibition.

Glutamate is probably secreted by the presynaptic terminals in many of the sensory pathways as well as in many areas of the cortex. It probably always causes excitation.

Serotonin is secreted by nuclei that originate in the median raphe of the brain stem and project to many brain areas, especially to the dorsal horns of the spinal cord and to the hypothalamus. Serotonin acts as an inhibitor of pain pathways in the cord, and it is also believed to help control the mood of the person, and perhaps even to cause sleep.

The Neuropeptides

The neuropeptides are an entirely different group of transmitters that are synthesized differently and whose actions are usually slow and in other ways quite different from those of the small-molecule transmitters.

The neuropeptides are not synthesized in the cytosol of the presynaptic terminals. Instead, they are synthesized as integral parts of large protein molecules by the ribosomes in the neuronal cell body. The protein molecules are transported immediately into the endoplasmic reticulum of the cell body, and the endoplasmic reticulum and subsequently the Golgi apparatus function together to do two things: First, they enzymatically split the original protein into smaller fragments and thereby release either the neuropeptide itself or a precursor of it. Second, the Golgi apparatus packages the neuropeptide into minute transmitter vesicles that are released into the cytoplasm. Then the transmitter vesicles are transported all the way to the tips of the nerve fibers by *axonal streaming* of the axon cytoplasm, traveling at the slow rate of only a few centimeters per day. Finally, these vesicles release their transmitter in response to action potentials in the same manner as for small-molecule transmitters. However, the vesicle is autolyzed and is not reused.

Because of this laborious method of forming the neuropeptides, much smaller quantities of these are usually released than for the small-molecule transmitters. However, this is partly compensated for by the fact that the neuropeptides are generally a thousand or more times as potent as the small-molecule transmitters. Another important characteristic of the neuropeptides is that they usually cause much more prolonged actions. Some of these actions include prolonged closure of calcium pores, prolonged changes in the metabolic machinery of cells, prolonged changes in activation or deactivation of specific genes in the cell nucleus, and prolonged alterations in numbers of excitatory or inhibitory receptors. Some of these effects can last for days or perhaps even months or years. Unfortunately, our knowledge of the functions of the neuropeptides is only at an early beginning.

Release of Only a Single Small-Molecule Transmitter by Each Type of Neuron

Almost invariably, only a single small-molecule type of transmitter is released by each type of neuron. However, the terminals of the same neuron may also release one or more neuropeptides at the same time. Yet whatever small-molecule transmitter and neuropeptides are released at one terminal of the neuron, these same transmitters will be released at all other terminals of the same neuron, whether these are few in number or many thousand and also wherever these terminate within the nervous system or in peripheral organs.

Removal of the Transmitter Substance From the Synapse

After a transmitter is released at a nerve ending, it is either destroyed or removed in some other way to prevent continued action forever thereafter. In the case of the neuropeptides, these are removed mainly by diffusion into the surrounding tissues, followed by destruction within a few minutes to several hours by specific or nonspecific enzymes. For the small-molecule, rapidly acting transmitters, removal usually occurs within a few milliseconds. This is achieved in three different ways:

1. By *diffusion* of the transmitter out of the cleft into the surrounding fluids.

2. By *enzymatic destruction* within the cleft itself. For instance, in the case of acetylcholine, the enzyme *cholinesterase* is present in the cleft, bound in the proteoglycan matrix that fills the space. Each molecule of this enzyme can split as many as 10 molecules of acetylcholine each millisecond, thus inactivating this transmitter substance. Similar effects occur for other transmitters.

3. By *active transport back into the presynaptic terminal itself* and reuse. This is called *transmitter re-uptake*. It occurs especially prominently at the presynaptic terminals of the sympathetic nervous system for the reuptake of norepinephrine, as we shall discuss in Chapter 22.

The degree to which each of these methods of removal is utilized is different for each type of transmitter.

ELECTRICAL EVENTS DURING NEURONAL EXCITATION

The electrical events in neuronal excitation have been studied especially in the large motor neurons of the anterior horns of the spinal cord. Therefore, the events to be described in the following few sections pertain essentially to these neurons. However, except for some quantitative differences, they apply to most other neurons of the nervous system as well.

The Resting Membrane Potential of the Neuronal Soma. Figure 7–7 illustrates the soma of a motor neuron showing the resting membrane potential to be about −65 mV. This is somewhat less than the −90 mV found in large peripheral nerve fibers and in skeletal muscle fibers; the lower voltage is important, however, because it allows both positive and negative control of the degree of excitability of the neuron. That is, decreasing the voltage to a less negative value makes the membrane of the neuron more excitable, whereas increasing this voltage to a more negative value makes the neuron less excitable. This is the basis of the two modes of function of the neuron—either excitation or inhibition—as is explained in detail in the following sections.

Concentration Differences of Ions Across the Neuronal Somal Membrane. Figure 7–7 also illustrates the concentration differences across the neuronal somal membrane of the three ions that are most important for neuronal function: sodium ions, potassium ions, and chloride ions.

At the top, the sodium ion concentration is shown to be very great in the extracellular fluid but low inside the neuron. This sodium concentration gradient is caused by a strong sodium pump that continually pumps sodium out of the neuron.

The figure also shows that the potassium ion concentration is large inside the neuronal soma but very low in the extracellular fluid. It illustrates that there is also a potassium pump (the other half of the Na$^+$-K$^+$ pump, as described in Chapter 5) that pumps potassium to the interior. However, potassium ions leak through the membrane ion channels at a rate sufficient

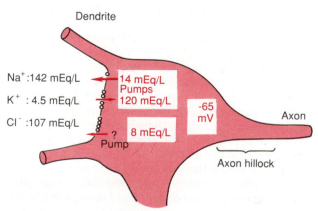

Figure 7–7. Distribution of sodium, potassium, and chloride ions across the neuronal somal membrane; origin of the intrasomal membrane potential.

to nullify much of the effectiveness of the potassium pump.

Figure 7–7 shows the chloride ion to be at high concentration in the extracellular fluid but low concentration inside the neuron. It also shows that the membrane is quite permeable to chloride ions and that there may be a weak chloride pump. Yet most of the reason for the low concentration of chloride ions inside the neuron is the −65-mV potential in the neuron. That is, this negative voltage repels the negatively charged chloride ions, forcing them outward through the pores until the concentration difference is much greater outside the membrane than inside.

Let us recall at this point what we learned in Chapters 5 and 6 about the relationship of ionic concentration differences to membrane potentials. It will be recalled that an electrical potential across the membrane can exactly oppose the movement of ions through a membrane despite concentration differences between the outside and inside of the membrane if the potential is of the proper polarity and magnitude. Such a potential that exactly opposes movement of each type of ion is called the Nernst potential for that ion; the equation for this is the following:

$$\text{EMF (mV)} = \pm 61 \times \log\left(\frac{\text{Concentration outside}}{\text{Concentration inside}}\right)$$

where EMF is the Nernst potential in millivolts on the *inside of the membrane*. The potential will be positive (+) for a positive ion and negative (−) for a negative ion.

Now, let us calculate the Nernst potential that will exactly oppose the movement of each of the three separate ions: sodium, potassium, and chloride.

For the sodium concentration difference shown in Figure 7–7, 142 mEq/L on the exterior and 14 mEq/L on the interior, the membrane potential that would exactly oppose sodium ion movement through the sodium channels would be +61 mV. However, the actual membrane potential is −65 mV, not +61 mV. Therefore, sodium ions normally diffuse inward through the sodium channels; however, not many sodium ions will diffuse because most of the sodium channels are normally closed. Furthermore, those sodium ions that do diffuse to the interior are normally pumped immediately back to the exterior by the sodium pump.

For potassium ions, the concentration gradient is 120 mEq/L inside the neuron and 4.5 mEq/L outside. This gives a Nernst potential of −86 mV inside the neuron, which is more negative than the −65 that actually exists. Therefore, there is a tendency for potassium ions to diffuse to the outside of the neuron, but this is opposed by the continual pumping of these potassium ions back to the interior.

Finally, the chloride ion gradient, 107 mEq/L outside and 8 mEq/L inside, yields a Nernst potential of −70 mV inside the neuron, which is slightly more negative than the actual value measured. Therefore, chloride ions tend normally to leak to the interior of the neuron, whereas those that do diffuse are moved back to the exterior, perhaps by an active chloride pump.

Keep these three Nernst potentials in mind and also remember the direction in which the different ions tend to diffuse, for this information will be important in understanding both excitation and inhibition of the neuron by synaptic activation of receptor ion channels.

Origin of the Resting Membrane Potential of the Neuronal Soma. The basic cause of the −65-mV resting membrane potential of the neuronal soma is the sodium-potassium pump. This pump causes the extrusion of more positively charged sodium ions to the exterior than potassium to the interior — three sodium ions outward for each two potassium ions inward. Because there are large numbers of negatively charged ions inside the soma that cannot diffuse through the membrane — protein ions, phosphate ions, and many others — extrusion of the excess positive ions to the exterior leaves some of these nondiffusible negative ions inside the cell unbalanced by positive ions. Therefore, the interior of the neuron becomes negatively charged as the result of the Na+-K+ pump. This principle was discussed in more detail in Chapter 6 in relation to the resting membrane potential of nerve fibers. In addition, as also explained in Chapter 6, diffusion of potassium ions outward through the membrane is another cause of intracellular negativity.

Uniform Distribution of the Potential Inside the Soma. The interior of the neuronal soma contains a very highly conductive electrolytic solution, the intracellular fluid of the neuron. Furthermore, the diameter of the neuronal soma is very large (from 10 to 80 μm in diameter), causing there to be almost no resistance to conduction of electrical current from one part of the somal interior to another part. Therefore, any change in potential in any part of the intrasomal fluid causes an almost exactly equal change in potential at all other points inside the soma. This is an important principle because it plays a major role in the summation of signals entering the neuron from multiple sources, as we shall see in subsequent sections of this chapter.

Effect of Synaptic Excitation on the Postsynaptic Membrane — The Excitatory Postsynaptic Potential. Figure 7–8A illustrates the resting neuron with an unexcited presynaptic terminal resting upon its surface. The resting membrane potential everywhere in the soma is −65 mV.

Figure 7–8B illustrates a presynaptic terminal that has secreted a transmitter into the cleft between the terminal and the neuronal somal membrane. This transmitter acts on a membrane excitatory receptor *to increase the membrane's permeability to* Na+. Because of the large electrochemical gradient that tends to move sodium inward, this large increase in membrane conductance for sodium ions allows these ions to rush to the inside of the membrane.

The rapid influx of the positively charged sodium ions to the interior neutralizes part of the negativity of the resting membrane potential. Thus, in Figure 7–8B the resting membrane potential has increased from −65 mV to −45 mV. This increase in voltage above the normal resting neuronal potential — that is, to a

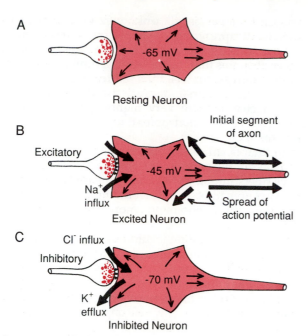

A
-65 mV
Resting Neuron

B
Initial segment of axon
Excitatory
-45 mV
Na⁺ influx
Spread of action potential
Excited Neuron

C
Cl⁻ influx
Inhibitory
-70 mV
K⁺ efflux
Inhibited Neuron

Figure 7–8. Three states of a neuron. *A,* A resting neuron. *B,* A neuron in an excited state, with more positive intraneuronal potential caused by sodium influx. *C,* A neuron in an inhibited state, with more negative intraneuronal membrane potential caused by potassium ion efflux, chloride ion influx, or both.

less negative value — is called the *excitatory postsynaptic potential* (or EPSP) because if this potential rises high enough it will elicit an action potential in the neuron, thus exciting it. In this case the EPSP is +20 mV.

However, we must issue a word of warning at this point. Discharge of a single presynaptic terminal can never increase the neuronal potential from −65 mV up to −45 mV. Instead, an increase of this magnitude requires the simultaneous discharge of many terminals — about 40 to 80 for the usual anterior motor neuron — at the same time or in rapid succession. This occurs by a process called *summation,* which will be discussed in detail in the following sections.

Generation of Action Potentials at the Initial Segment of the Axon Leaving the Neuron — Threshold for Excitation. When the excitatory postsynaptic potential rises high enough, there comes a point at which this initiates an action potential in the neuron. However, the action potential does not begin on the somal membrane adjacent to the excitatory synapses. Instead, it begins in the initial segment of the axon leaving the neuronal soma. The main reason for this point of origin of the action potential is that the soma has relatively few voltage-gated sodium channels in its membrane, which makes it difficult to open the required number of channels to elicit an action potential. On the other hand, the membrane of the initial segment has seven times as great a concentration of voltage-gated sodium channels and therefore can generate an action potential with much greater ease than can the soma. The excitatory postsynaptic potential that will elicit an action potential at the initial segment is be-

tween +15 and +20 mV. This is in contrast to the +30 mV or more required on the soma.

Once the action potential begins, it travels both peripherally along the axon and often also backward over the soma. In some instances, it travels backward into the dendrites, too, but not into all of them, because they, like the neuronal soma, also have very few voltage-gated sodium channels and therefore frequently cannot generate action potentials at all.

Thus, in Figure 7–8B, it is shown that under normal conditions the *threshold* for excitation of the neuron is about −45 mV, which represents an excitatory postsynaptic potential of +20 mV — that is, 20 mV more positive than the normal resting neuronal potential of −65 mV.

ELECTRICAL EVENTS IN NEURONAL INHIBITION

Effect of Inhibitory Synapses on the Postsynaptic Membrane — The Inhibitory Postsynaptic Potential. The inhibitory synapses open the potassium or the chloride channels, or both, instead of sodium channels, allowing easy passage of one or both of these ions. Now, to understand how the inhibitory synapses inhibit the postsynaptic neuron, we must recall what we learned about the Nernst potentials for both the potassium ions and the chloride ions. We calculated this potential for potassium ions to be about −86 mV and for chloride ions about −70 mV. Both of these potentials are more negative than the −65 mV normally present inside the resting neuronal membrane. Therefore, opening the potassium channels will allow positively charged potassium ions to move to the exterior, which will make the membrane potential more negative than normal; and opening the chloride channels will allow negatively charged chloride ions to move to the interior, which also will make the membrane potential more negative than usual. This increases the degree of intracellular negativity, which is called *hyperpolarization.* It obviously inhibits the neuron because the membrane potential is now farther away than ever from the threshold for excitation. Therefore, an increase in negativity beyond the normal resting membrane potential level is called the *inhibitory postsynaptic potential* (IPSP).

Thus, Figure 7–8C illustrates the effect on the membrane potential caused by activation of inhibitory synapses, allowing chloride influx into the cell or potassium efflux from the cell, with the membrane potential decreasing from its normal value of −65 mV to the more negative value of −70 mV. This membrane potential that is 5 mV more negative is the inhibitory postsynaptic potential. Thus the IPSP in this instance is −5 mV.

Inhibition of Neurons Without Causing an Inhibitory Postsynaptic Potential — "Short Circuiting" of the Membrane. Sometimes activation of the inhibitory synapses causes little or no inhibitory postsynaptic potential but nevertheless still inhibits the neuron.

The reason that the potential often does not change

is that in some neurons the concentration differences across the membrane for the potassium and chloride ions are only able to cause a Nernst equilibrium potential equal to the normal resting potential. Therefore, when the inhibitory channels open, there is no net flow of ions to cause an inhibitory postsynaptic potential. Yet the potassium or chloride ions, or both, do diffuse bidirectionally through the wide-open channels many times as rapidly as normally, and this high flux inhibits the neuron in the following way: When excitatory synapses cause sodium ions to flow into the neuron, the wide-open potassium or chloride channels cause far less excitatory postsynaptic potential than usual because any tendency for the membrane potential to change away from the resting potential is immediately opposed by rapid flux of potassium or chloride ions through the inhibitory channels to bring the potential back toward the negative Nernst equilibrium potentials for these two ions. Therefore, the influx of sodium ions required to overcome the potassium or chloride flux and therefore cause excitation may be as much as 5 to 20 times normal.

This tendency for the potassium and chloride ions to maintain the membrane potential near the resting value when the inhibitory channels are wide open is called "short circuiting" of the membrane, thus making the sodium current flow caused by excitatory synapses ineffective in exciting the cell.

To express the phenomenon of short circuiting more mathematically, one needs to recall the Goldman equation from Chapter 6. This equation shows that the membrane potential is determined by summation of the tendencies for the different ions to carry electrical charges through the membrane in the two directions. The membrane potential will approach the Nernst equilibrium potentials for those ions that permeate the membrane to the greatest extent. When the inhibitory channels are wide open, the chloride and potassium ions permeate the membrane greatly. Therefore, when the excitatory channels open, the summated effect of the inhibitory channels makes it difficult to raise the neuronal potential up to the threshold value for excitation.

Presynaptic Inhibition

In addition to the inhibition caused by inhibitory synapses operating at the neuronal membrane, which is called *postsynaptic inhibition,* another type of inhibition often occurs in the presynaptic terminals before the signal ever reaches the synapse. This type of inhibition, called *presynaptic inhibition,* is believed to occur in the following way:

In presynaptic inhibition, the inhibition is caused by "presynaptic" synapses that lie on the terminal nerve fibrils before they themselves terminate on the following neuron. It is believed that activation of these synapses on the presynaptic terminals decreases the ability of the calcium channels in the terminals to open. Because calcium ions must enter the presynaptic terminals before the vesicles can release transmitter at the neuronal synapse, the obvious result is to reduce neuronal excitation.

The cause of the reduced calcium entry into the pre-synaptic terminals is still unknown. One theory suggests that the presynaptic synapses release a transmitter that directly blocks calcium channels. Another theory proposes that the transmitter inhibits the opening of sodium channels, thus reducing the action potential in the terminal; because the voltage-activated calcium channels are very highly voltage-sensitive, any decrease in action potential greatly reduces calcium entry.

Presynaptic inhibition occurs in many of the sensory pathways in the nervous system. That is, the adjacent nerve fibers inhibit each other, which minimizes the sideways spread of signals from one fiber to the next. We discuss this phenomenon more fully in subsequent chapters.

Presynaptic inhibition is different from postsynaptic inhibition in its time sequence. It requires many milliseconds to develop; but once it does occur, it can last for as long as minutes or even hours. *Postsynaptic* inhibition, on the other hand, normally lasts for only a few milliseconds.

Summation of Postsynaptic Potentials

Time Course of Postsynaptic Potentials. When a synapse excites the anterior motor neuron the neuronal membrane remains highly permeable for only 1 to 2 msec. During this time sodium ions diffuse rapidly to the interior of the cell to increase the intraneuronal potential, thus creating the *excitatory postsynaptic potential,* illustrated by the two lower curves of Figure 7–9. This potential then slowly dissipates over the next 15 msec, because this is the time required for the positive charges to flow away from the excited synapses along the lengths of the dendrites and axon and for potassium ions to leak out or chloride ions to leak in to re-establish the normal resting membrane potential.

Precisely the opposite effect occurs for the inhibitory postsynaptic potential. That is, the inhibitory synapse increases the permeability of the membrane to potassium or chloride ions, or both, for 1 to 2 msec, and this decreases the intraneuronal potential to a more negative value than normal, thereby creating the *inhibitory postsynaptic potential.* This potential also persists for about 15 msec.

However, other types of transmitter substances acting on other neurons can excite or inhibit for hundreds of milliseconds or even for seconds, minutes, or hours.

Spatial Summation of the Postsynaptic Potentials — The Threshold for Firing

It has already been pointed out that excitation of a single presynaptic terminal on the surface of a neuron will almost never excite the neuron. The reason for this is that sufficient transmitter substance is released by a single terminal to cause an excitatory postsynaptic potential usually no more than 0.5 to 1 mV at most, instead of the required 10 to 20 mV to reach the usual threshold for excitation. However, during excitation in a neuronal pool of the nervous system, many presy-

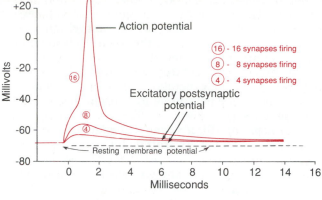

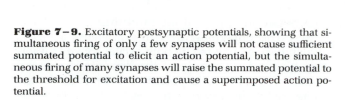

Figure 7–9. Excitatory postsynaptic potentials, showing that simultaneous firing of only a few synapses will not cause sufficient summated potential to elicit an action potential, but the simultaneous firing of many synapses will raise the summated potential to the threshold for excitation and cause a superimposed action potential.

naptic terminals are usually stimulated at the same time; and even though these terminals are spread over wide areas of the neuron, their effects can still summate. The reason for this is the following: It has already been pointed out that a change in the potential at any single point within the soma will cause the potential to change everywhere in the soma almost exactly equally. Therefore, for each excitatory synapse that discharges simultaneously, the intrasomal potential becomes more positive by as much as a fraction of a millivolt up to about 1 mV. When the excitatory postsynaptic potential becomes great enough, the *threshold for firing* will be reached, and an action potential will generate in the initial segment of the axon. This effect is illustrated in Figure 7–9, which shows several excitatory postsynaptic potentials. The bottom postsynaptic potential in the figure was caused by simultaneous stimulation of four synapses; the next higher potential was caused by stimulation of two times as many synapses; finally, a still higher excitatory postsynaptic potential was caused by stimulation of four times as many synapses. This time an action potential was generated in the initial axon segment.

This effect of summing simultaneous postsynaptic potentials by excitation of multiple terminals on widely spaced areas of the membrane is called *spatial summation.*

Temporal Summation

Each time a terminal fires, the released transmitter substance opens the membrane channels for a millisecond or so. Since the postsynaptic potential lasts up to 15 msec, a second opening of the same channels can increase the postsynaptic potential to a still greater level; therefore, the more rapid the rate of terminal stimulation, the greater the effective postsynaptic potential. Thus, successive postsynaptic potentials in a presynaptic terminal, if they occur rapidly enough, can summate in the same way that postsynaptic potentials can summate from widely distributed terminals over the surface of the neuron. This summation is called *temporal summation.*

Simultaneous Summation of Inhibitory and Excitatory Postsynaptic Potentials. Obviously, if an inhibitory postsynaptic potential is tending to decrease the membrane potential to a more negative value while an excitatory postsynaptic potential is tending to increase the potential at the same time, these two effects can either completely nullify each other or partially nullify each other. Also, inhibitory "short circuiting" of the membrane potential can nullify much of an excitatory potential. Thus, if a neuron is currently being excited by an excitatory postsynaptic potential, then an inhibitory signal from another source can easily reduce the postsynaptic potential to less than the threshold value for excitation, thus turning off the activity of the neuron.

Facilitation of Neurons. Often the summated postsynaptic potential is excitatory in nature but has not risen high enough to reach the threshold for excitation. When this happens the neuron is said to be *facilitated.* That is, its membrane potential is nearer the threshold for firing than normally but not yet to the firing level. Nevertheless, another signal entering the neuron from some other source can then excite the neuron very easily. Diffuse signals in the nervous system often facilitate large groups of neurons so that they can respond quickly and easily to signals arriving from second sources.

SPECIAL FUNCTIONS OF DENDRITES IN EXCITING NEURONS

The Large Spatial Field of Excitation of the Dendrites. The dendrites of the anterior motor neurons extend for 500 to 1000 μm in all directions from the neuronal soma. Therefore, these dendrites can receive signals from a large spatial area around the motor neuron. This provides vast opportunity for summation of signals from many separate presynaptic neurons.

It is also important that between 80 and 90 per cent of all the presynaptic terminals terminate on the dendrites of the anterior motor neuron, in contrast to only 10 to 20 per cent terminating on the neuronal soma. Therefore, the preponderant share of the excitation is provided by signals transmitted over the dendrites.

Many Dendrites Cannot Transmit Action Potentials — But They Can Transmit Signals by Electronic Conduction. Many dendrites fail to transmit action potentials because their membranes have relatively few voltage-gated sodium channels, so that their thresholds for excitation are too high for action potentials ever to occur. Yet they do transmit *electrotonic current* down the dendrites to the soma. Transmission of electrotonic current means the direct spread of current by electrical conduction in the fluids of the dendrites with no generation of action potentials. Stimulation of the neuron by this current has special characteristics, as follows:

Decrement of Electrotonic Conduction in the Dendrites — Greater Excitation by Synapses Near the Soma. In Figure 7–10 a number of excitatory and inhibitory synapses are shown stimulating the dendrites of a neuron. On the two dendrites to the left in the figure are shown excitatory effects near the ends of the dendrites; note the high levels of the excitatory postsynaptic potentials at these ends — that is, the less negative membrane potentials at these points. However, a large share of the excitatory postsynaptic potential is lost before it reaches the soma. The reason for this is that the dendrites are long and thin, and their membranes are also thin and excessively permeable to potassium and chloride ions, making them "leaky" to electrical current. Therefore, before the excitatory potentials can reach the soma, a large share of the potential is lost by leakage through the membrane. This decrease in membrane potential as it spreads electrotonically along dendrites toward the soma is called *decremental conduction*.

It is also obvious that the nearer the excitatory synapse is to the soma of the neuron, the less will be the decrement of conduction. Therefore, those synapses that lie near the soma have much more excitatory effect than those that lie far away from the soma.

Rapid Re-excitation of the Neuron by the Dendrites after the Neuron Fires. When an action potential is generated in a neuron, this action potential usually spreads back over the soma but not always over the dendrites. Therefore, the excitatory postsynaptic potentials in the dendrites often are only partially disturbed by the action potential, so that just as soon as the action potential is over, the potentials still existing in the dendrites are ready and waiting to excite the neuron again. Thus, the dendrites have a "holding capacity" for the excitatory signal from presynaptic sources.

Summation of Excitation and Inhibition in Dendrites. The uppermost dendrite of Figure 7–10 is shown to be stimulated by both excitatory and inhibitory synapses. At the tip of the dendrite is a strong excitatory postsynaptic potential, but nearer to the soma are two inhibitory synapses acting on the same dendrite. These inhibitory synapses provide a hyperpolarizing voltage that completely nullifies the excitatory effect and indeed transmits a small amount of inhibition by electrotonic conduction toward the soma. Thus, dendrites can summate excitatory and inhibitory postsynaptic potentials in the same way that the soma can.

Also shown in the figure are several inhibitory synapses located directly on the axon hillock and initial axon segment. This location provides especially powerful inhibition because it has the direct effect of increasing the threshold for excitation at the very point where the action potential is normally generated.

RELATION OF STATE OF EXCITATION OF THE NEURON TO THE RATE OF FIRING

The "Excitatory State." The "excitatory state" of a neuron is defined as the degree of excitatory drive to the neuron. If there is a higher degree of excitation than inhibition of the neuron at any given instant, then it is said that there is an *excitatory state*. On the other hand, if there is more inhibition than excitation, then it is said that there is an *inhibitory state*.

When the excitatory state of a neuron rises above the threshold for excitation, then the neuron will fire repetitively as long as the excitatory state remains at this level. However, *the rate at which it will fire is determined by how much* the excitatory state is above threshold. To explain this, we must first consider what happens to the neuronal somal potential during and following the action potential.

Changes in Neuronal Somal Potential During and Following the Action Potential. The lower curve of Figure 7–11 illustrates an action potential spreading backward over the neuronal soma after being initiated in the initial axon segment by an excitatory postsynaptic potential. After the spike portion of the action potential, there is a very long state of "hyperpolarization" lasting for many milliseconds. During this interval the somal membrane potential falls below the normal resting membrane potential. This is caused at least partly by a high degree of permeability of the neuronal membrane to potassium ions that persists for many milliseconds after the action potential is over.

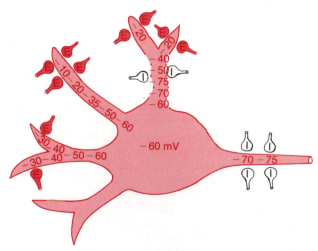

Figure 7–10. Stimulation of a neuron by presynaptic terminals located on dendrites, showing, especially, decremental conduction of excitatory (E) electrotonic potentials in the two dendrites to the left and inhibition (I) of dendritic excitation in the dendrite that is uppermost. A powerful effect of inhibitory synapses at the initial segment is also shown.

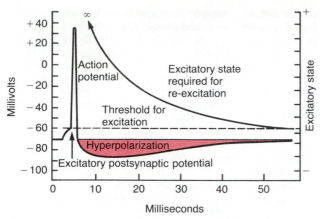

Figure 7–11. A neuronal action potential followed by a prolonged period of neuronal hyperpolarization. Also shown is the "excitatory state" required for re-excitation of the neuron at given intervals after the action potential is over.

The high membrane conductivity for potassium ions also short circuits excitatory potentials, as has already been explained.

The importance of this hyperpolarization as well as the short circuiting that occurs after the spike potential is over is that the neuron remains in an *inhibited state* during this period of time. Therefore, a far greater excitatory state is required during this time than normally to cause re-excitation of the neuron.

Relationship of Excitatory State to Frequency of Firing. The curve shown at the top of Figure 7–11, labeled "Excitatory state required for re-excitation," depicts the relative level of the excitatory state required at each instance after an action potential is over to re-excite the neuron. Note that very soon after an action potential is over, a very high excitatory state is required. That is, a very large number of excitatory synapses must be firing simultaneously. Then, after many milliseconds have passed and the state of hyperpolarization and the short circuiting of the neuron have begun to disappear, the excitatory state required becomes greatly reduced.

Therefore, when the excitatory state is high, a second action potential will appear very soon after the previous one. Then still a third action potential will appear soon after the second, and this process will continue indefinitely. Thus, at a very high excitatory state the frequency of firing of the neuron is great.

On the other hand, when the excitatory state is only barely above threshold, the neuron must recover almost completely from the hyperpolarization and short circuiting, which requires many milliseconds, before it will fire again. Therefore, the frequency of neuronal firing is low.

Response Characteristics of Different Neurons to Increasing Levels of Excitatory State. Histological study of the nervous system provides convincing evidence for the widely varying types of neurons in different parts of the nervous system. And, physiologically, the different types of neurons perform different functions. Therefore, as would be expected, the ability

to respond to stimulation by the synapses varies from one type of neuron to another.

Figure 7–12 illustrates theoretical responses of three different types of neurons to varying levels of the excitatory state. Note that neuron 1 has a low threshold for excitation, whereas neuron 3 has a high threshold. But note also that neuron 2 has the lowest maximum frequency of discharge, whereas neuron 3 has the highest maximum frequency.

Some neurons in the central nervous system fire continuously because even the normal excitatory state is above the threshold level. Their frequency of firing can usually be increased still more by further increasing their excitatory state. Or the frequency may be decreased, or firing even be stopped, by superimposing an inhibitory state on the neuron.

Thus, different neurons respond differently, have different thresholds for excitation, and have widely differing maximal frequencies of discharge. With a little imagination one can readily understand the importance of having neurons with many different types of response characteristics to perform the widely varying functions of the nervous system.

■ SOME SPECIAL CHARACTERISTICS OF SYNAPTIC TRANSMISSION

Fatigue of Synaptic Transmission. When excitatory synapses are repetitively stimulated at a rapid rate, the number of discharges by the postsynaptic neuron is at first very great, but it becomes progressively less in succeeding milliseconds or seconds. This is called *fatigue* of synaptic transmission.

Fatigue is an exceedingly important characteristic of synaptic function, for when areas of the nervous system become overexcited, fatigue causes them to lose this excess excitability after a while. For example, fatigue is probably the most important means by which the excess excitability of the brain during an epileptic convulsion is finally subdued so that the convulsion

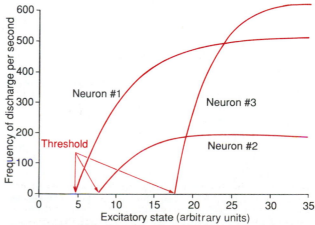

Figure 7–12. Response characteristics of different types of neurons to progressively increasing levels of excitatory state.

ceases. Thus, the development of fatigue is a protective mechanism against excess neuronal activity. This is discussed further in the description of reverberating neuronal circuits in the following chapter.

The mechanism of fatigue is mainly exhaustion of the stores of transmitter substance in the synaptic terminals, particularly because it has been calculated that the excitatory terminals on most neurons can store enough excitatory transmitter for only 10,000 normal synaptic transmissions, so that the transmitter can be exhausted in only a few seconds to a few minutes of rapid stimulation. However, part of the fatigue process probably also results from two other factors as well: (1) progressive inactivation of many of the postsynaptic membrane receptors; and (2) slow buildup of calcium ions inside the *postsynaptic* neuronal cell caused by the successive action potentials—these calcium ions in turn open calcium-activated potassium channels, which cause an inhibitory effect on the postsynaptic neuron.

Post-tetanic Facilitation. When a rapidly repetitive series of impulses stimulates an excitatory synapse for a period of time and then a rest period is allowed, the synapse will often become for a period of seconds or minutes even more responsive to subsequent stimulation than normally. This is called *post-tetanic facilitation.*

Experiments have shown that post-tetanic facilitation is caused mainly by the buildup of excess calcium ions in the *presynaptic* terminals because the calcium pump operates too slowly to remove all of these immediately after each action potential. These accumulated calcium ions cause more and more vesicular release of transmitter substance, occasionally increasing the release of transmitter to a rate two times normal.

The physiological significance of post-tetanic facilitation is still very doubtful, and it may have no real significance at all. However, neurons could possibly store information by this mechanism. Therefore post-tetanic facilitation might well be a mechanism of "short-term" memory in the central nervous system.

Effect of Acidosis and Alkalosis on Synaptic Transmission. The neurons are highly responsive to changes in pH of the surrounding interstitial fluids. *Alkalosis greatly increases neuronal excitability.* For instance, a rise in arterial pH from the 7.4 norm to 7.8 to 8.0 often causes cerebral convulsions because of increased excitability of the neurons. This can be demonstrated especially well by having a person who is predisposed to epileptic convulsions overbreathe. The overbreathing elevates the pH of the blood only momentarily, but even this short interval can often precipitate an epileptic attack.

On the other hand, *acidosis greatly depresses neuronal activity;* a fall in pH from 7.4 to below 7.0 usually causes a comatose state. For instance, in very severe diabetic or uremic acidosis, coma always develops.

Effect of Hypoxia on Synaptic Transmission. Neuronal excitability is also highly dependent on an adequate supply of oxygen. Cessation of oxygen for only a few seconds can cause complete inexcitability of the neurons. This is often seen when the cerebral circulation is temporarily interrupted, for within 3 to 5 sec the person becomes unconscious.

Effect of Drugs on Synaptic Transmission. Many different drugs are known to increase the excitability of neurons, and others are known to decrease the excitability. For instance, caffeine, theophylline, and theobromine, which are found in coffee, tea, and cocoa, respectively, all increase neuronal excitability, presumably by reducing the threshold for excitation of the neurons. Also, strychnine is one of the best known of all the agents that increase the excitability of neurons. However, it does not reduce the threshold for excitation of the neurons at all; instead, it *inhibits the action of some of the inhibitory transmitters* on the neurons, especially the inhibitory effect of glycine in the spinal cord. In consequence, the effects of the excitatory transmitters become overwhelming, and the neurons become so excited that they go into rapidly repetitive discharge, resulting in severe tonic muscle spasms.

Most anesthetics increase the membrane threshold for excitation and thereby decrease synaptic transmission at many points in the nervous system. Because most of the anesthetics are lipid-soluble, it has been reasoned that they might change the physical characteristics of the neuronal membranes, making them less responsive to excitatory agents.

Synaptic Delay. In transmission of an action potential from a presynaptic neuron to a postsynaptic neuron, a certain amount of time is consumed in the process of (1) discharge of the transmitter substance by the presynaptic terminal, (2) diffusion of the transmitter to the postsynaptic neuronal membrane, (3) action of the transmitter on the membrane receptor, (4) action of the receptor to increase the membrane permeability, and (5) inward diffusion of sodium to raise the excitatory postsynaptic potential to a high enough value to elicit an action potential. The *minimal* period of time required for all these events to take place, even when large numbers of excitatory synapses are stimulated simultaneously, is about 0.5 msec. This is called the *synaptic delay*. It is important for the following reason: Neurophysiologists can measure the *minimal* delay time between an input volley of impulses and an output volley and from this can estimate the number of series neurons in the circuit.

REFERENCES

Bahill, A. T., and Hamm, T. M.: Using open-loop experiments to study physiological systems, with examples from the human eye-movement systems. News Physiol. Sci., 4:104, 1989.

Barchi, R. L.: Probing the molecular structure of the voltage-dependent sodium channel. Annu. Rev. Neurosci., 11:455, 1988.

Berg, D. K.: New neuronal growth factors. Annu. Rev. Neurosci., 7:149, 1984.

Bloom, F. E.: Neurotransmitters: past, present, and future directions. FASEB J., 2:32, 1988.

Bousfield, D. (ed.): Neurotransmitters in Action. New York, Elsevier Science Publishing Co., 1985.

Byrn, J. H., and Schultz, S. G.: An Introduction to Membrane Transport and Bioelectricity. New York, Raven Press, 1988.

Changeux, J.-P., et al.: Acetylcholine receptor: An allosteric protein. Science, 225:1335, 1984.

Cotman, C. W., et al.: Excitatory animo acid neurotransmission: NMDA receptors and Hebb-type synaptic plasticity. Annu. Rev. Neurosci., 11:61, 1988.

Eldefrawi, A. T., and Eldefrawi, M. E.: Receptors for γ-aminobutyric acid and voltage-dependent chloride channels as targets for drugs and toxicants. FASEB J., 1:262, 987.

Eyzaguirre, C.: Physiology of the Nervous System. Chicago, Year Book Medical Publishers, 1985.

Grinnell, A. D., et al. (eds.): Calcium and Ion Channel Modulation. New York, Plenum Publishing Corp., 1988.

Grinvald, A., et al.: Optical imaging of neuronal activity. Physiol. Rev., 68:1285, 1988.

Hanin, I. (ed.): Dynamics of Neurotransmitter Function. New York, Raven Press, 1984.

Hansen, A. J.: Disturbed ion gradients in brain anoxia. News Physiol. Sci., 2:54, 1987.

Heinemann, S., and Patrick, J. (eds.): Molecular Neurobiology. New York, Plenum Publishing Corp., 1987.

Ito, M.: Where are neurophysiologists going? News Physiol. Sci., 1:30, 1986.

Iversen, L. L., and Goodman, E. C. (eds.): Fast and Slow Chemical Signalling in the Nervous System. New York, Oxford University Press, 1986.

Iversen, L. L.: Nonopioid neuropeptides in mammalian CNS. Annu. Rev. Pharmacol. Toxicol., 23:1, 1983.

Johnson, R. G., Jr.: Accumulation of biological amines into chromaffin granules: A model for hormone and neurotransmitter transport. Physiol. Rev., 68:232, 1988.

Kito, S., et al. (eds.): Neuroreceptors and Signal Transduction. New York, Plenum Publishing Corp., 1988.

Kostyuk, P. G.: Intracellular perfusion of nerve cells and its effect on membrane currents. Physiol. Rev., 64:435, 1984.

Krnjevic, K.: Ephaptic interactions: A significant mode of communications in the brain. News Physiol. Sci., 1:28, 1986.

Laduron, P. M.: Presynaptic heteroreceptors in regulation of neuronal transmission. Biochem. Pharmacol., 34:467, 1985.

Marx, J. L.: NMDA receptors trigger excitement. Science, 239:254, 1988.

McGeer, P. L., et al.: Molecular Neurobiology of the Mammalian Brain, 2nd Ed. New York, Plenum Publishing Corp., 1987.

McKay, R. D. G.: Molecular approach to the nervous system. Annu. Rev. Neurosci., 6:527, 1983.

Millhorn, D. E., and Hokfelt, T.: Chemical messengers and their coexistence in individual neurons. News Physiol. Sci., 3:1, 1988.

Nakanishi S.: Substance P precursor and kininogen: Their structures, gene organizations, and regulation. Physiol. Rev., 67:1117, 1987.

Narahashi, T. (ed.): Ion Channels. New York, Plenum Publishing Corp., 1988.

Nicholl, R. A.: The coupling of neurotransmitter receptors to ion channels in the brain. Science, 241:545, 1988.

Phillips, M. I.: Functions of angiotensin in the central nervous system. Annu. Rev. Physiol., 49:413, 1987.

Pickering, P. T., et al. (eds.): Neurosecretion. New York, Plenum Publishing Corp., 1988.

Popot, J.-L., and Changeux, J.-P.: Nicotinic receptor of acetylcholine: Structure of an oligomeric integral membrane protein. Physiol. Rev., 64:1162, 1984.

Purves, D., and Lichtman, J. W.: Specific connections between nerve cells. Annu. Rev. Physiol., 45:553, 1983.

Purves, D., et al.: Nerve terminal remodeling visualized in living mice by repeated examination of the same neuron. Science, 238:1122, 1987.

Reichardt, L. F.: Immunological approaches to the nervous system. Science, 225:1294, 1984.

Robinson, M. B., and Coyle, J. T.: Glutamate and related acidic excitatory neurotransmitters: From basic science to clinical application. FASEB J., 1:446, 1987.

Schubert, D.: Developmental Biology of Cultured Nerve, Muscle and Glia. New York, John Wiley & Sons, 1984.

Skok, V. I., et al. (eds.): Neuronal Acetylcholine Receptors. New York, Plenum Publishing Corp., 1989.

Snyder, S. H.: Neuronal receptors. Annu. Rev. Physiol., 48:461, 1986.

Starke, K.: Presynaptic autoregulation: Does it play a role? News Physiol. Sci., 4:1, 1989.

Stein, J. F.: Introduction to Neurophysiology. St. Louis, C. V. Mosby, Co., 1982.

Su, C.: Purinergic neurotransmission and neuromodulation. Annu. Rev. Pharmacol. Toxicol., 23:397, 1983.

Thompson, R. F.: The neurobiology of learning and memory. Science, 233:941, 1986.

Tucek, S.: Regulation of acetylcholine synthesis in the brain. J. Neurochem., 44:11, 1985.

White, J. D., et al.: Biochemistry of peptide-secreting neurons. Physiol. Rev., 65:553, 1985.

Williams, R. W., and Herrup, K.: The control of neuron number. Annu. Rev. Neurosci., 11:423, 1988.

Wurtman, R. J.: Presynaptic control of release of amine neurotransmitters by precursor levels. News Physiol. Sci., 3:158, 1988.

Zucker, R. S.: Short-term synaptic plasticity. Annu. Rev. Physiol., 12:13, 1989.

Sensory Receptors; Neuronal Circuits for Processing Information

■ TYPES OF SENSORY RECEPTORS AND THE SENSORY STIMULI THEY DETECT

Input to the nervous system is provided by the sensory receptors that detect such sensory stimuli as touch, sound, light, pain, cold, warmth, and so forth. The purpose of this chapter is to discuss the basic mechanisms by which these receptors change sensory stimuli into nerve signals and, also, how the information conveyed in the signals is processed in the nervous system.

Table 8–1 gives a list and classification of most of the body's sensory receptors. This table shows that there are basically five different types of sensory receptors: (1) *mechanoreceptors*, which detect mechanical deformation of the receptor or of cells adjacent to the receptor; (2) *thermoreceptors*, which detect changes in temperature, some receptors detecting cold and others warmth; (3) *nociceptors* (pain receptors), which detect damage in the tissues, whether physical damage or chemical damage; (4) *electromagnetic receptors*, which detect light on the retina of the eye; and (5) *chemoreceptors*, which detect taste in the mouth, smell in the nose, oxygen level in the arterial blood, osmolality of the body fluids, carbon dioxide concentration, and perhaps other factors that make up the chemistry of the body.

In this chapter we discuss the function of a few specific types of receptors, primarily peripheral mechanoreceptors, to illustrate some of the basic principles by which receptors in general operate. Other receptors will be discussed in relation to the sensory systems that they subserve.

Figure 8–1 illustrates some of the different types of mechanoreceptors found in the skin or in the deep structures of the body, and Table 8–1 gives their respective sensory functions. All these receptors are discussed in the following chapters in relation to the respective sensory systems.

DIFFERENTIAL SENSITIVITY OF RECEPTORS

The first question that must be answered is how different types of sensory receptors detect different types of sensory stimuli. The answer is by virtue of differential sensitivities. That is, each type of receptor is very highly sensitive to one type of stimulus for which it is designed, and yet is almost nonresponsive to normal intensities of the other types of sensory stimuli. Thus, the rods and cones are highly responsive to light but are almost completely nonresponsive to heat, cold, pressure on the eyeballs, or chemical changes in the blood. The osmoreceptors of the supraoptic nuclei in the hypothalamus detect minute changes in the osmolality of the body fluids but have never been known to respond to sound. Finally, pain receptors in the skin are almost never stimulated by usual touch or pressure stimuli but do become highly active the moment tactile stimuli become severe enough to damage the tissues.

Modality of Sensation — The "Labeled Line" Principle

Each of the principal types of sensation that we can experience — pain, touch, sight, sound, and so forth — is called a *modality* of sensation. Yet despite the fact that we experience these different modalities of sensation, nerve fibers transmit only impulses. Therefore, how is it that different nerve fibers transmit different modalities of sensation?

The answer to this is that each nerve tract terminates at a specific point in the central nervous system, and

TABLE 8–1 *Classification of Sensory Receptors*

Mechanoreceptors
  Skin tactile sensibilities (epidermis and dermis)
    Free nerve endings
    Expanded tip endings
      Merkel's discs
      Several other variants
    Spray endings
      Ruffini's endings
    Encapsulated endings
      Meissner's corpuscles
      Krause's corpuscles
    Hair end-organs
  Deep tissue sensibilities
    Free nerve endings
    Expanded tip endings
    Spray endings
      Ruffini's endings
    Encapsulated endings
      Pacinian corpuscles
      A few other variants
    Muscle endings
      Muscle spindles
      Golgi tendon receptors
  Hearing
    Sound receptors of cochlea
  Equilibrium
    Vestibular receptors
  Arterial pressure
    Baroreceptors of carotid sinuses and aorta
Thermoreceptors
  Cold
    Cold receptors
  Warmth
    Warm receptors
Nociceptors
  Pain
    Free nerve endings
Electromagnetic receptors
  Vision
    Rods
    Cones
Chemoreceptors
  Taste
    Receptors of taste buds
  Smell
    Receptors of olfactory epithelium
  Arterial oxygen
    Receptors of aortic and carotid bodies
  Osmolality
    Probably neurons in or near supraoptic nuclei
  Blood CO_2
    Receptors in or on surface of medulla and in aortic
      and carotid bodies
  Blood glucose, amino acids, fatty acids
    Receptors in hypothalamus

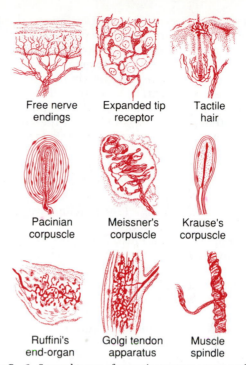

Figure 8–1. Several types of somatic sensory nerve endings.

the vision areas of the brain, fibers from the ear terminate in the auditory areas of the brain, and temperature fibers terminate in the temperature areas.

This specificity of nerve fibers for transmitting only one modality of sensation is called the *"labeled line" principle.*

■ TRANSDUCTION OF SENSORY STIMULI INTO NERVE IMPULSES

LOCAL CURRENTS AT NERVE ENDINGS—RECEPTOR POTENTIALS

All sensory receptors have one feature in common. Whatever the type of stimulus that excites the receptor, its immediate effect is to change the membrane potential of the receptor. This change in potential is called a *receptor potential.*

Mechanisms of Receptor Potentials. Different receptors can be excited in several different ways to cause receptor potentials: (1) by mechanical deformation of the receptor, which stretches the membrane and opens ion channels; (2) by application of a chemical to the membrane, which also opens ion channels; (3) by change of the temperature of the membrane, which alters the permeability of the membrane; and (4) by the effects of electromagnetic radiation such as light on the receptor, which either directly or indirectly changes the membrane characteristics and allows ions to flow through membrane channels. It will be recognized that these four different means of exciting receptors correspond in general with the different types of

the type of sensation felt when a nerve fiber is stimulated is determined by the point in the nervous system to which the fiber leads. For instance, if a pain fiber is stimulated, the person perceives pain regardless of what type of stimulus excites the fiber. The stimulus can be electricity, heat, crushing, or stimulation of the pain nerve ending by damage to the tissue cells. Yet the person still perceives pain. Likewise, if a touch fiber is stimulated by exciting a touch receptor electrically or in any other way, the person perceives touch because touch fibers lead to specific touch areas in the brain. Similarly, fibers from the retina of the eye terminate in known sensory receptors. In all instances, the basic cause of the change in membrane potential is a change in receptor membrane permeability, which allows ions to diffuse more or less readily through the membrane and thereby change the transmembrane potential.

The Receptor Potential Amplitude. The maximum amplitude of most sensory receptor potentials is around 100 mV. This is approximately the same maximum voltage recorded in action potentials and is also approximately the voltage at which the membrane becomes maximally permeable to sodium ions.

Relationship of the Receptor Potential to Action Potentials. When the receptor potential rises above the *threshold* for eliciting action potentials in the nerve fiber attached to the receptor, then action potentials begin to appear. This is illustrated in Figure 8–2. Note also that the more the receptor potential rises above the threshold level, the greater becomes the action potential frequency. Thus, the receptor potential stimulates the sensory nerve fiber in the same way that the excitatory postsynaptic potential in the central nervous system neuron stimulates the neuron's axon.

The Receptor Potential of the Pacinian Corpuscle — An Illustrative Example of Receptor Function

The student should at this point restudy the anatomical structure of the pacinian corpuscle illustrated in Figure 8–1. Note that the corpuscle has a central nerve fiber extending through its core. Surrounding this are multiple concentric capsule layers, so that compression anywhere on the outside of the corpuscle will elongate, indent, or otherwise deform the central fiber.

Now study Figure 8–3, which illustrates only the central fiber of the pacinian corpuscle after all capsule layers have been removed by microdissection. The very tip of the central fiber is unmyelinated, but it becomes myelinated shortly before leaving the corpuscle to enter the peripheral sensory nerve.

The figure also illustrates the mechanism by which the receptor potential is produced in the pacinian corpuscle. Observe the small area of the terminal fiber

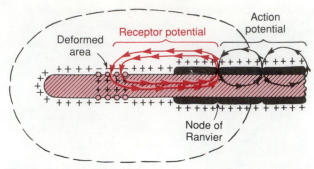

Figure 8–3. Excitation of a sensory nerve fiber by a receptor potential produced in a pacinian corpuscle. (Modified from Loëwenstein: Ann. N.Y. Acad. Sci., 94:510, 1961.)

that has been deformed by compression of the corpuscle, and note that ion channels have opened in the membrane, allowing positively charged sodium ions to diffuse to the interior of the fiber. This in turn creates increased positivity inside the fiber, which is the receptor potential. The receptor potential in turn induces a *local circuit* of current flow, illustrated by the red arrows, that spreads along the nerve fiber. At the first node of Ranvier, which itself lies inside the capsule of the pacinian corpuscle, the local current flow depolarizes the node, and this then sets off typical action potentials that are transmitted along the nerve fiber toward the central nervous system.

Relationship Between Stimulus Intensity and the Receptor Potential. Figure 8–4 illustrates the changing amplitude of the receptor potential caused by progressively stronger mechanical compression applied experimentally to the central core of a pacinian corpuscle. Note that the amplitude increases rapidly at first but then progressively less rapidly at high stimulus strength.

In general, the frequency of repetitive action potentials transmitted from sensory receptors increases approximately in proportion to the increase in receptor potential. Putting this information together with the data in Figure 8–4, one can see that even though a very

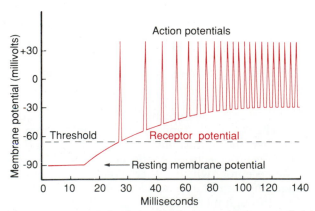

Figure 8–2. Typical relationship between receptor potential and action potentials when the receptor potential rises above the threshold level.

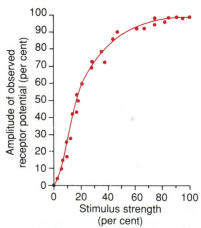

Figure 8–4. Relationship of amplitude of receptor potential to strength of a stimulus applied to a pacinian corpuscle. (From Loëwenstein: Ann. N.Y. Acad. Sci., 94:510, 1961.)

weak sensory stimulus can usually elicit at least some sensory signal, very intense stimulation of the receptor causes progressively less and less further increase in numbers of action potentials. This is an exceedingly important principle, employed by almost all sensory receptors. It allows the receptor to be very sensitive to weak sensory experience and yet not reach a maximum firing rate until the sensory experience is extreme. Obviously, this allows the receptor to have an extreme range of response, from very weak to very intense.

ADAPTATION OF RECEPTORS

A special characteristic of all sensory receptors is that they *adapt* either partially or completely to their stimuli after a period of time. That is, when a continuous sensory stimulus is applied, the receptors respond at a very high impulse rate at first, then at a progressively lower rate until finally many of them no longer respond at all.

Figure 8–5 illustrates typical adaptation of certain types of receptors. Note that the pacinian corpuscle adapts extremely rapidly and hair receptors adapt within a second or so, whereas joint capsule and muscle spindle receptors adapt very slowly.

Furthermore, some sensory receptors adapt to a far greater extent than others. For example, the pacinian corpuscles adapt to "extinction" within a few hundredths of a second, and the hair base receptors adapt to extinction within a second or more. It is probable that all other *mechanoreceptors* also adapt completely eventually, but some require hours or days to do so, for which reason they are frequently called "nonadapting" receptors. The longest measured time for complete adaptation of a mechanoreceptor is about two days for the carotid and aortic baroreceptors.

Some of the nonmechanoreceptors, the chemoreceptors and pain receptors for instance, probably never adapt completely.

Mechanisms by Which Receptors Adapt. Adaptation of receptors is an individual property of each type of receptor in much the same way that development of a receptor potential is an individual property. For instance, in the eye, the rods and cones adapt by changing the concentrations of their light-sensitive chemicals (which is discussed in Chapter 12).

In the case of the mechanoreceptors, the receptor that has been studied for adaptation in greatest detail is again the pacinian corpuscle. Adaptation occurs in this receptor in two ways. First, the pacinian corpuscle is a viscoelastic structure so that when a distorting force is suddenly applied to one side of the corpuscle, this force is instantly transmitted by the viscous component of the corpuscle directly to the same side of the central core, thus eliciting the receptor potential. However, within a few hundredths of a second the fluid within the corpuscle redistributes, so that the pressure becomes essentially equal all through the corpuscle; this now applies an even pressure on all sides of the central core fiber, so that the receptor potential is no longer elicited. Thus, the receptor potential appears at the onset of compression but then disappears within a small fraction of a second even though the compression continues.

Then, when the distorting force is removed from the corpuscle, essentially the reverse events occur. The sudden removal of the distortion from one side of the corpuscle allows rapid expansion on that side, and a corresponding distortion of the central core occurs once more. Again, within hundredths of a second, the pressure becomes equalized all through the corpuscle, and the stimulus is lost. Nevertheless, this disturbance of the central core fiber signals the offset of compression as well as signaling the onset of compression.

The second mechanism of adaptation of the pacinian corpuscle, but a much slower one, results from a process called *accommodation* that occurs in the nerve fiber itself. That is, even if by chance the central core fiber should continue to be distorted, as can be achieved after the capsule has been removed and the core compressed with a stylus, the tip of the nerve fiber itself gradually becomes "accommodated" to the stimulus. This probably results from "inactivation" of the sodium channels in the nerve fiber membrane, which means that the current flow itself through the channels in some way causes them gradually to close, as was explained in Chapter 6.

Presumably, these same two general mechanisms of adaptation apply also to the other types of mechanoreceptors. That is, part of the adaptation results from readjustments in the structure of the receptor itself, and part results from accommodation in the terminal nerve fibril.

Function of the Slowly Adapting Receptors to Detect Continuous Stimulus Strength — The "Tonic" Receptors. The slowly adapting receptors continue to transmit impulses to the brain as long as the stimulus is present (or at least for many minutes or hours). Therefore, they keep the brain constantly apprised of the status of the body and its relation to its surroundings. For instance, impulses from the muscle spindles and Golgi tendon apparatus allow the central nervous system to know the status of muscle contraction and the load on the muscle tendon at each instant.

Other types of slowly adapting receptors include the

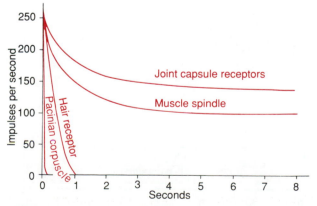

Figure 8–5. Adaptation of different types of receptors, showing rapid adaptation of some receptors and slow adaptation of others.

receptors of the macula in the vestibular apparatus, the pain receptors, the baroreceptors of the arterial tree, the chemoreceptors of the carotid and aortic bodies, and some of the tactile receptors, such as Ruffini's endings and Merkel's discs.

Because the slowly adapting receptors can continue to transmit information for many hours, they are also called *tonic* receptors. Many of these slowly adapting receptors will adapt to extinction if the intensity of the stimulus remains absolutely constant for several hours or days. Fortunately, because of our continually changing bodily state, these receptors almost never reach a state of complete adaptation.

Function of the Rapidly Adapting Receptors to Detect Change in Stimulus Strength—The "Rate Receptors" or "Movement Receptors" or "Phasic Receptors." Obviously, receptors that adapt rapidly cannot be used to transmit a continuous signal because these receptors are stimulated only when the stimulus strength changes. Yet they react strongly *while a change is actually taking place.* Furthermore, the number of impulses transmitted is directly related to the *rate at which the change takes place.* Therefore, these receptors are called *rate* receptors, *movement* receptors, or *phasic* receptors. Thus, in the case of the pacinian corpuscle, sudden pressure applied to the skin excites this receptor for a few milliseconds, and then its excitation is over even though the pressure continues. But later it transmits a signal again when the pressure is released. In other words, the pacinian corpuscle is exceedingly important for transmitting information about rapid changes in pressure against the body, but it is useless for transmitting information about constant pressure applied to the body.

Importance of the Rate Receptors — Their Predictive Function. If one knows the rate at which some change in bodily status is taking place, one can predict the state of the body a few seconds or even a few minutes later. For instance, the receptors of the semicircular canals in the vestibular apparatus of the ear detect the rate at which the head begins to turn when a person runs around a curve. Using this information, a person can predict how much he will turn within the next 2 sec and can adjust the motion of the limbs *ahead of time* to keep from losing balance. Likewise, receptors located in or near the joints help detect the rates of movement of the different parts of the body. Therefore, when a person is running, information from these receptors allows the nervous system to predict where the feet will be during any precise fraction of a second, and appropriate motor signals can be transmitted to the muscles of the legs to make any necessary anticipatory corrections in limb position so that the person will not fall. Loss of this predictive function makes it impossible for the person to run.

■ **THE NERVE FIBERS THAT TRANSMIT SIGNALS AND THEIR PHYSIOLOGICAL CLASSIFICATION**

Some signals need to be transmitted to the central nervous system extremely rapidly; otherwise the information would be useless. An example of this is the sensory signals that apprise the brain of the momentary positions of the limbs at each fraction of a second during running. Another example is the motor signals sent back to the muscles from the brain. At the other extreme, some types of sensory information, such as that depicting prolonged, aching pain, do not need to be transmitted rapidly at all, so that very slowly conducting fibers will suffice. Fortunately, nerve fibers come in all sizes between 0.2 and 20 μm in diameter — the larger the diameter, the greater the conducting velocity. The range of conducting velocities is between 0.5 and 120 m/sec.

Figure 8–6 gives two different classifications of nerve fibers that are in general use. One of these is a general classification that includes both sensory and motor fibers, including the autonomic nerve fibers as well. The other is a classification of sensory nerve fibers that is used primarily by sensory neurophysiologists.

In the general classification, the fibers are divided into types A and C, and the type A fibers are further subdivided into α, β, γ, and δ fibers.

Type A fibers are the typical myelinated fibers of spinal nerves. Type C fibers are the very small, unmyelinated nerve fibers that conduct impulses at low velocities. These constitute more than half the sensory fibers in most peripheral nerves and also all of the postganglionic autonomic fibers.

The sizes, velocities of conduction, and functions of the different nerve fiber types are given in the figure. Note that a

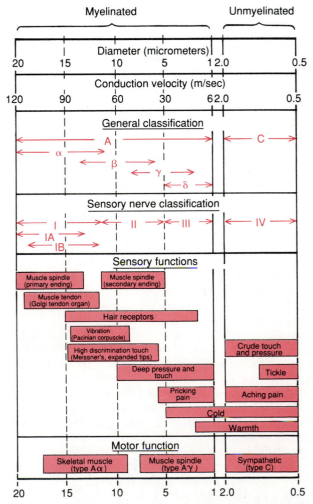

Figure 8–6. Physiological classifications and functions of nerve fibers.

few very large fibers can transmit impulses at velocities as great as 120 m/sec, a distance in 1 sec that is longer than a football field. On the other hand, the smallest fibers transmit impulses as slowly as 0.5 m/sec, requiring about 2 sec to go from the big toe to the spinal cord.

Alternate Classification Used by Sensory Physiologists. Certain recording techniques have made it possible to separate the type Aα fibers into two subgroups; and yet these same recording techniques cannot distinguish easily between Aβ and Aγ fibers. Therefore, the following classification is frequently used by sensory physiologists:

Group Ia. Fibers from the annulospiral endings of muscle spindles (average about 17 μ in diameter; these are type Aα fibers in the general classification).

Group Ib. Fibers from the Golgi tendon organs (average about 16 μm in diameter; these also are type Aα fibers).

Group II. Fibers from the discrete cutaneous tactile receptors and also from the flower-spray endings of the muscle spindles (average about 8 μm in diameter; these are type Aβ and type Aγ fibers in the other classification).

Group III. Fibers carrying temperature, crude touch, and pricking pain sensations (average about 3 μm in diameter; these are type Aδ fibers in the other classification).

Group IV. Unmyelinated fibers carrying pain, itch, temperature, and crude touch sensations (0.5 to 2 μ in diameter; these are called type C fibers in the other classification).

■ TRANSMISSION OF SIGNALS OF DIFFERENT INTENSITY IN NERVE TRACTS— SPATIAL AND TEMPORAL SUMMATION

One of the characteristics of each signal that always must be conveyed is its intensity, for instance, the intensity of pain. The different gradations of intensity can be transmitted either by utilizing increasing numbers of parallel fibers or by sending more impulses along a single fiber. These two mechanisms are called, respectively, spatial summation and temporal summation.

Figure 8–7 illustrates the phenomenon of *spatial summation*, whereby increasing signal strength is transmitted by using progressively greater numbers of fibers. This figure shows a section of skin innervated by a large number of parallel pain nerve fibers. Each of these arborizes into hundreds of minute *free nerve endings* that serve as pain receptors. The entire cluster of fibers from one pain fiber frequently covers an area of skin as large as 5 cm in diameter, and this area is called the *receptor field* of that fiber. The number of endings is large in the center of the field but diminishes toward the periphery. One can also see from the figure that the arborizing nerve fibrils overlap those from other pain fibers. Therefore, a pinprick of the skin usually stimulates endings from many different pain fibers simultaneously. When the pinprick is in the center of the receptive field of a particular pain fiber, however, the degree of stimulation of that fiber is far greater than when it is in the periphery of the field.

Thus, in the lower part of Figure 8–7 are shown three separate views of the cross section of the nerve bundle leading from the skin area. To the left is shown

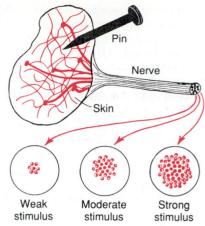

Figure 8–7. Pattern of stimulation of pain fibers in a nerve trunk leading from an area of skin pricked by a pin. This is an example of *spatial summation.*

the effect of a weak stimulus, with only a single nerve fiber in the middle of the bundle stimulated very strongly (represented by the solid fiber), whereas several adjacent fibers are stimulated weakly (half-solid fibers). The other two views of the nerve cross section show the effect respectively of a moderate stimulus and a strong stimulus, with progressively more fibers being stimulated. Thus, the stronger signals spread to more and more fibers. This is the phenomenon of spatial summation.

A second means for transmitting signals of increasing strength is by increasing the *frequency* of nerve impulses in each fiber, which is called *temporal summation.* Figure 8–8 illustrates this, showing in the upper part a changing strength of signal and in the lower part the actual impulses transmitted by the nerve fiber.

■ TRANSMISSION AND PROCESSING OF SIGNALS IN NEURONAL POOLS

The central nervous system is made up of literally hundreds or even thousands of separate neuronal pools,

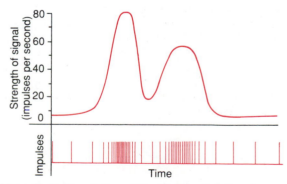

Figure 8–8. Translation of signal strength into a frequency-modulated series of nerve impulses, showing *above* the strength of signal and *below* the separate nerve impulses. This is an example of *temporal summation.*

some of which contain very few neurons while others hold vast numbers. For instance, the entire cerebral cortex could be considered to be a single large neuronal pool, or it could be considered to be a collection of smaller pools each observing separate functions. Other neuronal pools include the different basal ganglia, the specific nuclei in the thalamus, in the cerebellum, the mesencephalon, pons, and medulla. Also, the entire dorsal gray matter of the spinal cord could be considered to be one long pool of neurons, and the entire anterior gray matter another long neuronal pool.

Each pool has its own special characteristics of organization that cause it to process signals in its own special way, thus allowing these special characteristics to achieve the multitude of functions of the nervous system. Yet despite their differences in function, the pools also have many similar principles of function, described in the following pages.

RELAYING OF SIGNALS THROUGH NEURONAL POOLS

Organization of Neurons for Relaying Signals. Figure 8–9 is a schematic diagram of several neurons in a neuronal pool, showing "input" fibers to the left and "output" fibers to the right. Each input fiber divides hundreds to thousands of times, providing an average of a thousand or more terminal fibrils that spread over a large area in the pool to synapse with the dendrites or cell bodies of the neurons in the pool. The dendrites usually also arborize and spread for hundreds to thousands of micrometers in the pool. The neuronal area stimulated by each incoming nerve fiber is called its *stimulatory field*. Note that large numbers of the terminals from each input fiber lie on the center-most neuron in its "field," but progressively fewer terminals lie on the neurons farther from the center of the field.

Threshold and Subthreshold Stimuli—Facilitation. From the discussion of synaptic function in the previous chapter, it will be recalled that discharge of a single excitatory presynaptic terminal almost never stimulates the postsynaptic neuron. Instead, large numbers of terminals must discharge on the same neuron either simultaneously or in rapid succession to cause excitation. For instance, in Figure 8–9, let us assume that six separate terminals must discharge simultaneously to excite any one of the neurons. If the student will count the number of terminals on each one of the neurons from each input fiber, he or she will see that input *fiber 1* has more than enough terminals to cause *neuron a* to discharge. Therefore, the stimulus from input fiber 1 to this neuron is said to be an *excitatory stimulus*; it is also called a *suprathreshold stimulus* because it is above the threshold required for excitation.

Input fiber 1 also contributes terminals to neurons b and c, but not enough to cause excitation. Nevertheless, discharge of these terminals makes both these neurons more excitable to signals arriving through other incoming nerve fibers. Therefore, the stimulus to these neurons is said to be *subthreshold*, and the neurons are said to be *facilitated*.

Similarly, for input *fiber 2*, the stimulus to *neuron d* is a suprathreshold stimulus; and the stimulus to *neurons b* and *c* is a subthreshold, but facilitating, stimulus.

It must be recognized that Figure 8–9 represents a highly condensed version of a neuronal pool, for each input nerve fiber usually provides terminals to hundreds or thousands of separate neurons in its distribution "field," as illustrated in Figure 8–10. In the central portion of the field, almost all the neurons are stimulated by the incoming fiber, designated in Figure 8–10 by the darkened circle area. Therefore, this is said to be the *discharge* zone of the incoming fiber, also called *excited zone* or *liminal zone*. To either side, the neurons are facilitated but not excited, and these areas are called the *facilitated zone*, also called *subthreshold zone* or *subliminal zone*.

Inhibition of a Neuronal Pool. We must also remember that some incoming fibers inhibit neurons, rather than exciting them. This is exactly the opposite of facilitation, and the entire field of the inhibitory branches is called the *inhibitory zone*. The degree of

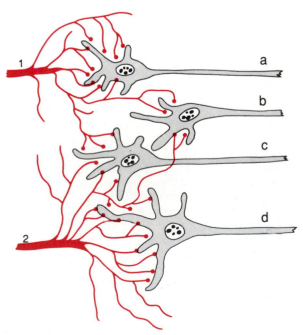

Figure 8–9. Basic organization of a neuronal pool.

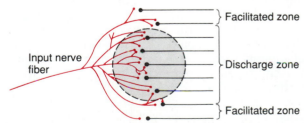

Figure 8–10. "Discharge" and "facilitated" zones of a neuronal pool.

inhibition in the center of this zone obviously is very great because of large numbers of endings in the center; it becomes progressively less toward its edges.

Divergence of Signals Passing Through Neuronal Pools

Often it is important for signals entering a neuronal pool to excite far greater numbers of nerve fibers leaving the pool. This phenomenon is called *divergence*. Two major types of divergence occur and have entirely different purposes:

An *amplifying* type of divergence is illustrated in Figure 8–11A. This means simply that an input signal spreads to an increasing number of neurons as it passes through successive orders of neurons in its path. This type of divergence is characteristic of the corticospinal pathway in its control of skeletal muscles, with a single large pyramidal cell in the motor cortex capable, under appropriate conditions, of exciting as many as 10,000 muscle fibers.

The second type of divergence, illustrated in Figure 8–11B, is *divergence into multiple tracts*. In this case, the signal is transmitted in two separate directions from the pool. For instance, information transmitted in the dorsal columns of the spinal cord takes two courses in the lower part of the brain: (1) into the cerebellum; and (2) on through the lower regions of the brain to the thalamus and cerebral cortex. Likewise, in the thalamus almost all sensory information is relayed both into deep structures of the thalamus and to discrete regions of the cerebral cortex.

Convergence of Signals

"Convergence" means the coming together of signals from multiple inputs to excite a single neuron. Figure 8–12A shows *convergence from a single source*. That is, multiple terminals from an incoming fiber tract terminate on the same neuron. The importance of this is that neurons are almost never excited by an action potential from a single input terminal. But action potentials from multiple input terminals will provide enough "spatial

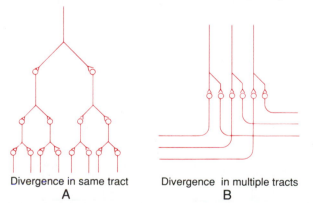

Figure 8–11. "Divergence" in neuronal pathways. *A*, Divergence within a pathway to cause "amplification" of the signal. *B*, Divergence into multiple tracts to transmit the signal to separate areas.

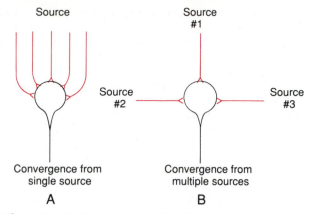

Figure 8–12. "Convergence" of multiple input fibers on a single neuron. *A*, Input fibers from a single source. *B*, Input fibers from multiple sources.

summation" to bring the neuron to the threshold required for discharge.

However, *convergence can also result from input signals* (excitatory or inhibitory) *from multiple sources*, as illustrated in Figure 8–12B. For instance, the interneurons of the spinal cord receive converging signals from (1) peripheral nerve fibers entering the cord, (2) propriospinal fibers passing from one segment of the cord to another, (3) corticospinal fibers from the cerebral cortex, and (4) several other long pathways descending from the brain into the spinal cord. Then the signals from the interneurons converge on the anterior motor neurons to control muscle function.

Such convergence allows summation of information from different sources, and the resulting response is a summated effect of all the different types of information. Obviously, therefore, convergence is one of the important means by which the central nervous system correlates, summates, and sorts different types of information.

Neuronal Circuit Causing Both Excitatory and Inhibitory Output Signals

Sometimes an incoming signal to a neuronal pool causes an output excitatory signal going in one direction and at the same time an inhibitory signal going elsewhere. For instance, at the same time that an excitatory signal is transmitted by one set of neurons in the spinal cord to cause forward movement of a leg, an inhibitory signal is transmitted simultaneously through a separate set of neurons to inhibit the muscles on the back of the leg so that they will not oppose the forward movement. This type of circuit is characteristic of control of all antagonistic pairs of muscles, and it is called the *reciprocal inhibition circuit*.

Figure 8–13 illustrates the means by which the inhibition is achieved. The input fiber directly excites the excitatory output pathway, but it stimulates an intermediate *inhibitory neuron* (neuron 2) which then inhibits the second output pathway from the pool. This type of circuit is also important in preventing overactivity in many parts of the brain.

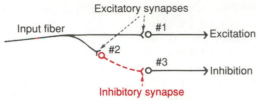

Figure 8–13. Inhibitory circuit. Neuron 2 is an inhibitory neuron.

PROLONGATION OF A SIGNAL BY A NEURONAL POOL— "AFTERDISCHARGE"

Thus far, we have considered signals that are merely relayed through neuronal pools. However, in many instances, a signal entering a pool causes a prolonged output discharge, called *afterdischarge*, even after the incoming signal is over, and lasting from a few milliseconds to as long as many minutes. The two most important mechanisms by which afterdischarge occurs are the following:

Synaptic Afterdischarge. When excitatory synapses discharge on the surfaces of dendrites or the soma of a neuron, a postsynaptic potential develops in the neuron that lasts for many milliseconds, especially so when some of the long-acting synaptic transmitter substances are involved. As long as this potential lasts, it can continue to excite the neuron, causing it to transmit a continuous train of output impulses, as was explained in the previous chapter. Thus, as a result of this synaptic "afterdischarge" mechanism alone, it is possible for a single instantaneous input to cause a sustained signal output (a series of repetitive discharges) lasting for many milliseconds.

The Reverberatory (Oscillatory) Circuit as a Cause of Signal Prolongation. One of the most important of all circuits in the entire nervous system is the *reverberatory*, or *oscillatory*, *circuit*. Such circuits are caused by positive feedback within the neuronal network. That is, the output of a neuronal circuit feeds back to re-excite the input of the same circuit. Consequently, once stimulated, the circuit discharges repetitively for a long time.

Several different possible varieties of reverberatory circuits are illustrated in Figure 8–14, the simplest—in Figure 8–14A—involving only a single neuron. In this case, the output neuron simply sends a collateral nerve fiber back to its own dendrites or soma to re-stimulate itself; therefore, once the neuron discharged, the feedback stimuli could help keep the neuron discharging for a long time thereafter.

Figure 8–14B illustrates a few additional neurons in the feedback circuit, which would give a longer period of time between the initial discharge and the feedback signal. Figure 8–14C illustrates a still more complex system in which both facilitatory and inhibitory fibers impinge on the reverberating circuit. A facilitatory signal enhances the intensity and frequency of reverberation, whereas an inhibitory signal depresses or stops the reverberation.

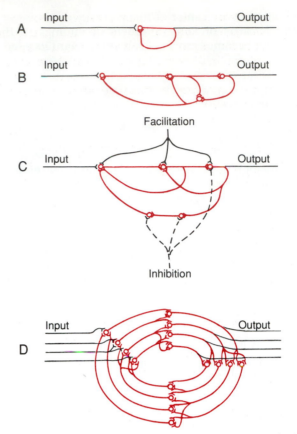

Figure 8–14. Reverberatory circuits of increasing complexity.

Figure 8–14D illustrates that most reverberating pathways are constituted of many parallel fibers, and at each cell station the terminal fibrils diffuse widely. In such a system the total reverberating signal can be either weak or strong, depending on how many parallel nerve fibers are momentarily involved in the reverberation.

Characteristics of Signal Prolongation From a Reverberatory Circuit. Figure 8–15 illustrates output signals from a typical reverberatory circuit. The input stimulus need last only 1 msec or so, and yet the output can last for many milliseconds or even minutes. The figure demonstrates that the intensity of the output signal usually increases to a high value early in the reverberation, then decreases to a critical point, at which it suddenly ceases entirely. The cause of this

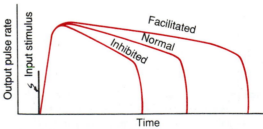

Figure 8–15. Typical pattern of the output signal from a reverberatory circuit following a single input stimulus, showing the effects of facilitation and inhibition.

sudden cessation of reverberation is fatigue of one or more of the synaptic junctions in the circuit, for fatigue beyond a certain critical level lowers the stimulation of the next neuron in the circuit below threshold level so that the circuit is suddenly broken. Obviously, the duration of the signal before cessation can also be controlled by signals from other parts of the brain that inhibit or facilitate the circuit.

Almost these exact patterns of output signals are recorded from the motor nerves exciting a muscle involved in the flexor reflex following pain stimulation of the foot (as illustrated in Figure 8–18).

Continuous Signal Output From Neuronal Circuits

Some neuronal circuits emit output signals continuously even without excitatory input signals. At least two different mechanisms can cause this effect: (1) intrinsic neuronal discharge and (2) continuous reverberatory signals.

Continuous Discharge Caused by Intrinsic Neuronal Excitability. Neurons, like other excitable tissues, discharge repetitively if their membrane potentials rise above certain threshold levels. The membrane potentials of many neurons even normally are high enough to cause them to emit impulses continually. This occurs especially in large numbers of the neurons of the cerebellum as well as in most of the interneurons of the spinal cord. The rates at which these cells emit impulses can be increased by facilitatory signals or decreased by inhibitory signals; the latter can sometimes decrease the rate to extinction.

Continuous Signals Emitted From Reverberating Circuits as a Means for Transmitting Information. Obviously, a reverberating circuit that never fatigues to extinction can also be a source of continual impulses. And facilitatory impulses entering the reverberating pool can increase the output signal, while inhibition can decrease or even extinguish the signal.

Figure 8–16 illustrates a continual output signal from a pool of neurons, which may emit impulses either because of intrinsic neuronal excitability or as a result of reverberation. Note that an excitatory (or fa-

cilitatory) input signal greatly increases the output signal, whereas an inhibitory input signal greatly decreases the output. Those students who are familiar with radio transmitters will recognize this to be a *carrier wave* type of information transmission. That is, the excitatory and inhibitory control signals are not the *cause* of the output signal, but they do *control* it. Note that this carrier wave system allows decrease in signal intensity as well as increase, whereas, up to this point, the types of information transmission that we have discussed have been only positive information, rather than negative information. This type of information transmission is used by the autonomic nervous system to control such functions as vascular tone, gut tone, degree of constriction of the iris, heart rate, and others.

RHYTHMIC SIGNAL OUTPUT

Many neuronal circuits emit rhythmic output signals—for instance, the rhythmic respiratory signal originating in the reticular substance of the medulla and pons. This repetitive rhythmic signal continues throughout life, while other rhythmic signals, such as those that cause scratching movements by the hind leg of a dog or the walking movements in an animal, require input stimuli into the respective circuits to initiate the signals.

Either all or almost all rhythmic signals that have been studied experimentally have been found to result from reverberating circuits or successive reverberating circuits that feed excitatory or inhibitory signals from one neuron to the next.

Obviously, facilitatory or inhibitory signals can affect rhythmic signal output in the same way that they can affect continuous signal outputs. Figure 8–17, for instance, illustrates the rhythmic respiratory signal in the phrenic nerve. However, when the carotid body is stimulated by arterial oxygen deficiency, the frequency and amplitude of the rhythmic signal pattern increase progressively.

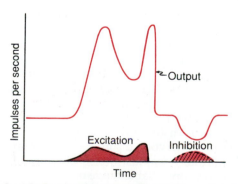

Figure 8–16. Continuous output from either a reverberating circuit or a pool of intrinsically discharging neurons. This figure also shows the effect of excitatory and inhibitory input signals.

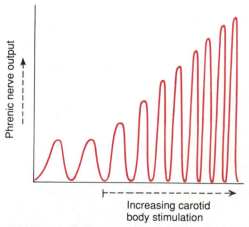

Figure 8–17. The rhythmic output from the respiratory center, showing that progressively increasing stimulation of the carotid body increases both the intensity and frequency of oscillation.

■ INSTABILITY AND STABILITY OF NEURONAL CIRCUITS

Almost every part of the brain connects either directly or indirectly with every other part, and this creates a serious problem. If the first part excites the second, the second the third, the third the fourth, and so on until finally the signal re-excites the first part, it is clear that an excitatory signal entering any part of the brain would set off a continuous cycle of re-excitation of all parts. If this should occur, the brain would be inundated by a mass of uncontrolled reverberating signals—signals that would be transmitting no information but, nevertheless, would be consuming the circuits of the brain so that none of the informational signals could be transmitted. Such an effect actually occurs in widespread areas of the brain during *epileptic convulsions*.

How does the central nervous system prevent this from happening all the time? The answer seems to lie in two basic mechanisms that function throughout the central nervous system: (1) inhibitory circuits; and (2) fatigue of synapses.

INHIBITORY CIRCUITS AS A MECHANISM FOR STABILIZING NERVOUS SYSTEM FUNCTION

Two types of inhibitory circuits in widespread areas of the brain help prevent excessive spread of signals: (1) inhibitory feedback circuits that return from the termini of pathways back to the initial excitatory neurons of the same pathways—these are believed to occur in all the sensory nervous pathways, and they inhibit the input neurons when the termini become overly excited; and (2) some neuronal pools that exert gross inhibitory control over widespread areas of the brain—for instance, many of the basal ganglia exert inhibitory influences throughout the motor control system.

SYNAPTIC FATIGUE AS A MEANS OF STABILIZING THE NERVOUS SYSTEM

Synaptic fatigue means simply that synaptic transmission becomes progressively weaker the more prolonged the period of excitation. Figure 8–18 illustrates three successive records of a flexor reflex elicited in an animal caused by inflicting pain in the footpad of the paw. Note in each record that the strength of contraction progressively "decrements"—that is, its strength diminishes; this is believed to be caused by *fatigue* of the synapses in the flexor reflex circuit. Furthermore, the shorter the interval between the successive flexor reflexes, the less is the intensity of the subsequent reflex response. Thus, in most neuronal circuits that are overused, the sensitivities of the circuits become depressed.

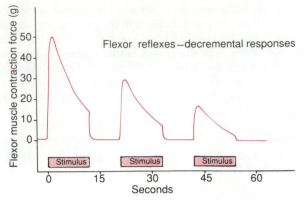

Figure 8–18. Successive flexor reflexes illustrating fatigue of conduction through the reflex pathway.

Automatic Short-term Adjustment of Pathway Sensitivity by the Fatigue Mechanism. Now let us apply this phenomenon of fatigue to multiple pathways in the brain. Those that are overused usually become fatigued, and so their sensitivities will be reduced. On the other hand, those that are underused will become rested, and their sensitivities will increase. Thus, fatigue and recovery from fatigue constitute an important short-term means of moderating the sensitivities of the different nervous system circuits, helping to keep them operating in a range of sensitivity that allows effective function.

Long-term Changes in Synaptic Sensitivity Caused by Automatic Downgrading or Upgrading of Synaptic Receptors. Recently it has been learned that the long-term sensitivities of synapses can be changed tremendously by downgrading the number of receptor proteins at the synaptic sites when there is overactivity and upgrading the receptors when there is underactivity. The mechanism for this is believed to be the following: Receptor proteins are being formed constantly by the endoplasmic reticulum–Golgi apparatus system and are constantly being inserted into the synaptic membrane. However, when the synapses are overused and excesses of transmitter substance combine with the receptor proteins, many of these proteins are inactivated permanently and presumably removed from the synaptic membrane. This is especially true when some of the "modulator" transmitter substances are released at the synapses.

It is indeed fortunate that fatigue and downgrading or upgrading of receptors, as well as other control mechanisms of the nervous system, continually adjust the sensitivity in each circuit to almost the exact level required for proper function. Think for a moment how serious it would be if the sensitivities of only a few of these circuits should be abnormally high; one might then expect almost continual muscle cramps, convulsions, psychotic disturbances, hallucinations, tension, or many other nervous disorders. But the automatic controls normally readjust the sensitivities of the circuits back to a controllable range of reactivity any time the circuits begin to be either too active or too depressed.

REFERENCES

An der Heiden, U.: Analysis of Neural Networks. New York, Springer-Verlag, 1980.

Baldissera, F., et al.: Integration in spinal neuronal systems. In Brooks, V. B. (ed.): Handbook of Physiology. Sec. 1, Vol. II. Bethesda, Md., American Physiological Society, 1981, p. 509.

Bjorklund, A., and Stenevi, U.: Intercerebral neural implants: Neuronal replacement and reconstruction of damaged circuitries. Annu. Rev. Neurosci., 7:279, 1984.

Bousfield, D. (ed.): Neurotransmitters in Action. New York, Elsevier Science Publishing Co., 1985.

Connor, J. A.: Neural pacemakers and rhythmicity. Annu. Rev. Physiol., 47:17, 1985.

Cotman, C. W., et al.: Synapse replacement in the nervous system of adult vertebrates. Physiol. Rev., 61:684, 1981.

Cowan, W. M.: The development of the brain. Sci. Am., 241(3):112, 1979.

Dumont, J. P. C., and Robertson R. M.: Neuronal circuits: An evolutionary perspective. Science, 233:849, 1986.

Faber, D. S., and Korn, H.: Electrical field effects: Their relevance in central neural networks. Physiol. Rev., 69:821, 1989.

Gilbert, C. D.: Microcircuitry of the visual cortex. Annu. Rev. Neurosci., 6:217, 1983.

Hemmings, H. C., Jr., et al.: Role of protein phosphorylation in neuronal signal transduction. FASEB J., 3:1583, 1989.

Henneman, E., and Mendell, L. M.: Functional organization of motoneuron pool and its inputs. In Brooks, V. B. (ed.): Handbook of Physiology. Sec. 1, Vol. II. Bethesda, Md., American Physiological Society, 1981, p. 423.

Hopfield, J. J., and Tank, D. W.: Computing with neural circuits: A model. Science, 233:625, 1986.

Kalia, M. P.: Anatomical organization of central respiratory neurons. Annu. Rev. Physiol., 43:105, 1981.

Laduron, P. M.: Presynaptic heteroreceptors in regulation of neuronal transmission. Biochem. Pharmacol., 34:467, 1985.

Llinas, R. R.: The intrinsic electrophysiological properties of mammalian neurons: Insights into central nervous system function. Science, 242:1654, 1988.

Mendell, L. M.: Modifiability of spinal synapses. Physiol. Rev., 64:260, 1984.

Mountcastle, V. B.: Central nervous mechanisms in mechanoreceptive sensibility. In Darian-Smith, I. (ed.): Handbook of Physiology. Sec. 1, Vol. III. Bethesda, Md., American Physiological Society, 1984, p. 789.

Pinsker, H. M., and Willis, W. D., Jr. (eds.): Information Processing in the Nervous System. New York, Raven Press, 1980.

Purves, D., and Lichtman, J. W.: Specific connections between nerve cells. Annu. Rev. Physiol., 45:553, 1983.

Robinson, D. A.: Integrating with neurons. Annu. Rev. Physiol., 12:33, 1989.

Sachs, M. B.: Neural coding of complex sounds: Speech. Annu. Rev. Physiol., 46:261, 1984.

Sejnowski, T. J., et al.: Computational neuroscience. Science, 241:1299, 1988.

Selverston, A. I., and Moulins, M.: Oscillatory neural networks. Annu. Rev. Physiol., 47:29, 1985.

Sherman, S. M., and Spear, P. D.: Organization of visual pathways in normal and visually deprived cats. Physiol. Rev., 62:738, 1982.

Starke, K., et al.: Modulation of neurotransmitter release by presynaptic autoreceptors. Physiol. Rev., 69:864, 1989.

Sterling, P.: Microcircuitry of the cat retina. Annu. Rev. Neurosci., 6:149, 1983.

Su, C.: Purinergic neurotransmission and neuromodulation. Annu. Rev. Pharmacol. Toxicol., 23:397, 1983.

Turek, F. W.: Circadian neural rhythms in mammals. Annu. Rev. Physiol., 47:49, 1985.

Wong, R. K., et al.: Local circuit interactions in synchronization of cortical neurons. J. Exp. Biol., 112:169, 1984.

9

Somatic Sensations:

I. General Organization; the Tactile and Position Senses

The *somatic senses* are the nervous mechanisms that collect sensory information from the body. These senses are in contradistinction to the *special senses,* which mean specifically vision, hearing, smell, taste, and equilibrium.

■ CLASSIFICATION OF SOMATIC SENSES

The somatic senses can be classified into three different physiological types: (1) the *mechanoreceptive somatic senses,* which include both *tactile* and *position* sensations that are stimulated by mechanical displacement of some tissue of the body; (2) the *thermoreceptive senses,* which detect heat and cold; and (3) the *pain sense,* which is activated by any factor that damages the tissues. This chapter deals with the mechanoreceptive tactile and position senses, and the following chapter discusses the thermoreceptive and pain senses.

The tactile senses include *touch, pressure, vibration,* and *tickle* senses, and the position senses include *static position* and *rate of movement* senses.

Other Classifications of Somatic Sensations. Somatic sensations are also often grouped together in other classes that are not necessarily mutually exclusive, as follows:

Exteroceptive sensations are those from the surface of the body. *Proprioceptive sensations* are those having to do with the physical state of the body, including position sensations, tendon and muscle sensations, pressure sensations from the bottom of the feet, and even the sensation of equilibrium, which is generally considered to be a "special" sensation rather than a somatic sensation.

Visceral sensations are those from the viscera of the body; in using this term one usually refers specifically to sensations from the internal organs.

The *deep sensations* are those that come from the deep tissues, such as from fasciae, muscles, bone, and so forth. These include mainly "deep" pressure, pain, and vibration.

■ DETECTION AND TRANSMISSION OF TACTILE SENSATIONS

Interrelationship Between the Tactile Sensations of Touch, Pressure, and Vibration. Though touch, pressure, and vibration are frequently classified as separate sensations, they are all detected by the same types of receptors. There are only three differences among them: (1) touch sensation generally results from stimulation of tactile receptors in the skin or in tissues immediately beneath the skin; (2) pressure sensation generally results from deformation of deeper tissues; and (3) vibration sensation results from rapidly repetitive sensory signals, but some of the same types of receptors as those for touch and pressure are utilized—specifically the very rapidly adapting types of receptors.

The Tactile Receptors. At least six entirely different types of tactile receptors are known, but many more similar to these also exist. Some of these receptors were illustrated in Figure 8–1, and their special characteristics are the following:

First, some *free nerve endings,* which are found everywhere in the skin and in many other tissues, can detect touch and pressure. For instance, even light contact with the cornea of the eye, which contains no other type of nerve ending besides free nerve endings, can nevertheless elicit touch and pressure sensations.

Second, a touch receptor of special sensitivity is *Meissner's corpuscle,* an elongated encapsulated nerve ending that excites a large (type Aβ) myelinated sensory nerve fiber. Inside the capsule are many whorls of terminal nerve filaments. These receptors are present in the nonhairy parts of the skin (called *glabrous skin*) and are particularly abundant in the fingertips, lips, and other areas of the skin where one's ability to discern spatial characteristics of touch sensations is highly developed. Meissner's corpuscles adapt in a fraction of a second after they are stimulated, which means that they are particularly sensitive to movement of very light objects over the surface of the skin and also to low frequency vibration.

Third, the fingertips and other areas that contain large numbers of Meissner's corpuscles also contain large numbers of *expanded tip tactile receptors*, one type of which is *Merkel's discs*, illustrated in Figure 9–1. The hairy parts of the skin also contain moderate numbers of expanded tip receptors, even though they have almost no Meissner's corpuscles. These receptors differ from Meissner's corpuscles in that they transmit an initially strong but partially adapting signal and then a continuing weaker signal that adapts only slowly. Therefore, they are responsible for giving steady state signals that allow one to determine continuous touch of objects against the skin. Merkel's discs are often grouped together in a single receptor organ called the *Iggo dome receptor*, which projects upward against the underside of the epithelium, as also illustrated in Figure 9–1. This causes the epithelium at this point to protrude outward, thus creating a dome and constituting an extremely sensitive receptor. Also note that the entire group of Merkel's discs is innervated by a single large type of myelinated nerve fiber (type Aβ). These receptors, along with the Meissner's corpuscles discussed above, play extremely important roles in localizing touch sensations to the specific surface areas of the body and also in determining the texture of what is felt.

Fourth, slight movement of any hair on the body stimulates the nerve fiber entwining its base. Thus,

each hair and its basal nerve fiber, called the *hair end-organ*, is also a touch receptor. This receptor adapts readily and, therefore, like Meissner's corpuscles, detects mainly movement of objects on the surface of the body or initial contact with the body.

Fifth, located in the deeper layers of the skin and also in deeper tissues are many *Ruffini's end-organs*, which are multibranched, encapsulated endings, as illustrated in Figure 8–1. These endings adapt very little and, therefore, are important for signaling continuous states of deformation of the skin and deeper tissues, such as heavy and continuous touch signals and pressure signals. They are also found in joint capsules and help signal the degree of joint rotation.

Sixth, *pacinian corpuscles*, which were discussed in detail in Chapter 8, lie both immediately beneath the skin and also deep in the fascial tissues of the body. These are stimulated only by very rapid movement of the tissues because they adapt in a few hundredths of a second. Therefore, they are particularly important for detecting tissue vibration or other extremely rapid changes in the mechanical state of the tissues.

Transmission of Tactile Sensations in Peripheral Nerve Fibers. Almost all the specialized sensory receptors, such as Meissner's corpuscles, Iggo dome receptors, hair receptors, pacinian corpuscles, and Ruffini's endings, transmit their signals in type Aβ nerve fibers, which have transmission velocities of 30 to 70 m/sec. On the other hand, free nerve ending tactile receptors transmit signals mainly via the small type Aδ myelinated fibers, which conduct at velocities of 5 to 30 m/sec. Some tactile free nerve endings transmit via type C unmyelinated fibers at velocities from a fraction of a meter up to 2 m/sec; these send signals into the spinal cord and lower brain stem, probably subserving mainly the sensation of tickle. Thus, the more critical types of sensory signals — those that help to determine precise localization on the skin, minute gradations of intensity, or rapid changes in sensory signal inten-

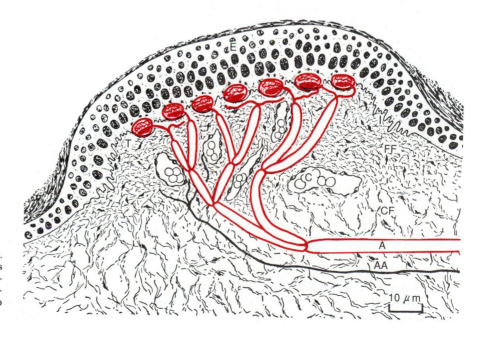

Figure 9–1. The Iggo dome receptor. Note the multiple numbers of Merkel's discs innervated by a single large myelinated fiber and abutting tightly the undersurface of the epithelium. (From Iggo and Muir: J. Physiol., 200:763, 1969.)

sity—are all transmitted in more rapidly conducting types of sensory nerve fibers. On the other hand, the cruder types of signals, such as crude pressure, poorly localized touch, and especially tickle, are transmitted via much slower nerve fibers that require much less space in the nerve bundle than the faster fibers.

DETECTION OF VIBRATION

All the different tactile receptors are involved in detection of vibration, though different receptors detect different frequencies of vibration. Pacinian corpuscles can signal vibrations from 30 to 800 cycles/sec, because they respond extremely rapidly to minute and rapid deformations of the tissues, and they also transmit their signals over type Aβ nerve fibers, which can transmit more than 1000 impulses/sec.

Low frequency vibrations up to 80 cycles/sec, on the other hand, stimulate other tactile receptors—especially Meissner's corpuscles, which are less rapidly adapting than pacinian corpuscles.

TICKLE AND ITCH

Recent neurophysiological studies have demonstrated the existence of very sensitive, rapidly adapting, mechanoreceptive free nerve endings that elicit only the tickle and itch sensations. Furthermore, these endings are found almost exclusively in the superficial layers of the skin, which is also the only tissue from which the tickle and itch sensations usually can be elicited. These sensations are transmitted by very small type C, unmyelinated fibers similar to those that transmit the aching, slow type of pain.

The purpose of the itch sensation is presumably to call attention to mild surface stimuli such as a flea crawling on the skin or a fly about to bite, and the elicited signals then excite the scratch reflex or other maneuvers that rid the host of the irritant.

Itch can be relieved by the process of scratching if this removes the irritant or if the scratch is strong enough to elicit pain. The pain signals are believed to suppress the itch signals in the cord by the process of lateral inhibition, which will be described later.

■ THE TWO SENSORY PATHWAYS FOR TRANSMISSION OF SOMATIC SIGNALS INTO THE CENTRAL NERVOUS SYSTEM

Almost all sensory information from the somatic segments of the body enters the spinal cord through the dorsal roots of the spinal nerves (with the exception of a few very small fibers of questionable importance that enter the ventral roots). However, from the entry point of the cord and then to the brain the sensory signals are carried through one of two alternate sensory pathways: (1) the *dorsal column – lemniscal system;* and (2) the *anterolateral system.* These two systems again come together partially at the level of the thalamus.

The dorsal column – lemniscal system, as its name implies, carries signals mainly in the *dorsal columns* of the cord and then, after crossing to the opposite side in the medulla, upward through the brain stem to the thalamus by way of the *medial lemniscus.* On the other hand, signals of the anterolateral system, after originating in the dorsal horns of the spinal gray matter, cross to the opposite side of the cord and ascend through the anterior and lateral white columns to terminate at all levels of the brain stem and also in the thalamus.

The dorsal column – lemniscal system is composed of large, myelinated nerve fibers that transmit signals to the brain at velocities of 30 to 110 m/sec, whereas the anterolateral system is composed of much smaller myelinated fibers (averaging 4 μm in diameter) that transmit signals at velocities ranging from a few meters per second up to 40 m/sec.

Another difference between the two systems is that the dorsal column – lemniscal system has a very high degree of spatial orientation of the nerve fibers with respect to their origin on the surface of the body, whereas the anterolateral system has a much smaller degree of spatial orientation.

These differences immediately characterize the types of sensory information that can be transmitted by the two systems. That is, sensory information that must be transmitted rapidly and with temporal and spatial fidelity is transmitted in the dorsal column – lemniscal system, while that which does not need to be transmitted rapidly nor with great spatial fidelity is transmitted mainly in the anterolateral system. On the other hand, the anterolateral system has a special capability that the dorsal system does not have: the ability to transmit a broad spectrum of sensory modalities—pain, warmth, cold, and crude tactile sensations; the dorsal system is limited to the more discrete types of mechanoreceptive sensations alone.

With this differentiation in mind we can now list the types of sensations transmitted in the two systems:

THE DORSAL COLUMN – LEMNISCAL SYSTEM

1. Touch sensations requiring a high degree of localization of the stimulus.
2. Touch sensations requiring transmission of fine gradations of intensity.
3. Phasic sensations, such as vibratory sensations.
4. Sensations that signal movement against the skin.
5. Position sensations.
6. Pressure sensations having to do with fine degrees of judgment of pressure intensity.

THE ANTEROLATERAL SYSTEM

1. Pain.
2. Thermal sensations, including both warm and cold sensations.
3. Crude touch and pressure sensations capable of only crude localizing ability on the surface of the body.
4. Tickle and itch sensations.
5. Sexual sensations.

■ TRANSMISSION IN THE DORSAL COLUMN–LEMNISCAL SYSTEM

ANATOMY OF THE DORSAL COLUMN–LEMNISCAL SYSTEM

On entering the spinal cord from the spinal nerve dorsal roots, the large myelinated fibers from specialized mechano-receptors pass medially into the lateral margin of the dorsal white columns. However, almost immediately each fiber divides to form a *medial branch* and a *lateral branch*, as illustrated by the medial fiber from the dorsal root in Figure 9–2. The medial branch turns upward in the dorsal column and proceeds by way of the dorsal column pathway to the brain.

The lateral branch enters the dorsal horn of the cord gray matter and then divides many times, synapsing with neurons in almost all parts of the intermediate and anterior portions of the cord gray matter. The neurons that are excited, in turn, serve three functions: (1) A few of them give off second-order fibers that re-enter the dorsal column, to make up about 15 per cent of all the dorsal column fibers; and a few other second-order fibers enter the posterolateral column, forming the *spinocervical tract*, which rejoins the dorsal column system in the neck and low medulla. (2) Many of the neurons elicit local spinal cord reflexes, which will be discussed in Chapter 16. (3) Others give rise to the spinocerebellar tracts, which we will discuss in Chapter 18 in relation to the function of the cerebellum.

The Dorsal Column–Medial Lemniscal Pathway. Note in Figure 9–3 that the nerve fibers entering the dorsal columns pass uninterrupted up to the medulla, where they synapse in the *dorsal column nuclei* (the *cuneate* and *gracile nuclei*). From here, *second-order neurons* decussate immediately

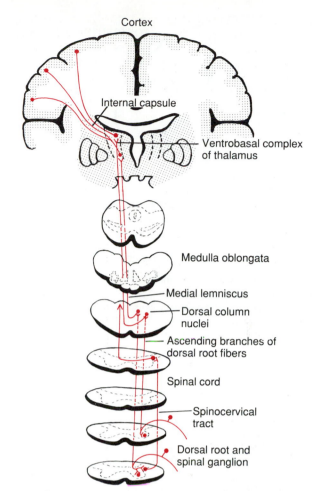

Figure 9–3. The dorsal column and spinocervical pathways for transmitting critical types of tactile signals. (Modified from Ranson and Clark: Anatomy of the Nervous System. Philadelphia, W. B. Saunders Company, 1959.)

to the opposite side and then continue upward to the thalamus through bilateral pathways called the *medial lemnisci*. In this pathway through the brain stem, the medial lemniscus is joined by additional fibers from the *main sensory nucleus of the trigeminal nerve* and from the *upper portion of its descending nuclei;* these fibers subserve the same sensory functions for the head that the dorsal column fibers subserve for the body.

In the thalamus, the medial lemniscal fibers from the dorsal columns terminate in the *ventral posterolateral nucleus,* whereas those from the trigeminal nuclei terminate in the *ventral posteromedial nucleus.* These two nuclei, along with the posterior thalamic nuclei, where some fibers from the anterolateral system terminate, are together called the *ventrobasal complex.* From the ventrobasal complex, *third-order nerve fibers* project, as shown in Figure 9–4, mainly to the *postcentral gyrus* of the *cerebral cortex,* which is called *somatic sensory area I (S-I area).* In addition, fewer fibers project to the lowermost lateral portion of each parietal lobe, an area called *somatic sensory area II (S-II area).*

Spatial Orientation of the Nerve Fibers in the Dorsal Column–Lemniscal System

One of the distinguishing features of the dorsal column–lemniscal system is a distinct spatial orienta-

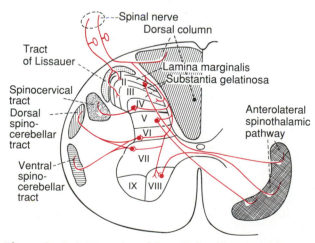

Figure 9–2. Cross-section of the spinal cord, showing the anatomical laminae I through IX of the cord gray matter and the ascending sensory tracts in the white columns of the spinal cord.

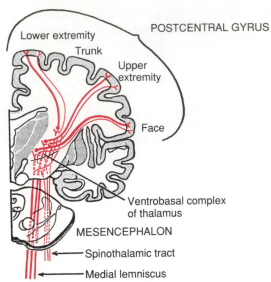

Figure 9–4. Projection of the dorsal column–lemniscal system from the thalamus to the somatic sensory cortex. (Modified from Brodal: Neurological Anatomy in Relation to Clinical Medicine. New York, Oxford University Press, 1969.)

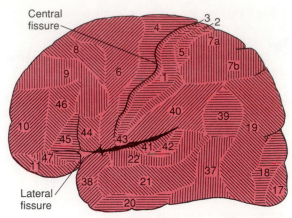

Figure 9–5. Structurally distinct areas, called "Brodmann areas," of the human cerebral cortex. (From Everett: Functional Neuroanatomy, 5th Ed. Philadelphia, Lea & Febiger, 1965. Modified from Brodmann.)

tion of nerve fibers from the individual parts of the body that is maintained throughout. For instance, in the dorsal columns, the fibers from the lower parts of the body lie toward the center, while those that enter the spinal cord at progressively higher segmental levels form successive layers laterally.

In the thalamus, the distinct spatial orientation is still maintained, with the tail end of the body represented by the most lateral portions of the ventrobasal complex and the head and face represented in the medial component of the complex. However, because of the crossing of the medial lemnisci in the medulla, the left side of the body is represented in the right side of the thalamus, and the right side of the body is represented in the left side of the thalamus.

THE SOMATIC SENSORY CORTEX

Before discussing the role of the cerebral cortex in somatic sensation, we need to give an orientation to the various areas of the cortex. Figure 9–5 is a map of the human cerebral cortex, showing that it is divided into about 50 distinct areas called *Brodmann areas* based on histological structural differences. The map itself is important because it has come to be used by virtually all neurophysiologists and neurologists in referring to the different functional areas of the human cortex.

Note in the figure the large *central fissure* (also called "central sulcus") that extends horizontally across the brain. In general, sensory signals from all modalities of sensation terminate in the cerebral cortex posterior to the central fissure. Most importantly, the *somatic sensory cortex* lies immediately behind the central fissure, located mainly in Brodmann areas 1, 2, 3, 5, 7, and 40. By and large, these constitute the *parietal lobe* of the cortex. In addition, visual signals terminate in the oc-

cipital lobe; and auditory signals, in the temporal lobe.

The portion of the cortex anterior to the central fissure is devoted to motor control of the body and to some aspects of analytical thought.

The two distinct and separate areas known to receive direct afferent nerve fibers from the somesthetic relay nuclei in the ventrobasal complex of the thalamus (S-I area and S-II area), are illustrated in Figure 9–6. However, somatic sensory area I is so much more important to the sensory functions of the body than is somatic sensory area II that in popular usage the term "somatic sensory cortex" most often is used to mean area I.

Projection of the Body in Somatic Sensory Area I. Somatic sensory area I lies in the postcentral gyrus of the human cerebral cortex (in Brodmann areas 3, 1, and 2). A distinct spatial orientation exists in this area for reception of nerve signals from the different areas of the body. Figure 9–7 illustrates a cross-section through the brain at the level of the postcentral gyrus, showing the representations of the different parts of the body in separate regions of somatic sensory area I. Note, however, that each side of the cortex receives sensory information exclusively from the opposite side

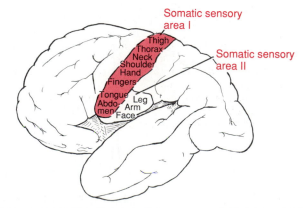

Figure 9–6. The two somatic sensory cortical areas, somatic sensory areas I and II.

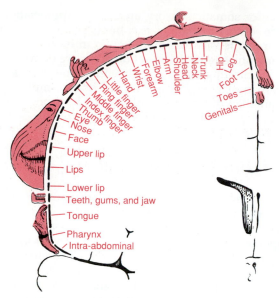

Figure 9–7. Representation of the different areas of the body in the somatic sensory area I of the cortex. (From Penfield and Rasmussen: Cerebral Cortex of Man: A Clinical Study of Localization of Function. New York, Macmillan Company, 1968.)

of the body (with the exception of a small amount of sensory information from the same side of the face).

Some areas of the body are represented by large areas in the somatic cortex — the lips the greatest of all, followed by the face and thumb — whereas the entire trunk and lower part of the body are represented by relatively small areas. The sizes of these areas are directly proportional to the number of specialized sensory receptors in each respective peripheral area of the body. For instance, a great number of specialized nerve endings are found in the lips and thumb, whereas only a few are present in the skin of the trunk.

Note also that the head is represented in the most lateral portion of somatic sensory area I, whereas the lower part of the body is represented medially.

Somatic Sensory Area II. The second cortical area to which thalamic somatic fibers project, somatic sensory area II, is much smaller and lies posterior and inferior to the lateral end of somatic sensory area I, as shown in Figure 9–6. The degree of localization of the different parts of the body is very poor in this area, compared with somatic sensory area I. The face is represented anteriorly, the arms centrally, and the legs posteriorly.

So little is known about the function of somatic sensory area II that it cannot be discussed intelligently. It is known that signals enter this area from both sides of the body, from somatic sensory area I, and also from other sensory areas of the brain, such as visual and auditory signals. Also, in lower animals, ablation of this area makes it difficult for the animal to learn to discriminate different shapes of objects.

The Layers of the Somatic Sensory Cortex and Their Function

The cerebral cortex contains *six* separate layers of neurons, beginning with layer I next to the surface and extending progressively deeper to layer VI, as illus-

trated in Figure 9–8. As would be expected, the neurons in each layer perform functions different from those in other layers. Some of these functions are as follows:

1. The incoming sensory signal excites mainly neuronal layer IV first; then the signal spreads toward the surface of the cortex and also toward the deeper layers.

2. Layers I and II receive a diffuse, nonspecific input from lower brain centers that can facilitate a whole region of the cortex at once; this system will be described in Chapter 19. This input perhaps controls the overall level of excitability of the region stimulated.

3. The neurons in layers II and III send axons to other closely related portions of the cerebral cortex.

4. The neurons in layers V and VI send axons to more distant parts of the nervous system. Those in layer V are generally larger and project to more distant areas. For instance, many of these pass all the way into the brain stem and spinal cord to provide control signals to these areas. From layer VI, especially large numbers of axons extend to the thalamus, providing feedback signals from the cerebral cortex to the thalamus.

Representation of the Different Sensory Modalities in the Somatic Sensory Cortex — The Vertical Columns of Neurons

Functionally, the neurons of the somatic sensory cortex are arranged in vertical columns extending all the

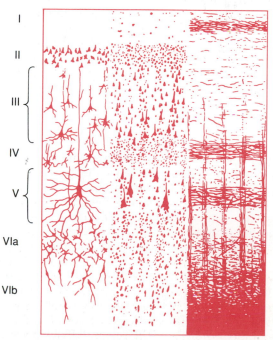

Figure 9–8. Structure of the cerebral cortex, illustrating *I*, molecular layer; *II*, external granular layer; *III*, layer of pyramidal cells; *IV*, internal granular layer; *V*, large pyramidal cell layer; and *VI*, layer of fusiform or polymorphic cells. (From Ranson and Clark [after Brodmann]: Anatomy of the Nervous System. Philadelphia, W. B. Saunders Company, 1959.)

way through the six layers of the cortex, each column having a diameter of 0.3 to 0.5 mm and containing perhaps 10,000 neuronal cell bodies. Each of these columns serves a single specific sensory modality, some columns responding to stretch receptors around joints, some to stimulation of tactile hairs, others to discrete localized pressure points on the skin, and so forth. Furthermore, the columns for the different modalities are interspersed among each other. At layer IV, where the signals first enter the cord, the columns of neurons function almost entirely separately from each other. However, at other levels of the columns interactions occur that allow beginning analysis of the meanings of the sensory signals.

In the most anterior portion of the postcentral gyrus, located deep in the central fissure, in Brodmann area 3a, a disproportionately large share of the vertical columns respond to muscle, tendon, or joint stretch receptors. Many of the signals from these in turn spread directly to the motor cortex located immediately anterior to the central fissure and help control muscle function. As one proceeds more posteriorly in somatic sensory cortex I, more and more of the vertical columns respond to the slowly adapting cutaneous receptors, and then still farther posteriorly greater numbers of the columns are sensitive to deep pressure.

In the most posterior portion of somatic sensory area I, about 6 per cent of the vertical columns respond only when a stimulus moves across the skin in a particular direction. Thus, this is a still higher order of interpretation of sensory signals; and the process becomes even more complex still farther posteriorly in the parietal cortex, which is called the *somatic association area*, as we discuss subsequently.

Functions of Somatic Sensory Area I

The functional capabilities of different areas of the somatic sensory cortex have been determined by selective excision of the different portions. Widespread excision of somatic sensory area I causes loss of the following types of sensory judgment:

1. The person is unable to localize discretely the different sensations in the different parts of the body. However, he or she can localize these sensations very crudely, such as to a particular hand, which indicates that the thalamus or parts of the cerebral cortex not normally considered to be concerned with somatic sensations can perform some degree of localization.

2. He is unable to judge critical degrees of pressure against his body.

3. He is unable to judge exactly the weights of objects.

4. He is unable to judge shapes or forms of objects. This is called *astereognosis.*

5. He is unable to judge texture of materials, for this type of judgment depends on highly critical sensations caused by movement of the skin over the surface to be judged.

Note in the list that nothing has been said about loss of pain and temperature sense. However, in the absence of somatic sensory area I, the appreciation of these sensory modalities may be altered either in quality or in intensity. But more importantly, the pain and temperature sensations that do occur are poorly localized, indicating that both pain and temperature localization probably depend mainly upon simultaneous stimulation of tactile stimuli, using the topographical map of the body in somatic sensory area I to localize the source.

SOMATIC ASSOCIATION AREAS

Brodmann areas 5 and 7 of the cerebral cortex, which are located in the parietal cortex behind somatic sensory area I and above somatic sensory area II, play important roles in deciphering the sensory information that enters the somatic sensory areas. Therefore, these areas are called the *somatic association areas.*

Electrical stimulation in the somatic association area can occasionally cause a person to experience a complex somatic sensation, sometimes even the "feeling" of an object such as a knife or a ball. Therefore, it seems clear that the somatic association area combines information from multiple points in the somatic sensory area to decipher its meaning. This also fits with the anatomical arrangement of the neuronal tracts that enter the somatic association area, for it receives signals from (1) somatic sensory area I, (2) the ventrobasal nuclei of the thalamus, (3) other areas of the thalamus, (4) the visual cortex, and (5) the auditory cortex.

Effect of Removing the Somatic Association Area — Amorphosynthesis. When the somatic association area is removed, the person loses the ability to recognize complex objects and complex forms by the process of feeling them. In addition, he or she loses most of the sense of form of his or her own body. Especially interesting, the person is mainly oblivious to the opposite side of the body — that is, forgets that it is there. Therefore, he also often forgets to use the other side for motor functions as well. Likewise, when feeling objects, the person will tend to feel only one side of the object and to forget that the other side even exists. This complex sensory deficit is called *amorphosynthesis.*

OVERALL CHARACTERISTICS OF SIGNAL TRANSMISSION AND ANALYSIS IN THE DORSAL COLUMN – LEMNISCAL SYSTEM

Basic Neuronal Circuit and Discharge Cortical "Field" in the Dorsal Column – Lemniscal System. The lower part of Figure 9 – 9 illustrates the basic organization of the neuronal circuit of the dorsal column pathway, showing that at each synaptic stage divergence occurs. However, the upper part of the figure shows that a single receptor stimulus on the skin does not cause all the cortical neurons with which that receptor connects to discharge at the same rate. Instead, the cortical neurons that discharge to the greatest extent are those in a central part of the cortical "field" for

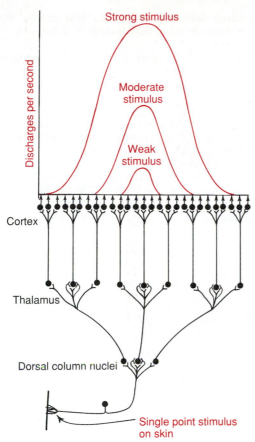

Figure 9–9. Transmission of a pinpoint stimulus signal to the cortex.

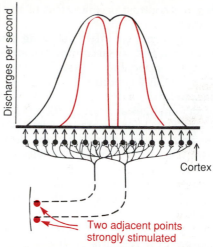

Figure 9–10. Transmission of signals to the cortex from two adjacent pinpoint stimuli. The solid black curve represents the pattern of cortical stimulation without "surround" inhibition, and the two colored curves represent the pattern with "surround" inhibition.

each respective receptor. Thus, a weak stimulus causes only the centralmost neurons to fire. A stronger stimulus causes still more neurons to fire, but those in the center still discharge at a considerably more rapid rate than do those farther away from the center.

Two-Point Discrimination. A method frequently used to test tactile capabilities is to determine a person's so-called "two-point discriminatory ability." In this test, two needles are pressed lightly against the skin, and the subject determines whether two points of stimulus are felt or one point. On the tips of the fingers a person can distinguish two separate points even when the needles are as close together as 1 to 2 mm. However, on the person's back, the needles must usually be as far apart as 30 to 70 mm before two separate points can be detected. The reason for this difference is the different numbers of specialized tactile receptors in the two areas.

Figure 9–10 illustrates the mechanism by which the dorsal column pathway, and all other sensory pathways as well, transmit two-point discriminatory information. This figure shows two adjacent points on the skin that are strongly stimulated, and it also shows the area of the somatic sensory cortex (greatly enlarged) that is excited by signals from the two stimulated points. The solid black curve shows the spatial pattern of cortical excitation when both skin points are stimulated simultaneously. Note that the resultant zone of

excitation has two separate peaks. It is these two peaks, separated by a valley, that allow the sensory cortex to detect the presence of two stimulatory points, rather than a single point. However, the capability of the sensorium to distinguish between two points of stimulation is strongly influenced by another mechanism, *lateral inhibition,* as explained in the following section.

Effect of Lateral Inhibition in Increasing the Degree of Contrast in the Perceived Spatial Pattern. It was pointed out in Chapter 8 that virtually every sensory pathway, when excited, gives rise simultaneously to lateral inhibitory signals; these spread to the sides of the excitatory signal and inhibit adjacent neurons. For instance, consider an excited neuron in a dorsal column nucleus. Aside from the central excitatory signal, short collateral fibers transmit inhibitory signals to the surrounding neurons. Some of these pass through an additional interneuron that secretes an inhibitory transmitter, and others pass directly to presynaptic terminals on the adjacent neurons and inhibit them by the mechanism of presynaptic inhibition.

The importance of *lateral inhibition* is that it blocks lateral spread of the excitatory signals and therefore increases the degree of contrast in the sensory pattern perceived in the cerebral cortex.

In the case of the dorsal column system, lateral inhibitory signals occur at each synaptic level, for instance in the dorsal column nuclei, in the ventrobasal nuclei of the thalamus, and in the cortex itself. At each of these levels, the lateral inhibition helps to block lateral spread of the excitatory signal. As a result, the peaks of excitation stand out, and much of the surrounding diffuse stimulation is blocked. This effect is illustrated by the two colored curves in Figure 9–10, showing complete separation of the peaks when the intensity of the lateral inhibition, also called *surround inhibition,* is very great. Obviously, this mechanism accentuates the contrast between the areas of peak stimulation and the surrounding areas, thus greatly increasing the contrast or sharpness of the perceived spatial pattern.

Transmission of Rapidly Changing and Repetitive Sensations. The dorsal column system is of particular value for apprising the sensorium of rapidly changing peripheral conditions. This system can "follow" changing stimuli up to at least 400 cycles/sec and can "detect" changes as high as 700 cycles.

Vibratory Sensation. Vibratory signals are rapidly repetitive and can be detected as vibration up to 700 cycles/sec. The higher frequency vibratory signals originate from the pacinian corpuscles, but lower frequency (below about 100/sec) signals can originate from Meissner's corpuscles as well. These signals are transmitted only in the dorsal column pathway. For this reason, application of vibration with a tuning fork to different peripheral parts of the body is an important tool used by the neurologist for testing the functional integrity of the dorsal columns.

PSYCHIC INTERPRETATION OF SENSORY STIMULUS INTENSITY

The ultimate goal of most sensory stimulation is to apprise the mind of the state of the body and its surroundings. Therefore, it is important that we discuss briefly some of the principles related to the transmission of sensory stimulus intensity to the higher levels of the nervous system.

The first question that comes to mind is: How is it possible for the sensory system to transmit sensory experiences of tremendously varying intensities? For instance, the auditory system can detect the weakest possible whisper but can also discern the meanings of an explosive sound only a few feet away, even though the sound intensities of these two experiences can vary by more than 10 billion times; the eyes can see visual images with light intensities that vary by as much as a half million times; or the skin can detect pressure differences of 10,000 to 100,000 times.

As a partial explanation of these effects, Figure 8–4 in the previous chapter showed the relationship of the receptor potential produced by the pacinian corpuscle to the intensity of the sensory stimulus. At low stimulus strength, very slight changes in strength increase the potential markedly, whereas at high levels of stimulus strength, further increases in receptor potential are very slight. Thus, the pacinian corpuscle is capable of accurately measuring extremely minute changes in stimulus strength at low intensity levels, but at high intensity levels the change in stimulus strength must be much greater to cause the same amount of change in receptor potential.

The transduction mechanism for detecting sound by the cochlea of the ear illustrates still another method for separating gradations of stimulus intensity. When sound causes vibration at a specific point on the basilar membrane, weak vibration stimulates only those hair cells at the point of maximum vibration. But as the vibration intensity increases, not only do these hair cells become more intensely stimulated, but still many more hair cells in each direction farther away from the maximum vibratory point also become stimulated. Thus, signals transmitted over progressively increasing numbers of nerve fibers is another mechanism by which stimulus strength is transmitted into the central nervous system. This mechanism, plus the direct effect of stimulus strength on impulse rate in each nerve fiber, as well as several other mechanisms, makes it possible for most sensory systems to operate reasonably faithfully at stimulus intensity levels changing by more than hundreds of thousands to billions of times.

Importance of the Tremendous Intensity Range of Sensory Reception. Were it not for the tremendous intensity range of sensory reception that we can experience, the various sensory systems would more often than not operate in the wrong range. This is illustrated by the attempts of most persons to adjust the light exposure on a camera without using a light meter. Left to intuitive judgment of light intensity, a person almost always overexposes the film on very bright days and greatly underexposes the film at twilight. Yet that person's eyes are capable of discriminating with great detail the visual objects in both very bright sunlight and at twilight; the camera cannot do this because of the narrow critical range of light intensity required for proper exposure of film.

JUDGMENT OF STIMULUS INTENSITY

Physiopsychologists have evolved numerous methods for testing one's judgment of sensory stimulus intensity, but only rarely do the results from the different methods agree with each other. Yet the basic principle of decreasing intensity discrimination as the sensory intensity increases is applicable to virtually all sensory modalities. Two formulations of this principle are widely discussed in the physiopsychology field of sensory interpretation: the *Weber-Fechner principle* and the *power principle.*

The Weber-Fechner Principle — Detection of "Ratio" of Stimulus Strength. In the mid-1800s, Weber first and Fechner later proposed the principle that *gradations of stimulus strength are discriminated approximately in proportion to the logarithm of stimulus strength.* That is, a typical test of this principle might show that a person can barely detect a 1-g increase in weight when holding 30 g or a 10-g increase when holding 300 g. Thus, the *ratio* of the change in stimulus strength required for detection of a change remains essentially constant, about 1 to 30, which is what the logarithmic principle means. To express this mathematically,

$$\text{Interpreted signal strength} = \log (\text{Stimulus}) + \text{Constant}$$

More recently it has become evident that the Weber-Fechner principle is quantitatively accurate only for the higher intensities of visual, auditory, and cutaneous sensory experience and applies only poorly to most other types of sensory experience.

Yet the Weber-Fechner principle is still a good one to remember because it emphasizes that the greater the background sensory intensity, the greater also must be the additional change in stimulus strength in order for the mind to detect the change.

The Power Law. Another attempt by physiopsychologists to find a good mathematical relationship is the following formula, known as the power law:

$$\text{Interpreted signal strength} = K \cdot (\text{Stimulus} - k)^y$$

In this formula the exponent y and the constants K and k are different for each type of sensation.

When this power law relationship is plotted on a graph using double logarithmic coordinates, as illustrated in Figure 9–11, a linear relationship can be attained between interpreted stimulus strength and actual stimulus strength over a large range for almost any type of sensory perception. How-

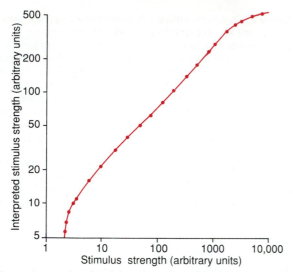

Figure 9–11. Graphical demonstration of the "power law" relationship between actual stimulus strength and strength that the psyche interprets it to be. Note that the power law does not hold at either very weak or very strong stimulus strengths.

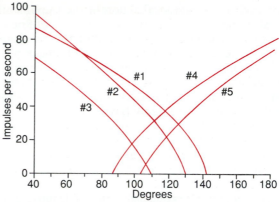

Figure 9–12. Typical responses of five different neurons in the knee joint receptor field of the thalamic ventrobasal complex when the knee joint is moved through its range of motion. (The curves were constructed from data in Mountcastle et al.: J. Neurophysiol., 26:807, 1963.)

ever, as illustrated in the figure, even this power law relationship fails to hold satisfactorily at both very low and very high stimulus strengths.

THE POSITION SENSES

The *position senses* are frequently also called *proprioceptive senses.* They can be divided into two subtypes: (1) *static position sense,* which means conscious orientation of the different parts of the body with respect to each other; and (2) *rate of movement sense,* also called *kinesthesia* or *dynamic proprioception.*

The Position Sensory Receptors. Knowledge of position, both static and dynamic, depends upon knowing the degrees of angulation of all joints in all planes and their rates of change. Therefore, multiple different types of receptors help to determine joint angulation and are used together for position sense. Furthermore, both skin tactile receptors and deep receptors near the joints are also used. In the case of the fingers, where skin receptors are in great abundance, as much as half of position recognition is probably detected through the skin receptors. On the other hand, for most of the larger joints of the body, deep receptors are more important.

For determining joint angulation in midranges of motion, the most important receptors are believed to be the *muscle spindles.* These are also exceedingly important in helping to control muscle movement, as we see in Chapter 16. When the angle of a joint is changing, some muscles are being stretched while others are loosened, and the stretch information from the spindles is passed into the computational system of the spinal cord and higher regions of the dorsal column system for deciphering the complex interrelations of joint angulations.

At the extremes of joint angulation, the stretch of the ligaments and deep tissues around the joints is an additional important factor in determining position.

Some types of endings used for this are the pacinian corpuscles, Ruffini's endings, and receptors similar to the Golgi tendon receptors found in muscle tendons.

The pacinian corpuscles and muscle spindles are especially adapted for detecting rapid rates of change. Therefore, it is likely that these are the receptors most responsible for detecting rate of movement.

Processing of Position Sense Information in the Dorsal Column–Lemniscal Pathway. Despite the usual faithfulness of transmission of signals from the periphery to the sensory cortex in the dorsal column–lemniscal system, there nevertheless seems to be some processing of position sense signals before they reach the cerebral cortex. For instance, individual joint receptors are stimulated maximally at specific degrees of rotation of the joint, with the intensity of stimulation decreasing on either side of the maximal point for each receptor. However, the static position signal for joint rotation is quite different at the level of the thalamus, as can be seen by referring to Figure 9–12. This figure shows that the thalamic neurons that respond to joint rotation are of two types: (1) those maximally stimulated when the joint is at full rotation; and (2) those maximally stimulated when the joint is at minimal rotation. In each case, as the degree of rotation changes, the rate of stimulation of the neuron either decreases or increases, depending on the direction in which the joint is being rotated. Furthermore, the intensity of neuronal excitation changes over angles of 40 to 60 degrees of angulation, in contrast to 20 to 30 degrees for the individual joint receptors. Thus, the signals from the individual joint receptors have been integrated in the space domain by the time they reach the thalamic neurons, illustrating some degree of processing of the signals either in the cord or in the thalamus.

■ TRANSMISSION IN THE ANTEROLATERAL SYSTEM

The anterolateral system, in contrast to the dorsal column system, transmits sensory signals that do not require highly discrete localization of the signal source and also that do not require discrimination of fine gradations of intensity. These include pain, heat, cold, crude tactile, tickle and itch, and sexual sensations. In the following chapter pain and temperature sensations are discussed; the present chapter is still concerned principally with transmission of the tactile sensations, but now with the less acute types.

ANATOMY OF THE ANTEROLATERAL PATHWAY

The anterolateral fibers originate mainly in laminae I, IV, V, and VI (see Fig. 9–2) in the dorsal horns, where many of the dorsal root sensory nerve fibers terminate after entering the cord. Then, as illustrated in Figure 9–13, the fibers cross in the anterior commissure of the cord to the opposite anterior and lateral white columns, where they turn upward toward the brain. These fibers ascend rather diffusely throughout the anterolateral columns. However, anatomical studies suggest a partial differentiation of this pathway into an anterior division, called the *anterior spinothalamic tract*, and a lateral division, called the *lateral spinothalamic tract*. Also encompassed in the anterolateral pathway are a *spinoreticular pathway* (to the reticular substance of the brain stem) and a *spinotectal tract* (to the tectum of the mesencephalon).

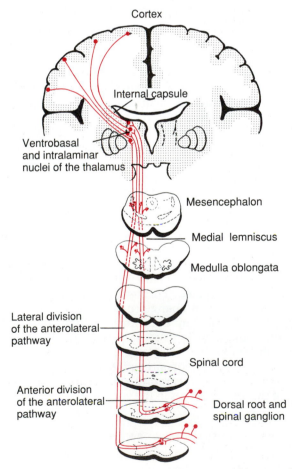

Cortex

Internal capsule

Ventrobasal and intralaminar nuclei of the thalamus

Mesencephalon

Medial lemniscus

Medulla oblongata

Lateral division of the anterolateral pathway

Spinal cord

Anterior division of the anterolateral pathway

Dorsal root and spinal ganglion

Figure 9–13. The anterior and lateral divisions of the anterolateral pathway.

However, it has been difficult to make these differentiations using electrical recording techniques.

The upper terminus of the anterolateral pathway is mainly twofold: (1) throughout the *reticular nuclei of the brain stem;* and (2) in two different nuclear complexes of the thalamus, the *ventrobasal complex* and the *intralaminar nuclei.* In general, the tactile signals are transmitted mainly into the ventrobasal complex, terminating in the same *ventral posterior lateral* and *medial* nuclei as the dorsal column system, and this is probably also true for the temperature signals. From here, the tactile signals are transmitted to the somatosensory cortex along with the signals from the dorsal columns. On the other hand, only part of the pain signals project to this complex. Instead, most of these enter the reticular nuclei of the brain stem and via relay from the brain stem to the intralaminar nuclei of the thalamus, as is discussed in greater detail in the following chapter.

Characteristics of Transmission in the Anterolateral Pathway. In general, the same principles apply to transmission in the anterolateral pathway as in the dorsal column–lemniscal system except for the following differences: (1) the velocities of transmission are only one-third to one-half those in the dorsal column–lemniscal system, ranging between 8 and 40 m/sec; (2) the degree of spatial localization of signals is poor, especially in the pain pathways; (3) the gradations of intensities are also far less accurate, most of the sensations being recognized in 10 to 20 gradations of strength, rather than as many as 100 gradations for the dorsal column system; and (4) the ability to transmit rapidly repetitive signals is poor.

Thus, it is evident that the anterolateral system is a cruder type of transmission system than the dorsal column–lemniscal system. Even so, certain modalities of sensation are transmitted only in this system and not at all in the dorsal column–lemniscal system. These are pain, thermal, tickle and itch, and sexual sensations in addition to crude touch and pressure.

▪ SOME SPECIAL ASPECTS OF SOMATIC SENSORY FUNCTION

Function of the Thalamus in Somatic Sensation

When the somatosensory cortex of a human being is destroyed, that person loses most critical tactile sensibilities, but a slight degree of crude tactile sensibility does return. Therefore, it must be assumed that the thalamus (as well as other lower centers) has a slight ability to discriminate tactile sensation even though the thalamus normally functions mainly to relay this type of information to the cortex.

On the other hand, loss of the somatosensory cortex has little effect on one's perception of pain sensation and only a moderate effect on the perception of temperature. Therefore, there is much reason to believe that the brain stem, the thalamus, and other associated basal regions of the brain play perhaps the dominant role in discrimination of these sensibilities. It is interesting that these sensibilities appeared very early in the phylogenetic development of animalhood, whereas the critical tactile sensibilities were a late development.

CORTICAL CONTROL OF SENSORY SENSITIVITY — "CORTICOFUGAL" SIGNALS

In addition to somatic sensory signals transmitted from the periphery to the brain, "corticofugal" signals are transmitted in the backward direction from the cerebral cortex to the lower sensory relay stations of the thalamus, medulla, and spinal cord; these control the sensitivity of the sensory input. The corticofugal signals are inhibitory, so that when the input intensity becomes too great, the corticofugal signals automatically decrease the transmission in the relay nuclei. Obviously, this does two things: First, it decreases lateral spread of the sensory signals into adjacent neurons and therefore increases the contrast of the signal pattern. Second, it keeps the sensory system operating in a range of sensitivity that is not so low that the signals are ineffectual nor so high that the system is swamped beyond its capacity to differentiate sensory patterns.

This principle of corticofugal sensory control is utilized by all of the different sensory systems, not only the somatic system, as we shall see in subsequent chapters.

SEGMENTAL FIELDS OF SENSATION — THE DERMATOMES

Each spinal nerve innervates a "segmental field" of the skin called a *dermatome*. The different dermatomes are illustrated in Figure 9–14. However, these are shown as if there were distinct borders between the adjacent dermatomes, which is far from true, because much overlap exists from segment to segment.

The figure shows that the anal region of the body lies in the dermatome of the most distal cord segment. In the embryo, this is the tail region and is the most distal portion of the body. The legs develop from the lumbar and upper sacral segments, rather than from the distal sacral segments, which is evident from the dermatomal map. Obviously, one can use a dermatomal map such as that illustrated in Figure 9–14 to determine the level in the spinal cord at which various cord injuries may have occurred when the peripheral sensations are disturbed.

REFERENCES

Akil, H., and Lewis, J. W. (eds.): Neurotransmitters and Pain Control. New York, S. Karger Publishers, Inc., 1987.
Akil, H., et al.: Endogenous opioids: Etiology and function. Annu. Rev. Neurosci., 7:223, 1984.
American Physiological Society: Sensory Processes. Washington, D.C., American Physiological Society, 1984.
Amit, Z., and Galina, Z. H.: Stress-induced analgesia: Adaptive pain suppression. Physiol. Rev., 68:1091, 1988.
Basbaum, A. I., and Fields, H. L.: Endogenous pain control systems: Brainstem spinal pathways and endorphin circuitry. Annu. Rev. Neurosci., 7:309, 1984.
Berger, P. A., et al.: Behavioral pharmacology of the endorphins. Annu. Rev. Med., 33:397, 1982.
Besson, J. M., and Chaouch, A.: Peripheral and spinal mechanisms of nociception. Physiol. Rev., 67:67, 1987.
Bond, M. R.: Pain — Its Nature, Analysis, and Treatment. New York, Churchill Livingstone, 1984.
Darian-Smith, I.: The sense of touch: Performance and peripheral neural processes. In Darian-Smith I. (ed.): Handbook of Physiology. Sec. 1, Vol. III. Bethesda, Md., American Physiological Society, 1984, p. 739.
Darian-Smith, I.: Thermal sensibility. In Darian-Smith, I. (ed.): Handbook of Physiology. Sec. 1, Vol. III. Bethesda, Md., American Physiological Society, 1984, p. 879.
Dubner, R., and Bennett, G. J.: Spinal and trigeminal mechanisms of nociception. Annu. Rev. Neurosci., 6:381, 1983.
Emmers, R.: Somesthetic System of the Rat. New York, Raven Press, 1988.
Fields, H. L. (ed.): Pain: Mechanisms and Management. New York, McGraw-Hill Book Co., 1987.
Foreman, R. D., and Blair, R. W.: Central organization of sympathetic cardiovascular response to pain. Annu. Rev. Physiol., 50:607, 1988.
Friedhoff, A. J., and Miller, J. C.: Clinical implications of receptor sensitivity modification. Annu. Rev. Neurosci., 6:121, 1983.
Gelmers, H. J.: Calcium-channel blockers in the treatment of migraine. Am. J. Cardiol., 55:139B, 1985.
Goldman-Rakic, P. S.: Topography of cognition: Parallel distributed networks in primate association cortex. Annu. Rev. Neurosci., 11:137, 1988.
Goldstein, E. B.: Sensation and Perception. Belmont, Calif., Wadsworth Publishing Co., 1980.
Guyton, A. C., and Reeder, R. C.: Pain and contracture in poliomyelitis. Arch. Neurol. Psychiatr., 63:954, 1950.
Haft, J. I. (ed.): Differential Diagnosis of Chest Pain and Other Cardiac Symptoms. Mt. Kisco, N.Y., Futura Publishing Co., 1983.
Han, J. S., and Terenius, L.: Neurochemical basis of acupuncture analgesia. Annu. Rev. Pharmacol. Toxicol., 22:193, 1982.
Hnik, P., et al. (eds.): Mechanoreceptors. Development, Structure and Function. New York, Plenum Publishing Corp., 1988.
Hochberg, J.: Perception. In Darian-Smith, I. (ed.): Handbook of Physiology. Sec. 1, Vol. III. Bethesda, Md., American Physiological Society, 1984, p. 75.
Hyvarinen, J.: Posterior parietal lobe of the primate brain. Physiol. Rev., 62:1060, 1982.
Iggo, A., et al. (eds.): Nociception and Pain. New York, Cambridge University Press, 1986.
Jung, R.: Sensory research in historical perspective: Some philosophical foundations of perception. In Darian-Smith, I. (ed.): Handbook of Physiology. Sec. 1, Vol. III. Bethesda, Md., American Physiological Society, 1984, p. 1.
Kaas, J. H.: What, if anything, is SI? Organization of first somatosensory area of cortex. Physiol. Rev., 63:206, 1983.
Kaas, J. H., et al.: The reorganization of the somatosensory cortex following peripheral nerve damage in adult and developing mammals. Annu. Rev. Neurosci., 6:325, 1983.
Kruger, L. (ed.): Neural Mechanisms of Pain. New York, Raven Press, 1984.
Lewis, R. V., and Stern, A. S.: Biosynthesis of the enkephalins and enkephalin-containing polypeptides. Annu. Rev. Pharmacol. Toxicol., 23:353, 1983.
Loewenstein, E. R.: Excitation and inactivation in the receptor membrane. Ann. N.Y. Acad. Sci., 94:510, 1961.
Lucente, F. E., and Cooper, B. C.: Management of Facial, Head and Neck Pain. Philadelphia, W. B. Saunders Co., 1989.
Lund, J. S. (ed.): Sensory Processing in the Mammalian Brain. New York, Oxford University Press, 1988.

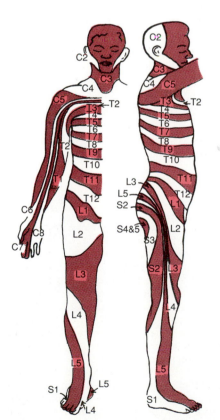

Figure 9–14. The dermatomes. (Modified from Grinker and Sahs: Neurology. Springfield, Ill., Charles C Thomas, 1966.)

McCloskey, D. I.: Kinesthetic sensibility. Physiol. Rev., 58:763, 1978.

Mountcastle, V. B.: Central nervous mechanisms in mechanoreceptive sensibility. In Darian-Smith, I. (ed.): Handbook of Physiology. Sec. 1, Vol. III. Bethesda, Md., American Physiological Society, 1984, p. 789.

Nakanishi, S.: Substance P precursor and kininogen: Their structures, gene organizations, and regulation. Physiol. Rev., 67:1117, 1987.

Neff, W. D. (ed.): Contributions to Sensory Physiology. New York, Academic Press, 1982.

Paintal, A. S.: The visceral sensations—some basic mechanisms. Prog. Brain Res., 67:3, 1986.

Paris, P. M., and Stewart, R. D.: Pain Management in Emergency Medicine. East Norwalk, Conn., Appleton & Lange, 1988.

Perl, E. R.: Pain and nociception. In Darian-Smith, I. (ed.): Handbook of Physiology. Sec. 1, Vol. III. Bethesda, Md., American Physiological Society, 1984, p. 915.

Porter, R. (ed.): Studies in Neurophysiology. New York, Cambridge University Press, 1978.

Price, D. D.: Psychological and Neural Mechanisms of Pain. New York, Raven Press, 1988.

Saper, J. P. (ed.): Controversies and Clinical Variants of Migraine. New York, Pergamon Press, 1987.

Scheibel, A. B.: The brain stem reticular core and sensory function: In Darian-Smith, I. (ed.): Handbook of Physiology. Sec. 1, Vol. III. Bethesda, Md., American Physiological Society, 1984, p. 213.

Stebbins, W. C., et al.: Sensory function in animals. In Darian-Smith, I. (ed.): Handbook of Physiology. Sec. 1, Vol. III. Bethesda, Md., American Physiological Society, 1984, p. 123.

Tollison, C. D., et al.: Handbook of Chronic Pain Management. Baltimore, Williams & Wilkins, 1988.

Udin, S. B., and Fawcett, J. W.: Formation of topographic maps. Annu. Rev. Neurosci., 11:289, 1988.

Weiss, T. F.: Relation of receptor potentials of cochlear hair cells to spike discharges of cochlear neurons. Annu. Rev. Physiol., 46:247, 1984.

10

Somatic Sensations:

II. Pain, Headache, and Thermal Sensations

Many, if not most, ailments of the body cause pain. Furthermore, the ability to diagnose different diseases depends to a great extent on a doctor's knowledge of the different qualities of pain. For these reasons, the present chapter is devoted mainly to pain and to the physiologic basis of some of the associated clinical phenomena.

The Purpose of Pain. Pain is a protective mechanism for the body; it occurs whenever any tissues are being damaged, and it causes the individual to react to remove the pain stimulus. Even such simple activities as sitting for a long time on the ischia can cause tissue destruction because of lack of blood flow to the skin where the skin is compressed by the weight of the body. When the skin becomes painful as a result of the ischemia, the person normally shifts weight unconsciously. But a person who has lost the pain sense, such as after spinal cord injury, fails to feel the pain and therefore fails to shift. This very soon results in ulceration at the areas of pressure.

■ THE TWO TYPES OF PAIN AND THEIR QUALITIES— FAST PAIN AND SLOW PAIN

Pain has been classified into two different major types: *fast pain* and *slow pain*. Fast pain occurs within about 0.1 sec when a pain stimulus is applied, whereas slow pain begins only after a second or more and then increases slowly over many seconds and sometimes even minutes. During the course of this chapter we shall see that the conduction pathways for these two types of pain are different and that each of them has specific qualities.

Fast pain is also described by many alternate names, such as *sharp pain, pricking pain, acute pain, electric pain,* and others. This type of pain is felt when a needle is stuck into the skin or when the skin is cut with a knife, and this pain is also felt when the skin is subjected to electric shock. Fast, sharp pain is not felt in most of the deeper tissues of the body.

Slow pain also goes by multiple additional names such as *burning pain, aching pain, throbbing pain, nauseous pain,* and *chronic pain*. This type of pain is usually associated with *tissue destruction*. It can become excruciating and can lead to prolonged, unbearable suffering. It can occur both in the skin and in almost any deep tissue or organ.

We will learn later that the fast type of pain is transmitted through type $A\delta$ pain fibers, whereas the slow type of pain results from stimulation of the more primitive type C fibers.

■ THE PAIN RECEPTORS AND THEIR STIMULATION

All Pain Receptors Are Free Nerve Endings. The pain receptors in the skin and other tissues are all free nerve endings. They are widespread in the superficial layers of the *skin* and also in certain internal tissues, such as the *periosteum,* the *arterial walls,* the *joint surfaces,* and the *falx* and *tentorium* of the cranial vault. Most of the other deep tissues are not extensively supplied with pain endings but are weakly supplied; nevertheless, any widespread tissue damage can still summate to cause the slow-chronic-aching type of pain in these areas.

Three Different Types of Stimuli Excite Pain Receptors—Mechanical, Thermal, and Chemical. Most pain fibers can be excited by multiple types of stimuli. However, some fibers are more likely to respond to excessive mechanical stretch, others to extremes of heat or cold, and still others to specific chemicals in the tissues. These are classified respectively as

127

mechanical, thermal, and *chemical pain receptors.* In general, fast pain is elicited by the mechanical and thermal types of receptors, whereas slow pain can be elicited by all three types.

Some of the chemicals that excite the chemical type of pain receptors include *bradykinin, serotonin, histamine, potassium ions, acids, acetylcholine,* and *proteolytic enzymes.* In addition, *prostaglandins* enhance the sensitivity of pain endings, but do not directly excite them. The chemical substances are especially important in stimulating the slow, suffering type of pain that occurs following tissue injury.

Nonadapting Nature of Pain Receptors. In contrast to most other sensory receptors of the body, the pain receptors adapt very little and sometimes not at all. In fact, under some conditions, the excitation of the pain fibers becomes progressively greater as the pain stimulus continues. This increase in sensitivity of the pain receptors is called *hyperalgesia.*

One can readily understand the importance of this failure of pain receptors to adapt, for it allows them to keep the person apprised of a damaging stimulus that causes the pain as long as it persists.

Rate of Tissue Damage as a Cause of Pain. The average person first begins to perceive pain when the skin is heated above 45°C, as illustrated in Figure 10–1. This is also the temperature at which the tissues begin to be damaged by heat; indeed, the tissues are eventually completely destroyed if the temperature remains above this level indefinitely. Therefore, it is immediately apparent that pain resulting from heat is closely correlated with the ability of heat to damage the tissues.

Furthermore, the intensity of pain has also been closely correlated with the rate of tissue damage from causes other than heat—bacterial infection, tissue ischemia, tissue contusion, and so forth.

Special Importance of Chemical Pain Stimuli During Tissue Damage. Extracts from damaged tissues cause intense pain when injected beneath the normal skin. All the chemicals listed above that excite the chemical pain receptors are found in these extracts. However, the chemical that seems to be most painful of all is *bradykinin.* Therefore, many research workers have suggested that bradykinin might be the single agent most responsible for causing the pain associated with tissue damage. Also, the intensity of the pain felt correlates with the local increase in potassium ion concentration as well. And it should be remembered, too, that proteolytic enzymes can directly attack the nerve endings and excite pain by making their membranes more permeable to ions.

Release of the various chemical pain excitants not only stimulates the chemosensitive pain endings but also greatly decreases the threshold for stimulation of the mechanosensitive and thermosensitive pain receptors as well. A widely known example of this is the extreme pain caused by slight mechanical or heat stimuli following tissue damage by sunburn.

Tissue Ischemia as a Cause of Pain. When blood flow to a tissue is blocked, the tissue becomes very painful within a few minutes. And the greater the rate of metabolism of the tissue, the more rapidly the pain appears. For instance, if a blood pressure cuff is placed around the upper arm and inflated until the arterial blood flow ceases, exercise of the forearm muscles can cause severe muscle pain within 15 to 20 sec. In the absence of muscle exercise, the pain will not appear for 3 to 4 min.

One of the suggested causes of pain in ischemia is accumulation of large amounts of lactic acid in the tissues, formed as a consequence of the anaerobic metabolism (metabolism without oxygen) that occurs during ischemia. However, it is also possible that other chemical agents, such as bradykinin, proteolytic enzymes, and so forth, are formed in the tissues because of cell damage and that these, rather than lactic acid, stimulate the pain nerve endings.

Muscle Spasm as a Cause of Pain. Muscle spasm is also a very common cause of pain, and it is the basis of many clinical pain syndromes. This pain probably results partially from the direct effect of muscle spasm in stimulating mechanosensitive pain receptors. However, it possibly results also from the indirect effect of muscle spasm to compress the blood vessels and cause ischemia. Also, the spasm increases the rate of metabolism in the muscle tissue at the same time, thus making the relative ischemia even greater, creating ideal conditions for release of chemical pain-inducing substances.

■ THE DUAL TRANSMISSION OF PAIN SIGNALS INTO THE CENTRAL NERVOUS SYSTEM

Even though all pain endings are free nerve endings, these endings utilize two separate pathways for trans-

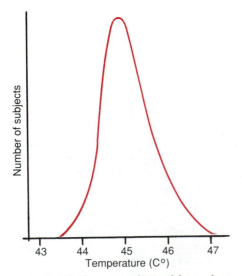

Figure 10–1. Distribution curve obtained from a large number of subjects for the minimal skin temperature that causes pain. (Modified from Hardy: J. Chronic Dis., 4:22, 1956.)

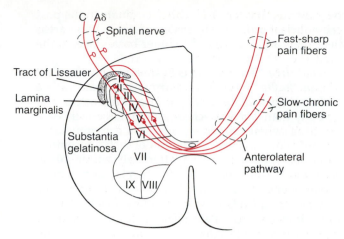

Figure 10–2. Transmission of both "acute-sharp" and "slow-chronic" pain signals into and through the spinal cord on the way to the brain stem.

mitting pain signals into the central nervous system. The two pathways correspond to the two different types of pain, a *fast-sharp pain pathway* and a *slow-chronic pain pathway*.

The Peripheral Pain Fibers — "Fast" and "Slow" Fibers. The fast-sharp pain signals are transmitted in the peripheral nerves to the spinal cord by small type $A\delta$ fibers at velocities of between 6 and 30 m/sec. On the other hand, the slow-chronic type of pain is transmitted by type C fibers at velocities of between 0.5 and 2 m/sec. When the type $A\delta$ fibers are blocked without blocking the C fibers by moderate compression of the nerve trunk, the fast-sharp pain disappears. On the other hand, when the type C fibers are blocked without blocking the δ fibers by low concentrations of local anesthetic, the slow-chronic-aching type of pain disappears.

Because of this double system of pain innervation, a sudden onset of painful stimulus gives a "double" pain sensation: a fast-sharp pain followed a second or so later by a slow, burning pain. The sharp pain apprises the person very rapidly of a damaging influence and, therefore, plays an important role in making the person react immediately to remove himself or herself from the stimulus. On the other hand, the slow, burning sensation tends to become more and more painful over a period of time. This sensation eventually gives one the intolerable suffering of long-continued pain.

On entering the spinal cord from the dorsal spinal roots, the pain fibers ascend or descend one to three segments in the *tract of Lissauer*, which lies immediately posterior to the dorsal horn of the cord gray matter. Then they terminate on neurons in the dorsal horns. However, here again, there are two systems for processing the pain signals on their way to the brain, as illustrated in Figures 10–2 and 10–3. These are as follows:

DUAL PAIN PATHWAYS IN THE CORD AND BRAIN STEM — THE NEOSPINOTHALAMIC TRACT AND THE PALEOSPINOTHALAMIC TRACT

On entering the spinal cord, the pain signals take two different pathways to the brain, through the *neospi-nothalamic tract* and through the *paleospinothalamic tract*.

The Neospinothalamic Tract for Fast Pain. The "fast" type $A\delta$ pain fibers transmit mainly mechanical and thermal pain. They terminate mainly in lamina I (lamina marginalis) of the dorsal horns and there excite second-order neurons of the neospinothalamic tract. These give rise to long fibers that cross immediately to the opposite side of the cord through the anterior commissure and then pass upward to the brain in the anterolateral columns.

Termination of the Neospinothalamic Tract in the Brain Stem and Thalamus. A few fibers of the neo-spinothalamic tract terminate in the reticular areas of the brain stem, but most pass all the way to the thalamus, terminating in the *ventrobasal complex* along with the dorsal column – medial lemniscal tract discussed in

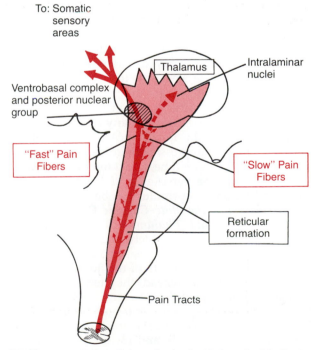

Figure 10–3. Transmission of pain signals into the hindbrain, thalamus, and cortex via the fast "pricking pain" pathway and the slow "burning pain" pathway.

the previous chapter. A few also terminate in the posterior nuclear group of the thalamus. From these areas the signals are transmitted to other basal areas of the brain and to the somatic sensory cortex.

Capability of the Nervous System to Localize Fast Pain in the Body. The fast-sharp type of pain can be localized much more exactly in the different parts of the body than can slow-chronic pain. However, even fast pain, when only pain receptors are stimulated without simultaneous stimulation of tactile receptors, is still quite poorly localized, often only within 10 cm or so of the stimulated area. Yet when tactile receptors are also stimulated, the localization can be very exact.

The Paleospinothalamic Pathway for Transmitting Slow-Chronic Pain. The paleospinothalamic pathway is a much older system, and transmits pain mainly carried in the peripheral slow-suffering type C pain fibers, though it does also transmit some signals from type Aδ fibers as well. In this pathway, the peripheral fibers terminate almost entirely in laminae II and III of the dorsal horns, which together are called the _substantia gelatinosa_, as illustrated by the lateralmost dorsal root fiber in Figure 10–2. Most of the signals then pass through one or more additional short-fiber neurons within the dorsal horns themselves before entering mainly lamina V, also in the dorsal horn. Here the last neuron in the series gives rise to long axons that mostly join the fibers from the fast pathway, passing through the anterior commissure to the opposite side of the cord, then upward to the brain in the same anterolateral pathway. However, a few of these fibers do not cross but instead pass ipsilaterally to the brain.

Substance P, the Probable Neurotransmitter of the Type C Nerve Endings. Where the type C fibers synapse in the dorsal horns of the spinal cord, they are believed to release substance P as the synaptic transmitter. Substance P is a neuropeptide; and as is true of all neuropeptides, it is slow to build up at the synapse and also slow to be destroyed. Therefore, its concentration at the synapse is believed to increase for at least several seconds, and perhaps much longer, after pain stimulation begins. After the pain is over, the substance P probably persists for many more seconds or perhaps minutes. The importance of this is that it might explain the progressive increase in intensity of slow-chronic pain with time, and it might also explain at least partially the persistence of this type of pain even after the painful stimulus has been removed.

Termination of the Slow-Chronic Pain Signals in the Brain Stem and Thalamus. The slow-chronic pathway terminates very widely in the brain stem, in the large pink-shaded area illustrated in Figure 10–3. Only one-tenth to one-fourth of the fibers pass all the way to the thalamus. Instead, they terminate principally in one of three different areas: (1) the _reticular nuclei_ of the medulla, pons, and mesencephalon; (2) the _tectal area_ of the mesencephalon deep to the superior and inferior colliculi; and (3) the _periaqueductal gray region_ surrounding the aqueduct of Sylvius. These lower regions of the brain appear to be very important in the appreciation of the suffering types of pain, for

animals with their brains sectioned above the mesencephalon evince undeniable evidence of suffering when any part of the body is traumatized.

From the reticular area of the brain stem, multiple short-fiber neurons relay the pain signals upward into the intralaminar nuclei of the thalamus and also into certain portions of the hypothalamus and other adjacent regions of the basal brain.

Capability of the Nervous System to Localize Pain Transmitted in the Slow-Chronic Pathway. Localization of pain transmitted in the paleospinothalamic pathway is very poor. In fact, electrophysiological studies suggest that the localization is often only to a major part of the body such as to one limb but not to a detailed point on the limb. This is in keeping with the multisynaptic, diffuse connectivity to the brain. It also explains why patients often have serious difficulty in localizing the source of some chronic types of pain.

Function of the Reticular Formation, Thalamus, and Cerebral Cortex in the Appreciation of Pain. Complete removal of the somatic sensory areas of the cerebral cortex does not destroy one's ability to perceive pain. Therefore, it is likely that pain impulses entering the reticular formation, thalamus, and other lower centers can cause conscious perception of pain. However, this does not mean that the cerebral cortex has nothing to do with normal pain appreciation; indeed, electrical stimulation of the cortical somatic sensory areas causes a person to perceive mild pain in approximately 3 per cent of the different points stimulated. It is believed that the cortex plays an important role in interpreting the quality of pain even though pain perception might be a function of lower centers.

Special Capability of Pain Signals to Arouse the Nervous System. Electrical stimulation in the reticular areas of the brain stem and also in the intralaminar nuclei of the thalamus, the areas where the slow-suffering type of pain terminates, has a strong arousal effect on nervous activity throughout the brain. In fact, these two areas are parts of the brain's principal arousal system, which is discussed in Chapter 21. This explains why a person with severe pain is frequently strongly aroused, and it also explains why it is almost impossible for a person to sleep when he or she is subjected to pain.

Surgical Interruption of Pain Pathways. Often a person has such severe and intractable pain (often resulting from rapidly spreading cancer) that it is necessary to relieve the pain. To do this the pain pathway can be destroyed at any one of several different points. If the pain is in the lower part of the body, a _cordotomy_ in the upper thoracic region often relieves the pain for a few weeks to a few months. To do this, the spinal cord on the side opposite the pain is sectioned almost entirely through its anterolateral quadrant, which interrupts the anterolateral sensory pathway.

Unfortunately, though, the cordotomy is not always successful in relieving the pain for two reasons. First, many of the pain fibers from the upper part of the body do not cross to the opposite side of the spinal cord until they have reached the brain, so that the cordotomy does not transect these fibers. Second, pain frequently returns several months later, perhaps caused partly by sensitization of other pain pathways and partly by stimulation by fibrous tissue from the remaining fibers. This new pain is often even more objectionable than the original pain.

Another operative procedure to relieve pain is to place lesions in the intralaminar nuclei in the thalamus, which often relieve the suffering type of pain while leaving intact one's appreciation of "acute" pain, an important protective mechanism.

■ A PAIN CONTROL ("ANALGESIA") SYSTEM IN THE BRAIN AND SPINAL CORD

The degree to which each person reacts to pain varies tremendously. This results partly from the capability of the brain itself to control the degree of input of pain signals to the nervous system by activation of a pain control system, called an *analgesia system.*

The analgesia system is illustrated in Figure 10–4. It consists of three major components (plus other accessory components): (1) the *periaqueductal gray area* of the mesencephalon and upper pons surrounding the aqueduct of Sylvius; neurons from this area send their signals to (2) the *raphe magnus nucleus,* a thin midline nucleus located in the lower pons and upper medulla. From here the signals are transmitted down the dorsolateral columns in the spinal cord to (3) a *pain inhibitory complex located in the dorsal horns of the spinal cord.* At this point the analgesia signals can block the pain before it is relayed on to the brain.

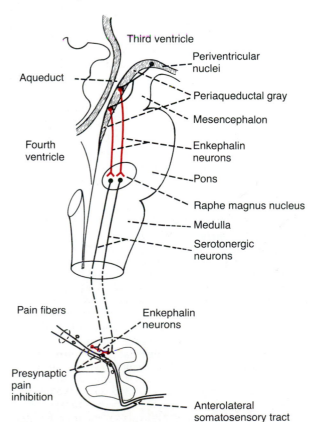

Figure 10–4. The analgesia system of the brain and spinal cord, showing inhibition of incoming pain signals at the cord level.

Electrical stimulation either in the periaqueductal gray area or in the raphe magnus nucleus can almost completely suppress many very strong pain signals entering by way of the dorsal spinal roots. Also, stimulation of areas at still higher levels of the brain that in turn excite the periaqueductal gray, especially the *periventricular nuclei in the hypothalamus* lying adjacent to the third ventricle and to a lesser extent the *medial forebrain bundle* also in the hypothalamus, can suppress pain, though perhaps not by quite so much.

Several different transmitter substances are involved in the analgesia system; especially involved are *enkephalin* and *serotonin.* Many of the nerve fibers derived from both periventricular nuclei and the periaqueductal gray area secrete enkephalin at their endings. Thus, as shown in Figure 10–4, the endings of many of the fibers in the raphe magnus nucleus release enkephalin. The fibers originating in this nucleus but terminating in the dorsal horns of the spinal cord secrete serotonin at their endings. The serotonin in turn acts on still another set of local cord neurons that are believed to secrete enkephalin. Enkephalin is believed to cause *presynaptic inhibition* of both incoming type C and type Aδ pain fibers where they synapse in the dorsal horns. It probably does this by blocking calcium channels in the membranes of the nerve terminals. Because it is calcium ions that cause release of transmitter at the synapse, such calcium blockage would obviously result in presynaptic inhibition. Furthermore, the blockage appears to last for prolonged periods of time, because after activating the analgesia system, analgesia often lasts for many minutes or even for hours.

Thus, the analgesia system can block pain signals at the initial entry point to the spinal cord. In fact, it can also block many of the local cord reflexes that result from pain signals, especially the withdrawal reflexes, which are described in Chapter 16.

It is probable that this analgesia system can also inhibit pain transmission at other points in the pain pathway, especially in the reticular nuclei in the brain stem and in the intralaminar nuclei of the thalamus.

The Brain's Opiate System — The Endorphins and Enkephalins

More than 20 years ago it was discovered that injection of extremely minute quantities of morphine either into the periventricular nucleus around the third ventricle of the diencephalon or into the periaqueductal gray area of the brain stem will cause an extreme degree of analgesia. In subsequent studies, it has now been found that morphine acts at still many other points in the analgesia system, including the dorsal horns of the spinal cord. Because most drugs that alter the excitability of neurons do so by acting on synaptic receptors, it was assumed that the "morphine receptors" of the analgesia system must in fact be receptors for some morphine-like neurotransmitter that is naturally secreted in the brain. Therefore, an extensive search was set into motion for a natural opiate of the brain. About a dozen such opiate-like substances have now been found in different points of the nervous system, but all are breakdown products of three large protein molecules: *proopiomelanocortin, proenkephalin,* and *prodynorphin.* Furthermore, multiple areas of the brain have been shown to have opiate receptors, especially the areas in the analgesia system. Among the more important of the opiate substances are *β-endorphin, met-enkephalin, leu-enkephalin,* and *dynorphin.*

The two enkephalins are found in the portions of the analgesia system described earlier, and β-endorphin is present both in the hypothalamus and in the pituitary gland. Dynorphin, though found in only minute quantities in nervous tissue, is important because it is an extremely powerful opiate, having 200 times as much pain-killing effect as morphine when injected directly into the analgesia system.

Thus, although all the fine details of the brain's opiate system are not yet entirely understood, nevertheless activation of the analgesia system either by nervous signals entering the periaqueductal gray area or by morphine-like drugs can totally or almost totally suppress many pain signals entering through the peripheral nerves.

INHIBITION OF PAIN TRANSMISSION BY TACTILE SENSORY SIGNALS

Another important landmark in the saga of pain control was the discovery that stimulation of large sensory fibers from the peripheral tactile receptors depresses the transmission of pain signals either from the same area of the body or even from areas sometimes located many segments away. This effect presumably results from a type of local lateral inhibition. It explains why such simple maneuvers as rubbing the skin near painful areas is often very effective in relieving pain. And it probably also explains why liniments are often useful in the relief of pain. This mechanism and simultaneous psychogenic excitation of the central analgesia system are probably also the basis of pain relief by acupuncture.

TREATMENT OF PAIN BY ELECTRICAL STIMULATION

Several clinical procedures have been developed recently for suppressing pain by electrical stimulation of large sensory nerve fibers. The stimulating electrodes are placed on selected areas of the skin, or on occasion they have been implanted over the spinal cord to stimulate the dorsal sensory columns.

In a few patients, electrodes have also been placed stereotactically in the intralaminar nuclei of the thalamus or in the periventricular or periaqueductal area of the diencephalon. The patient can then personally control the degree of stimulation. Dramatic relief has been reported in some instances. Also, the pain relief often lasts for as long as 24 hours after only a few minutes of stimulation.

■ REFERRED PAIN

Often a person feels pain in a part of his or her body that is considerably removed from the tissues causing the pain. This pain is called *referred pain.* Usually the

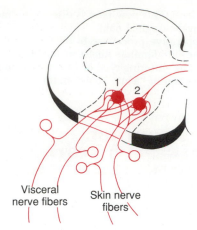

Figure 10–5. Mechanism of referred pain and referred hyperalgesia.

pain is initiated in one of the visceral organs and referred to an area on the body surface. Also, pain may be referred to another deep area of the body not exactly coincident with the location of the viscus producing the pain. A knowledge of these different types of referred pain is extremely important in clinical diagnosis because many visceral ailments cause no other signs except referred pain.

Mechanism of Referred Pain. Figure 10–5 illustrates the most likely mechanism by which most pain is referred. In the figure, branches of visceral pain fibers are shown to synapse in the spinal cord with some of the same second-order neurons that receive pain fibers from the skin. When the visceral pain fibers are stimulated, pain signals from the viscera are then conducted through at least some of the same neurons that conduct pain signals from the skin, and the person has the feeling that the sensations actually originate in the skin itself.

■ VISCERAL PAIN

In clinical diagnosis, pain from the different viscera of the abdomen and chest is one of the few criteria that can be used for diagnosing visceral inflammation, disease, and other ailments. In general, the viscera have sensory receptors for no other modalities of sensation besides pain. Also, visceral pain differs from surface pain in several important aspects.

One of the most important differences between surface pain and visceral pain is that highly localized types of damage to the viscera rarely cause severe pain. For instance, a surgeon can cut the gut entirely in two in a patient who is awake without causing significant pain. On the other hand, any stimulus that causes *diffuse stimulation of pain nerve endings* throughout a viscus causes pain that can be extremely severe. For instance, ischemia caused by occluding the blood supply to a large area of gut stimulates many diffuse pain fibers at the same time and can result in extreme pain.

CAUSES OF TRUE VISCERAL PAIN

Any stimulus that excites pain nerve endings in diffuse areas of the viscera causes visceral pain. Such stimuli include ischemia of visceral tissue, chemical damage to the surfaces of the viscera, spasm of the smooth muscle in a hollow viscus, distension of a hollow viscus, or stretching of the ligaments.

Essentially all the true visceral pain originating in the thoracic and abdominal cavities is transmitted through sensory nerve fibers that run in the sympathetic nerves. These fibers are small type C fibers and, therefore, can transmit only the chronic-aching-suffering type of pain.

Ischemia. Ischemia causes visceral pain in exactly the same way that it does in other tissues, presumably because of the formation of acidic metabolic end products or tissue-degenerative products, such as bradykinin, proteolytic enzymes, or others that stimulate the pain nerve endings.

Chemical Stimuli. On occasion, damaging substances leak from the gastrointestinal tract into the peritoneal cavity. For instance, proteolytic acidic gastric juice often leaks through a ruptured gastric or duodenal ulcer. This juice causes widespread digestion of the visceral peritoneum, thus stimulating extremely broad areas of pain fibers. The pain is usually extremely severe.

Spasm of a Hollow Viscus. Spasm of the gut, the gallbladder, a bile duct, the ureter, or any other hollow viscus can cause pain, possibly by mechanical stimulation of the pain endings. Or its cause might be diminished blood flow to the muscle combined with increased metabolic need of the muscle for nutrients. Thus, *relative* ischemia could develop, which causes severe pain.

Often, pain from a spastic viscus occurs in the form of *cramps,* the pain increasing to a high degree of severity and then subsiding, this process continuing rhythmically once every few minutes. The rhythmic cycles result from rhythmic contraction of smooth muscle. For instance, each time a peristaltic wave travels along an overly excitable spastic gut, a cramp occurs. The cramping type of pain frequently occurs in gastroenteritis, constipation, menstruation, parturition, gallbladder disease, or ureteral obstruction.

Overdistension of a Hollow Viscus. Extreme overfilling of a hollow viscus also results in pain, presumably because of overstretch of the tissues themselves. However, overdistension can also collapse the blood vessels that encircle the viscus, or that pass into its wall, thus perhaps promoting ischemic pain.

Insensitive Viscera

A few visceral areas are almost entirely insensitive to pain of any type. These include the parenchyma of the liver and the alveoli of the lungs. Yet the liver *capsule* is extremely sensitive to both direct trauma and stretch, and the *bile ducts* are also sensitive to pain. In the lungs, even though the alveoli are insensitive, the *bronchi* and the *parietal pleura* are both very sensitive to pain.

PARIETAL PAIN CAUSED BY VISCERAL DAMAGE

In addition to true visceral pain, pain sensations are also transmitted from the viscera through nonvisceral nerve fibers that innervate the parietal peritoneum, pleura, or pericardium.

When a disease affects a viscus, it often spreads to the parietal wall of the visceral cavity. This wall, like the skin, is supplied with extensive innervation from the spinal nerves, not from the sympathetic nerves. Therefore, pain from the parietal wall overlying the viscus is frequently very sharp. To emphasize the difference between this pain and true visceral

pain: a knife incision through the *parietal* peritoneum is very painful, even though a similar cut through the visceral peritoneum or through the gut is not very painful if at all.

LOCALIZATION OF VISCERAL PAIN — THE "VISCERAL" AND THE "PARIETAL" TRANSMISSION PATHWAYS

Pain from the different viscera is frequently difficult to localize for a number of reasons. First, the brain does not know from firsthand experience that the different organs exist, and, therefore, any pain that originates internally can be localized only generally. Second, sensations from the abdomen and thorax are transmitted through two separate pathways to the central nervous system — the *true visceral pathway* and the *parietal pathway.* The true visceral pain is transmitted via sensory fibers of the autonomic nervous system (both sympathetic and parasympathetic), and the sensations are *referred* to surface areas of the body often far from the painful organ. On the other hand, parietal sensations are conducted *directly* into the local spinal nerves from the parietal peritoneum, pleura, or pericardium, and the sensations are usually *localized directly over the painful area.*

Localization of Referred Pain Transmitted in the Visceral Pathways. When visceral pain is referred to the surface of the body, the person generally localizes it in the dermatomal segment from which the visceral organ originated in the embryo. For instance the heart originated in the neck and upper thorax, so that the heart's visceral pain fibers pass into the cord between segments C-3 and T-5. Therefore, as illustrated in Figure 10–6, pain from the heart is referred to the side of the neck, over the shoulder, over the pectoral muscles, down the arm, and into the substernal area of the chest. Most frequently, the pain is on the left side, rather than on the right — because the left side of the heart is much more frequently involved in coronary disease than the right.

The stomach originated approximately from the seventh to the ninth thoracic segments of the embryo. Therefore, stomach pain is referred to the anterior epigastrium above the umbilicus, which is the surface area of the body subserved by the seventh through ninth thoracic segments. And Figure 10–6 shows several other surface areas to which vis-

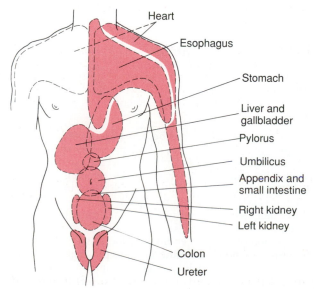

Figure 10–6. Surface areas of referred pain from different visceral organs.

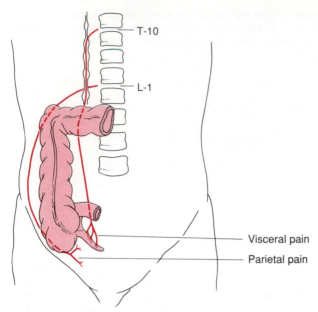

Figure 10 – 7. Visceral and parietal transmission of pain from the appendix.

ceral pain is referred from other organs, representing in general the areas in the embryo from which the respective organ originated.

The Parietal Pathway for Transmission of Abdominal and Thoracic Pain. Pain from the viscera is frequently localized to two surface areas of the body at the same time because of the dual transmission of pain through the referred visceral pathway and the direct parietal pathway. Thus, Figure 10 – 7 illustrates dual transmission from an inflamed appendix. Impulses pass from the appendix through the sympathetic visceral pain fibers into the sympathetic chain and then into the spinal cord at approximately T-10 or T-11; this pain is referred to an area around the umbilicus and is of the aching, cramping type. On the other hand, pain impulses also often originate in the parietal peritoneum where the inflamed appendix touches or is adherent to the abdominal wall. These cause pain of the sharp type directly over the irritated peritoneum in the right lower quadrant of the abdomen.

■ SOME CLINICAL ABNORMALITIES OF PAIN AND OTHER SOMATIC SENSATIONS

HYPERALGESIA

A pain pathway sometimes becomes excessively excitable; this gives rise to *hyperalgesia,* which means hypersensitivity to pain. The basic causes of hyperalgesia are excessive sensitivity of the pain receptors themselves, which is called *primary hyperalgesia,* and facilitation of sensory transmission, which is called *secondary hyperalgesia.*

An example of primary hyperalgesia is the extreme sensitivity of sunburned skin, which is believed to result from sensitization of the pain endings by local tissue products of the burn — perhaps histamine, perhaps prostaglandins, perhaps others. Secondary hyperalgesia frequently results from lesions in the spinal cord or in the thalamus. Several of these are discussed in subsequent sections.

THE THALAMIC SYNDROME

Occasionally the posterolateral branch of the posterior cerebral artery, a small artery supplying the posteroventral portion of the thalamus, becomes blocked by thrombosis, so that the nuclei of this area of the thalamus degenerate, while the medial and anterior nuclei of the thalamus remain intact. The patient suffers a series of abnormalities, as follows: First, loss of almost all sensations from the opposite side of the body occurs because of destruction of the relay nuclei. Second, ataxia (inability to control movements precisely) may be evident because of loss of position and kinesthetic signals normally relayed through the thalamus to the cortex. Third, after a few weeks to a few months some sensory perception in the opposite side of the body returns, but strong stimuli are usually necessary to elicit this. When the sensations do occur, they are poorly — if at all — localized and are almost always very painful, sometimes lancinating, regardless of the type of stimulus applied to the body. Fourth, the person is likely to perceive many affective sensations of extreme unpleasantness or, rarely, extreme pleasantness; the unpleasant ones are often associated with emotional tirades.

The medial nuclei of the thalamus are not destroyed by thrombosis of the artery. Therefore, it is believed that these nuclei become facilitated and give rise to the enhanced sensitivity to pain transmitted through the reticular system as well as to the affective perceptions.

HERPES ZOSTER (SHINGLES)

Occasionally a herpes virus infects a dorsal root ganglion. This causes severe pain in the dermatomal segment normally subserved by the ganglion, thus eliciting a segmental type of pain that circles halfway around the body. The disease is called *herpes zoster,* or "shingles" because of the eruption described in the next paragraph.

The cause of the pain is presumably excitation of the neuronal cells of the dorsal root ganglion by the virus infection. Aside from causing pain, the virus is also carried by neuronal cytoplasmic flow outward through the peripheral axons to their cutaneous terminals. Here the virus causes a rash that vesiculates within a few days and within another few days crusts over, all of this occurring within the dermatomal area served by the infected dorsal root.

TIC DOULOUREUX

Lancinating pains occur in some persons over one side of the face in part of the sensory distribution area of the fifth or ninth nerve; this phenomenon is called *tic douloureux* (or *trigeminal neuralgia* or *glossopharyngeal neuralgia*). The pains feel like sudden electric shocks, and they may appear for only a few seconds at a time or they may be almost continuous. Often, they are set off by exceedingly sensitive "trigger areas" on the surface of the face, in the mouth, or in the throat — almost always by a mechanoreceptive stimulus instead of a pain stimulus. For instance, when the patient swallows a bolus of food, as the food touches a tonsil it might set off a severe lancinating pain in the mandibular portion of the fifth nerve.

The pain of tic douloureux can usually be blocked by cutting the peripheral nerve from the hypersensitive area. The

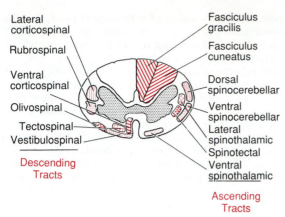

Figure 10–8. Cross-section of the spinal cord, showing principal ascending tracts on the right and principal descending tracts on the left.

sensory portion of the fifth nerve is often sectioned immediately inside the cranium, where the motor and sensory roots of the fifth nerve can be separated so that the motor portions, which are needed for many of the jaw movements, are spared while the sensory elements are destroyed. Obviously, this operation leaves the side of the face anesthetic, which in itself may be annoying. Furthermore, it is sometimes unsuccessful, indicating that the lesion that causes the pain is in the sensory nucleus in the brain stem and not in the peripheral nerves.

THE BROWN-SÉQUARD SYNDROME

Obviously, if the spinal cord is transected entirely, all sensations and motor functions distal to the segments of transection are blocked, but if only one half of the spinal cord is transected on a single side, the so-called Brown-Séquard syndrome occurs. The following effects of such a transection occur, and these can be predicted from a knowledge of the cord fiber tracts illustrated in Figure 10–8: All motor functions are blocked on the side of the transection in all segments below the level of the transection. Yet only some of the modalities of sensation are lost on the transected side, and others are lost on the opposite side. The sensations of pain, heat, and cold are lost on the opposite side of the body in all dermatomes two to six segments below the level of the transection. The sensations that are transmitted only in the dorsal and dorsolateral columns—kinesthetic and position sensations, vibration sensation, discrete localization, and two-point discrimination—are lost entirely on the side of the transection in all dermatomes below the level of the transection. Touch is impaired on the side of the transection because the principal pathway for transmission of light touch, the dorsal columns, is transected. Yet "crude touch," which is poorly localized, still persists because of transmission in the opposite ventral spinothalamic tract.

■ HEADACHE

Headaches are actually referred pain to the surface of the head from the deep structures. Many headaches result from pain stimuli arising inside the cranium, but others result from pain arising outside the cranium, such as from the nasal sinuses.

HEADACHE OF INTRACRANIAL ORIGIN

Pain-Sensitive Areas in the Cranial Vault. The brain itself is almost totally insensitive to pain. Even cutting or electrically stimulating the somatic sensory areas of the cortex only occasionally causes pain; instead, it causes pins-and-needles types of paresthesias on the area of the body represented by the portion of the sensory cortex stimulated. Therefore, it is likely that much or most of the pain of headache is not caused by damage within the brain itself.

On the other hand, *tugging on the venous sinuses, damaging the tentorium,* or *stretching the dura at the base of the brain* can all cause intense pain that is recognized as headache. Also, almost any type of traumatizing, crushing, or stretching stimulus to the *blood vessels of the dura* can cause headache. A very sensitive structure is the middle meningeal artery, and neurosurgeons are careful to anesthetize this artery specifically when performing brain operations under local anesthesia.

Areas of the Head to Which Intracranial Headache Is Referred. Stimulation of pain receptors in the intracranial vault above the tentorium, including the upper surface of the tentorium itself, initiates impulses in the fifth nerve and, therefore, causes referred headache to the front half of the head in the area supplied by the fifth cranial nerve, as illustrated in Figure 10–9.

On the other hand, pain impulses from beneath the tentorium enter the central nervous system mainly through the second cervical nerve, which also supplies the scalp behind the ear. Therefore, subtentorial pain stimuli cause "occipital headache" referred to the posterior part of the head as shown in Figure 10–9.

Types of Intracranial Headache. *Headache of Meningitis.* One of the most severe headaches of all is that resulting from meningitis, which causes inflammation of all of the meninges, including the sensitive areas of the dura and the sensitive areas around the venous sinuses. Such intense damage as this can cause extreme headache pain referred over the entire head.

Headache Caused by Low Cerebrospinal Fluid Pressure. Removing as little as 20 ml of fluid from the spinal canal, particularly if the person remains in the upright position, often causes intense intracranial headache. Removing this quantity of fluid removes the flotation for the brain that is normally provided by the cerebrospinal fluid. Therefore, the

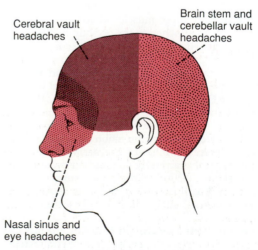

Figure 10–9. Areas of headache resulting from different causes.

weight of the brain stretches and otherwise distorts the various dural surfaces and thereby elicits the pain that causes the headache.

Migraine Headache. Migraine headache is a special type of headache that is thought to result from abnormal vascular phenomena, though the exact mechanism is unknown.

Migraine headaches often begin with various prodromal sensations, such as nausea, loss of vision in part of the field of vision, visual aura, or other types of sensory hallucinations. Ordinarily, the prodromal symptoms begin half an hour to an hour prior to the beginning of the headache itself. Therefore, any theory that explains migraine headache must also explain these prodromal symptoms.

One of the *theories* of the cause of migraine headaches is that prolonged emotion or tension causes reflex vasospasm of some of the arteries of the head, including arteries that supply the brain itself. The vasospasm theoretically produces ischemia of portions of the brain, and this is responsible for the prodromal symptoms. Then, as a result of the intense ischemia, something happens to the vascular wall to allow it to become flaccid and incapable of maintaining vascular tone for 24 to 48 hours. The blood pressure in the vessels causes them to dilate and pulsate intensely, and it is postulated that the excessive stretching of the walls of the arteries—including some extracranial arteries as well, such as the temporal artery—causes the actual pain of migraine headaches. However, it is possible that diffuse aftereffects of ischemia in the brain itself are at least partially if not mainly responsible for this type of headache.

Alcoholic Headache. As many people have experienced, a headache usually follows an alcoholic binge. It is most likely that alcohol, because it is toxic to tissues, directly irritates the meninges and causes intracranial pain.

Headache Caused by Constipation. Constipation causes headache in many persons. Because it has been shown that constipation headache can occur in persons whose spinal cords have been cut, we know that this headache is not caused by nervous impulses from the colon. Therefore, it possibly results from absorbed toxic products or from changes in the circulatory system resulting from loss of fluid into the gut.

EXTRACRANIAL TYPES OF HEADACHE

Headache Resulting from Muscular Spasm. Emotional tension often causes many of the muscles of the head, including especially those muscles attached to the scalp and the neck muscles attached to the occiput, to become spastic, and it is postulated that this is one of the common causes of headache. The pain of the spastic head muscles supposedly is referred to the overlying areas of the head and produces the same type of headache as do intracranial lesions.

Headache Caused by Irritation of the Nasal and Accessory Nasal Structures. The mucous membranes of the nose and also of all the nasal sinuses are sensitive to pain, but not intensely so. Nevertheless, infection or other irritative processes in widespread areas of the nasal structures usually cause headache that is referred behind the eyes or, in the case of frontal sinus infection, to the frontal surfaces of the forehead and scalp, as illustrated in Figure 10–9. Also, pain from the lower sinuses—such as the maxillary sinuses—can be felt in the face.

Headache Caused by Eye Disorders. Difficulty in focusing one's eyes clearly may cause excessive contraction of the ciliary muscles in an attempt to gain clear vision. Even though these muscles are extremely small, tonic contraction of them can be the cause of retro-orbital headache. Also, excessive attempts to focus the eyes can result in reflex spasm in various facial and extraocular muscles, which is also a possible cause of headache.

A second type of headache originating in the eyes occurs when the eyes are exposed to excessive irradiation by light rays, especially ultraviolet light. Watching the sun or the arc of an arc-welder for even a few seconds may result in headache that lasts from 24 to 48 hours. The headache sometimes results from "actinic" irritation of the conjunctivae, and the pain is referred to the surface of the head or retro-orbitally. However, intense light from an arc or the sun focused on the retina can actually burn the retina, and this could result in headache.

■ THERMAL SENSATIONS

THERMAL RECEPTORS AND THEIR EXCITATION

The human being can perceive different gradations of cold and heat, progressing from *freezing cold* to *cold* to *cool* to *indifferent* to *warm* to *hot* to *burning hot.*

Thermal gradations are discriminated by at least three different types of sensory receptors: the cold receptors, the warmth receptors, and pain receptors. The pain receptors are stimulated only by extreme degrees of heat or cold and therefore are responsible, along with the cold and warmth receptors, for "freezing cold" and "burning hot" sensations.

The cold and warmth receptors are located immediately under the skin at discrete but separated points, each having a stimulatory diameter of about 1 mm. In most areas of the body there are three to ten times as many cold receptors as warmth receptors, and the number in different areas of the body varies from as great as 15 to 25 cold points/cm² in the lips, to 3 to 5 cold points/cm² in the finger, to less than 1 cold point/cm² in some broad surface areas of the trunk. There are correspondingly fewer numbers of warmth points.

Although it is quite certain, on the basis of psychological tests, that there are distinctive warmth nerve endings, these have not yet been identified histologically. They are presumed to be free nerve endings because warmth signals are transmitted mainly over type C nerve fibers at transmission velocities of only 0.4 to 2 m/sec.

On the other hand, a definitive cold receptor has been identified. It is a special, small, type Aδ myelinated nerve ending that branches a number of times, the tips of which protrude into the bottom surfaces of basal epidermal cells. Signals are transmitted from these receptors via delta nerve fibers at velocities of up to about 20 m/sec. However, some cold sensations are also transmitted in type C nerve fibers, which suggests that some free nerve endings also might function as cold receptors.

Stimulation of Thermal Receptors—Sensations of Cold, Cool, Indifferent, Warm, and Hot. Figure 10–10 illustrates the effects of different temperatures on the responses of four different nerve fibers: (1) a

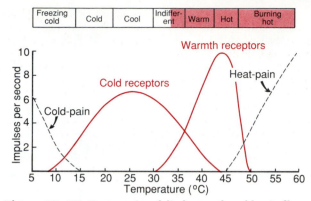

Figure 10–10. Frequencies of discharge of a cold-pain fiber, a cold fiber, a warmth fiber, and a heat-pain fiber. (The responses of these fibers are drawn from original data collected in separate experiments by Zotterman, Hensel, and Kenshalo.)

pain fiber stimulated by cold; (2) a cold fiber; (3) a warmth fiber; and (4) a pain fiber stimulated by heat. Note especially that these fibers respond differently at different levels of temperature. For instance, in the *very* cold region only the pain fibers are stimulated (if the skin becomes even colder so that it nearly freezes or actually does freeze, even these fibers cannot be stimulated). As the temperature rises to 10° to 15°C, pain impulses cease, but the cold receptors begin to be stimulated. Then, above about 30°C the warmth receptors become stimulated, and the cold receptors fade out at about 43°C. Finally, at around 45°C, pain fibers begin to be stimulated by heat.

One can understand from Figure 10–10, therefore, that a person determines the different gradations of thermal sensations by the relative degrees of stimulation of the different types of endings. One can understand also from this figure why extreme degrees of cold or heat can be painful and why both these sensations, when intense enough, may give almost exactly the same quality of sensation—that is, freezing cold and burning hot sensations feel almost alike; they are both very painful.

Stimulatory Effects of Rising and Falling Temperature — Adaptation of Thermal Receptors. When a cold receptor is suddenly subjected to an abrupt fall in temperature, it becomes strongly stimulated at first, but this stimulation fades rapidly during the first few seconds and progressively more slowly during the next half hour or more. In other words, the receptor "adapts" to a very great extent but does not appear to adapt 100 per cent.

Thus, it is evident that the thermal senses respond markedly to *changes in temperature* in addition to being able to respond to steady states of temperature. This means, therefore, that when the temperature of the skin is actively falling, a person feels much colder than when the temperature remains at the same level. Conversely, if the temperature is actively rising the person feels much warmer than he or she would at the same temperature if it were constant.

The response to changes in temperature explains the extreme degree of heat that one feels on first entering a tub of hot water and the extreme degree of cold felt on going from a heated room to the out-of-doors on a cold day.

Mechanism of Stimulation of the Thermal Receptors

It is believed that the cold and warmth receptors are stimulated by changes in their metabolic rates, these changes resulting from the fact that temperature alters the rates of intracellular chemical reactions more than twofold for each 10°C change. In other words, thermal detection probably results not from direct physical effects of heat or cold on the nerve endings, but instead from chemical stimulation of the endings as modified by the temperature.

Spatial Summation of Thermal Sensations. Because the number of cold or warmth endings in any one surface area of the body is very slight, it is difficult to judge gradations of temperature when small areas are stimulated. However, when a large area of the body is stimulated all at once, the thermal signals from the entire area summate. For instance, rapid changes in temperature of as little as 0.01°C can be detected if this change affects the entire surface of the body simultaneously. On the other hand, temperature changes 100 times this great might not be detected when the skin surface affected is only a square centimeter or so in size.

TRANSMISSION OF THERMAL SIGNALS IN THE NERVOUS SYSTEM

In general, thermal signals are transmitted in almost parallel, but not the same, pathways as pain signals. On entering the spinal cord, the signals travel for a few segments upward or downward in the *tract of Lissauer* and then terminate mainly in laminae I, II, and III of the dorsal horns—the same as for pain. After a small amount of processing by one or more cord neurons, the signals enter long, ascending thermal fibers that cross to the opposite anterolateral sensory tract and terminate in (1) the reticular areas of the brain stem and (2) the ventrobasal complex of the thalamus. A few thermal signals are also relayed to the somatic sensory cortex from the ventrobasal complex. Occasionally, a neuron in somatic sensory area I has been found by microelectrode studies to be directly responsive to either cold or warm stimuli in specific areas of the skin. Furthermore, it is known that removal of the postcentral gyrus in the human being reduces the ability to distinguish gradations of temperature.

REFERENCES

See references for Chapter 9.

V

THE CENTRAL NERVOUS SYSTEM:

B. The Special Senses

The Eye:

I. Optics of Vision

Before it is possible to understand the optical system of the eye, the student must be thoroughly familiar with the basic physical principles of optics, including the physics of refraction, a knowledge of focusing, depth of focus, and so forth. Therefore, a brief review of these physical principles is first presented, and then the optics of the eye is discussed.

REFRACTION OF LIGHT

The Refractive Index of a Transparent Substance. Light rays travel through air at a velocity of approximately 300,000 km/sec but much slower through transparent solids and liquids. The refractive index of a transparent substance is the *ratio* of the velocity of light in air to the velocity in the substance. Obviously, the refractive index of air itself is 1.00.

If light travels through a particular type of glass at a velocity of 200,000 km/sec, the refractive index of this glass is 300,000 divided by 200,000, or 1.50.

Refraction of Light Rays at an Interface Between Two Media With Different Refractive Indices. When light waves traveling forward in a beam, as shown in the upper part of Figure 11–1, strike an interface that is perpendicular to the beam, the waves enter the second refractive medium without deviating from their course. The only effect that occurs is decreased velocity of transmission and shorter wavelength. On the other hand, as illustrated in the lower part of the figure, if the light waves strike an angulated interface, the light waves bend if the refractive indices of the two media are different from each other. In this particular figure the light waves are leaving air, which has a refractive index of 1.00, and are entering a block of glass having a refractive index of 1.50. When the beam first strikes the angulated interface, the lower edge of the beam enters the glass ahead of the upper edge. The wave front in the upper portion of the beam continues to travel at a velocity of 300,000 km/sec, whereas that which has entered the glass travels at a velocity of 200,000 km/sec. This causes the upper portion of the wave front to move ahead of the lower portion so that the wave front is no longer vertical but is angulated to the right. Because *the direction in which light travels is always perpendicular to the plane of the wave front,* the direction of travel of the light beam now bends downward.

The bending of light rays at an angulated interface is known as *refraction.* Note particularly that the degree of refraction increases as a function of (1) the ratio of the two refractive indices of the two transparent media and (2) the degree of angulation between the interface and the entering wave front.

APPLICATION OF REFRACTIVE PRINCIPLES TO LENSES

The Convex Lens – Focusing of Light Rays. Figure 11–2 shows parallel light rays entering a convex lens. The light rays passing through the center of the lens strike the lens exactly perpendicular to the lens surface and therefore pass through the lens without being refracted at all. Toward either edge of the lens, however, the light rays strike a progressively more angulated interface. Therefore, the outer rays bend more and more toward the center. Half the bending occurs when the rays enter the lens and half as they exit from the opposite side. (At this time the student should pause and analyze why the rays still bend toward the center upon leaving the lens.)

Finally, if the lens is ground with exactly the proper curvature, parallel light rays passing through each part of the lens will be bent exactly enough so that all the rays will pass through a single point, which is called the *focal point.*

The Concave Lens. Figure 11–3 shows the effect of a concave lens on parallel light rays. The rays that enter the very center of the lens strike an interface that is absolutely perpendicular to the beam and, therefore, do not refract at all. The rays at the edge of the lens enter the lens ahead of the rays toward the center. This is opposite to the effect in the convex lens, and it causes the peripheral light rays to *diverge* away from the light rays that pass through the center of the lens.

Thus, the concave lens *diverges* light rays, whereas the convex lens *converges* light rays.

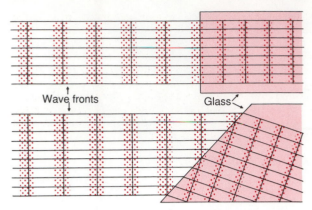

Figure 11–1. Wave fronts entering *(top)* a glass surface perpendicular to the light rays and *(bottom)* a glass surface angulated to the light rays. This figure illustrates that the distance between waves after they enter the glass is shortened to approximately two thirds that in air. It also illustrates that light rays striking an angulated glass surface are bent.

Cylindrical Lenses—Comparison With Spherical Lenses. Figure 11–4 illustrates both a convex *spherical* lens and a convex *cylindrical* lens. Note that the cylindrical lens bends light rays from the two sides of the lens but not from either the top or the bottom. Therefore, parallel light rays are bent to a *focal line*. On the other hand, light rays that pass through the spherical lens are refracted at all edges of the lens toward the central ray, and all the rays come to a *focal point*.

The cylindrical lens is well illustrated by a test tube full of water. If the test tube is placed in a beam of sunlight and a piece of paper is brought progressively closer to the opposite side of the tube, a certain distance will be found at which the light rays come to a *focal line*. On the other hand, the spherical lens is illustrated by an ordinary magnifying glass. If such a lens is placed in a beam of sunlight and a piece of paper is brought progressively closer to the lens, the light rays will impinge on a common focal point at an appropriate distance.

Concave cylindrical lenses *diverge* light rays in only one plane in the same manner that *convex* cylindrical lenses *converge* light rays in one plane.

Combination of Two Cylindrical Lenses to Equal a Spherical Lens. Figure 11–5 shows two convex cylindrical lenses at right angles to each other. The vertical cylindrical lens causes convergence of the light rays that pass through the two sides of the lens, and the horizontal lens converges the top and bottom rays. Thus, all the light rays come to a single-point focus. In other words, *two cylindrical lenses crossed at right angles to each other perform the same function as one spherical lens of the same refractive power.*

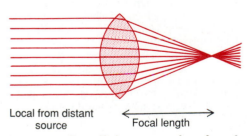

Figure 11–2. Bending of light rays at each surface of a convex spherical lens, showing that parallel light rays are focused to a point focus.

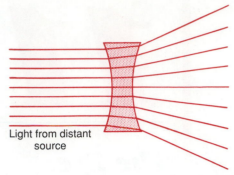

Figure 11–3. Bending of light rays at each surface of a concave spherical lens, illustrating that parallel light rays are diverged by a concave lens.

FOCAL LENGTH OF A LENS

The distance beyond a convex lens at which *parallel* rays converge to a common focal point is called the *focal length* of the lens. The diagram at the top of Figure 11–6 illustrates this focusing of parallel light rays.

In the middle diagram, the light rays that enter the convex lens are not parallel but are diverging because the origin of the light is a point source not far away from the lens itself. Because these rays are diverging outward from the point source, it can be seen from the diagram that they do not come to a point focus at the same distance away from the lens as do

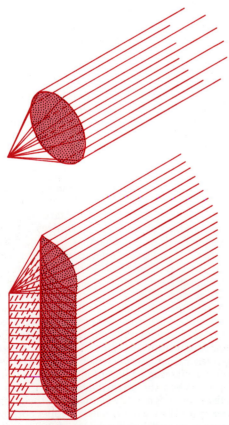

Figure 11–4. *Top:* Point focus of parallel light rays by a spherical convex lens. *Bottom:* Line focus of parallel light rays by a cylindrical convex lens.

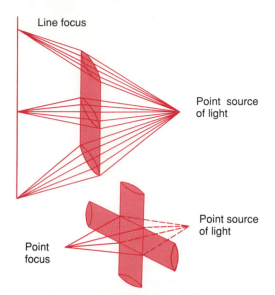

Figure 11−5. Two cylindrical convex lenses at right angles to each other, illustrating that one lens converges light rays in one plane and the other lens converges light rays in the plane at right angles. The two lenses combined give the same point focus as that obtained with a spherical convex lens.

parallel rays. In other words, when rays of light that are already diverging enter a convex lens, the distance of focus on the other side of the lens is farther from the lens than is the case when the entering rays are parallel.

In the lower diagram of Figure 11−6 are shown light rays that are diverging toward a convex lens with far greater curvature than that of the upper two lenses of the figure. In this diagram the distance from the lens at which the light rays come to a focus is exactly the same as that from the lens in the first diagram, in which the lens was less convex but the rays entering it were parallel. This illustrates that both parallel rays and diverging rays can be focused at the same distance behind a lens provided the lens changes its convexity.

The relationship of focal length of the lens, distance of the point source of light, and distance of focus is expressed by the following formula:

$$\frac{1}{f} = \frac{1}{a} + \frac{1}{b}$$

in which *f* is the focal length of the lens, *a* the distance of the point source of light from the lens, and *b* the distance of focus from the lens.

FORMATION OF AN IMAGE BY A CONVEX LENS

The upper drawing of Figure 11−7 illustrates a convex lens with two point sources of light to the left. Because light rays pass through the center of a convex lens without being refracted in either direction, the light rays from each point source of light are shown to come to a point focus on the opposite side of the lens *directly in line with the point source and the center of the lens.*

Any object in front of the lens is in reality a mosaic of point sources of light. Some of these points are very bright, some are very weak, and they vary in color. And each point source

of light on the object comes to a separate point focus on the opposite side of the lens in line with the lens center. Furthermore, all the focal points behind the lens will fall in a common plane a certain distance behind the lens. If a white piece of paper is placed at this distance, one can see an image of the object, as illustrated in the lower portion of Figure 11−7. However, this image is upside down with respect to the original object, and the two lateral sides of the image are reversed as well. This is the method by which the lens of a camera focuses images on the camera film.

MEASUREMENT OF THE REFRACTIVE POWER OF A LENS — THE DIOPTER

The more a lens bends light rays, the greater is its "refractive power." This refractive power is measured in terms of *diopters*. The refractive power of a convex lens is equal to 1 meter divided by its focal length. Thus, a spherical lens that converges parallel light rays to a focal point 1 meter beyond the lens has a refractive power of +1 diopter, as illustrated in Figure 11−8. If the lens is capable of bending parallel light rays twice as much as a lens with a power of +1 diopter, it is said to have a strength of +2 diopters, and the light rays come to a focal point 0.5 m beyond the lens. A lens capable of converging parallel light rays to a focal point only 10 cm (0.10 m) beyond the lens has a refractive power of +10 diopters.

The refractive power of concave lenses cannot be stated in terms of the focal distance beyond the lens because the light rays diverge, rather than focusing to a point. However, if a concave lens diverges light rays the same amount that a 1 diopter convex lens converges them, the concave lens is said to have a dioptric strength of −1. Likewise, if the concave lens diverges the light rays as much as a +10 diopter lens converges them, it is said to have a strength of −10 diopters.

Note particularly that concave lenses "neutralize" the refractive power of convex lenses. Thus, placing a 1 diopter

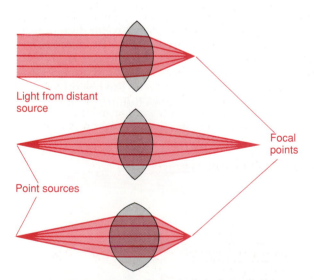

Figure 11−6. The upper two lenses of this figure have the same strength, but the light rays entering the top lens are parallel, whereas those entering the second lens are diverging; the effect of parallel versus diverging rays on the focal distance is illustrated. The bottom lens has far more refractive power than either of the other two lenses, illustrating that the stronger the lens the nearer to the lens is the point focus.

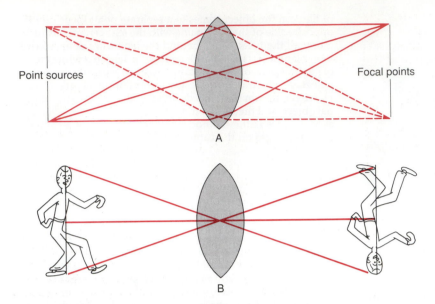

Figure 11–7. *A*, Two point sources of light focused at two separate points on the opposite side of the lens. *B*, Formation of an image by a convex spherical lens.

concave lens immediately in front of a 1 diopter convex lens results in a lens system with zero refractive power.

The strengths of cylindrical lenses are computed in the same manner as the strengths of spherical lenses. If a cylindrical lens focuses parallel light rays to a line focus 1 m beyond the lens, it has a strength of +1 diopter. On the other hand, if a cylindrical lens of a concave type *diverges* light rays as much as a +1 diopter cylindrical lens *converges* them, it has a strength of −1 diopter. However, the *axis* of the cylindrical lens must also be stated in addition to its strength.

■ THE OPTICS OF THE EYE

THE EYE AS A CAMERA

The eye, illustrated in Figure 11–9, is optically equivalent to the usual photographic camera, for it has a lens system, a variable aperture system (the pupil), and a retina that corresponds to the film. The lens system of the eye is composed of four refractive interfaces: (1) the interface between air and the anterior surface of the cornea, (2) the interface between the posterior surface of the cornea and the aqueous humor, (3) the interface

between the aqueous humor and the anterior surface of the crystalline lens of the eye, and (4) the interface between the posterior surface of the lens and the vitreous humor. The refractive index of air is 1; the cornea, 1.38; the aqueous humor, 1.33; the crystalline lens (on the average), 1.40; and the vitreous humor, 1.34.

The Reduced Eye. If all the refractive surfaces of the eye are algebraically added together and then considered to be one single lens, the optics of the normal eye may be simplified and represented schematically as a "reduced eye." This is useful in simple calculations. In the reduced eye, a single refractive surface is considered to exist with its central point 17 mm in front of the retina and to have a total refractive power of approximately 59 diopters when the lens is accommodated for distant vision.

Most of the refractive power of the eye is provided not by the crystalline lens but instead by the anterior surface of the cornea. The principal reason for this is that the refractive index of the cornea is markedly different from that of air.

On the other hand, the total refractive power of the crystalline lens of the eye, as it normally lies in the eye surrounded by fluid on each side, is only 20 diopters, about one third the total refractive power of the eye's lens system. If this lens were removed from the eye and then surrounded by air, its refractive power would be about six times as great. The reason for this difference

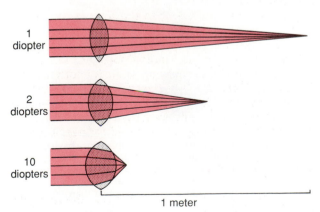

Figure 11–8. Effect of lens strength on the focal distance.

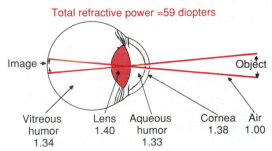

Figure 11–9. The eye as a camera. The numbers are the refractive indices.

is that the fluids surrounding the lens have refractive indices not greatly different from the refractive index of the lens itself, and the smallness of the differences greatly decreases the amount of light refraction at the lens interfaces. But the importance of the crystalline lens is that its curvature can be increased markedly to provide "accommodation," which will be discussed later in the chapter.

Formation of an Image on the Retina. In exactly the same manner that a glass lens can focus an image on a sheet of paper, the lens system of the eye can focus an image on the retina. The image is inverted and reversed with respect to the object. However, the mind perceives objects in the upright position despite the upside-down orientation on the retina because the brain is trained to consider an inverted image as the normal.

THE MECHANISM OF ACCOMMODATION

The refractive power of the crystalline lens of the eye can be voluntarily increased from 20 diopters to approximately 34 diopters in young children; this is a total "accommodation" of 14 diopters. To do this, the shape of the lens is changed from that of a moderately convex lens to that of a very convex lens. The mechanism of this is the following:

In the young person, the lens is composed of a strong elastic capsule filled with viscous, proteinaceous, but transparent fibers. When the lens is in a relaxed state, with no tension on its capsule, it assumes an almost spherical shape, owing entirely to the elasticity of the lens capsule. However, as illustrated in Figure 11–10, approximately 70 ligaments (called *zonules*) attach radially around the lens, pulling the lens edges toward the anterior edges of the retina. These ligaments are constantly tensed by the elastic pull of their attachments to the ciliary body at the anterior border of the choroid. The tension on the ligaments causes the lens to remain relatively flat under normal resting condi-

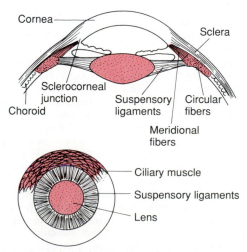

Figure 11–10. Mechanism of accommodation (focusing).

tions of the eye. At the insertions of the ligaments in the ciliary body is the *ciliary muscle*, which has two sets of smooth muscle fibers, the *meridional fibers* and the *circular fibers*. The meridional fibers extend to the corneoscleral junction. When these muscle fibers contract, the peripheral insertions of the lens ligaments are pulled forward, thereby releasing a certain amount of tension on the lens. The circular fibers are arranged circularly all the way around the eye so that when they contract, a sphincter-like action occurs, decreasing the diameter of the circle of ligament attachments and also allowing the ligaments to pull less on the lens capsule.

Thus, contraction of both sets of smooth muscle fibers in the ciliary muscle relaxes the ligaments to the lens capsule, and the lens assumes a more spherical shape, like that of a balloon, because of the natural elasticity of its capsule. Therefore, when the ciliary muscle is completely relaxed, the dioptric strength of the lens is as weak as it can become. On the other hand, when the ciliary muscle contracts as strongly as possible, the dioptric strength of the lens becomes maximal.

Autonomic Control of Accommodation. The ciliary muscle is controlled almost entirely by the parasympathetic nervous system. Stimulation of the parasympathetic nerves contracts the ciliary muscle, which relaxes the lens ligaments and increases the refractive power. With an increased refractive power, the eye is capable of focusing on objects nearer at hand than when the eye has less refractive power. Consequently, as a distant object moves toward the eye, the number of parasympathetic impulses impinging on the ciliary muscle must be progressively increased for the eye to keep the object constantly in focus. (Sympathetic stimulation has a weak effect in relaxing the ciliary muscle, but this plays almost no role in the normal accommodation mechanism, the neurology of which will be discussed in Chapter 13.)

Presbyopia. As a person grows older, the lens grows larger and thicker, and it also becomes less elastic, partly because of progressive denaturation of the lens proteins. Therefore, the ability of the lens to change shape progressively decreases, and the power of accommodation decreases from approximately 14 diopters in the young child to less than 2 diopters at the age of 45 to 50 and to about 0 at age 70. Thereafter, the lens is then almost totally nonaccommodating, a condition known as "presbyopia."

Once a person has reached the state of presbyopia, each eye remains focused permanently at an almost constant distance; this distance depends on the physical characteristics of each individual's eyes. Obviously, the eyes can no longer accommodate for both near and far vision. Therefore, to see clearly both in the distance and nearby, an older person must wear bifocal glasses with the upper segment normally focused for far-seeing and the lower segment focused for near-seeing.

THE PUPILLARY APERTURE

A major function of the iris is to increase the amount of light that enters the eye during darkness and to decrease the light in bright light. The reflexes for controlling this mechanism will be considered in the discussion of the neurology of the eye in Chapter 13. The amount of light that enters the eye through the pupil is proportional to the *area* of the pupil or to the *square of the diameter* of the pupil. The pupil of the human eye can become as small as approximately 1.5 mm and as large as 8 mm in diameter. Therefore, the quantity of light entering the eye may vary approximately 30 times as a result of changes in pupillary aperture.

Depth of Focus of the Lens System of the Eye. Figure 11–11 illustrates two separate eyes that are exactly alike except for the diameters of the pupillary apertures. In the upper eye the pupillary aperture is small, and in the lower eye the aperture is large. In front of each of these two eyes are two small point sources of light, and light from each passes through the pupillary aperture and focuses on the retina. Consequently, in both eyes the retina sees two spots of light in perfect focus. It is evident from the diagrams, however, that if the retina is moved forward or backward to an out-of-focus position, the size of each spot will not change much in the upper eye, but in the lower eye the size of each spot will increase greatly, becoming a "blur circle." In other words, the upper lens system has far greater *depth of focus* than the bottom lens system. When a lens system has great depth of focus, the retina can be considerably displaced from the focal plane and still the image remains in sharp focus; whereas when a lens system has a shallow depth of focus, moving the retina only slightly away from the focal plane causes extreme blurring.

The greatest possible depth of focus occurs when the pupil is extremely small. The reason for this is that with a very small aperture all light rays must pass through the center of the lens, and the centralmost rays are always in focus, as was explained earlier.

ERRORS OF REFRACTION

Emmetropia. As shown in Figure 11–12, the eye is considered to be normal, or "emmetropic," if parallel light rays

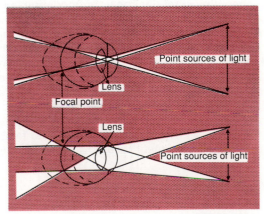

Figure 11–11. Effect of small and large pupillary apertures on the depth of focus.

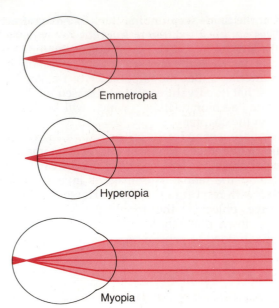

Figure 11–12. Parallel light rays focus on the retina in emmetropia, behind the retina in hyperopia, and in front of the retina in myopia.

from distant objects are in sharp focus on the retina *when the ciliary muscle is completely relaxed.* This means that the emmetropic eye can see all distant objects clearly, with its ciliary muscle relaxed, but to focus objects at close range it must contract its ciliary muscle and thereby provide various degrees of accommodation.

Hyperopia. Hyperopia, which is also known as "far-sightedness," is usually due either to an eyeball that is too short or occasionally to a lens system that is too weak when the ciliary muscle is relaxed. In this condition, as seen in the middle panel of Figure 11–12, parallel light rays are not bent sufficiently by the lens system to come to a focus by the time they reach the retina. To overcome this abnormality, the ciliary muscle may contract to increase the strength of the lens. Therefore, the far-sighted person is capable, by using his mechanism of accommodation, of focusing distant objects on the retina. If he has used only a small amount of strength in his ciliary muscle to accommodate for the distant objects, then he still has much accommodative power left, and objects closer and closer to the eye can also be focused sharply until the ciliary muscle has contracted to its limit.

In old age, when the lens becomes presbyopic, the far-sighted person often is not able to accommodate his or her lens sufficiently to focus even distant objects, much less to focus near objects.

Myopia. In myopia, or "near-sightedness," when the ciliary muscle is completely relaxed the light rays coming from distant objects are focused in front of the retina, as shown in the lower panel of Figure 11–12. This is usually due to too long an eyeball but it can occasionally result from too much refractive power of the lens system of the eye.

No mechanism exists by which the eye can decrease the strength of its lens to less than that which exists when the ciliary muscle is completely relaxed. Therefore, the myopic person has no mechanism by which he can ever focus distant objects sharply on his retina. However, as an object comes nearer to his eye, it finally comes near enough that its image will focus. Then, when the object comes still closer to the eye, the person can use his mechanism of accommodation to keep the image focused clearly. Therefore, a myopic person has a definite limiting "far point" for clear vision.

Correction of Myopia and Hyperopia by Use of Lenses. It will be recalled that light rays passing through a concave lens diverge. Therefore, if the refractive surfaces of the eye have too much refractive power, as in *myopia*, some of this excessive refractive power can be neutralized by placing in front of the eye a concave spherical lens, which will diverge rays.

On the other hand, in a person who has *hyperopia*—that is, someone who has too weak a lens system—the abnormal vision can be corrected by adding refractive power with a convex lens in front of the eye. These corrections are illustrated in Figure 11–13.

One usually determines the strength of the concave or convex lens needed for clear vision by "trial and error"—that is, by trying first a strong lens and then a stronger or weaker lens until the one that gives the best visual acuity is found.

Astigmatism. Astigmatism is a refractive error of the lens system, caused usually by an oblong shape of the cornea or, rarely, an oblong shape of the lens. A lens surface like the side of an egg lying sidewise to the incoming light would be an example of an astigmatic lens. The degree of curvature in the plane through the long axis of the egg is not nearly so great as the degree of curvature in the plane through the short axis.

Because the curvature of the astigmatic lens along one plane is less than the curvature along the other plane, light rays striking the peripheral portions of the lens in one plane are not bent nearly so much as are rays striking the peripheral portions of the other plane. This is illustrated in Figure 11–14, which shows rays of light emanating from a point source and passing through an oblong, astigmatic lens. The light rays in the vertical plane, indicated by plane BD, are refracted greatly by the astigmatic lens because of the greater curvature in the vertical direction than in the horizontal direction. However, the light rays in the horizontal plane, indicated by plane AC, are bent not nearly so much as the light rays in the vertical plane. It is obvious, therefore, that the light rays passing through an astigmatic lens do not all come to a common focal point, because the light rays passing through one plane focus far in front of those passing through the other plane.

The accommodative powers of the eyes can never compensate for astigmatism because, during accommodation, the curvature of the eye lens changes equally in both planes.

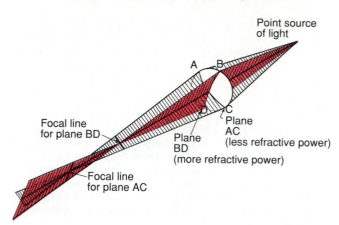

Figure 11–14. Astigmatism, illustrating that light rays focus at one focal distance in one focal plane and at another focal distance in the plane at right angles.

Therefore, when the accommodation corrects the refractive error in one plane, the error in the other plane is not corrected. That is, each of the two planes requires a different degree of accommodation to be corrected, so that the two planes are never corrected at the same time without the help of glasses. Thus, vision never occurs with a sharp focus.

Correction of Astigmatism With a Cylindrical Lens. One may consider an astigmatic eye as having a lens system made up of two cylindrical lenses of different strengths and placed at right angles to each other. Therefore, to correct for astigmatism the usual procedure is to find a spherical lens by "trial and error" that corrects the focus in one of the two planes of the astigmatic lens. Then an additional cylindrical lens is used to correct the error in the remaining plane. To do this, both the *axis* and the *strength* of the required cylindrical lens must be determined.

There are several methods for determining the axis of the abnormal cylindrical component of the lens system of an eye. One of these methods is based on the use of parallel black bars of the type shown in Figure 11–15. Some of these parallel bars are vertical, some horizontal, and some at various angles to the vertical and horizontal axes. After placing

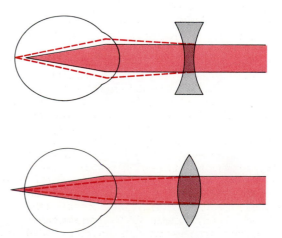

Figure 11–13. Correction of myopia with a concave lens, and correction of hyperopia with a convex lens.

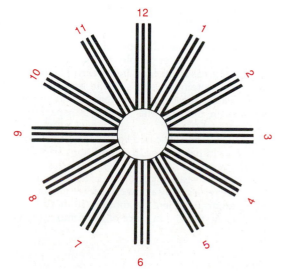

Figure 11–15. Chart composed of parallel black bars for determining the axis of astigmatism.

various spherical lenses in front of the astigmatic eye by trial and error, a strength of lens will usually be found that will cause sharp focus of one set of these parallel bars on the retina of the astigmatic eye but will not correct the fuzziness of the set of bars at right angles to the sharp bars. It can be shown from the physical principles of optics discussed earlier in this chapter that the axis of the *out-of-focus* cylindrical component of the optical system is parallel to the bars that are fuzzy. Once this axis is found, the examiner tries progressively stronger and weaker positive or negative cylindrical lenses, the axes of which are placed parallel to the out-of-focus bars, until the patient sees all the crossed bars with equal clarity. When this has been accomplished, the examiner directs the optician to grind a special lens having the spherical correction as well as the cylindrical correction at the appropriate axis.

Correction of Optical Abnormalities by Use of Contact Lenses. In recent years, either glass or plastic contact lenses have been fitted snugly against the anterior surface of the cornea. These lenses are held in place by a thin layer of tears that fills the space between the contact lens and the anterior eye surface.

A special feature of the contact lens is that it nullifies almost entirely the refraction that normally occurs at the anterior surface of the cornea. The reason for this is that the tears between the contact lens and the cornea have a refractive index almost equal to that of the cornea, so that no longer does the anterior surface of the cornea play a significant role in the eye's optical system. Instead, the anterior surface of the contact lens now plays the major role and its posterior surface a minor role. Thus, the refraction of this lens substitutes for the cornea's usual refraction. This is especially important in persons whose eye refractive errors are caused by an abnormally shaped cornea, such as persons who have an odd-shaped, bulging cornea—a condition called *keratoconus*. Without the contact lens the bulging cornea causes such severe abnormality of vision that almost no glasses can correct the vision satisfactorily; when a contact lens is used, however, the corneal refraction is neutralized, and normal refraction by the anterior surface of the contact lens is substituted in its place.

The contact lens has several other advantages as well, including (1) the lens turns with the eye and gives a broader field of clear vision than do usual glasses, and (2) the contact lens has little effect on the size of the object that the person sees through the lens; on the other hand, lenses placed several centimeters in front of the eye do affect the size of the image in addition to correcting the focus.

Cataracts. Cataracts are an especially common eye abnormality that occurs in older people. A cataract is a cloudy or opaque area or areas in the lens. In the early stage of cataract formation the proteins in some of the lens fibers become denatured. Later, these same proteins coagulate to form opaque areas in place of the normal transparent protein fibers.

When a cataract has obscured light transmission so greatly that it seriously impairs vision, the condition can be corrected by surgical removal of the entire lens. When this is done, however, the eye loses a large portion of its refractive power, which must be replaced by a powerful convex lens in front of the eye, or an artificial lens of about +20 diopters may be implanted inside the eye in place of the removed lens.

VISUAL ACUITY

Theoretically, light from a distant point source, when focused on the retina, should be infinitely small. How-

ever, since the lens system of the eye is not perfect, such a retinal spot ordinarily has a total diameter of about 11 μm even with maximal resolution of the optical system. However, it is brightest in its very center and shades off gradually toward the edges, as illustrated by the two-point images in Figure 11–16.

The average diameter of cones *in the fovea* of the retina, the central part of the retina where vision is most highly developed, is approximately 1.5 μm, which is one seventh the diameter of the spot of light. Nevertheless, since the spot of light has a bright center point and shaded edges, a person can distinguish two separate points if their centers lie approximately 2 μm apart on the retina, which is slightly greater than the width of a foveal cone. This discrimination between points is also illustrated in Figure 11–16.

The normal visual acuity of the human eye for discriminating between point sources of light is about 45 sec of arc. That is, when light rays from two separate points strike the eye with an angle of at least 45 sec between them, they can usually be recognized as two points instead of one. This means that a person with normal acuity looking at two bright pinpoint spots of light 10 m away can barely distinguish the spots as separate entities when they are 1.5 to 2 mm apart.

The fovea is less than one half a millimeter (less than 500 μm) in diameter, which means that maximum visual acuity occurs in only 3 degrees of the visual field. Outside this foveal area the visual acuity is reduced five- to tenfold, and it becomes progressively poorer as the periphery is approached. This is caused by the connection of many rods and cones to the same optic nerve fiber in the nonfoveal parts of the retina, as will be discussed in Chapter 13.

Clinical Method for Stating Visual Acuity. Usually the test chart for testing eyes is placed 20 ft away from the tested person, and if the person can see the letters of the size that he should be able to see at 20 ft, he is said to have 20/20 vision: that is, normal vision. If he can see only letters that he should be able to see at 200 ft, he is said to have 20/200 vision. In other words, the clinical method for expressing visual acuity is to use a mathematical fraction that expresses the ratio of two distances, which is also the ratio of one's visual acuity to that of the normal person.

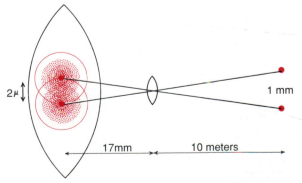

Figure 11–16. Maximum visual acuity for two point sources of light.

DETERMINATION OF DISTANCE OF AN OBJECT FROM THE EYE—DEPTH PERCEPTION

The visual apparatus normally perceives distance by three major means. This phenomenon is known as *depth perception*. These means are (1) the size of the image of known objects on the retina, (2) the phenomenon of moving parallax, and (3) the phenomenon of stereopsis.

Determination of Distance by Sizes of Retinal Images of Known Objects. If one knows that a man whom he is viewing is 6 ft tall, he can determine how far away the man is simply by the size of the man's image on his retina. He does not consciously think about the size, but his brain has learned to calculate automatically from image sizes the distances of objects when the dimensions are known.

Determination of Distance by Moving Parallax. Another important means by which the eyes determine distance is that of moving parallax. If a person looks off into the distance with his eyes completely still, he perceives no moving parallax, but when he moves his head to one side or the other, the images of objects close to him move rapidly across his retinas while the images of distant objects remain rather stationary. For instance, if he moves his head 1 in. and an object is only 1 in. in front of his eye, the image moves almost all the way across his retinas, whereas the image of an object 200 ft away from his eyes does not move perceptibly. Thus, by this mechanism of moving parallax, one can tell the *relative distances* of different objects even though only one eye is used.

Determination of Distance by Stereopsis—Binocular Vision. Another method by which one perceives parallax is that of binocular vision. Because one eye is a little more than 2 in. to one side of the other eye, the images on the two retinas are different one from the other—that is, an object that is 1 in. in front of the bridge of the nose forms an image on the temporal portion of the retina of each eye, whereas a small object 20 ft in front of the nose has its image at closely corresponding points in the middle of each retina. This type of parallax is illustrated in Figure 11–17, which shows the images of a black spot and a square actually reversed on the two retinas because they are at differ-

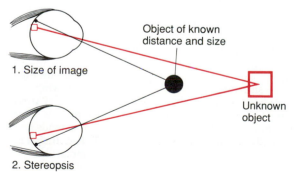

Figure 11–17. Perception of distance (1) by the size of the image on the retina and (2) as a result of stereopsis.

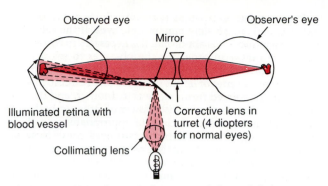

Figure 11–18. The optical system of the ophthalmoscope.

ent distances in front of the eyes. This gives a type of parallax that is present all the time when both eyes are being used. It is almost entirely this binocular parallax (or stereopsis) that gives a person with two eyes far greater ability to judge relative distances *when objects are nearby* than a person who has only one eye. However, stereopsis is virtually useless for depth perception at distances beyond 200 ft.

■ OPTICAL INSTRUMENTS

THE OPHTHALMOSCOPE

The ophthalmoscope is an instrument through which an observer can look into another person's eye and see the retina with clarity. Though the ophthalmoscope appears to be a relatively complicated instrument, its principles are simple. The basic components are illustrated in Figure 11–18 and may be explained as follows.

If a bright spot of light is on the retina of an *emmetropic eye*, light rays from this spot diverge toward the lens system of the eye, and, after passing through the lens system, they are parallel with each other because the retina is located exactly one focal length distance behind the lens. Then, when these parallel rays pass into an emmetropic eye of another person, they focus back again to a point focus on the retina of the second person because his retina is also one focal length distance behind the lens. Therefore, any spot of light on the retina of the observed eye comes to a focal spot on the retina of the observing eye. Likewise, when the bright spot of light is moved to different points on the observed retina, the focal spot on the retina of the observer also moves an equal amount. Thus, if the retina of one person is made to emit light, the image of his retina will be focused on the retina of the observer provided the two eyes are simply looking into each other. These principles, of course, apply only to completely emmetropic eyes.

To make an ophthalmoscope, one need only devise a means for illuminating the retina to be examined. Then, the reflected light from that retina can be seen by the observer simply by putting the two eyes close to each other. To illuminate the retina of the observed eye, an angulated mirror or a segment of a prism is placed in front of the observed eye in such a manner, as illustrated in Figure 11–18, that light from a bulb is reflected into the observed eye. Thus, the retina is illuminated through the pupil, and the observer sees into the subject's pupil by looking over the edge of the mirror or prism, or *through* an appropriately designed prism so that the light will not have to enter the pupil at an angle.

It was noted above that these principles apply only to persons with completely emmetropic eyes. If the refractive power of either eye is abnormal, it is necessary to correct this refractive power in order for the observer to see a sharp image of the observed retina. Therefore, the usual ophthalmoscope has a series of lenses mounted on a turret so that the turret can be rotated from one lens to another, and the correction for abnormal refractive power of either or both eyes can be made by selecting a lens of appropriate strength. In normal young adults, when the two eyes come close together, a natural accommodative reflex occurs that causes an approximate +2 diopter increase in the strength of the lens of each eye. To correct for this, it is necessary that the lens turret be rotated to an approximately −4 diopter correction.

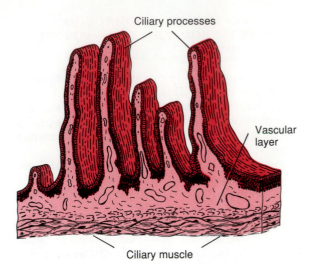

Figure 11–20. Anatomy of the ciliary processes.

■ THE FLUID SYSTEM OF THE EYE—THE INTRAOCULAR FLUID

The eye is filled with *intraocular fluid*, which maintains sufficient pressure in the eyeball to keep it distended. Figure 11–19 illustrates that this fluid can be divided into two portions, the *aqueous humor*, which lies in front and to the sides of the lens, and the fluid of the *vitreous humor*, which lies between the lens and the retina. The aqueous humor is a freely flowing fluid, whereas the vitreous humor, sometimes called the *vitreous body*, is a gelatinous mass held together by a fine fibrillar network composed primarily of large proteoglycan molecules. Substances can *diffuse* slowly in the vitreous humor, but there is little *flow* of fluid.

Aqueous humor is continually being formed and reabsorbed. The balance between formation and reabsorption of aqueous humor regulates the total volume and pressure of the intraocular fluid.

FORMATION OF AQUEOUS HUMOR BY THE CILIARY BODY

Aqueous humor is formed in the eye *at an average rate of 2 to 3 μL each minute.* Essentially all of this is secreted by the

ciliary processes, which are linear folds projecting from the *ciliary body* into the space behind the iris where the lens ligaments also attach to the eyeball. A cross-section of these ciliary processes is illustrated in Figure 11–20, and their relationship to the fluid chambers of the eye can be seen in Figure 11–19. Because of their folded architecture, the total surface area of the ciliary processes is approximately 6 cm² in each eye—a large area, considering the small size of the ciliary body. The surfaces of these processes are covered by epithelial cells, and immediately beneath these is a highly vascular area.

Aqueous humor is formed almost entirely as an active secretion of the epithelium lining the ciliary processes. Secretion begins with active transport of sodium ions into the spaces between the epithelial cells. The sodium ions in turn pull chloride and bicarbonate ions along with them to maintain electrical neutrality. Then all these ions together cause osmosis of water from the sublying tissue into the same epithelial intercellular spaces, and the resulting solution washes from the spaces onto the surfaces of the ciliary processes. In addition, several nutrients are transported across the epithelium by active transport or facilitated diffusion; these include amino acids, ascorbic acid, and probably also glucose.

OUTFLOW OF AQUEOUS HUMOR FROM THE EYE

After aqueous humor is formed by the ciliary processes, it flows, as shown in Figure 11–19, *between the ligaments of the lens,* then *through the pupil,* and finally *into the anterior chamber of the eye.* Here, the fluid flows into the *angle between the cornea and the iris* and thence through a meshwork of *trabeculae,* finally entering the *canal of Schlemm,* which empties into extraocular veins. Figure 11–21 illustrates the anatomical structures at the iridocorneal angle, showing that the spaces between the trabeculae extend all the way from the anterior chamber to the canal of Schlemm. The canal of Schlemm in turn is a thin-walled vein that extends circumferentially all the way around the eye. Its endothelial membrane is so porous that even large protein molecules, as well as small particulate matter up to the size of red blood cells, can pass from the anterior chamber into the canal of

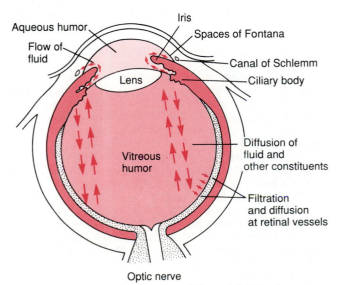

Figure 11–19. Formation and flow of fluid in the eye.

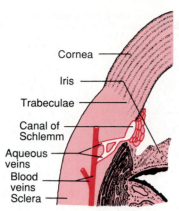

Figure 11–21. Anatomy of the iridocorneal angle, showing the system for outflow of aqueous humor into the conjunctival veins.

Schlemm. Even though the canal of Schlemm is actually a venous blood vessel, so much aqueous humor normally flows into it that it is filled only with aqueous humor rather than with blood. Also, the small veins that lead from the canal of Schlemm to the larger veins of the eye usually contain only aqueous humor, and these are called *aqueous veins*.

INTRAOCULAR PRESSURE

The average normal intraocular pressure is approximately 15 mm Hg, with a range of from 12 to 20.

Tonometry. Because it is impractical to pass a needle into a patient's eye for measurement of intraocular pressure, this pressure is measured clinically by means of a tonometer, the principle of which is illustrated in Figure 11–22. The cornea of the eye is anesthetized with a local anesthetic, and the footplate of the tonometer is placed on the cornea. A small force is then applied to a central plunger, causing the part of the cornea beneath the plunger to be displaced inward. The amount of displacement is recorded on the scale of the tonometer, and this in turn is calibrated in terms of intraocular pressure.

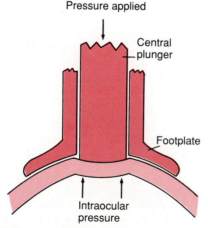

Figure 11–22. Principles of the tonometer.

Regulation of Intraocular Pressure. Intraocular pressure remains very constant in the normal eye, normally within about ±2 mm Hg. The level of this pressure is determined mainly by the resistance to outflow of aqueous humor from the anterior chamber into the canal of Schlemm. This outflow resistance results from a meshwork of trabeculae through which the fluid must percolate on its way from the lateral angles of the anterior chamber to the wall of the canal of Schlemm. These trabeculae have minute openings of only 2 to 3 μm. The rate of fluid flow into the canal increases markedly as the pressure rises. At approximately 15 mm Hg in the normal eye the amount of fluid leaving the eye by way of the canal of Schlemm averages 2.5 μL/min and exactly equals the inflow of fluid from the ciliary body. Therefore, normally the pressure remains at approximately this level of 15 mm Hg.

Cleansing of the Trabecular Spaces and of the Intraocular Fluid. When large amounts of debris occur in the aqueous humor, as occurs following hemorrhage into the eye or during intraocular infection, the debris is likely to accumulate in the trabecular spaces leading to the canal of Schlemm, therefore preventing adequate reabsorption of fluid from the anterior chamber and sometimes causing glaucoma, as explained subsequently. However, on the surfaces of the trabecular plates are large numbers of phagocytic cells. Also, immediately outside the canal of Schlemm is a layer of interstitial gel containing large numbers of reticuloendothelial cells that have an extremely high capacity for both engulfing debris and degrading it into small molecular substances that can then be absorbed. Thus, this phagocytic system keeps the trabecular spaces cleaned.

In addition, the surface of the iris and other surfaces of the eye behind the iris are covered with an epithelium that is capable of phagocytizing proteins and small particles from the aqueous humor, thereby helping maintain a perfectly clear fluid.

Glaucoma. Glaucoma is one of the most common causes of blindness. It is a disease of the eye in which the intraocular pressure becomes pathologically high, sometimes rising to as high as 60 to 70 mm Hg. Pressures rising above as little as 20 to 30 mm Hg can cause loss of vision when maintained for long periods of time. And the extremely high pressures can cause blindness within days or even hours. As the pressure rises, the axons of the optic nerve are compressed where they leave the eyeball at the optic disk. This compression is believed to block axonal flow of cytoplasm from the neuronal cell bodies in the retina to the peripheral optic nerve fibers entering the brain. The result is lack of appropriate nutrition, which eventually causes death of the involved neurons. It is possible that compression of the retinal artery, which also enters the eyeball at the optic disk, adds as well to the neuronal damage by reducing nutrition to the retina.

In most cases of glaucoma the abnormally high pressure results from increased resistance to fluid outflow through the trabecular spaces into the canal of Schlemm at the iridocorneal junction. For instance, in acute eye inflammation, white blood cells and tissue debris can block these spaces and cause acute increase in intraocular pressure. In chronic conditions, especially in older age, fibrous occlusion of the trabecular spaces appears to be the likely culprit.

Glaucoma can sometimes be treated by placing drops in the eye containing a drug that diffuses into the eyeball and causes reduced secretion or increased absorption of aqueous humor. However, when drug therapy fails, operative techniques to open the spaces of the trabeculae or to make channels directly between the fluid space of the eyeball and the subconjunctival space outside the eyeball can often effectively reduce the pressure.

REFERENCES

Apple, D. J., et al.: Intraocular Lenses: Evolution, Design, Complications and Pathology. Baltimore, Williams & Wilkins, 1988.

Bentley, P. J.: The crystalline lens of the eye: An optical microcosm. News in Physiol. Sci., 1:195, 1986.

Bill, A.: Blood circulation and fluid dynamics in the eye. Physiol. Rev., 55:383, 1975.

Bill, A.: Circulation in the eye. In Renkin, E. M., and Michel, C. C. (eds.): Handbook of Physiology. Sec. 2, Vol. IV. Bethesda, Md., American Physiological Society, 1984, p. 1001.

Caldwell, D. R.: Cataracts. New York, Raven Press, 1988.

Cavanagh, H. D.: The Cornea: Transactions of the World Congress on the Cornea III. New York, Raven Press, 1988.

Collins, R., and Van der Werff, T. J.: Mathematical Models of the Dynamics of the Human Eye. New York, Springer-Verlag, 1980.

Duncan, G., and Jacob, T. J.: Calcium and the physiology of cataract. Ciba Found. Symp., 106:132, 1984.

Elliot, R. H.: A Treatise on Glaucoma. Huntington, N.Y., R. E. Krieger, 1979.

Fischbarg, J., and Lim, J. J.: Fluid and electrolyte transports across corneal endothelium. Curr. Top. Eye Res., 4:201, 1984.

Guyton, D. L.: Sights and Sounds in Ophthalmology: Ocular Motility and Binocular Vision. St. Louis, C. V. Mosby Co., 1989.

Jaffe, N. S.: Cataract Surgery and Its Complications. St. Louis, C. V. Mosby Co., 1983.

Kavner, R. S., and Dusky, L.: Total Vision. New York, A & W Publishers, 1980.

Koretz, J. F., and Handelman, G. H.: How the human eye focuses. Sci. Am., July, 1988, p. 92.

Kuszak, J. R., et al.: Sutures of the crystalline lens: A review. Scan. Electron Miscrosc., 3:1369, 1984.

Lee, J. R.: Contact Lens Handbook. Philadelphia, W. B. Saunders Co., 1986.

Lesperace: Ophthalmic Lasers. Photocoagulation, Photoradiation and Surgery. St. Louis, C. V. Mosby, 1983.

Leydhecker, W., and Krieglstein, G. K. (eds.): Recent Advances in Glaucoma. New York, Springer-Verlag, 1979.

Michaels, D. D.: Basic Refraction Techniques. New York, Raven Press, 1988.

Moses, R. A.: Adler's Physiology of the Eye; Clinical Application, 7th Ed. St. Louis, C. V. Mosby, 1981.

Piatigorsky, J.: Lens crystallins and their genes: diversity and tissue-specific expression. FASEB J., 3:1933, 1989.

Ritch, R., et al. (eds.): The Glaucomas. St. Louis, C. V. Mosby Co., 1989.

Roth, H. W., and Roth-Wittig, M.: Contact Lenses. Hagerstown, Md., Harper & Row, 1980.

Safir, A. (ed.): Refraction and Clinical Optics. Hagerstown, Md., Harper & Row, 1980.

Shields, M. B.: Textbook of Glaucoma, 2nd Ed. Baltimore, Williams & Wilkins, 1986.

Stenson, S. M.: Contact Lenses: Guide to Selection, Fitting, and Management of Complications. East Norwalk, Conn., Appleton & Lange, 1987.

Toates, F. M.: Accommodation function of the human eye. Physiol. Rev., 52:828, 1972.

Whitnall, S. E. The Anatomy of the Human Orbit and Accessory Organs of Vision. Huntington, N. Y., R. E. Kreiger Publishing Co., 1979.

Wiederholt M.: Ion transport by the cornea. News Physiol. Sci., 3:97, 1988.

Yellott, J. I., Jr., et al.: The beginnings of visual perception: The retinal image and its initial encoding. In Darian-Smith, I. (ed.): Handbook of Physiology. Sec. 1, Vol. III. Bethesda Md., American Physiological Society, 1984, p. 257.

The Eye:

II. Receptor and Neural Function of the Retina

The retina is the light-sensitive portion of the eye, containing the cones, which are responsible for color vision, and the rods, which are mainly responsible for vision in the dark. When the rods and cones are excited, signals are transmitted through successive neurons in the retina itself and finally into the optic nerve fibers and cerebral cortex. The purpose of the present chapter is to explain specifically the mechanisms by which the rods and cones detect both white and colored light and then convert the visual image into nerve impulses.

ANATOMY AND FUNCTION OF THE STRUCTURAL ELEMENTS OF THE RETINA

The Layers of the Retina. Figure 12–1 shows the functional components of the retina arranged in layers from the outside to the inside as follows: (1) pigment layer, (2) layer of rods and cones projecting into the pigment, (3) outer limiting membrane, (4) outer nuclear layer containing the cell bodies of the rods and cones, (5) outer plexiform layer, (6) inner nuclear layer, (7) inner plexiform layer, (8) ganglionic layer, (9) layer of optic nerve fibers, and (10) inner limiting membrane.

After light passes through the lens system of the eye and then through the vitreous humor, it enters the retina from the inside (see Figure 12–1); that is, it passes through the ganglion cells, the plexiform layers, the nuclear layer, and the limiting membranes before it finally reaches the layer of rods and cones located all the way on the outer side of the retina. This distance is a thickness of several hundred micrometers; visual acuity is obviously decreased by this passage through such nonhomogeneous tissue. However, in the central region of the retina, as will be discussed, the initial layers are pulled aside for prevention of this loss of acuity.

The Foveal Region of the Retina and Its Importance in Acute Vision. A minute area in the center of the retina, illustrated in Figure 12–2, called the *macula* and occupying a total area of less than 1 mm², is especially capable of acute and detailed vision. The central portion of the macula, only 0.4 mm in diameter, is called the *fovea*; this area is composed entirely of cones, and the cones have a special structure that aids their detection of detail in the visual image, especially a long slender body, in contradiction to much larger cones located further peripherally in the retina. Also, in this region the blood vessels, the ganglion cells, the inner nuclear layer of cells, and the plexiform layers are all displaced to one side rather than resting directly on top of the cones. This allows light to pass unimpeded to the cones.

The Rods and Cones. Figure 12–3 is a diagrammatic representation of a photoreceptor (either a rod or a cone), though the cones are distinguished by having a conical upper end (the outer segment) as shown in Figure 12–4. In general, the rods are narrower and longer than the cones, but this is not always the case. In the peripheral portions of the retina the rods are 2 to 5 μm in diameter, whereas the cones are 5 to 8 μm in diameter; in the central part of the retina, in the fovea, the cones have a diameter of only 1.5 μm.

To the right in Figure 12–3 are labeled the four major functional segments of either a rod or a cone: (1) the *outer segment,* (2) the *inner segment,* (3) the *nucleus,* and (4) the *synaptic body.* In the outer segment the light-sensitive photochemical is found. In the case of the rods, this is *rhodopsin,* and in the cones it is one of several "color" photochemicals that function almost exactly the same as rhodopsin except for differences in spectral sensitivity.

Note in both Figures 12–3 and 12–4 the large numbers of discs in both the rods and the cones. In the cones, each of the discs is actually an infolded shelf of cell membrane; in the rods this is also true near the base of the rod. However, toward the tip of the rod the discs separate from the membrane and are flat sacs lying totally inside the cell. There are as many as 1000 discs in each rod or cone.

Both rhodopsin and the color photochemicals are conjugated proteins. These are incorporated into the membranes of the discs in the form of transmembrane proteins. The concentrations of these photosensitive pigments in the disc are so great that they constitute approximately 40 per cent of the entire mass of the outer segment.

The inner segment contains the usual cytoplasm of the cell with the usual cytoplasmic organelles. Particularly impor-

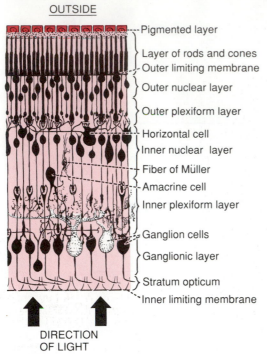

OUTSIDE

Pigmented layer
Layer of rods and cones
Outer limiting membrane
Outer nuclear layer
Outer plexiform layer
Horizontal cell
Inner nuclear layer
Fiber of Müller
Amacrine cell
Inner plexiform layer
Ganglion cells
Ganglionic layer
Stratum opticum
Inner limiting membrane

DIRECTION
OF LIGHT

Figure 12–1. Plan of the retinal neurons. (Modified from Polyak: The Retina. Copyright 1941 by The University of Chicago. All rights reserved.)

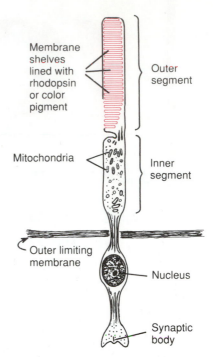

Membrane shelves lined with rhodopsin or color pigment

Outer segment

Mitochondria

Inner segment

Outer limiting membrane

Nucleus

Synaptic body

Figure 12–3. Schematic drawing of the functional parts of the rods and cones.

tant are the mitochondria; we see later that the mitochondria in this segment play an important role in providing the energy for function of the photoreceptors.

The synaptic body is the portion of the rod and cone that connects with the subsequent neuronal cells, the horizontal and bipolar cells, that represent the next stages in the vision chain.

The Pigment Layer of the Retina. The black pigment *melanin* in the pigment layer prevents light reflection throughout the globe of the eyeball; this is extremely important for clear vision. This pigment performs the same function in the eye as the black coloring inside the bellows of a camera. Without it, light rays would be reflected in all direc-

tions within the eyeball and would cause diffuse lighting of the retina rather than the contrast between dark and light spots required for formation of precise images.

The importance of melanin in the pigment layer and choroid is well illustrated by its absence in *albinos*, persons hereditarily lacking in melanin pigment in all parts of their bodies. When an albino enters a bright area, light that impinges on the retina is reflected in all directions by the white unpigmented surfaces so that a single discrete spot of light that would normally excite only a few rods or cones is reflected everywhere and excites many of the receptors. Therefore, the visual acuity of albinos, even with the best of optical correction, is rarely better than 20/100 to 20/200.

Figure 12–2. Photomicrograph of the macula and of the fovea in its center. Note that the inner layers of the retina are pulled to the side to decrease the interference with light transmission. (From Fawcett: Bloom and Fawcett: A Textbook of Histology. 11th Ed. Philadelphia, W. B. Saunders Company, 1986; courtesy of H. Mizoguchi.)

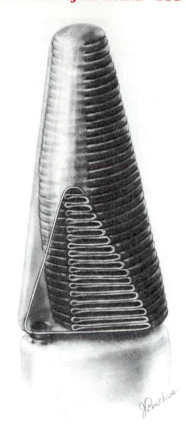

Figure 12–4. Membranous structures of the outer segments of a rod (left) and a cone (right). (Courtesy of Dr. Richard Young.)

The pigment layer also stores large quantities of *vitamin A.* This vitamin A is exchanged back and forth through the membranes of the outer segments of the rods and cones, which themselves are embedded in the pigment layers. We shall see later that vitamin A is an important precursor of the photosensitive pigments and that this interchange of vitamin A is very important for adjustment of the light sensitivity of the receptors.

The Blood Supply of the Retina — The Arterial System and the Choroid. The nutrient blood supply for the inner layers of the retina is derived from the central retinal artery, which enters the eyeball along with the optic nerve and then divides to supply the entire inner retinal surface. Thus, to a great extent, the retina has its own blood supply independent of the other structures of the eye.

However, the outer surface of the retina is adherent to the *choroid,* which is a highly vascular tissue between the retina and the sclera. The outer layers of the retina, including the outer segments of the rods and cones, depend mainly on diffusion from the choroid vessels for their nutrition, especially for their oxygen.

Retinal Detachment. The neural retina occasionally detaches from the pigment epithelium. In some instances the cause of such detachment is injury to the eyeball that allows fluid or blood to collect between the retina and the pigment epithelium, but often it is also caused by contracture of fine collagenous fibrils in the vitreous humor, which pull the retina unevenly toward the interior of the globe.

Fortunately, partly because of diffusion across the detachment gap and partly because of the independent blood supply to the retina through the retinal artery, the detached retina can resist degeneration for days and can become functional once again if surgically replaced in its normal relationship with the pigment epithelium. But, if not replaced soon, the retina finally is destroyed and is then unable to function even after surgical repair.

■ PHOTOCHEMISTRY OF VISION

Both the rods and cones contain chemicals that decompose on exposure to light and, in the process, excite the nerve fibers leading from the eye. The chemical in the *rods* is called *rhodopsin,* and the light-sensitive chemicals in the *cones* have compositions only slightly different from that of rhodopsin.

In the present section we discuss principally the photochemistry of rhodopsin, but we can apply almost exactly the same principles to the photochemistry of the cones.

THE RHODOPSIN-RETINAL VISUAL CYCLE, AND EXCITATION OF THE RODS

Rhodopsin and Its Decomposition by Light Energy. The outer segment of the rod that projects into the pigment layer of the retina has a concentration of about 40 per cent of the light-sensitive pigment called *rhodopsin,* or *visual purple.* This substance is a combination of the protein *scotopsin* and the carotenoid pigment *retinal* (also called "retinene"). Furthermore, the retinal is a particular type called 11-*cis* retinal. This *cis* form of the retinal is important because only this form can bind with scotopsin to synthesize rhodopsin.

When light energy is absorbed by rhodopsin, the rhodopsin begins within trillionths of a second to decompose, as shown at the top of Figure 12–5. The cause of this is photoactivation of electrons in the reti-

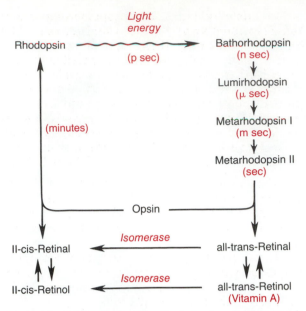

Figure 12–5. The rhodopsin-retinal visual cycle, showing decomposition of rhodopsin during exposure to light and subsequent slow reformation of rhodopsin by the chemical processes of the rod.

nal portion of the rhodopsin, which leads to an instantaneous change (on the order of trillionths of a second) of the *cis* form of retinal into an all-*trans* form, which still has the same chemical structure as the *cis* form but has a different physical structure — a straight molecule rather than a curved molecule. Because the three-dimensional orientation of the reactive sites of the all-*trans* retinal no longer fits with that of the reactive sites on the protein scotopsin, it begins to pull away from the scotopsin. The immediate product is *bathorhodopsin*, which is a partially split combination of the all-*trans* retinal and scotopsin. Bathorhodopsin itself is an extremely unstable compound and decays in nanoseconds to *lumirhodopsin*. This then decays in microseconds to *metarhodopsin I*, then in about a millisecond to *metarhodopsin II*, and, finally, much more slowly (in seconds) into the completely split products: *scotopsin* and *all-trans retinal*. It is the metarhodopsin II, also called *activated rhodopsin*, that excites electrical changes in the rods that then transmit the visual image into the central nervous system, as we discuss later.

Reformation of Rhodopsin. The first stage in reformation of rhodopsin, as shown in Figure 12–5, is to reconvert the all-*trans* retinal into 11-*cis* retinal. In the dark this process is catalyzed by the enzyme *retinal isomerase*. Once the 11-*cis* retinal is formed, it automatically recombines with the scotopsin to reform rhodopsin, which then remains stable until its decomposition is again triggered by absorption of light energy.

The Role of Vitamin A in the Formation of Rhodopsin. Note in Figure 12–5 that there is a second chemical route by which all-*trans* retinal can be converted into 11-*cis* retinal. This is by conversion of the all-*trans* retinal first into *all-trans retinol*, which is one form of vitamin A. Then, the all-*trans* retinol is con-

verted into 11-*cis* retinol under the influence of the enzyme isomerase. And, finally, the 11-*cis* retinol is converted into 11-*cis* retinal.

Vitamin A is present both in the cytoplasm of the rods and in the pigment layer of the retina as well. Therefore, vitamin A is normally always available to form new retinal when needed. On the other hand, when there is excess retinal in the retina, the excess is converted back into vitamin A, thus reducing the amount of light-sensitive pigment in the retina. We shall see later that this interconversion between retinal and vitamin A is especially important in long-term adaptation of the retina to different light intensities.

Night Blindness. Night blindness occurs in severe vitamin A deficiency. The simple reason for this is that not enough vitamin A is then available to form adequate quantities of retinal. Therefore, the amounts of rhodopsin that can be formed in the rods, as well as the amounts of color-photosensitive chemicals in the cones, are all depressed. This condition is called night blindness because the amount of light available at night is then too little to permit adequate vision, though in daylight the cones especially can still be excited despite their reduction in photochemical substances.

For night blindness to occur, a person usually must remain on a vitamin A–deficient diet for months, because large quantities of vitamin A are normally stored in the liver and can be made available to the eyes. However, once night blindness does develop, it can sometimes be completely cured in less than an hour by intravenous injection of vitamin A.

Excitation of the Rod When Rhodopsin Is Activated

The Rod Receptor Potential Is Hyperpolarizing, Not Depolarizing. The rod receptor potential is different from the receptor potentials in almost all other sensory receptors. That is, excitation of the rod causes *increased negativity* of the membrane potential, which is a state of *hyperpolarization*, rather than decreased negativity, which is the process of "depolarization" that is characteristic of almost all other sensory receptors.

But how does the activation of rhodopsin cause hyperpolarization? The answer to this is that *when rhodopsin decomposes, it decreases the membrane conductance for sodium ions in the outer segment of the rod.* And this causes hyperpolarization of the entire rod membrane in the following way:

Figure 12–6 illustrates movement of sodium ions in a complete electrical circuit through the inner and outer segments of the rod. The inner segment continually pumps sodium from inside the rod to the outside, thereby creating a negative potential on the inside of the entire cell. However, the membrane of the outer segment, in the *dark* state, is very leaky to sodium. Therefore, sodium continually leaks back to the inside of the rod and thereby neutralizes much of the negativity on the inside of the entire cell. Thus, under normal conditions, when the rod is not excited there is a reduced amount of electronegativity inside the membrane of the rod, normally about -40 mV.

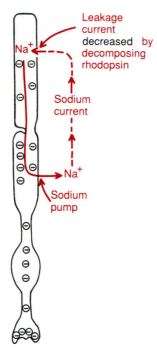

Figure 12–6. Theoretical basis for the generation of a hyperpolarization receptor potential caused by rhodopsin decomposition.

When the rhodopsin in the outer segment of the rod is exposed to light and begins to decompose, this *decreases* the outer segment conductance of sodium to the interior of the rod even though sodium ions continue to be pumped out of the inner segment. Thus, more sodium ions now leave the rod than leak back in. Because these are positive ions, their loss from inside the rod creates increased negativity inside the membrane; and the greater the amount of light energy striking the rod, the greater the electronegativity — that is, the greater the degree of *hyperpolarization*. At maximum light intensity, the membrane potential approaches −70 to −80 mV, which is near the equilibrium potential for potassium ions across the membrane.

Duration of the Receptor Potential and Logarithmic Relationship of the Receptor Potential to Light Intensity. When a sudden pulse of light strikes the retina, the transient hyperpolarization that occurs in rods — that is, the receptor potential that occurs — reaches a peak in about 0.3 sec and lasts for more than a second. In cones these changes occur four times as fast. Therefore, a visual image impinged on the retina for only a millionth of a second nevertheless can cause the sensation of seeing the image sometimes for longer than a second.

Another characteristic of the receptor potential is that it is approximately proportional to the logarithm of the light intensity. This is exceedingly important, because it allows the eye to discriminate light intensities through a range many thousand times as great as would be possible otherwise.

Mechanism by Which Rhodopsin Decomposition Decreases Membrane Sodium Conductance — The Excitation "Cascade." Under optimal conditions, a single photon of light, the smallest possible quantal unit of light energy, can cause a measurable receptor potential in a rod of about 1 mV. Only 30 photons of light will cause half saturation of the rod. How can such a small amount of light cause such great excitation? The answer is that the photoreceptors have an extremely sensitive chemical cascade that amplifies the stimulatory effects about a millionfold, as follows:

1. The *photon activates an electron* in the *11-cis retinal* portion of the rhodopsin; this leads to the formation of *metarhodopsin II*, which is the active form of rhodopsin, as already discussed and illustrated in Figure 12–5.

2. The *activated rhodopsin* functions as an enzyme to activate many molecules of *transducin*, a protein present in an inactive form in the membranes of the discs and cell membrane of the rod.

3. The *activated transducin* in turn activates many more molecules of *phosphodiesterase*.

4. *Activated phosphodiesterase* is another enzyme; it immediately hydrolyzes many, many molecules of *cyclic* guanosine monophosphate (cGMP), thus destroying it. Before being destroyed, the cGMP had been bound with the sodium channel protein in a way to "splint" it in the open state, allowing continued rapid influx of sodium ions during dark conditions. But in light, when phosphodiesterase hydrolyzes the cGMP, this removes the splinting and causes the sodium channels to close. Several hundred channels close for each originally activated molecule of rhodopsin. Because the sodium flux through each of these channels is extremely rapid, flow of more than a million sodium ions is blocked by the channel closure before the channel opens again. This diminishment of sodium ion flow is what excites the rod, as already discussed.

5. Within a small fraction of a second, another enzyme, *rhodopsin kinase*, that is always present in the rod, inactivates the activated rhodopsin, and the entire cascade reverses back to the normal state with open sodium channels.

Thus, the rods have invented an important chemical cascade that amplifies the effect of a single photon of light to cause movement of millions of sodium ions. This explains the extreme sensitivity of the rods under dark conditions.

The cones are about 300 times less sensitive than the rods, but even this allows color vision in any light greater than very dim twilight.

PHOTOCHEMISTRY OF COLOR VISION BY THE CONES

It was pointed out at the outset of this discussion that the photochemicals in the cones have almost exactly the same chemical composition as that of rhodopsin in the rods. The only difference is that the protein

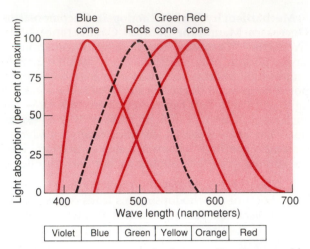

Figure 12–7. Light absorption by the respective pigments of the three color-receptive cones of the human retina. (Drawn from curves recorded by Marks, Dobelle, and MacNichol, Jr.: Science, 143:1181, 1964, and by Brown and Wald: Science, 144:45, 1964. Copyright 1964 by the American Association for the Advancement of Science.)

portions, the opsins, called *photopsins* in the cones, are different from the scotopsin of the rods. The retinal portion is exactly the same in the cones as in the rods. The color-sensitive pigments of the cones, therefore, are combinations of retinal and photopsins.

In the discussion of color vision later in the chapter, it will become evident that three different types of photochemicals are present in different cones, thus making these cones selectively sensitive to the different colors of blue, green, and red. These photochemicals are called, respectively, *blue-sensitive pigment*, *green-sensitive pigment*, and *red-sensitive pigment*. The absorption characteristics of the pigments in the three types of cones show peak absorbancies at light wavelengths, respectively, of 445, 535, and 570 nm. These are also the wavelengths for peak light sensitivity for each type of cone, which begins to explain how the retina differentiates the colors. The approximate absorption curves for these three pigments are shown in Figure 12–7. Also shown is the absorption curve for the rhodopsin of the rods, having a peak at 505 nm.

AUTOMATIC REGULATION OF RETINAL SENSITIVITY — DARK AND LIGHT ADAPTATION

Relationship of Sensitivity to Pigment Concentration. The sensitivity of rods is approximately proportional to the antilogarithm of the rhodopsin concentration, and it is assumed that this relationship also holds true in the cones. Therefore, the sensitivity of the rods and cones can be altered up or down tremendously by only slight changes in concentrations of the photosensitive chemicals.

Light and Dark Adaptation. If a person has been in bright light for a long time, large proportions of the photochemicals in both the rods and cones have been reduced to retinal and opsins. Furthermore, much of

the retinal of both the rods and cones has also been converted into vitamin A. Because of these two effects, the concentrations of the photosensitive chemicals are considerably reduced, and the sensitivity of the eye to light is even more reduced. This is called *light adaptation.*

On the other hand, if the person remains in darkness for a long time, the retinal and opsins in the rods and cones are converted back into the light-sensitive pigments. Furthermore, vitamin A is reconverted back into retinal to give still additional light-sensitive pigments, the final limit being determined by the amount of opsins in the rods and cones. This is called *dark adaptation.*

Figure 12–8 illustrates the course of dark adaptation when a person is exposed to total darkness after having been exposed to bright light for several hours. Note that sensitivity of the retina is very low on first entering the darkness, but within 1 min the sensitivity has increased tenfold—that is, the retina can respond to light of one tenth the previously required intensity. At the end of 20 min the sensitivity has increased about 6000-fold, and at the end of 40 min it has increased about 25,000-fold.

The resulting curve of Figure 12–8 is called the *dark adaptation curve.* Note, however, the inflection in the curve. The early portion of the curve is caused by adaptation of the cones, for all of the chemical events of vision occur about 4 times as rapidly in cones as in rods. On the other hand, the cones do not achieve anywhere near the same degree of sensitivity as the rods. Therefore, despite rapid adaptation by the cones, they cease adapting after only a few minutes, while the slowly adapting rods continue to adapt for many minutes and even hours, their sensitivity increasing tremendously. In addition, a large share of the greater sensitivity of the rods is also caused by convergence of as many as 100 or more rods onto a single ganglion cell in the retina; these rods summate to increase their sensitivity, as will be discussed later in the chapter.

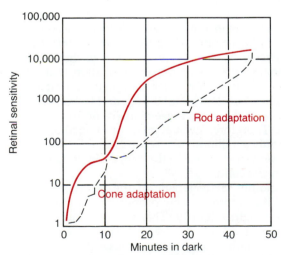

Figure 12–8. Dark adaptation, illustrating the relationship of cone adaptation to rod adaptation.

Other Mechanisms of Light and Dark Adaptation. In addition to adaptation caused by changes in concentrations of rhodopsin or color photochemicals, the eye has two other mechanisms for light and dark adaptation. The first of these is a *change in pupillary size,* which was discussed in the previous chapter. This can cause adaptation of approximately 30-fold because of changes in the amount of light allowed through the pupillary opening.

The other mechanism is *neural adaptation,* involving the neurons in the successive stages of the visual chain in the retina itself. That is, when the light intensity first increases, the intensities of the signals transmitted by the bipolar cells, the horizontal cells, the amacrine cells, and the ganglion cells are all very intense. However, the intensities of most of these signals decrease rapidly. Although the degree of this adaptation is only a fewfold rather than the many thousand-fold that occurs during adaptation of the photochemical system, this neural adaptation occurs in a fraction of a second, in contrast to the many minutes required for full adaptation by the photochemicals.

Value of Light and Dark Adaptation in Vision. Between the limits of maximal dark adaptation and maximal light adaptation, the eye can change its sensitivity to light by as much as 500,000 to 1,000,000 times, the sensitivity automatically adjusting to changes in illumination.

Since the registration of images by the retina requires detection of both dark and light spots in the image, it is essential that the sensitivity of the retina always be adjusted so that the receptors respond to the lighter areas but not to the darker areas. An example of maladjustment of the retina occurs when a person leaves a movie theater and enters the bright sunlight, for even the dark spots in the images then seem exceedingly bright, and, as a consequence, the entire visual image is bleached, having little contrast between its different parts. Obviously, this is poor vision, and it remains poor until the retina has adapted sufficiently for the darker areas of the image no longer to stimulate the receptors excessively.

Conversely, when a person enters darkness, the sensitivity of the retina is usually so slight that even the light spots in the image cannot excite the retina. After dark adaptation, however, the light spots begin to register. As an example of the extremes of light and dark adaptation, the intensity of sunlight is approximately 10 billion times that of starlight; yet the eye can function both in bright sunlight and in starlight.

■ COLOR VISION

From the preceding sections, we know that different cones are sensitive to different colors of light. The present section is a discussion of the mechanisms by which the retina detects the different gradations of color in the visual spectrum.

THE TRICOLOR MECHANISM OF COLOR DETECTION

All the theories of color vision are based on the well-known observation that the human eye can detect almost all gradations of colors when red, green, and blue monochromatic lights are appropriately mixed in different combinations.

Spectral Sensitivities of the Three Types of Cones. On the basis of color vision tests, the spectral sensitivities of the three different types of cones in human beings have been proved to be essentially the same as the light absorption curves for the three types of pigment found in the respective cones. These were illustrated in Figure 12–7 and are also shown in Figure 12–9. These curves can explain most but not all the phenomena of color vision.

Interpretation of Color in the Nervous System. Referring to Figure 12–9, one can see that an orange monochromatic light with a wavelength of 580 nm stimulates the red cones to a stimulus value of approximately 99 (99 per cent of the peak stimulation at optimum wavelength), whereas it stimulates the green cones to a stimulus value of approximately 42 and the blue cones not at all. Thus, the ratios of stimulation of the three different types of cones in this instance are 99:42:0. The nervous system interprets this set of ratios as the sensation of orange. On the other hand, a monochromatic blue light with a wavelength of 450 nm stimulates the red cones to a stimulus value of 0, the green cones to a value of 0, and the blue cones to a value of 97. This set of ratios—0:0:97—is interpreted by the nervous system as blue. Likewise, ratios of 83:83:0 are interpreted as yellow, and 31:67:36 as green.

Perception of White Light. Approximately equal stimulation of all the red, green, and blue cones gives one the sensation of seeing white. Yet there is no wavelength of light corresponding to white; instead, white is a combination of all the wavelengths of the spectrum. Furthermore, the sensation of white can be achieved by stimulating the retina with a proper combination of only three chosen colors that stimulate the respective types of cones approximately equally.

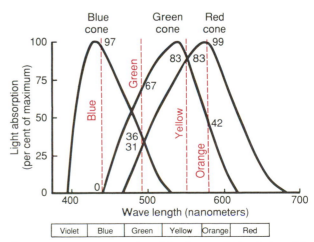

Figure 12–9. Demonstration of the degree of stimulation of the different color-sensitive cones by monochromatic lights of four separate colors: blue, green, yellow, and orange.

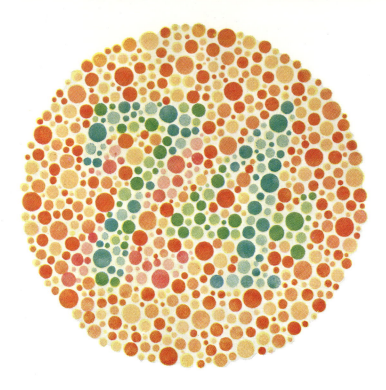

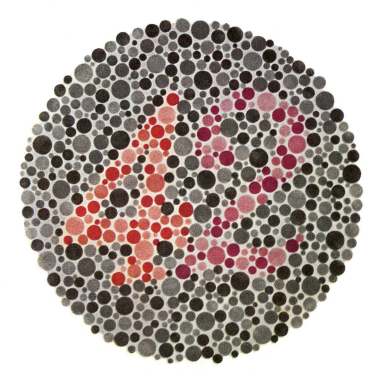

Figure 12–10. Two Ishihara charts. *Upper:* In this chart, the normal person reads "74," whereas the red-green color blind person reads "21." *Lower:* In this chart, the red-blind person (protanope) reads "2," whereas the green-blind person (deuteranope) reads "4." The normal person reads "42." (From Ishihara: Tests for Colour-Blindness. 6th Ed. Tokyo, Kanehara and Co.)

Failure of Changes in the Color of an Illuminating Light to Alter the Perceived Colors of a Visual Scene — The Phenomenon of Color Constancy. When Edwin Land was developing the Polaroid color camera, he noted that changing the color of a light illuminating a scene altered the hue of a color picture taken by the camera but did not significantly alter the hue of the scene as observed by the human eye under the same changed light conditions. This phenomenon is called *color constancy;* to this day it has not been completely explained. What is believed to occur is the following:

First, the brain computes from all the colors in the scene the overall hue of the entire vision. This computation is helped when some areas in the picture are known by the person to be white. Using this information of overall hue, the brain then adjusts mathematically for the changed color of the illuminating light, though the exact neural mechanism for doing this has not been explained. Spotted throughout the primary visual cortex of the brain are irregular peg-shaped blocks of cells called by the simple name "blobs" that demonstrate color constancy when the illuminating light changes its wavelength. Therefore, it is believed that somewhere in the neighborhood of these blobs is located the computational machinery that allows for this phenomenon of color constancy.

Obviously, color constancy is of survival value to the foraging animal when he must distinguish nutritious food from poisonous plants both in bright sunlight and in the rose-colored dawn of day.

COLOR BLINDNESS

Red-Green Color Blindness. When a single group of color receptive cones is missing from the eye, the person is unable to distinguish some colors from others. For instance, one can see in Figure 12–9 that green, yellow, orange, and red colors, which are the colors between the wavelengths of 525 and 675 nm, are normally distinguished one from the other entirely by the red and the green cones. If either of these two cones is missing, one no longer can use this mechanism for distinguishing these four colors; the person is especially unable to distinguish red from green and therefore is said to have *red-green color blindness.*

The person with loss of red cones is called a *protanope;* his overall visual spectrum is noticeably shortened at the long wavelength end because of lack of the red cones. The color blind person who lacks green cones is called a *deuteranope;* this person has a perfectly normal visual spectral width because the absent green cones operate in the middle of the spectrum.

Red-green color blindness is a genetic disease in males that is transmitted through the female. That is, genes in the female X chromosome code for the respective cones. Yet color blindness almost never occurs in the female because at least one of her two X chromosomes will almost always have normal genes for all the cones. But the male has only one X chromosome, so that a missing gene will lead to color blindness in him.

Since the X chromosome in the male is always inherited from the mother, never from the father, color blindness is passed from mother to son, and the mother is said to be a *color blindness* carrier; this is the case for about 8 per cent of all women.

Blue Weakness. Only rarely are blue cones missing, though sometimes they are underrepresented, which is a state also genetically inherited, giving rise to the phenomenon called blue weakness.

Color Test Charts. A rapid method for determining color blindness is based on the use of spot-charts such as those illustrated in Figure 12–10. These charts are arranged with a confusion of spots of several different colors. In the top chart, the normal person reads "74," while the red-green color blind person reads "21." In the bottom chart, the normal person reads "42," while the red blind "protanope" reads "2," and the green blind "deuteranope" reads "4."

If one will study these charts while at the same time observing the spectral sensitivity curves of the different cones in Figure 12–9, it can be readily understood how excessive emphasis can be placed on spots of certain colors by color blind persons in comparison with normal persons.

■ THE NEURAL FUNCTION OF THE RETINA

THE NEURAL CIRCUITRY OF THE RETINA

The first figure of this chapter, Figure 12–1, illustrated the tremendous complexity of neural organization in the retina. To simplify this, Figure 12–11 presents the basic essentials of the retina's neural connections. The different neuronal cell types are:

1. The photoreceptors themselves: the *rods* and *cones.*

2. The *horizontal cells,* which transmit signals horizontally in the outer plexiform layer from the rods and cones to the bipolar cell dendrites.

3. The *bipolar cells,* which transmit signals from the rods, cones, and horizontal cells to the inner plexiform layer, where they synapse with either amacrine cells or ganglion cells.

4. The *amacrine cells,* which transmit signals in two directions, either directly from bipolar cells to ganglion cells or horizontally within the inner plexiform layer between the axons of the bipolar cells, the dendrites of the ganglion cells, and/or other amacrine cells.

5. The *ganglion cells,* which transmit output signals from the retina through the optic nerve into the brain.

Still a sixth type of neuronal cell in the retina is the *interplexiform* cell. This cell transmits signals in the retrograde direction from the inner plexiform layer to the outer plexiform layer. These signals are all inhibitory and are believed to control the lateral spread of visual signals by the horizontal cells in the outer plexiform layer. Their role possibly is to control the degree of contrast in the visual image.

The Direct Visual Pathways From the Receptors to the Ganglion Cells. As is true of many of our sensory systems, the retina has both a very old type of vision based on rod vision and a new type of vision based on cone vision. The neurons and nerve fibers that conduct the visual signals for cone vision are considerably larger than those for rod vision, and the signals are conducted to the brain two to five times as rapidly. Also, the circuitries for the two systems are slightly different as follows:

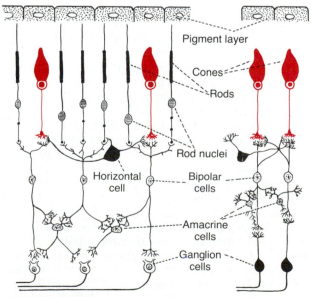

Figure 12–11. Neural organization of the retina: peripheral area to the left, foveal area to the right.

To the far right of Figure 12–11 is illustrated the visual pathway from the foveal portion of the retina, representing the new, fast system. This shows three neurons in the direct pathway: (1) cones, (2) bipolar cells, and (3) ganglion cells. In addition, horizontal cells transmit inhibitory signals laterally in the outer plexiform layer, and amacrine cells transmit signals laterally in the inner plexiform layer.

To the left in Figure 12–11 is illustrated the neural connections for the peripheral retina where both rods and cones are present. Three bipolar cells are shown; the middle of these connects only to rods, representing the old visual system. In this case, the output from the bipolar cell passes only to amacrine cells, and these in turn relay the signals to the ganglion cells. Thus, for pure rod vision there are four neurons in the direct visual pathway: (1) rods, (2) bipolar cells, (3) amacrine cells, and (4) ganglion cells. Also, both horizontal and amacrine cells provide lateral connectivity.

The other two bipolar cells illustrated in the peripheral retinal circuitry of Figure 12–11 connect with both rods and cones; the outputs of these bipolar cells pass both directly to ganglion cells and also by way of amacrine cells.

Neurotransmitters Released by Retinal Neurons. The neurotransmitters employed for synaptic transmission in the retina still have not all been delineated clearly. However, it is believed that both the rods and the cones release *glutamate*, an excitatory transmitter, at their synapses with the bipolar and horizontal cells. And histological and pharmacological studies have shown there to be many different types of amacrine cells secreting at least five different types of transmitter substances: *gamma-aminobutyric acid (GABA), glycine, dopamine, acetylcholine,* and *indolamine,* all of which normally function as inhibitory transmitters. The transmitters of the bipolar, horizontal, and interplexiform cells are still unknown.

Transmission of Most Signals Occurs in the Retina by Electrotonic Conduction, Not by Action Potentials. The only retinal neurons that always transmit visual signals by means of action potentials are the ganglion cells; and these send their signals all the way to the brain. Occasionally, though, action potentials have also been recorded in amacrine cells, though the importance of these action potentials is questionable. Otherwise, all the retinal neurons conduct their visual signals by *electrotonic conduction,* which can be explained as follows:

Electrotonic conduction means direct flow of electrical current, not action potentials, in the neuronal cytoplasm from the point of excitation all the way to the output synapses. In fact, even in the rods and cones, conduction from their outer segments where the visual signals are generated to the synaptic bodies is by electrotonic conduction. That is, when hyperpolarization occurs in response to light in the outer segment, approximately the same degree of hyperpolarization is conducted by direct electrical current flow to the synaptic body, and no action potential at all occurs. Then, when the transmitter from a rod or cone stimulates a bipolar cell or horizontal cell, once again the signal is

transmitted from the input to the output by direct electrical current flow, not by action potentials. Electrotonic conduction is also the means of signal transmission in most, if not all, of the different types of amacrine cells as well.

The importance of electrotonic conduction is that it allows *graded conduction* of signal strength. Thus, for the rods and cones, the hyperpolarizing output signal is directly related to the intensity of illumination; the signal is not all-or-none, as would be the case for action potential conduction.

LATERAL INHIBITION TO ENHANCE VISUAL CONTRAST—FUNCTION OF THE HORIZONTAL CELLS

The horizontal cells, illustrated in Figure 12–11, connect laterally between the synaptic bodies of the rods and cones and also with the dendrites of the bipolar cells. The outputs of the horizontal cells are always inhibitory. Therefore, this lateral connection provides the same phenomenon of lateral inhibition that is important in all other sensory systems, that is, allowing faithful transmission of visual patterns into the central nervous system. This phenomenon is illustrated in Figure 12–12, which shows a very minute spot of light focused on the retina. The visual pathway from the centralmost area where the light strikes is excited, whereas the area to the side, called the "surround," is inhibited. In other words, instead of the excitatory signal spreading widely in the retina because of the spreading dendritic and axonal trees in the plexiform layers, transmission through the horizontal cells puts a stop to this by providing lateral inhibition in the surrounding area. This is an essential mechanism allowing high visual accuracy in transmitting contrast borders in the visual image. It is probable that some of the amacrine cells provide additional lateral inhibition and further enhancement of visual contrast in the inner plexiform layer of the retina as well.

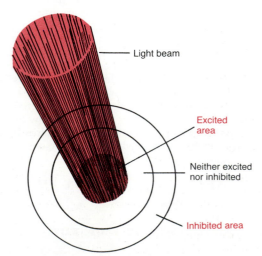

Light beam

Excited area

Neither excited nor inhibited

Inhibited area

Figure 12–12. Excitation and inhibition of a retinal area caused by a small beam of light.

EXCITATION OF SOME BIPOLAR CELLS AND INHIBITION OF OTHERS —THE DEPOLARIZING AND HYPERPOLARIZING BIPOLAR CELLS

Two different types of bipolar cells provide opposing excitatory and inhibitory signals in the visual pathway, the *depolarizing bipolar cell* and the *hyperpolarizing bipolar cell.* That is, some bipolar cells depolarize when the rods and cones are excited, and others hyperpolarize.

There are two possible explanations for this difference in response of the two different types of bipolar cells. One explanation is that the two bipolar cells are of entirely different types, one responding to the glutamate neurotransmitter released by the rods and cones by depolarizing and the other responding by hyperpolarizing. The other possibility is that one of the bipolar cells receives direct excitation from the rods and cones, whereas the other receives its signal indirectly through a horizontal cell. Because the horizontal cell is an inhibitory cell, this would reverse the polarity of the electrical response.

Regardless of the mechanism for the two different types of bipolar responses, the importance of this phenomenon is that it allows half of the bipolar cells to transmit positive signals and the other half to transmit negative signals. We shall see later that both positive and negative signals are used in transmitting visual information to the brain.

Another importance of this reciprocal relationship between depolarizing and hyperpolarizing bipolar cells is that it provides a second mechanism for lateral inhibition in addition to the horizontal cell mechanism. Since depolarizing and hyperpolarizing bipolar cells lie immediately against each other, this gives an extremely acute mechanism for separating contrast borders in the visual image even when the border lies exactly between two adjacent photoreceptors.

THE AMACRINE CELLS AND THEIR FUNCTIONS

About 30 different types of amacrine cells have been identified by morphological or histochemical means. The functions of a half dozen different types of amacrine cells have been characterized, and all of these are different from each other. It is probable that other amacrine cells have many additional functions yet to be determined.

One type of amacrine cell is part of the direct pathway for rod vision — that is, from rod to bipolar cells to amacrine cells to ganglion cells.

Another type of amacrine cell responds very strongly at the onset of a visual signal, but the response dies out rapidly. Other amacrine cells respond very strongly at the offset of visual signals, but again the response dies quickly. Finally, still other amacrine cells respond both when a light is turned on or off, signalling simply a change in illumination irrespective of direction.

Still another type of amacrine cell responds to movement of a spot across the retina in a specific direction; therefore, these amacrine cells are said to be *directional sensitive.*

In a sense, then, amacrine cells are types of interneurons that help in the beginning analysis of visual signals before they ever leave the retina.

THE GANGLION CELLS

Connectivity of the Ganglion Cells With Cones in the Fovea and With Rods and Cones in the Peripheral Retina. Each retina contains about 100,000,000 rods and 3,000,000 cones; yet the number of ganglion cells is only about 1,600,000. Thus, an average of 60 rods and 2 cones converge on each optic nerve fiber.

However, major differences exist between the peripheral retina and the central retina. As one approaches the fovea, fewer rods and cones converge on each optic fiber, and the rods and cones both become slenderer. These two effects progressively increase the acuity of vision toward the central retina. And in the very center, in the *fovea* itself, there are only slender cones, about 35,000 of them, and no rods at all. Also, the number of optic nerve fibers leading from this part of the retina is almost equal to the number of cones, as illustrated to the right in Figure 12–11. This mainly explains the high degree of visual acuity in the central retina in comparison with much poorer acuity peripherally.

Another difference between the peripheral and central portions of the retina is a much greater sensitivity of the peripheral retina to weak light. This results partly from the fact that rods are about 300 times more sensitive to light than are cones, but it is further magnified by the fact that as many as 200 rods converge on the same optic nerve fiber in the more peripheral portions of the retina, so that the signals from the rods summate to give even more intense stimulation of the peripheral ganglion cells.

Three Different Types of Retinal Ganglion Cells and Their Respective Fields

There are three distinct groups of ganglion cells designated as W, X, and Y cells. Each of these serves a different function:

Transmission of Rod Vision by the W Cells. The W cells, constituting about 40 per cent of all the ganglion cells, are small, having a diameter less than 10 μm and transmitting signals in their optic nerve fibers at the slow velocity of only 8 m/sec. These ganglion cells receive most of their excitation from rods, transmitted by way of small bipolar cells and amacrine cells. They have very broad fields in the retina because their dendrites spread widely in the inner plexiform layer, receiving signals from broad areas.

On the basis of histology as well as physiological experiments, it appears that the W cells are especially sensitive for detecting directional movement anywhere in the field of vision, and they probably also are important for much of our rod vision under dark conditions.

Transmission of the Visual Image and Color by the X Cells. The most numerous of the ganglion cells are the X cells, representing 55 per cent of the total. They are of medium diameter, between 10 and 15 μm, and transmit signals in their optic nerve fibers at about 14 m/sec.

The X cells have very small fields because their dendrites do not spread widely in the retina. Because of this, the signals represent rather discrete retinal locations. Therefore, it is through the X cells that the visual image itself is mainly transmitted. Also, because every X cell receives input from at least one cone, X cell transmission is probably responsible for all color vision as well.

Function of the Y Cells to Transmit Instantaneous Changes in the Visual Image. The Y cells are the largest of all, up to 35 μm in diameter, and they transmit their signals to the brain faster than 50 m/sec. However, they are also the fewest of all the ganglion cells, representing only 5 per cent of the total. Yet they have very broad dendritic fields, so that signals are picked up by these cells from widespread retinal areas.

The Y ganglion cells respond like many of the amacrine cells to rapid changes in the visual image, either rapid movement or rapid change in light intensity, sending bursts of signals for only a fraction of a second before the signal dies out. Therefore, these ganglion cells undoubtedly apprise the central nervous system almost instantaneously when an abnormal visual event occurs anywhere in the visual field, but without specifying with great accuracy the location of the event other than to give appropriate clues for moving the eyes toward the exciting vision.

Excitation of the Ganglion Cells

Spontaneous, Continuous Action Potentials in the Ganglion Cells. It is from the ganglion cells that the long fibers of the optic nerve lead into the brain. Because of the distance involved, the electrotonic method of conduction is no longer appropriate; and, true enough, ganglion cells transmit their signals by means of action potentials instead. Furthermore, even when unstimulated they still transmit continuous impulses at rates varying between 5 and 40 per second, with the larger nerve fibers, in general, firing more rapidly. The visual signals, in turn, are superimposed onto this background ganglion cell firing.

Transmission of Changes in Light Intensity — The On-Off Response. Many ganglion cells are especially excited by *changes* in light intensity. This is illustrated by the records of nerve impulses in Figure 12–13, showing in the upper panel strong excitation for a fraction of a second when a light was first turned on; then in another fraction of a second the level of excitation diminished. The lower tracing is from a ganglion cell located in the dark area lateral to the spot of light; this cell was markedly inhibited when the light

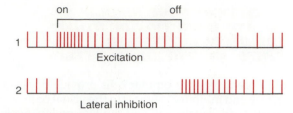

Figure 12–13. Responses of ganglion cells to light in (1) an area excited by a spot of light and (2) an area immediately adjacent to the excited spot; the ganglion cells in this area are inhibited by the mechanism of lateral inhibition. (Modified from Granit: Receptors and Sensory Perception: A Discussion of Aims, Means, and Results of Electrophysiological Research into the Process of Reception. New Haven, Conn., Yale University Press, 1955.)

was turned on because of lateral inhibition. Then, when the light was turned off, exactly the opposite effects occurred. Thus, these records are called "on-off" and "off-on" responses. The opposite directions of these responses to light are caused, respectively, by the depolarizing and hyperpolarizing bipolar cells, and the transient nature of the responses was probably generated by the amacrine cells, many of which also have similar transient responses themselves.

This capability of the eyes to detect change in light intensity is equally developed in the peripheral retina as in the central retina. For instance, a minute gnat flying across the peripheral field of vision is instantaneously detected. On the other hand, the same gnat sitting quietly remains entirely below the threshold of visual detection.

Transmission of Signals Depicting Contrasts in the Visual Scene — The Role of Lateral Inhibition

Most of the ganglion cells do not respond to the actual level of illumination of the scene; instead they respond mainly to contrast borders in the scene. Since it seems that this is the major means by which the form of the scene is transmitted to the brain, let us explain how this process occurs.

When flat light is applied to the entire retina — that is, when all the photoreceptors are stimulated equally by the incident light — the contrast type of ganglion cell is neither stimulated nor inhibited. The reason for this is that the signals transmitted *directly* from the photoreceptors through the depolarizing bipolar cells are excitatory, whereas the signals transmitted *laterally* through the horizontal cells and hyperpolarizing bipolar cells are inhibitory. Thus, the direct excitatory signal through one pathway is likely to be completely neutralized by the inhibitory signals through the lateral pathways. One circuit for this is illustrated in Figure 12–14, which shows three photoreceptors; the central one of these receptors excites a depolarizing bipolar cell. However, the two receptors on either side are connected to the same bipolar cell through inhibitory horizontal cells that neutralize the direct excitatory signal if these receptors are also stimulated by light.

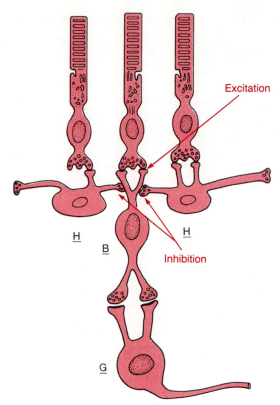

Figure 12–14. Typical arrangement of rods, horizontal cells (H), a bipolar cell (B), and a ganglion cell (G) in the retina, showing excitation at the synapses between the rods and the horizontal cells but inhibition between the horizontal cells and the bipolar cells.

Now, let us examine what happens when a contrast border occurs in the visual scene. Referring again to Figure 12–14, let us assume that the central photoreceptor is stimulated by a bright spot of light while one of the two lateral receptors is in the dark. The bright spot of light will excite the direct pathway through the bipolar cell. Then, in addition, the fact that one of the lateral photoreceptors is in the dark causes one of the horizontal cells to be inhibited. In turn, this cell loses its inhibitory effect on the bipolar cell, and this allows still more excitation of the bipolar cell. Thus, when light is everywhere, the excitatory and inhibitory signals to the bipolar cells mainly neutralize each other, but where contrasts occur the signals through the direct and lateral pathways actually accentuate each other.

Thus, the mechanism of lateral inhibition functions in the eye in the same way that it functions in most other sensory systems as well—that is, to provide contrast detection and enhancement.

Transmission of Color Signals by the Ganglion Cells

A single ganglion cell may be stimulated by a number of cones or by only a very few. When all three types of cones—the red, blue, and green types—stimulate the same ganglion cell, the signal transmitted through the ganglion cell is the same for any color of the spectrum.

Therefore, this signal plays no role in the detection of the different colors. Instead, it is a "white" signal.

On the other hand, some of the ganglion cells are excited by only one color type of cone but inhibited by a second type. For instance, this frequently occurs for the red and green cones, red causing excitation and green causing inhibition—or vice versa, with green causing excitation and red, inhibition. The same type of reciprocal effect also occurs between blue cones on the one hand and a combination of red and green cones on the other hand, giving a reciprocal excitation inhibition relationship between the blue and yellow colors.

The mechanism of this opposing effect of colors is the following: One color-type cone excites the ganglion cell by the direct excitatory route through a depolarizing bipolar cell, while the other color type inhibits the ganglion cell by the indirect inhibitory route through a horizontal cell or a hyperpolarizing bipolar cell.

The importance of these color-contrast mechanisms is that they represent a mechanism by which the retina itself begins to differentiate colors. Thus each color-contrast type of ganglion cell is excited by one color but inhibited by the "opponent color." Therefore, the process of color analysis begins in the retina and is not entirely a function of the brain.

REFERENCES

Allansmith, M. R.: The Eye and Immunology. St. Louis, C. V. Mosby Co., 1983.
Benson, W. E., et al.: Diabetes and Its Ocular Complications. Philadelphia, W. B. Saunders Co., 1988.
Cunha-Vaz, J. G. (ed.): The Blood-Retinal Barriers. New York, Plenum Press, 1980.
Dacey, D. M.: Dopamine-accumulating retinal neurons revealed by in vitro fluorescence display a unique morphology. Science, 240:1196, 1988.
Daw, N. W., et al.: The function of synaptic transmitters in the retina. Annu. Rev. Neurosci., 12:205, 1989.
DeValois, R. L., and Jacobs, G. H.: Neural mechanisms of color vision. In Darian-Smith, I. (ed.): Handbook of Physiology. Sec. 1, Vol. III. Bethesda, Md., American Physiological Society, 1984, p. 525.
Dowling, J. E., and Dubin, M. W.: The vertebrate retina. In Darian-Smith, I. (ed.): Handbook of Physiology. Sec. 1, Vol. III. Bethesda, Md., American Physiological Society, 1984, p. 317.
Fine, B. S., and Yanoff, M.: Ocular Histology: A Text and Atlas. Hagerstown, Md., Harper & Row, 1979.
Finlay, B. L., and Sengelaub, D. R. (eds.): Development of the Vertebrate Retina. New York, Plenum Publishing Corp., 1989.
Gurney, A. M., and Lester, H. A.: Light-flash physiology with synthetic photosensitive compounds. Physiol. Rev., 67:583, 1987.
Hillman, P., et al.: Transduction in invertebrate photoreceptors: Role of pigment bistability. Physiol. Rev., 63:668, 1983.
Huismans, H., et al.: The Photographed Fundus. Baltimore, Williams & Wilkins, 1988.
Hurley, J. G.: Molecular properties of the cGMP cascade of vertebrate photoreceptors. Annu. Rev. Physiol., 49:793, 1987.
Kaneko, A.: Physiology of the retina. Annu. Rev. Neurosci., 2:169, 1979.
Kanski, J. J. (ed.): BIMR Ophthalmology. Vol. 1. Disorders of the Vitreous, Retina, and Choroid. Woburn, Mass., Butterworths, 1983.
Land, E. H.: The retinex theory of color vision. Sci. Am., 237(6):108, 1977.
Liebman, P. A., et al.: The molecular mechanism of visual excitation and its relation to the structure and composition of the rod outer segment. Annu. Rev. Physiol., 49:765, 1987.
MacNichol, E. F., Jr.: Three-pigment color vision. Sci. Am., 211:48, 1964.
Marks, W. B., et al.: Visual pigments of single primate cones. Science, 143:1181, 1964.
Michaelson, I. C.: Textbook of the Fundus of the Eye. New York, Churchill Livingstone, 1980.
Ming, A. L. S., and Constable, I. J.: Colour Atlas of Ophthalmology. Boston, Houghton Mifflin, 1979.
Montgomery, G.: Seeing With the Brain. Discover, December 1988, p. 52.

Newsome, D. A.: Retinal Dystrophies and Degenerations. New York, Raven Press, 1988.

Owen, W. G.: Ionic conductances in rod photoreceptors. Annu. Rev. Physiol., 49:743, 1987.

Pugh, E. N., Jr.: The nature and identity of the internal excitational transmitter of vertebrate phototransduction. Annu. Rev. Physiol., 49:715, 1987.

Rushton, W. A. H.: Visual pigments and color blindness. Sci. Am., 232(3):64, 1975.

Ryan, S. J., et al. (eds.): Retina. St. Louis, C. V. Mosby Co., 1989.

Saibil, H. R.: From photon to receptor potential: The biochemistry of vision. News Physiol. Sci., 1:122, 1986.

Schepens, C. L.: Retinal Detachment and Allied Diseases. Philadelphia, W. B. Saunders Co., 1983.

Sherman, S. M., and Spear, P. D.: Organization of visual pathways in normal and visually deprived cats. Physiol. Rev., 62:738, 1982.

Spaeth, G. L. (ed.): Ophthalmic Surgery: Principles and Practice, 2nd Ed. Philadelphia, W. B. Saunders Co., 1989.

Stillman, A. J.: Current concepts in photoreceptor physiology. Physiologist, 28:122, 1985.

Wolf, G.: Multiple functions of vitamin A. Physiol. Rev., 64:738, 1982.

Yannuzzi, L. A., et al. (eds.): The Macula: A Comprehensive Text and Atlas. Baltimore, Williams & Wilkins, 1978.

13

The Eye:
III. Central Neurophysiology of Vision

■ THE VISUAL PATHWAYS

Figure 13–1 illustrates the principal visual pathways from the two retinas to the *visual cortex*. After nerve impulses leave the retinas they pass backward through the *optic nerves*. At the *optic chiasm* all the fibers from the nasal halves of the retinas cross to the opposite side, where they join the fibers from the opposite temporal retinas to form the *optic tracts*. The fibers of each optic tract synapse in the *dorsal lateral geniculate nucleus*, and from here the *geniculocalcarine fibers* pass by way of the *optic radiation*, or *geniculocalcarine tract*, to the *primary visual cortex* in the calcarine area of the occipital lobe.

In addition, visual fibers also pass to older precortical areas of the brain: (1) from the optic tracts to the *suprachiasmatic nucleus of the hypothalamus*, presumably for controlling circadian rhythms; (2) into the *pretectal nuclei*, for eliciting some reflex movements of the eyes focused on objects of importance and also for activating the pupillary light reflex; (3) into the *superior colliculus*, for control of rapid directional movements of the two eyes; and (4) into the *ventral lateral geniculate nucleus* of the thalamus and thence into surrounding basal regions of the brain, presumably to help control some of the body's behavioral functions.

Thus, the visual pathways can be divided roughly into an *old system* to the midbrain and base of the forebrain and a *new system* for direct transmission into the visual cortex. The new system is responsible in man for the perception of virtually all aspects of visual form, colors, and other conscious vision. On the other hand, in many lower animals, even visual form is detected by the older system, using the superior colliculus in the same manner that the visual cortex is used in mammals.

FUNCTION OF THE DORSAL LATERAL GENICULATE NUCLEUS

The optic nerve fibers of the new visual system all terminate in the *dorsal lateral geniculate nucleus*, located at the dorsal end of the thalamus and frequently also called the *lateral geniculate body*. The dorsal lateral geniculate nucleus serves two principal functions: First, it serves as a relay station to relay visual information from the optic tract to the *visual cortex* by way of the *geniculocalcarine tract*. This relay function is very accurate, so much so that there is exact point-to-point transmission with a high degree of spatial fidelity all the way from the retina to the visual cortex.

It will be recalled that half of the fibers in each optic tract after passing the optic chiasm are derived from one eye and half from the other eye, representing corresponding points on the two retinas. However, the signals from the two eyes are kept apart in the dorsal lateral geniculate nucleus. This nucleus is composed of six nuclear layers. Layers II, III, and V (from ventral to dorsal) receive signals from the temporal portion of the ipsilateral retina, whereas layers I, IV, and VI receive signals from the nasal retina of the opposite eye. The respective retinal areas of the two eyes connect with neurons that are approximately superimposed over each other in the paired layers, and similar parallel transmission is preserved all the way back to the visual cortex.

The second major function of the dorsal lateral geniculate nucleus is to "gate" the transmission of signals to the visual cortex, that is, to control how much of the signal is allowed to pass to the cortex. The nucleus receives gating control signals from two major sources, (1) *corticofugal fibers* returning in a backward direction from the primary visual cortex to the lateral geniculate

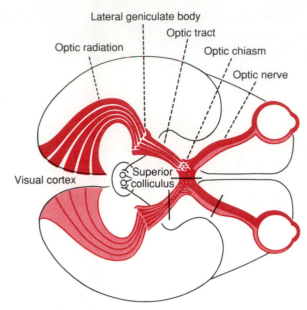

Figure 13-1. The principal visual pathways from the eyes to the visual cortex. (Modified from Polyak: The Retina. Copyright 1941 by The University of Chicago. All rights reserved.)

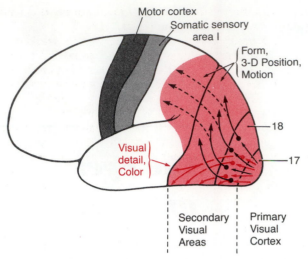

Figure 13-2. Transmission of visual signals from the primary visual cortex into secondary visual areas. Note that the signals representing form, third dimensional position, and motion are transmitted mainly superiorly into the superior portions of the occipital lobe and the posterior parietal lobe. By contrast, the signals for visual detail and color are transmitted mainly into the anteroventral portion of the occipital lobe and ventral portion of the posterior temporal lobe.

nucleus and (2) the reticular areas of the mesencephalon. Both of these are inhibitory and, when stimulated, can literally turn off transmission through selected portions of the dorsal lateral geniculate nucleus. Therefore, it is assumed that both of these gating circuits help to control the visual information that is allowed to pass.

Finally, the dorsal lateral geniculate nucleus is divided in another way: (1) Layers I and II are called *magnocellular layers* because they contain very large neurons. These receive their usual input almost entirely from the large type Y retinal ganglion cells. This magnocellular system provides a very rapidly conducting pathway to the visual cortex. On the other hand, it is "color blind," transmitting only black and white information. Also, its point-to-point transmission is poor, for there are not so many Y ganglion cells, and their dendrites spread widely in the retina. (2) Layers III through VI are called *parvocellular layers* because they contain large numbers of small- to medium-sized neurons. These receive their input almost entirely from the type X retinal ganglion cells that transmit color and also convey accurate point-to-point spatial information but at only a moderate velocity of conduction, rather than high velocity.

■ ORGANIZATION AND FUNCTION OF THE VISUAL CORTEX

Figures 13-2 and 13-3 show that the *visual cortex* is located primarily in the occipital lobes. Like the cortical representations of the other sensory systems, the visual cortex is divided into a *primary visual cortex* and *secondary visual areas.*

The Primary Visual Cortex. The primary visual cortex (Figure 13-3) lies in the *calcarine fissure area* and extends to the *occipital pole* on the medial aspect of each occipital cortex. This area is the terminus of the most direct visual signals from the eyes. Signals from the macular area of the retina terminate near the occipital pole, while signals from the more peripheral retina terminate in concentric circles anterior to the pole and along the calcarine fissure. The upper portion of the retina is represented superiorly and the lower portion inferiorly. Note in the figure the especially large area that represents the macula. It is to this region that the fovea transmits its signals. The fovea is responsible for the highest degree of visual acuity. Based on retinal area, the fovea has several hundred times as much

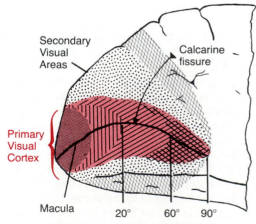

Figure 13-3. The visual cortex.

representation in the primary visual cortex as do the peripheral portions of the retina.

The primary visual cortex is coextensive with *Brodmann cortical area 17* (see the diagram of the Brodmann areas in Figure 9–5 in Chapter 9). It is frequently also called *visual area I* or simply *V-1*. Still another name for the primary visual cortex is the *striate cortex* because this area has a grossly striated appearance.

The Secondary Visual Areas. The secondary visual areas, also called *visual association areas,* lie anterior, superior, and inferior to the primary visual cortex. Secondary signals are transmitted to these areas for further analysis of visual meanings. For instance, on all sides of the primary visual cortex is *Brodmann area 18* (Figure 13–2), which is the association area where virtually all signals from the primary visual cortex pass next. Therefore, Brodmann area 18 is called *visual area II* or simply *V-2*. The other more distant secondary visual areas have specific designations V-3, V-4, and so forth. Various aspects of the visual image are progressively dissected and analyzed in separate areas.

THE LAYERED STRUCTURE OF THE PRIMARY VISUAL CORTEX

Like almost all other portions of the cerebral cortex, the primary visual cortex has six distinct layers, as illustrated in Figure 13–4. As is true for the other sensory systems, the geniculocalcarine fibers terminate mainly in layer IV. But this layer, too, is organized in subdivisions. The rapidly conducted signals from the Y retinal ganglion cells terminate in layer IVcα and from here are relayed vertically both outward toward the cortical surface and inward toward deeper levels.

The visual signals from the medium-sized optic nerve fibers, derived from the X ganglion cells in the retina, also terminate in layer IV but at points different from the Y signals, in layers IVa and IVcβ, the shallowest and deepest portions of layer IV. From here, these signals again are transmitted vertically both toward the surface of the cortex and to deeper layers. It is these X ganglion pathways that transmit the very accurate point-to-point type of vision and also color vision.

The Vertical Neuronal Columns in the Visual Cortex. The visual cortex is organized structurally into several million vertical columns of neuronal cells, each column having a diameter of 30 to 50 μm. This same vertical columnar organization is found throughout the cerebral cortex. Each column represents a functional unit. One can calculate from rough data that the number of neurons in each of the visual vertical columns is around 1000.

After the optic signals terminate in layer IV, they are further processed as they spread both outward and inward along each vertical column unit. This processing is believed to decipher separate bits of visual information at successive stations along the pathway. The signals that pass outward to layers I, II, and III eventu-

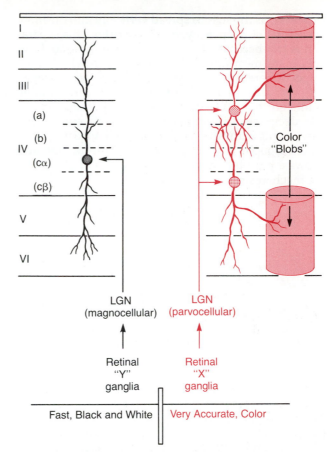

Figure 13–4. The six layers of the primary visual cortex. The connections to the left transmit very rapidly changing black and white visual signals. The pathways to the right transmit signals depicting very accurate detail and also color. Note especially small areas of the visual cortex called "color blobs" that are necessary for detection of color.

ally transmit higher orders of signals for short distances laterally in the cortex. On the other hand, the signals that pass inward to layers V and VI excite neurons that transmit signals much greater distances.

The "Color Blobs" in the Visual Cortex. Interspersed among the primary visual columns are special column-like areas called *color blobs.* These receive lateral signals from the adjacent visual columns and respond specifically to color signals. Therefore, it is presumed that these blobs are the primary areas for deciphering color. Also, in certain secondary visual areas additional color blobs are found, which presumably perform still higher levels of color deciphering.

Interaction of Visual Signals From the Separate Eyes. Recall that the visual signals from the two separate eyes are relayed through separate neuronal layers in the lateral geniculate nucleus. And these signals still remain separated from each other when they arrive in layer IV of the primary visual cortex. In fact, layer IV is interlaced with horizontal zebra-like stripes of neuronal columns, each stripe about 0.5 mm wide; the signals from one eye enter the columns of every other stripe, alternating with the signals from the other eye.

However, as the signals spread vertically into the more superficial or deeper layers of the cortex, this separation is lost because of lateral spread of the visual signals. In the meantime the cortex deciphers whether the respective areas of the two visual images are "in register" with each other, that is, whether corresponding points on the two retinas fit with each other. In turn, this deciphered information is used to control the movements of the eyes so that they will fuse with each other (brought into "register"). The information also allows a person to distinguish distances of objects by the mechanism of stereopsis.

TWO MAJOR PATHWAYS FOR ANALYSIS OF VISUAL INFORMATION — THE FAST "POSITION" AND "MOTION" PATHWAY; THE ACCURATE COLOR PATHWAY

Figure 13–2 shows that after leaving the primary visual cortex, the visual information is analyzed in two major pathways in the secondary visual areas.

1. Analysis of Three-Dimensional Position, Gross Form, and Motion of Objects. One of the analytical pathways, illustrated in Figure 13–2 by the broad black arrows, analyzes the three-dimensional positions of visual objects in the coordinates of space around the body. From this information, this pathway also analyzes the overall form of the visual scene as well as motion in the scene. In other words, this pathway tells "where" every object is at each instant and whether it is moving. After leaving the primary visual cortex (Brodmann area 17), the signals of this pathway next synapse in visual area 2 (Brodmann area 18), then flow generally into the posterior midtemporal area, and thence upward into the broad occipitoparietal cortex. At the anterior border of this last area, the signals overlap with signals from the posterior somatic association areas that analyze form and three-dimensional aspects of somatic sensory signals. The signals transmitted in this position-form-motion pathway are mainly from the large Y optic nerve fibers of the retinal Y ganglion cells, transmitting rapid signals but only black and white signals.

2. Analysis of Visual Detail and Color. The red arrows in Figure 13–2, passing from the primary visual cortex (Brodmann area 17) into visual area 2 (Brodmann area 18) and thence into the inferior ventral and medial regions of the occipital and temporal cortex, illustrate the principal pathway for analysis of visual detail. Also, separate portions of this pathway specifically dissect out color as well. Therefore, this pathway is concerned with such visual feats as recognizing letters, reading, determining the texture of surfaces, determining detailed colors of objects, and deciphering from all this information "what" the object is and its meaning.

■ NEURONAL PATTERNS OF STIMULATION DURING ANALYSIS OF THE VISUAL IMAGE

Analysis of Contrasts in the Visual Image. If a person looks at a blank wall, only a few neurons in the primary visual cortex will be stimulated whether the illumination of the wall is bright or weak. Therefore, the question must be asked, What does the visual cortex detect? To answer this, let us now place on the wall a large solid cross as illustrated to the left in Figure 13–5. To the right is illustrated the spatial pattern of the greater majority of the excited neurons in the visual cortex. *Note that the areas of maximum excitation occur along the sharp borders of the visual pattern.* Thus, the visual signal in the primary visual cortex is concerned mainly with the *contrasts* in the visual scene, rather than with the flat areas. We saw in the previous chapter that this is true of most of the retinal ganglion cells as well, because equally stimulated adjacent retinal receptors mutually inhibit each other. But at any border in the visual scene where there is a change from dark to light or light to dark, mutual inhibition does not occur, and the intensity of stimulation is proportional to the *gradient of contrast*—that is, the greater the sharpness of contrast and the greater the intensity difference between the light and dark areas, the greater the degree of stimulation.

Detection of Orientation of Lines and Borders — The "Simple" Cells. Not only does the visual cortex detect the existence of lines and borders in the different areas of the retinal image, but it also detects the orientation of each line or border—that is, whether it is vertical or horizontal or lies at some degree of inclination. This is believed to result from linear organizations of mutually inhibiting cells that excite second order neurons when mutual inhibition falls all along a line of cells, that is, where there is a contrast edge. Thus, for each such orientation of a line, a specific neuronal cell is stimulated. And a line oriented in a different direction excites a different cell. These neuronal cells are called *simple cells.* They are found mainly in layer IV of the primary visual cortex.

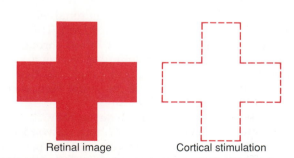

Retinal image   Cortical stimulation

Figure 13–5. Pattern of excitation occurring in the visual cortex in response to a retinal image of a dark cross.

Detection of Line Orientation When the Line Is Displaced Laterally or Vertically in the Visual Field—"Complex" Cells. As the signal progresses farther away from layer IV, some neurons now respond to lines still oriented in the same direction but not position-specific. That is, the line can be displaced moderate distances laterally or vertically in any direction in the field, and still the neuron will be stimulated if the line has the same direction. These cells are called *complex cells.*

Detection of Lines of Specific Lengths, Angles, or Other Shapes. Many neurons in the outer layers of the primary visual columns, as well as neurons in some secondary visual areas, are stimulated only by lines or borders of specific lengths, or by specific angulated shapes, or by images having other characteristics. Thus, these neurons detect still higher orders of information from the visual scene; therefore, they are called *hypercomplex cells.*

Thus, as one goes farther into the analytical pathway of the visual cortex, progressively more characteristics of each area of the visual scene are deciphered.

DETECTION OF COLOR

Color is detected in much the same way that lines are detected: by means of color contrast. The contrasts are between cones that lie immediately adjacent to each other or cones that lie far apart. For instance, a red area is often contrasted against a green area, or a blue area against a red, or a green area against a yellow. All of these colors can also be contrasted against a white area within the visual scene. In fact, it is this contrasting against white that is believed to be mainly responsible for the phenomenon called color constancy that was discussed in the previous chapter; that is, when the color of the illuminating light changes, the color of the "white" changes with the light, and appropriate computation in the brain allows red to be interpreted as red even though the illuminating light has actually changed the color spectrum entering the eyes.

The mechanism of color contrast analysis depends on the fact that contrasting colors, called opponent colors, mutually excite certain of the neuronal cells. It is presumed that the initial details of color contrast are detected by simple cells, whereas more complex contrasts are detected by complex and hypercomplex cells.

SERIAL ANALYSIS OF THE VISUAL IMAGE VERSUS PARALLEL ANALYSIS

From the foregoing discussion, it should by now be clear that the visual image is deciphered and analyzed by both serial and parallel pathways. The sequence from simple to complex to hypercomplex cells is serial analysis, with more and more details being deciphered. The transmission of different types of visual information into different brain locations represents parallel processing. It is the combination of both types of these analyses that gives one full interpretation of a visual scene. However, the highest levels of analysis are still mainly beyond present physiological understanding.

EFFECT OF REMOVING THE PRIMARY VISUAL CORTEX

Removal of the primary visual cortex in the human being causes loss of conscious vision. However, psychological studies demonstrate that such persons can still react subconsciously to changes in light intensity, to movement, and even to some gross patterns of vision. These reactions include turning the eyes, turning the head, avoidance, and so on. This vision is believed to be subserved by neuronal pathways that pass from the optic tracts mainly into the superior colliculi and other portions of the older visual system.

■ THE FIELDS OF VISION; PERIMETRY

The *field of vision* is the area seen by an eye at a given instant. The area seen to the nasal side is called the *nasal field of vision,* and the area seen to the lateral side is called the *temporal field of vision.*

To diagnose blindness in specific portions of the retinas, one charts the field of vision for each eye by a process known as *perimetry.* This is done by having the subject look with one eye toward a central spot directly in front of the eye. Then a small dot of light or a small object is moved back and forth in all areas of the field of vision, and the person indicates when the spot of light or object can be seen and when it cannot. Thus, the field of vision is plotted as illustrated in Figure 13–6.

In all perimetry charts, a *blind spot* caused by lack of rods and cones in the retina over the *optic disc* is found approxi-

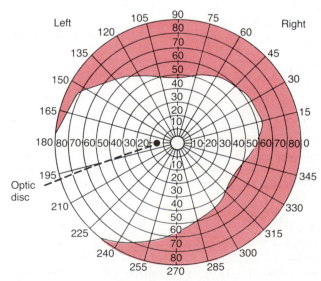

Figure 13–6. A perimetry chart, showing the field of vision for the left eye.

mately 15 degrees lateral to the central point of vision, as illustrated in the figure.

Abnormalities in the Fields of Vision. Occasionally blind spots are found in portions of the field of vision other than the optic disc area. Such blind spots are called *scotomata;* they frequently result from allergic reactions in the retina or from toxic conditions, such as lead poisoning or excessive use of tobacco.

Still another condition that can be diagnosed by perimetry is *retinitis pigmentosa*. In this disease, portions of the retina degenerate and excessive melanin pigment deposits in the degenerated areas. Retinitis pigmentosa generally causes blindness in the peripheral field of vision first and then gradually encroaches on the central areas.

Effect of Lesions in the Optic Pathway on the Fields of Vision. Destruction of an entire *optic nerve* obviously causes blindness of the respective eye. Destruction of the *optic chiasm*, as shown by the longitudinal line across the chiasm in Figure 13–1, prevents the passage of impulses from the nasal halves of the two retinae to the opposite optic tracts. Therefore, the nasal halves are both blinded, which means that the person is blind in both temporal fields of vision *because the image of the field of vision is inverted on the retina;* this condition is called *bitemporal hemianopsia.* Such lesions frequently result from tumors of the adenohypophysis pressing upward on the optic chiasm.

Interruption of an *optic tract,* which is shown by another line in Figure 13–1, denervates the corresponding half of each retina on the same side as the lesion, and, as a result, neither eye can see objects to the opposite side. This condition is known as *homonymous hemianopsia.* Destruction of the *optic radiation* or the *visual cortex* of one side also causes homonymous hemianopsia. A common condition that destroys the visual cortex is thrombosis of the posterior cerebral artery, which infarcts the occipital cortex except for part of the foveal area, thus often sparing central vision.

One can differentiate a lesion in the optic tract from a lesion in the geniculocalcarine tract or visual cortex by determining whether impulses can still be transmitted into the pretectal nuclei to initiate a pupillary light reflex.

■ EYE MOVEMENTS AND THEIR CONTROL

To make use of the abilities of the eye, almost equally as important as the system for interpretation of the visual signals from the eyes is the cerebral control system for directing the eyes toward the object to be viewed.

Muscular Control of Eye Movements. The eye movements are controlled by three separate pairs of muscles, shown in Figure 13–7: (1) the *medial* and *lateral recti,* (2) the *superior* and *inferior recti,* and (3) the *superior* and *inferior obliques.* The medial and lateral recti contract reciprocally mainly to move the eyes from side to side. The superior and inferior recti contract reciprocally to move the eyes mainly upward or downward. And the oblique muscles function mainly rotate the eyeballs to keep the visual fields in the upright position.

Neural Pathways for Control of Eye Movements. Figure 13–7 also illustrates the nuclei of the third, fourth, and sixth cranial nerves and their innervation of the ocular muscles. Shown, too, are the interconnections among these three nuclei through the *medial longitudinal fasciculus.* Either by way of this fasciculus or by way of other closely associated pathways, each of the three sets of muscles to each eye is *reciprocally* innervated so that one muscle of the pair relaxes while the other contracts.

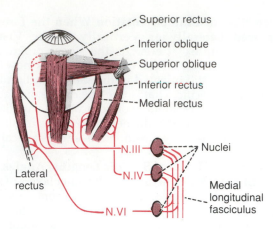

Figure 13 – 7. The extraocular muscles of the eye and their innervation.

Figure 13–8 illustrates cortical control of the oculomotor apparatus, showing spread of signals from the occipital visual areas through occipitotectal and occipitocollicular tracts into the pretectal and superior colliculus areas of the brain stem. In addition, a frontotectal tract passes from the frontal cortex into the pretectal area. From both the pretectal and the superior colliculus areas, the oculomotor control signals then pass to the nuclei of the oculomotor nerves. Strong signals are also transmitted into the oculomotor system from the vestibular nuclei by way of the medial longitudinal fasciculus.

FIXATION MOVEMENTS OF THE EYES

Perhaps the most important movements of the eyes are those that cause the eyes to "fix" on a discrete portion of the field of vision.

Fixation movements are controlled by two different neuronal mechanisms. The first of these allows the person to move his eyes voluntarily to find the object upon which he wishes to fix his vision; this is called the *voluntary fixation mechanism.* The second is an involuntary mechanism that holds the eyes firmly on the object once it has been found; this is called the *involuntary fixation mechanism.*

The voluntary fixation movements are controlled by a small cortical field located bilaterally in the premotor cortical regions of the frontal lobes, as illustrated in Figure 13–8. Bilateral dysfunction or destruction of these areas makes it difficult or almost impossible for the person to "unlock" the eyes from one point of fixation and then move them to another point. It is usually necessary for the person to blink the eyes or put a hand over the eyes for a short time, which then allows the eyes to be moved.

On the other hand, the fixation mechanism that causes the eyes to "lock" on the object of attention once it is found is controlled by *secondary visual areas of the occipital cortex*—mainly Brodmann area 19 located anterior to visual areas V-1 and V-2 (Brodmann areas 17 and 18). When this area is destroyed bilaterally, an animal has difficulty keeping its eyes directed toward a given fixation point or becomes completely unable to do so.

To summarize, the posterior eye fields automatically "lock" the eyes on a given spot of the visual field and thereby prevent movement of the image across the retina. To unlock

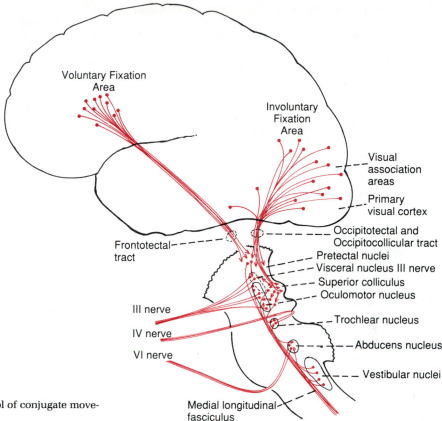

Figure 13–8. Neural pathways for control of conjugate movement of the eyes.

this visual fixation, voluntary impulses must be transmitted from the "voluntary" eye fields located in the frontal areas.

Mechanism of the Involuntary Locking Fixation — Role of the Superior Colliculi. The involuntary locking type of fixation discussed in the previous section results from a negative feedback mechanism that prevents the object of attention from leaving the foveal portion of the retina. The eyes even normally have three types of continuous but almost imperceptible movements: (1) a *continuous tremor* at a rate of 30 to 80 cycles/sec caused by successive contractions of the motor units in the ocular muscles, (2) a *slow drift* of the eyeballs in one direction or another, and (3) sudden *flicking movements* that are controlled by the involuntary fixation mechanism. When a spot of light has become fixed on the foveal region of the retina, the tremorous movements cause the spot to move back and forth at a rapid rate across the cones, and the drifting movements cause it to drift slowly across the cones. However, each time the spot of light drifts as far as the edge of the fovea, a sudden reflex reaction occurs, producing a flicking movement that moves the spot away from this edge back toward the center. Thus, an automatic response moves the image back toward the central portion of the fovea. These drifting and flicking motions are illustrated in Figure 13–9, which shows by the dashed lines the slow drifting across the retina and by the solid lines the flicks that keep the image from leaving the foveal region.

This involuntary fixation capability is mostly lost when the superior colliculi are destroyed. After the signals for fixation originate in the visual fixation areas of the occipital cortex, they pass to the superior colliculi, probably from there to reticular areas around the oculomotor nuclei, and thence into the motor nuclei themselves.

Saccadic Movement of the Eyes — a Mechanism of Successive Fixation Points. When the visual scene is moving continually before the eyes, such as when a person is riding in a car or turning around, the eyes fix on one highlight after another in the visual field, jumping from one to the next at a rate of two to three jumps per second. The jumps are called *saccades*, and the movements are called *opticokinetic movements*. The saccades occur so rapidly that not more than 10 per cent of the total time is spent in moving the eyes, 90 per cent of the time being allocated to the fixation sites. Also, the brain suppresses the visual image during the saccades so that one is completely unconscious of the movements from point to point.

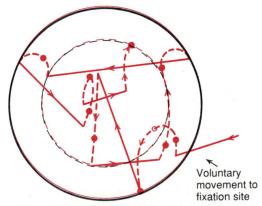

Figure 13–9. Movements of a spot of light on the fovea, showing sudden "flicking" movements to move the spot back toward the center of the fovea whenever it drifts to the foveal edge. (The dashed lines represent slow drifting movements, and the solid lines represent sudden flicking movements.) (Modified from Whitteridge: Handbook of Physiology. Vol. 2, Sec. 1. Baltimore, Williams & Wilkins, 1960.)

Saccadic Movements During Reading. During the process of reading, a person usually makes several saccadic movements of the eyes for each line. In this case the visual scene is not moving past the eyes, but the eyes are trained to scan across the visual scene to extract the important information. Similar saccades occur when a person observes a painting, except that the saccades occur in one direction after another from one highlight of the painting to another, then another, and so forth.

Fixation on Moving Objects—"Pursuit Movements." The eyes can also remain fixed on a moving object, which is called *pursuit movement*. A highly developed cortical mechanism automatically detects the course of movement of an object and then gradually develops a similar course of movement of the eyes. For instance, if an object is moving up and down in a wavelike form at a rate of several times per second, the eyes at first may be completely unable to fixate on it. However, after a second or so the eyes begin to jump coarsely in approximately the same pattern of movement as that of the object. Then after a few more seconds, the eyes develop progressively smoother and smoother movements and finally follow the course of movement almost exactly. This represents a high degree of automatic, subconscious computational ability by the cerebral cortex.

The Superior Colliculi Are Mainly Responsible for Turning the Eyes and Head Toward a Visual Disturbance. Even after the visual cortex has been destroyed, a sudden visual disturbance in a lateral area of the visual field will cause immediate turning of the eyes in that direction. This will not occur if the superior colliculi have also been destroyed. To support this function, the various points of the retina are represented topologically in the superior colliculi in the same way as in the primary visual cortex, though with less accuracy. Even so, the principal direction of a flash of light in a peripheral retinal field is mapped by the colliculi, and secondary signals are then transmitted to the oculomotor nuclei to turn the eyes.

The optic nerve fibers from the eyes to the colliculi that are responsible for these rapid turning movements are branches from the rapidly conducting Y fibers, with one branch going to the visual cortex and the other going to the superior colliculi. (The superior colliculi and other regions of the brain stem are also strongly supplied with visual signals transmitted in type W optic nerve fibers. These represent the older visual pathway, but their function is unclear.)

In addition to causing the eyes to turn toward the visual disturbance, signals are also relayed from the superior colliculi through the *medial longitudinal fasciculus* to other levels of the brain stem to cause turning of the whole head and even of the whole body toward the direction of the disturbance. Also, other types of disturbances besides visual, such as strong sounds or even stroking of the side of the body, will cause similar turning of the eyes, head, and body, but only if the superior colliculi are intact. Therefore, the superior colliculi play a global role in orienting the eyes, the head, and the body with respect to external disturbances whether visual, auditory, or somatic.

FUSION OF THE VISUAL IMAGES FROM THE TWO EYES

To make the visual perceptions more meaningful, the visual images in the two eyes normally *fuse* with each other on "corresponding points" of the two retinas.

The visual cortex plays a very important role in fusion. It was pointed out earlier in the chapter that corresponding points of the two retinas transmit visual signals to different neuronal layers of the lateral geniculate body, and these signals in turn are relayed to parallel stripes of neurons in the visual cortex. Interactions occur between the stripes of cortical neurons; these cause *interference patterns of excitation* in some of the local neuronal cells when the two visual images are not precisely "in register"—that is, not precisely fused. This excitation presumably provides the signal that is transmitted to the oculomotor apparatus to cause convergence or divergence or rotation of the eyes so that fusion can be reestablished. Once the corresponding points of the retinas are precisely in register with each other, the excitation of the specific cells in the visual cortex is greatly diminished or disappears.

The Neural Mechanism of Stereopsis for Judging Distances of Visual Objects

In Chapter 11 it was pointed out that because the two eyes are more than 2 in. apart the images on the two retinas are not exactly the same. That is, the right eye sees a little more of the right-hand side of the object and the left eye a little more of the left-hand side, and the closer the object the greater the disparity. Therefore, even when the two eyes are fused with each other, it is still impossible for all corresponding points in the two visual images to be absolutely in register at the same time. Furthermore, the nearer the object is to the eyes, the less the degree of register. This degree of non-register provides the mechanism for *stereopsis,* a very important mechanism for judging distances of visual objects up to distances of about 100 m.

The neuronal cellular mechanism for stereopsis is based on the fact that some of the fiber pathways from the retinas to the visual cortex stray 1 to 2 degrees on either side of the central pathway. Therefore, some optic pathways from the two eyes will be exactly in register for objects 2 m away; and still another set of pathways will be in register for objects 75 m away. Thus, the distance is determined by which set of pathways interact with each other. This phenomenon is called *depth perception,* which is another name for stereopsis.

Strabismus

Strabismus, which is also called *squint* or *cross-eyedness,* means lack of fusion of the eyes in one or more of the coordinates described above. Three basic types of strabismus are illustrated in Figure 13–10: *horizontal strabismus, vertical strabismus,* and *torsional strabismus.* However, combinations of two or even of all three of the different types of strabismus often occur.

Strabismus is often caused by an abnormal "set" of the fusion mechanism of the visual system. That is, in the early efforts of the child to fixate the two eyes on the same object, one of the eyes fixates satisfactorily while the other fails to

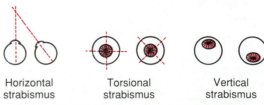

Figure 13 – 10. The three basic types of strabismus.

fixate, or they both fixate satisfactorily but never simultaneously. Soon, the patterns of conjugate movements of the eyes become abnormally "set" so that the eyes never fuse.

Frequently, some abnormality of the eyes contributes to the failure of the two eyes to fixate on the same point. For instance, in hyperopic infants, intense impulses must be transmitted to the ciliary muscles to focus the eyes, and some of these impulses overflow into the oculomotor nuclei to cause simultaneous convergence of the eyes, as will be discussed later. As a result, the child's fusion mechanism becomes "set" for continual inward deviation of the eyes.

Suppression of the Visual Image From a Repressed Eye. In a few patients with strabismus the eyes alternate in fixing on the object of attention. However, in other patients, one eye alone is used all the time while the other eye becomes repressed and is never used for vision. The vision in the repressed eye develops only slightly, usually remaining 20/400 or less. If the dominant eye then becomes blinded, vision in the repressed eye can develop only to a slight extent in the adult but far more in young children. This illustrates that visual acuity is highly dependent on proper development of the central synaptic connections from the eyes. In fact, even the numbers of neuronal connections diminish in the cortical stripes that receive signals from the repressed eye.

■ AUTONOMIC CONTROL OF ACCOMMODATION AND PUPILLARY APERTURE

The Autonomic Nerves to the Eyes. The eye is innervated by both parasympathetic and sympathetic fibers, as illustrated in Figure 13–11. The parasympathetic preganglionic fibers arise in the *Edinger-Westphal nucleus* (the visceral nucleus of the third nerve) and then pass in the *third nerve* to

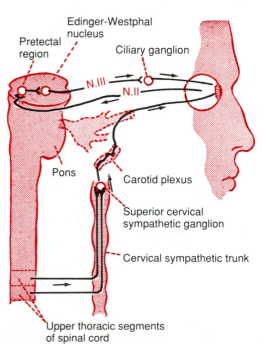

Figure 13–11. Autonomic innervation of the eye, showing also the reflex arc of the light reflex. (Modified from Ranson and Clark: Anatomy of the Nervous System. Philadelphia, W. B. Saunders Company, 1959.)

the *ciliary ganglion*, which lies immediately behind the eye. Here the preganglionic fibers synapse with postganglionic parasympathetic neurons that, in turn, send fibers through the *ciliary nerves* into the eyeball. These nerves excite the ciliary muscle and the sphincter of the iris.

The sympathetic innervation of the eye originates in the *intermediolateral horn cells* of the first thoracic segment of the spinal cord. From here, sympathetic fibers enter the sympathetic chain and pass upward to the *superior cervical ganglion*, where they synapse with postganglionic neurons. Fibers from these spread along the carotid artery and successively smaller arteries until they reach the eye. There the sympathetic fibers innervate the radial fibers of the iris as well as several extraocular structures around the eye, which are discussed shortly in relation to Horner's syndrome. Also, they supply very weak innervation to the ciliary muscle.

CONTROL OF ACCOMMODATION (FOCUSING THE EYES)

The accommodation mechanism—that is, the mechanism that focuses the lens system of the eye—is essential for a high degree of visual acuity. Accommodation results from contraction or relaxation of the ciliary muscle, contraction causing increased strength of the lens system, as explained in Chapter 11, and relaxation causing decreased strength. The question that must be answered now is, How does a person adjust accommodation to keep the eyes in focus all the time?

Accommodation of the lens is regulated by a negative feedback mechanism that automatically adjusts the focal power of the lens for the highest degree of visual acuity. When the eyes have been fixed on some far object and then suddenly fix on a near object, the lens accommodates for maximum acuity of vision usually within less than 1 sec. Though the precise control mechanism that causes this rapid and accurate focusing of the eye is still unclear, some of the known features are the following:

First, when the eyes suddenly change the distance of the fixation point, the lens always changes its strength in the proper direction to achieve a new state of focus. In other words, the lens does not make a mistake and change its strength in the wrong direction in an attempt to find the focus.

Second, different types of clues that can help the lens change its strength in the proper direction include the following: (1) *Chromatic aberration* appears to be important. That is, the red light rays focus slightly posteriorly to the blue light rays. The eyes appear to be able to detect which of these two types of rays is in better focus, and this clue relays information to the accommodation mechanism whether to make the lens stronger or weaker. (2) When the eyes fixate on a near object they also converge toward each other. The neural mechanisms for *convergence cause a simultaneous signal to strengthen the lens of the eye.* (3) *Because the fovea lies in a hollowed-out depression that is deeper than the remainder of the retina, the clarity of focus in the depth of the fovea versus the clarity of focus on the edges will be different.* It has been suggested that this also gives clues as to which way the strength of the lens needs to be changed. (4) It has been found that *the degree of accommodation of the lens oscillates slightly* all of the time, at a frequency up to two times per second. It has been suggested that the visual image becomes clearer when the oscillation of the lens strength is changing in the appropriate direction and poorer when the lens strength is changing in the wrong direction. This could give a rapid cue as to which way the strength of the lens needs to change to provide appropriate focus.

It is presumed that the cortical areas that control accommodation closely parallel those that control fixation movements of the eyes, with final integration of the visual signals in Brodmann areas 18 and 19 and transmission of motor signals to the ciliary muscle through the pretectal area and Edinger-Westphal nucleus.

CONTROL OF THE PUPILLARY DIAMETER

Stimulation of the parasympathetic nerves excites the pupillary sphincter muscle, thereby decreasing the pupillary aperture; this is called *miosis*. On the other hand, stimulation of the sympathetic nerves excites the radial fibers of the iris and causes pupillary dilatation, which is called *mydriasis*.

The Pupillary Light Reflex. When light is shone into the eyes, the pupils constrict, a reaction called the *pupillary light reflex*. The neuronal pathway for this reflex is illustrated in Figure 13–11. When light impinges on the retina, the resulting impulses pass through the optic nerves and optic tracts to the pretectal nuclei. From here, impulses pass to the *Edinger-Westphal nucleus* and finally back through the *parasympathetic nerves* to constrict the sphincter of the iris. In darkness, the reflex becomes inhibited, which results in dilatation of the pupil.

The function of the light reflex is to help the eye adapt extremely rapidly to changing light conditions, as explained in the previous chapter. The limits of pupillary diameter are about 1.5 mm on the small side and 8 mm on the large side. Therefore, the range of light adaptation that can be effected by the pupillary reflex is about 30 to 1.

Pupillary Reflexes in Central Nervous System Disease. Certain central nervous system diseases block the transmission of visual signals from the retinas to the Edinger-Westphal nucleus. Such blocks frequently occur as a result of *central nervous system syphilis, alcoholism, encephalitis,* and so forth. The block usually occurs in the pretectal region of the brain stem, though it can also result from destruction of the small afferent fibers in the optic nerves.

The final nerve fibers in the pathway through the pretectal area to the Edinger-Westphal nucleus are of the inhibitory type. Therefore, when their inhibitory effect is lost, the nucleus becomes chronically active, causing the pupils thereafter to remain partially constricted in addition to their failure to respond to light.

Yet the pupils can still constrict some more if the Edinger-Westphal nucleus is stimulated through some other pathway. For instance, when the eyes fixate on a near object, the signals that cause accommodation of the lens and also those that cause convergence of the two eyes cause a mild degree of pupillary constriction at the same time. This is called the *accommodation reflex*. Such a pupil that fails to respond to light but does respond to accommodation and also is very small (an *Argyll Robertson pupil*) is an important diagnostic sign of central nervous system disease—very often syphilis.

Horner's Syndrome. The sympathetic nerves to the eye are occasionally interrupted, and this interruption frequently occurs in the cervical sympathetic chain. This results in *Horner's syndrome,* which consists of the following effects: First, because of interruption of fibers to the pupillary dilator muscle, the pupil remains persistently constricted to a smaller diameter than that of the pupil of the opposite eye. Second, the superior eyelid droops because this eyelid is normally maintained in an open position during the waking hours partly by contraction of a smooth muscle embedded in the lid and innervated by the sympathetics. Therefore, destruction of the sympathetics makes it impossible to open the superior eyelid nearly as widely as normally. Third, the blood vessels on the corresponding side of the face and head become persistently dilated. And fourth, sweating cannot occur on the side of the face and head affected by Horner's syndrome.

REFERENCES

Andersen, R. A.: Visual and eye movement functions of the posterior parietal cortex. Annu. Rev. Neurosci., 12:377, 1989.

Anderson, D. R.: Testing the Field of Vision. St. Louis, C. V. Mosby Co., 1983.

Bahill, A. T., and Hamm, T. M.: Using open-loop experiments to study physiological systems, with examples from the human eye-movement systems. News Physiol. Sci., 4:104, 1989.

Bishop, P. O.: Processing of visual information within the retinostriate system. In Darian-Smith, I. (ed.): Handbook of Physiology. Sec. 1, Vol. III. Bethesda, Md., American Physiological Society, 1984, p. 341.

Blasdel, G. G.: Visualization of neuronal activity in monkey striate cortex. Annu. Rev. Physiol., 51:561, 1989.

Buttner, E. J. (ed.): Neuroanatomy of the Oculomotor System. New York, Elsevier Science Publishing Co., 1984.

DeValois, R. L., and DeValois, K. K.: Spatial Vision. New York, Oxford University Press, 1988.

DeValois, R. L., and Jacobs, G. H.: Neural mechanisms of color vision. In Darian-Smith, I. (ed.): Handbook of Physiology. Sec. 1, Vol. III. Bethesda, Md., American Physiological Society, 1984, p. 525.

Eckmiller, R.: Neural control of pursuit eye movements. Physiol. Rev., 67:797, 1987.

Fregnac, Y., and Imbert, M.: Development of neuronal selectivity in primary visual cortex of cat. Physiol. Rev., 64:325, 1984.

Gilbert, C. C.: Microcircuitry of the visual cortex. Annu. Rev. Neurosci., 6:217, 1983.

Hubel, D. H., and Wiesel, T. N.: Brain mechanisms of vision. Sci. Am., 241(3):150, 1979.

Hubel, D. H., and Wiesel, T. N.: Cortical and callosal connections concerned with vertical meridian of visual fields in the cat. J. Neurophysiol., 30:1561, 1967.

Hubel, D. H., and Wiesel, T. N.: Receptive fields of cells in striate cortex of very young, visually inexperienced kittens. J. Neurophysiol., 26:994, 1963.

Jones, G. M.: The remarkable vestibuloocular reflex. News Physiol. Sci., 2:85, 1987.

Lennerstrand, G., et al. (eds.): Strabismus and Amblyopia. New York, Plenum Publishing Corp., 1988.

Livingstone, M., and Hubel, D.: Segregation of form, color, movement, and depth: Anatomy, physiology, and perception. Science, 240:740, 1988.

Lund, J. S.: Anatomical organization of Macaque monkey striate visual cortex. Annu. Rev. Neurosci., 11:253, 1988.

Mitchell, D. E., and Timney, B.: Postnatal development of function in the mammalian visual system. In Darian-Smith, I. (ed.): Handbook of Physiology. Sec. 1, Vol. III. Bethesda, Md., American Physiological Society, 1984, p. 507.

Moses, R. A.: Adler's Physiology of the Eye: Clinical Application, 7th Ed. St. Louis, C. V. Mosby Co., 1981.

Peters, A., and Jones, E. G. (eds.): Visual Cortex, New York, Plenum Publishing Corp., 1985.

Poggio, G. F., and Poggio, T.: The analysis of stereopsis. Annu. Rev. Neurosci., 7:379, 1984.

Reinecke, R. D., and Parks, M. M.: Strabismus, 3rd Ed. East Norwalk, Conn., Appleton & Lange, 1987.

Robinson, D. A.: Control of eye movements. In Brooks, V. B. (ed.): Handbook of Physiology. Sec. 1, Vol. II. Bethesda, Md., American Physiological Society, 1981, p. 1275.

Schiller, P. H.: The superior colliculus and visual function. In Darian-Smith, I. (ed.): Handbook of Physiology. Sec. 1, Vol. III. Bethesda, Md., American Physiological Society, 1984, p. 457.

Schor, C. M. (ed.): Vergence Eye Movements: Basic and Clinical Aspects. Woburn, Mass., Butterworth, 1982.

Sherman, S. M., and Spear, P. D.: Organization of visual pathways in normal and visually deprived cats. Physiol. Rev., 62:738, 1982.

Simpson, J. I.: The accessory optic system. Annu. Rev. Neurosci., 7:13, 1984.

Song, P.-S.: Protozoan and related photoreceptors: Molecular aspects. Annu. Rev. Biophys. Bioeng., 12:35, 1983.

Sparks, D. L.: Translation of sensory signals into commands for control of saccadic eye movements: Role of primate superior colliculus. Physiol. Rev., 66:118, 1986.

Sterling, P.: Microcircuitry of the cat retina. Annu. Rev. Neurosci., 6:149, 1983.

Wolfe, J. M. (ed.): The Mind's Eye. New York, W. H. Freeman and Company, 1986.

Woolsey, C. N. (ed.): Cortical Sensory Organization. Multiple Visual Areas. Clifton, N.J., Humana Press, 1981.

Wurtz, R. H., and Albano, J. E.: Visual-motor function of the primate superior colliculus. Annu. Rev. Neurosci., 3:189, 1980.

14

The Sense of Hearing

The purpose of this chapter is to describe and explain the mechanism by which the ear receives sound waves, discriminates their frequencies, and finally transmits auditory information into the central nervous system where its meaning is deciphered.

■ THE TYMPANIC MEMBRANE AND THE OSSICULAR SYSTEM

CONDUCTION OF SOUND FROM THE TYMPANIC MEMBRANE TO THE COCHLEA

Figure 14–1 illustrates the *tympanic membrane* (commonly called the *eardrum)* and the *ossicular system,* which conducts sound through the middle ear. The tympanic membrane is cone-shaped, with its concavity facing downward and outward toward the auditory canal. Attached to the very center of the tympanic membrane is the *handle* of the *malleus.* At its other end the malleus is tightly bound to the *incus* by ligaments so that whenever the malleus moves the incus moves with it. The opposite end of the incus in turn articulates with the stem of the *stapes,* and the *faceplate* of the stapes lies against the membranous labyrinth in the opening of the oval window where sound waves are conducted into the inner ear, the *cochlea.*

The ossicles of the middle ear are suspended by ligaments in such a way that the combined malleus and incus act as a single lever having its fulcrum approximately at the border of the tympanic membrane. The large *head* of the malleus, which is on the opposite side of the fulcrum from the handle, almost exactly balances the other end of the lever.

The articulation of the incus with the stapes causes the stapes to push forward on the cochlear fluid every time the handle of the malleus moves inward and to pull backward on the fluid every time the malleus moves outward, which promotes inward and outward motion of the faceplate at the oval window.

The handle of the malleus is constantly pulled inward by the *tensor tympani muscle,* which keeps the tympanic membrane tensed. This allows sound vibrations on *any* portion of the tympanic membrane to be transmitted to the malleus, which would not be true if the membrane were lax.

Impedance Matching by the Ossicular System. The amplitude of movement of the stapes faceplate with each sound vibration is only three fourths as much as the amplitude of the handle of the malleus. Therefore, the ossicular lever system does not amplify the movement distance of the stapes, as is commonly believed. Instead, the system actually reduces the amplitude but increases the *force* of movement about 1.3 times. However, the surface area of the tympanic membrane is approximately 55 mm^2, whereas the surface area of the stapes averages 3.2 mm^2. This 17-fold difference times the 1.3-fold ratio of the lever system allows energy of a sound wave impinging on the tympanic membrane to be applied to the small faceplate of the stapes, causing approximately 22 times as much *pressure* on the fluid of the cochlea as is exerted by the sound wave against the tympanic membrane. Because fluid has far greater inertia than air, it is easily understood that increased amounts of pressure are needed to cause vibration in the fluid. Therefore, the tympanic membrane and ossicular system provide *impedance matching* between the sound waves in air and the sound vibrations in the fluid of the cochlea. Indeed, the impedance matching is about 50 to 75 per cent of perfect for sound frequencies between 300 and 3000 cycles/sec, which allows utilization of most of the energy in the incoming sound waves.

In the absence of the ossicular system and tympanum, sound waves can travel directly through the air of the middle ear and can enter the cochlea at the oval window. However, the sensitivity for hearing is then 15 to 20 db less than for ossicular transmission — equivalent to a decrease from a medium voice to a barely perceptible voice level.

Attenuation of Sound by Contraction of the Stapedius and Tensor Tympani Muscles. When loud sounds are transmitted through the ossicular system

177

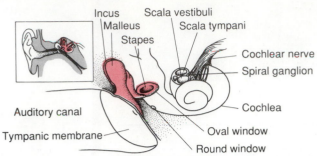

Figure 14 – 1. The tympanic membrane, the ossicular system of the middle ear, and the inner ear.

into the central nervous system, a reflex occurs after a latent period of 40 to 80 msec to cause contraction of the *stapedius* and *tensor tympani muscles.* The tensor tympani muscle pulls the handle of the malleus inward while the stapedius muscle pulls the stapes outward. These two forces oppose each other and thereby cause the entire ossicular system to develop a high degree of rigidity, thus greatly reducing the ossicular conduction of low frequency sound, mainly frequencies below 1000 cycles/sec.

This *attenuation reflex* can reduce the intensity of sound transmission by as much as 30 to 40 db, which is about the same difference as that between a loud voice and the sound of a whisper. The function of this mechanism is probably twofold:

1. To *protect* the cochlea from damaging vibrations caused by excessively loud sound.

2. To *mask* low frequency sounds in loud environments. This usually removes a major share of the background noise and allows a person to concentrate on sounds above 1000 cycles/sec, where most of the pertinent information in voice communication is transmitted.

Another function of the tensor tympani and stapedius muscles is to decrease a person's hearing sensitivity to his or her own speech. This effect is activated by collateral signals transmitted to these muscles at the same time that the brain activates the voice mechanism.

TRANSMISSION OF SOUND THROUGH BONE

Because the inner ear, the *cochlea,* is embedded in a bony cavity in the temporal bone called the bony labyrinth, vibrations of the entire skull can cause fluid vibrations in the cochlea itself. Therefore, under appropriate conditions, a tuning fork or an electronic vibrator placed on any bony protuberance of the skull, but especially on the mastoid process, causes the person to hear the sound. Unfortunately, the energy available even in very loud sound in the air is not sufficient to cause hearing through the bone except when a special electromechanical sound-transmitting device is applied directly to the bone.

■ THE COCHLEA

FUNCTIONAL ANATOMY OF THE COCHLEA

The cochlea is a system of coiled tubes, shown in Figure 14 – 1 and in cross-section in Figures 14 – 2 and 14 – 3. It consists of three different tubes coiled side by side: the *scala vestibuli,* the *scala media,* and the *scala tympani.* The scala vestibuli and scala media are separated from each other by *Reissner's membrane* (also called the *vestibular membrane*), shown in Figure 14 – 3; and the scala tympani and scala media are separated from each other by the *basilar membrane.* On the surface of the basilar membrane lies a structure, the *organ of Corti,* which contains a series of electromechanically sensitive cells, the *hair cells.* These are the receptive end-organs that generate nerve impulses in response to sound vibrations.

Figure 14 – 4 diagrams the functional parts of the uncoiled cochlea for conduction of sound vibrations. First, note that Reissner's membrane is missing from this figure. This membrane is so thin and so easily moved that it does not obstruct the passage of sound vibrations from the scala vestibuli into the scala media at all. Therefore, so far as the conduction of sound is concerned, the scala vestibuli and scala media are considered to be a single chamber. The importance of Reissner's membrane is to maintain a special fluid in the scala media that is required for normal function of the sound-receptive hair cells, as discussed later in the chapter.

Sound vibrations enter the scala vestibuli from the faceplate of the stapes at the oval window. The faceplate covers this window and is connected with the window's edges by a relatively loose annular ligament so that it can move inward and outward with the sound vibrations. Inward movement causes the fluid to move into the scala vestibuli and scala media, and outward movement causes the fluid to move backward.

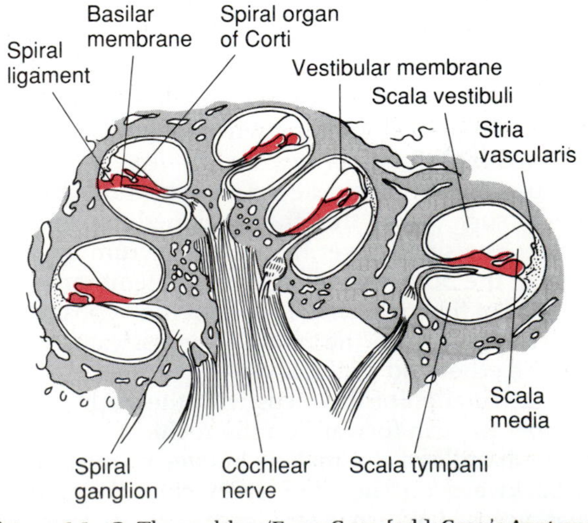

Figure 14 – 2. The cochlea. (From Goss, [ed.]: Gray's Anatomy of the Human Body. 35th Ed. Philadelphia, Lea & Febiger, 1948.)

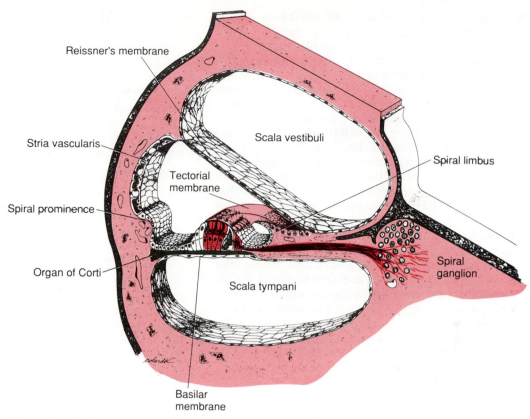

Figure 14–3. A section through one of the turns of the cochlea. (Drawn by Sylvia Colard Keene. From Fawcett: A Textbook of Histology, 11th Ed. Philadelphia, W. B. Saunders Company, 1986.)

The Basilar Membrane and Resonance in the Cochlea. The basilar membrane is a fibrous membrane that separates the scala media and the scala tympani. It contains 20,000 to 30,000 *basilar fibers* that project from the bony center of the cochlea, the *modiolus,* toward the outer wall. These fibers are stiff, elastic, reedlike structures that are fixed at their basal ends in the central bony structure of the cochlea (the modiolus) but not fixed at their distal ends, except that the distal ends are embedded in the loose basilar membrane. Because the fibers are stiff and also free at one end, they can vibrate like reeds of a harmonica.

The lengths of the basilar fibers increase progressively as one goes from the base of the cochlea to its apex, from a length of approximately 0.04 mm near the oval and round windows to 0.5 mm at the tip of the cochlea, a 12-fold increase in length.

The diameters of the fibers, on the other hand, decrease from the base to the helicotrema, so that their overall stiffness decreases more than 100-fold. As a result, the stiff, short fibers near the oval window of the cochlea will vibrate at a high frequency, whereas the long, limber fibers near the tip of the cochlea will vibrate at a low frequency.

Thus, high frequency resonance of the basilar membrane occurs near the base, where the sound waves enter the cochlea through the oval window; and low frequency resonance occurs near the apex mainly because of difference in stiffness of the fibers but also because of increased "loading" of the basilar membrane with extra amounts of fluid that must vibrate with the membrane at the apex.

TRANSMISSION OF SOUND WAVES IN THE COCHLEA — THE "TRAVELING WAVE"

If the foot of the stapes moves inward instantaneously, the round window must also bulge outward instantaneously because the cochlea is bounded on all sides by bony walls. Therefore, the initial effect is to cause the basilar membrane at the very base of the cochlea to bulge in the direction of the round window. However, the elastic tension that is built up in the basilar fibers as they bend toward the round window initiates a wave that "travels" along the basilar membrane toward the

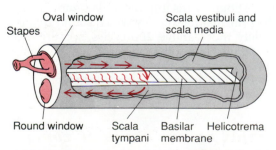

Figure 14–4. Movement of fluid in the cochlea following forward thrust of the stapes.

helicotrema, as illustrated in Figure 14–5. Figure 14–5A shows movement of a high frequency wave down the basilar membrane; Figure 14–5B, a medium frequency wave; and Figure 14–5C, a very low frequency wave. Movement of the wave along the basilar membrane is comparable to the movement of a pressure wave along the arterial walls, or it is also comparable to the wave that travels along the surface of a pond.

Pattern of Vibration of the Basilar Membrane for Different Sound Frequencies. Note in Figure 14–5 the different patterns of transmission for sound waves of different frequencies. Each wave is relatively weak at the outset but becomes strong when it reaches that portion of the basilar membrane that has a natural resonant frequency equal to the respective sound frequency. At this point the basilar membrane can vibrate back and forth with such great ease that the energy in the wave is completely dissipated. Consequently, the wave dies out at this point and fails to travel the remaining distance along the basilar membrane. Thus, a high frequency sound wave travels only a short distance along the basilar membrane before it reaches its resonant point and dies out; a medium frequency sound wave travels about halfway and then dies out; and finally, a very low frequency sound wave travels the entire distance along the membrane.

Another feature of the traveling wave is that it travels fast along the initial portion of the basilar membrane but progressively more slowly as it goes farther and farther into the cochlea. The cause of this is the high coefficient of elasticity of the basilar fibers near the stapes and a progressively decreasing coefficient farther along the membrane. This rapid initial transmission of the wave allows the high frequency sounds to travel far enough into the cochlea to spread out and separate from each other on the basilar membrane. Without this, all the high frequency waves

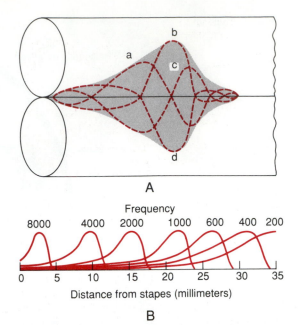

Figure 14–6. *A,* Amplitude pattern of vibration of the basilar membrane for a medium frequency sound. *B,* Amplitude patterns for sounds of all frequencies between 200 and 8000 per second, showing the points of maximum amplitude (the resonance points) on the basilar membrane for the different frequencies.

would be bunched together within the first millimeter or so of the basilar membrane, and their frequencies could not be discriminated one from the other.

Amplitude Pattern of Vibration of the Basilar Membrane. The dashed curves of Figure 14–6A show the position of a sound wave on the basilar membrane when the stapes (*a*) is all the way inward, (*b*) has moved back to the neutral point, (*c*) is all the way outward, and (*d*) has moved back again to the neutral point but is moving inward. The shaded area around these different waves shows the extent of vibration of the basilar membrane during a complete vibratory cycle. This is the *amplitude pattern of vibration* of the basilar membrane for this particular sound frequency.

Figure 14–6B shows the amplitude patterns of vibration for different frequencies, showing that the maximum amplitude for 8000 cycles occurs near the base of the cochlea, whereas that for frequencies less than 200 cycles/sec is all the way at the tip of the basilar membrane near the helicotrema where the scala vestibuli opens into the scala tympani.

The principal method by which sound frequencies, especially those above 200 cycles/sec, are discriminated from one another is based on the "place" of maximum stimulation of the nerve fibers from the organ of Corti lying on the basilar membrane, as is explained in the following section.

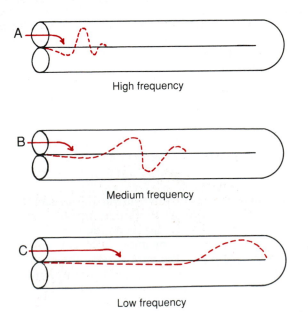

High frequency

Medium frequency

Low frequency

Figure 14–5. "Traveling waves" along the basilar membrane for high, medium, and low frequency sounds.

FUNCTION OF THE ORGAN OF CORTI

The organ of Corti, illustrated in Figures 14–2, 14–3, and 14–7, is the receptor organ that generates nerve

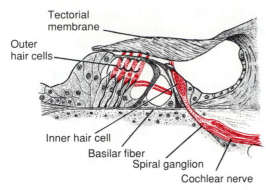

Figure 14 – 7. The organ of Corti, showing especially the hair cells and the tectorial membrane against the projecting hairs.

impulses in response to vibration of the basilar membrane. Note that the organ of Corti lies on the surface of the basilar fibers and basilar membrane. The actual sensory receptors in the organ of Corti are two types of *hair cells,* a single row of *internal hair cells,* numbering about 3500 and measuring about 12 μm in diameter, and three to four rows of *external hair cells,* numbering about 15,000 and having diameters of only about 8 μm. The bases and sides of the hair cells synapse with a network of cochlear nerve endings. These lead to the *spiral ganglion of Corti,* which lies in the modiolus (the center) of the cochlea. The spiral ganglion in turn sends axons into the *cochlear nerve* and thence into the central nervous system at the level of the upper medulla. The relationship of the organ of Corti to the spiral ganglion and to the cochlear nerve is illustrated in Figure 14 – 2.

Excitation of the Hair Cells. Note in Figure 14 – 7 that minute hairs, or *stereocilia,* project upward from the hair cells and either touch or are embedded in the surface gel coating of the *tectorial membrane,* which lies above the stereocilia in the scala media. These hair cells are similar to the hair cells found in the macula and cristae ampullaris of the vestibular apparatus, which are discussed in Chapter 17. Bending of the hairs in one direction depolarizes the hair cells, and bending them in the opposite direction hyperpolarizes them. This in turn excites the nerve fibers synapsing with their bases.

Figure 14 – 8 illustrates the mechanism by which vibration of the basilar membrane excites the hair endings. The upper ends of the hair cells are fixed tightly in a rigid structure composed of a flat plate, called the *reticular lamina,* supported by triangular *rods of Corti,* which in turn are attached tightly to the basilar fibers. Therefore, the basilar fiber, the rods of Corti, and the reticular lamina all move as a rigid unit.

Upward movement of the basilar fiber rocks the reticular lamina upward and *inward.* Then, when the basilar membrane moves downward, the reticular lamina rocks downward and *outward.* The inward and outward motion causes the hairs to shear back and forth against the tectorial membrane; or, in the case of the internal hair cells, the hairs of which do not necessarily touch the tectorial membrane, fluid rushes back

and forth over the hairs and bends them. Thus, the hair cells are excited whenever the basilar membrane vibrates.

Hair Cell Receptor Potentials and Excitation of Auditory Nerve Fibers. The stereocilia are stiff structures because each of these has a rigid internal structural protein framework as is true for all cilia in the body. Each hair cell has about 100 stereocilia on its apical border. These become progressively longer on the side away from the modiolus, and the tops of the shorter stereocilia are each attached by a thin filament to the side of its adjacent longer stereocilium. Therefore, whenever the cilia are bent in the direction of the longer ones, the tips of the smaller stereocilia are tugged outward from the surface of the hair cell. This causes a mechanical transduction that opens as many as 200 to 300 cation conducting channels, allowing rapid movement of positively charged potassium ions into the tips of the stereocilia, which in turn causes depolarization of the entire hair cell membrane.

Thus, when the basilar fibers bend toward the scala vestibuli, the hair cells depolarize, and in the opposite direction they hyperpolarize, thus generating an alternating hair cell receptor potential. This in turn stimulates the cochlear nerve endings that synapse with the bases of the hair cells. It is believed that a rapidly acting neurotransmitter is released by the hair cells at these synapses during depolarization. It is possible that the transmitter substance is glutamate, but this is not certain.

The Endocochlear Potential. To explain even more fully the electrical potentials generated by the hair cells, we need to explain another electrical phenomenon called the endocochlear potential: The scala media is filled with a fluid called *endolymph,* in contradistinction to the *perilymph* present in the scala vestibuli and scala tympani. The scala vestibuli and scala tympani communicate directly with the subarachnoid space around the brain, so that the perilymph is almost identical with cerebrospinal fluid. On the other hand, the endolymph that fills the scala media is an entirely different fluid secreted by the *stria vascularis,* a highly vascular area on the outer wall of the scala media. Endolymph contains a very high concentration of potassium and a very low concentration of sodium, which is exactly opposite to the perilymph.

An electrical potential of approximately +80 mV exists all the time between the endolymph and the perilymph, with

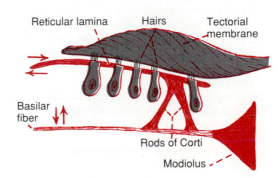

Figure 14 – 8. Stimulation of the hair cells by the to-and-fro movement of the hairs in the tectorial membrane.

positivity inside the scala media and negativity outside. This is called the *endocochlear potential,* and it is believed to be generated by continual transport of positive potassium ions from the perilymph into the scala media by the stria vascularis.

The importance of the endocochlear potential is that the tops of the hair cells project through the reticular lamina and are bathed by the endolymph of the scala media, whereas perilymph bathes the lower bodies of the hair cells. Furthermore, the hair cells have a negative intracellular potential of -60 mV with respect to the perilymph, but -140 mV with respect to the endolymph at their upper surfaces, where the hairs project into the endolymph. It is believed that this high electrical potential at the tips of the stereocilia greatly sensitizes the cell, thereby increasing its ability to respond to the slightest sound.

DETERMINATION OF SOUND FREQUENCY — THE "PLACE" PRINCIPLE

From earlier discussions in this chapter it is already apparent that low frequency sounds cause maximal activation of the basilar membrane near the apex of the cochlea, sounds of high frequency activate the basilar membrane near the base of the cochlea, and intermediate frequencies activate the membrane at intermediate distances between these two extremes. Furthermore, there is spatial organization of the nerve fibers in the cochlear pathway all the way from the cochlea to the cerebral cortex. And recording of signals from the auditory tracts in the brain stem and from the auditory receptive fields in the cerebral cortex shows that specific neurons are activated by specific sound frequencies. Therefore, the major method used by the nervous system to detect different frequencies is to determine the position along the basilar membrane that is most stimulated. This is called the *place principle* for determination of frequency (or of sound "pitch").

Yet, referring again to Figure 14–6, one can see that the distal end of the basilar membrane at the helicotrema is stimulated by all sound frequencies below 200 cycles/sec. Therefore, it has been difficult to understand from the place principle how one can differentiate between very low sound frequencies, from 200 down to 20. It is postulated that these low frequencies are discriminated mainly by the so-called *volley* or *frequency principle.* That is, low frequency sounds, from 20 up to 2000 to 4000 cycles/sec, can cause volleys of impulses at the same low frequencies to be transmitted by the cochlear nerve into the cochlear nuclei. It is believed that the cochlear nuclei then distinguish the different frequencies. In fact, destruction of the entire apical half of the cochlea, which destroys the basilar membrane where all the lower frequency sounds are normally detected, still does not completely eliminate the discrimination of low frequency sounds.

DETERMINATION OF LOUDNESS

Loudness is determined by the auditory system in at least three different ways: First, as the sound becomes louder, the amplitude of vibration of the basilar membrane and hair cells also increases, so that the hair cells excite the nerve endings at more rapid rates. Second, as the amplitude of vibration increases, it causes more and more of the hair cells on the fringes of the resonating portion of the basilar membrane to become stimulated, thus causing *spatial summation* of impulses — that is, transmission through many nerve fibers, rather than through a few. Third, certain hair cells do not become stimulated until the vibration of the basilar membrane reaches a relatively high intensity, and it is believed that stimulation of these cells in some way apprises the nervous system that the sound is then very loud.

Detection of Changes in Loudness — The Power Law. It was pointed out in Chapter 8 that a person interprets changes in intensity of sensory stimuli approximately in proportion to a power function of the actual intensity. In the case of sound, the interpreted sensation changes approximately in proportion to the cube root of the actual sound intensity. To express this another way, the ear can discriminate differences in sound intensity from the softest whisper to the loudest possible noise, representing an *approximate 1 trillion times* increase in sound energy or 1 million times increase in amplitude of movement of the basilar membrane. Yet the ear interprets this much difference in sound level as approximately a 10,000-fold change. Thus, the scale of intensity is greatly "compressed" by the sound perception mechanisms of the auditory system. This obviously allows a person to interpret differences in sound intensities over an extremely wide range, a far broader range than would be possible were it not for compression of the scale.

The Decibel Unit. Because of the extreme changes in sound intensities that the ear can detect and discriminate, sound intensities are usually expressed in terms of the logarithm of their actual intensities. A 10-fold increase in sound energy (or a $\sqrt{10}$-fold increase in sound pressure, because energy is proportional to the square of pressure) is called 1 *bel,* and 0.1 bel is called 1 *decibel* (db). One decibel represents an actual increase in sound energy of 1.26 times.

Another reason for using the decibel system in expressing changes in loudness is that, in the usual sound intensity range for communication, the ears can barely distinguish approximately a 1 db *change* in sound intensity.

Threshold for Hearing Sound at Different Frequencies. Figure 14–9 shows the pressure thresholds at which sounds of different frequencies can barely be heard by the ear. This figure illustrates that a 3000 cycle/sec sound can be heard even when its intensity is as low as 70 db below 1 dyne/cm² sound pressure level, which is one ten-millionth microwatt per square centimeter. On the other hand, a 100 cycle/sec sound can be detected only if its intensity is 10,000 times as great as this.

Frequency Range of Hearing. The frequencies of sound that a young person can hear, before aging has occurred in the ears, is generally stated to be between 20 and 20,000 cycles/sec. However, referring again to Figure 14–9, we see that the sound range depends to a great extent on intensity. If the intensity is 60 db below the 1 dyne/cm² sound pressure

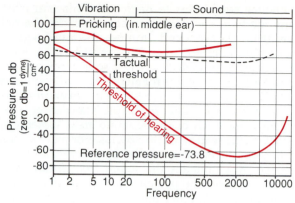

Figure 14 – 9. Relationship of the threshold of hearing and the threshold of somesthetic perception to the sound energy level at each sound frequency. (Modified from Stevens and Davis: Hearing. New York, John Wiley & Sons.)

level, the sound range is 500 to 5000 cycles/sec, and only with intense sounds can the complete range of 20 to 20,000 cycles be achieved. In old age, the frequency range falls to 50 to 8000 cycles/sec or less, as discussed later in the chapter.

■ CENTRAL AUDITORY MECHANISMS

THE AUDITORY PATHWAY

Figure 14 – 10 illustrates the major auditory pathways. It shows that nerve fibers from the *spiral ganglion of Corti* enter the *dorsal* and *ventral cochlear nuclei* located in the upper part of the medulla. At this point, all the fibers synapse, and second-order neurons pass mainly to the opposite side of the brain stem through the *trapezoid body* to the *superior olivary nucleus.* However,

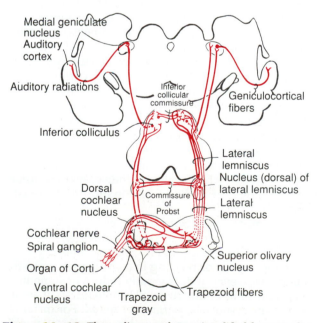

Figure 14 – 10. The auditory pathway. (Modified from Crosby, Humphrey, and Lauer.: Correlative Anatomy of the Nervous System. New York, Macmillan Publishing Co., 1962. Copyright 1962 by Macmillan Publishing Co. Reprinted with permission.)

some second-order fibers also pass ipsilaterally to the superior olivary nucleus on the same side. From the superior olivary nucleus the auditory pathway then passes upward through the *lateral lemniscus;* and some, but not all, of the fibers terminate in the *nucleus of the lateral lemniscus.* Many bypass this nucleus and pass on to the inferior colliculus, where either all or almost all of them terminate. From here, the pathway passes to the *medial geniculate nucleus,* where all the fibers again synapse. And, finally, the auditory pathway proceeds by way of the *auditory radiations* to the *auditory cortex,* located mainly in the superior gyrus of the temporal lobe.

Several points of importance in relation to the auditory pathway should be noted. First, signals from both ears are transmitted through the pathways of both sides of the brain with only slight preponderance of transmission in the contralateral pathway. In at least three different places in the brain stem crossing-over occurs between the two pathways: (1) in the trapezoid body, (2) in the commissure of Probst between the two nuclei of the lateral lemnisci, and (3) in the commissure connecting the two inferior colliculi.

Second, many collateral fibers from the auditory tracts pass directly into the *reticular activating system of the brain stem.* This system projects diffusely upward into the cerebral cortex and downward into the spinal cord and activates the entire nervous system in response to a loud sound. Other collaterals go to the *vermis of the cerebellum,* which is also activated instantaneously in the event of a sudden noise.

Third, a high degree of spatial orientation is maintained in the fiber tracts from the cochlea all the way to the cortex. In fact, there are *three* different spatial representations of sound frequencies in the cochlear nuclei, *two* representations in the inferior colliculi, *one precise* representation for discrete sound frequencies in the auditory cortex, and *at least five other less precise* representations in the auditory cortex and auditory association areas.

Firing Rates at Different Levels of the Auditory Pathway. Single nerve fibers entering the cochlear nuclei from the auditory nerve can fire at rates up to at least 1000/sec, the rate being determined mainly by the loudness of the sound. At sound frequencies up to 2000 to 4000 cycles/sec, the auditory nerve impulses are often synchronized with the sound waves, but they do not necessarily occur with every wave.

In the auditory tracts of the brain stem, the firing is usually no longer synchronized with the sound frequency except at sound frequencies below 200 cycles/sec. And above the level of the inferior colliculi, even this synchronization is mainly lost. These findings demonstrate that the sound signals are not transmitted unchanged directly from the ear to the higher levels of the brain; instead, information from the sound signals begins to be dissected from the impulse traffic at levels as low as the cochlear nuclei. We will have more to say about this later, especially in relation to perception of direction from which sound comes.

Another significant feature of the auditory pathways is that low rates of impulse firing continue even in the absence of sound all the way from the cochlear nerve fibers to the auditory cortex. When the basilar membrane moves toward the scala vestibuli, the impulse traffic increases; when the basilar membrane moves toward the scala tympani, the impulse traffic decreases. Thus, the presence of this background signal allows information to be transmitted from the basilar membrane when the membrane moves in either direction: positive information in one direction and negative information in the opposite direction. Were it not for the background signal, only the positive half of the information could be transmitted. This type of so-called ''carrier wave'' method for transmitting information is utilized in many parts of the brain, as discussed in several of the succeeding chapters.

FUNCTION OF THE CEREBRAL CORTEX IN HEARING

The projection areas of the auditory pathway to the cerebral cortex are illustrated in Figure 14–11, which shows that the auditory cortex lies principally on the *supratemporal plane of the superior temporal gyrus* but also extends over the *lateral border of the temporal lobe,* over much of the *insular cortex,* and even into the most lateral portion of the *parietal operculum.*

Two separate areas are shown in Figure 14–11: the *primary auditory cortex* and the *auditory association cortex* (also called the *secondary auditory cortex*). The primary auditory cortex is directly excited by projections from the medial geniculate body, whereas the auditory association areas are excited secondarily by impulses from the primary auditory cortex and by projections from thalamic association areas adjacent to the medial geniculate body.

Sound Frequency Perception in the Primary Auditory Cortex. At least six different *tonotopic maps* have been found in the primary auditory cortex and auditory association areas. In each of these maps, high

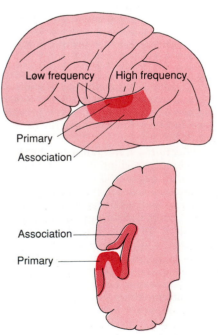

Figure 14–11. The auditory cortex.

frequency sounds excite neurons at one end of the map, whereas low frequency sounds excite the neurons at the opposite end. In most, the low frequency sounds are located anteriorly, as shown in Figure 14–11; and the high frequency sounds, posteriorly. However, this is not true for all the maps. The question that one must ask is why does the auditory cortex have so many different tonotopic maps? The answer is presumably that each of the separate areas dissects out some specific feature of the sounds. For instance, one of the large maps in the primary auditory cortex almost certainly discriminates the sound frequencies themselves and gives the person the psychic sensation of sound pitches. Another one of the maps probably is used to detect the direction from which the sound comes.

The frequency range to which each individual neuron in the auditory cortex responds is much narrower than that in the cochlear and brain stem relay nuclei. Referring back to Figure 14–6B, we note that the basilar membrane near the base of the cochlea is stimulated by all frequency sounds, and in the cochlear nuclei this same breadth of sound representation is found. Yet by the time the excitation has reached the cerebral cortex, most sound-responsive neurons respond only to a narrow range of frequencies, rather than a broad range. Therefore, somewhere along the pathway, processing mechanisms ''sharpen'' the frequency response. It is believed that this sharpening effect is caused mainly by the phenomenon of lateral inhibition, which was discussed in Chapter 8 in relation to mechanisms for transmitting information in nerves. That is, stimulation of the cochlea at one frequency causes inhibition of signals caused by sound frequencies on either side of the stimulated frequency, this resulting from collateral fibers angling off the primary signal pathway and exerting inhibitory influences on adjacent pathways. The same effect has also been demonstrated to be important in sharpening patterns of somesthetic images, visual images, and other types of sensations.

A large share of the neurons in the auditory cortex, especially in the auditory association cortex, do not respond to specific sound frequencies in the ear. It is believed that these neurons ''associate'' different sound frequencies with each other or associate sound information with information from other sensory areas of the cortex. Indeed, the parietal portion of the auditory association cortex partly overlaps somatic sensory area II, which could provide easy opportunity for association of auditory information with somatic sensory information.

Discrimination of Sound "Patterns" by the Auditory Cortex. Complete bilateral removal of the auditory cortex does not prevent a cat or monkey from detecting sounds or reacting in a crude manner to the sounds. However, it does greatly reduce or sometimes even abolish its ability to discriminate different sound pitches and especially *patterns of sound.* For instance, an animal that has been trained to recognize a combination or sequence of tones, one following the other in a particular pattern, loses this ability when the auditory cortex is destroyed; and, furthermore, it cannot relearn this type of response. Therefore, the auditory cortex is important in the discrimination of *tonal* and *sequential sound patterns.*

Total destruction of both primary auditory cortices in the human being is said to reduce greatly one's sensitivity for hearing, which is quite different from the effect in lower animals. However, this information is not clear. On the other hand, destruction of the primary auditory cortex on only one side in the human being has little effect on hearing because of the many crossover connections from side to side in the neural pathway. Yet this does affect one's ability to localize the source of sound, because comparative signals in both cortices are required for this localization function.

Lesions in the human being affecting the auditory association areas but not affecting the primary auditory cortex do not decrease the person's ability to hear and differentiate sound tones and to interpret at least simple patterns of sound. However, he or she will often be unable to interpret the *meaning* of the sound heard. For instance, lesions in the posterior portion of the superior temporal gyrus, which is called Wernicke's area and is also part of the auditory association cortex, often make it impossible for the person to interpret the meanings of words even though he hears them perfectly well and can even repeat them. These functions of the auditory association areas and their relationship to the overall intellectual functions of the brain are discussed in detail in Chapter 19.

DISCRIMINATION OF THE DIRECTION FROM WHICH SOUND EMANATES

A person determines the direction from which sound emanates by two principal mechanisms: (1) by the time lag between the entry of sound into one ear and into the opposite ear and (2) by the difference between the intensities of the sounds in the two ears. The first mechanism functions best at frequencies below 3000 cycles/sec, and the intensity mechanism operates best at higher frequencies because the head acts as a sound barrier at these frequencies. The time lag mechanism discriminates direction much more exactly than the intensity mechanism, for the time lag mechanism does not depend on extraneous factors but only on an exact interval of time between two acoustical signals. If a person is looking straight toward the sound, the sound reaches both ears at exactly the same instant, while, if the right ear is closer to the sound than the left ear, the sound signals from the right ear enter the brain ahead of those from the left ear.

Neural Mechanisms for Detecting Sound Direction. Destruction of the auditory cortex on both sides of the brain, in either human beings or lower mammals, causes loss of almost all ability to detect the direction from which sound comes. Yet the mechanism for this detection process begins in the superior olivary nuclei, even though it requires the neural pathways all the way from these nuclei to the cortex for interpretation of the signals. The mechanism is believed to be the following:

First, the superior olivary nucleus is divided into two sections, (1) the *medial superior olivary nucleus* and (2) the *lateral superior olivary nucleus.* The lateral nucleus is concerned with detecting the direction from which the sound is coming by the *difference in intensities of the sound* reaching the two ears, presumably by simply comparing the two intensities and sending an appropriate signal to the auditory cortex to estimate the direction.

The *medial superior olivary nucleus,* on the other hand, has a very specific mechanism for *detecting the time-lag between acoustic signals entering the two ears.* This nucleus contains large numbers of neurons that have two major dendrites, one projecting to the right and the other to the left. The acoustical signal from the right ear impinges on the right dendrite, and the signal from the left ear impinges on the left dendrite. The intensity of excitation of each of these neurons is highly sensitive to a specific time-lag between the two acoustical signals from the two ears. That is, the neurons near one border of the nucleus respond maximally to a short time-lag; whereas those near the opposite border respond to a very long time-lag; and those between, to intermediate time-lags. Thus, a spatial pattern of the neuronal stimulation develops in the medial superior olivary nucleus, with sound from directly in front of the head stimulating one set of olivary neurons maximally and sounds from different side angles stimulating other sets of neurons on opposite sides of the straight front neurons. This spatial orientation of signals is then transmitted all the way to the auditory cortex, where sound direction is determined by the locus in the cortex that is stimulated maximally. It is believed that the signals for determining sound direction are transmitted through a different pathway and that this pathway terminates in the cerebral cortex in a different locus from the transmission pathway and termination locus for the tonal patterns of sound.

This mechanism for detection of sound direction indicates again how information in sensory signals is dissected out as the signals pass through different levels of neuronal activity. In this case, the "quality" of sound direction is separated from the "quality" of sound tones at the level of the superior olivary nuclei.

CENTRIFUGAL SIGNALS FROM THE CENTRAL NERVOUS SYSTEM TO LOWER AUDITORY CENTERS

Retrograde pathways have been demonstrated at each level of the nervous system from the auditory cortex to the cochlea. The final pathway is mainly from the superior olivary nucleus to the hair cells themselves in the organ of Corti.

These retrograde fibers are inhibitory. Indeed, direct stimulation of discrete points in the olivary nucleus has been shown to inhibit specific areas of the organ of Corti, reducing their sound sensitivities as much as 15 to 20 db. One can readily understand how this could allow a person to direct attention to sounds of particular qualities while rejecting sounds of other qualities. This is readily demonstrated when one listens to a single instrument in a symphony orchestra.

■ HEARING ABNORMALITIES

TYPES OF DEAFNESS

Deafness is usually divided into two types: first, that caused by impairment of the cochlea or auditory nerve, which is usually classed as "nerve deafness," and, second, that caused by impairment of the mechanisms for transmitting sound into the cochlea, which is usually called "conduction deafness." Obviously, if either the cochlea or the auditory nerve is completely destroyed, the person is permanently deaf. However, if the cochlea and nerve are still intact but the ossicular system has been destroyed or ankylosed ("frozen" in place by fibrosis or calcification), sound waves can still be conducted into the cochlea by means of bone conduction.

The Audiometer. To determine the nature of hearing disabilities, the audiometer is used. Simply an earphone connected to an electronic oscillator capable of emitting pure tones ranging from low frequencies to high frequencies, the instrument is calibrated so that the zero intensity level of sound at each frequency is the loudness that can barely be heard by the normal person, based on previous studies of normal persons. However, a calibrated volume control can increase or decrease the loudness of each tone above or below the zero level. If the loudness of a tone must be increased to 30 db above normal before it can be heard, the person is said to have a *hearing loss* of 30 db for that particular tone.

In performing a hearing test using an audiometer, one tests approximately 8 to 10 frequencies covering the auditory spectrum, and the hearing loss is determined for each of these frequencies. Then the so-called "audiogram" is plotted as shown in Figures 14–12 and 14–13, depicting the hearing loss for each of the frequencies in the auditory spectrum.

The audiometer, in addition to being equipped with an earphone for testing air conduction by the ear, is also equipped with an electronic vibrator for testing bone conduction from the mastoid process into the cochlea.

The Audiogram in Nerve Deafness. In nerve deafness— this term including damage to the cochlea, to the auditory nerve, or to the central nervous system circuits from the ear—the person has decreased or total loss of ability to hear sound as tested by both air conduction and bone conduction. An audiogram depicting partial nerve deafness is illustrated in Figure 14–12. In this figure the deafness is mainly for high frequency sound. Such deafness could be caused by damage

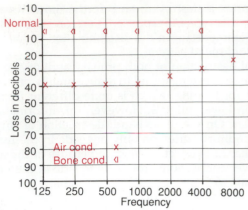

Figure 14–13. Audiogram of deafness resulting from middle ear sclerosis.

to the base of the cochlea. This type of deafness occurs to some extent in almost all older persons.

Other patterns of nerve deafness frequently occur as follows: (1) deafness for low frequency sounds caused by excessive and prolonged exposure to very loud sounds (a rock band or a jet airplane engine) because low frequency sounds are usually louder and more damaging to the organ of Corti, and (2) deafness for all frequencies caused by drug sensitivity of the organ of Corti, especially sensitivity to some antibiotics such as streptomycin, kanamycin, and chloramphenicol.

The Audiogram in Conduction Deafness. A frequent type of deafness is that caused by fibrosis of the middle ear following repeated infection in the middle ear or fibrosis occurring in the hereditary disease called *otosclerosis.* In this instance the sound waves cannot be transmitted easily through the ossicles from the tympanic membrane to the oval window. Figure 14–13 illustrates an audiogram from a person with "middle ear deafness" of this type. In this case the bone conduction is essentially normal, but air conduction is greatly depressed at all frequencies, more so at the low frequencies. In this type of deafness, the faceplate of the stapes frequently becomes "ankylosed" by bony overgrowth to the edges of the oval window. In this case, the person becomes totally deaf for air conduction but can be made to hear again almost normally by removing the stapes and replacing it with a minute Teflon or metal prosthesis that transmits the sound from the incus to the oval window.

REFERENCES

Aitkin, L. M., et al.: Central neural mechanisms of hearing. In Darian-Smith, I. (ed.): Handbook of Physiology. Sec. 1, Vol. III. Bethesda, Md., American Physiological Society, 1984, p. 675.

Altschuler, R. A., et al.: Neurobiology of Hearing: The Cochlea. New York, Raven Press, 1986.

Ballenger, J. J. (ed.): Diseases of the Nose, Throat, Ear, Head, and Neck. Philadelphia, Lea & Febiger, 1985.

Becker, W.: Atlas of Ear, Nose and Throat Diseases, Including Bronchoesophagology. Philadelphia, W. B. Saunders Co., 1984.

Borg, E., and Counter, S. A.: The middle-ear muscles. Sci. Am., August, 1989, p. 74.

Brugge, J. F., and Geisler, C. D.: Auditory mechanisms of the lower brainstem. Annu. Rev. Neurosci., 1:363, 1978.

Dallos, P.: Peripheral mechanisms of hearing. In Darian-Smith, I. (ed.): Handbook of Physiology. Sec. 1, Vol. III. Bethesda, Md., American Physiological Society, 1984, p. 595.

Fujimura, O.: Vocal Physiology: Voice Production, Mechanisms and Functions. New York, Raven Press, 1988.

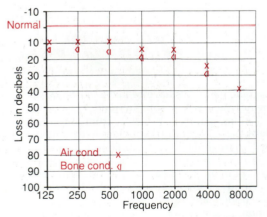

Figure 14–12. Audiogram of the old-age type of nerve deafness.

Glasscock, M., III: Shambaugh's Surgery of the Ear, 4th Ed. Philadelphia, W. B. Saunders Co., 1989.

Green, D. M., and Wier, C. C.: Auditory perception. In Darian-Smith, I. (ed.): Handbook of Physiology. Sec. 1, Vol. III. Bethesda, Md., American Physiological Society, 1984, p. 557.

Guth, P. S., and Melamed, B.: Neurotransmission in the auditory system: A primer for pharmacologists. Annu. Rev. Pharmacol. Toxicol., 22:383, 1982.

Hawke, M., et al.: Diseases of the Ear: Clinical and Pathologic Aspects. Philadelphia, Lea & Febiger, 1987.

Hudspeth, A. J.: Mechanoelectrical transduction by hair cells in the acoustico-lateralis sensory system. Annu. Rev. Neurosci., 6:187, 1983.

Hudspeth, A. J.: The cellular basis of hearing: The biophysics of hair cells. Science, 230:745, 1985.

Imig, T. J., and Morel, A.: Organization of the thalamocortical auditory system in the cat. Annu. Rev. Neurosci., 6:95, 1983.

Kay, R. H.: Hearing of modulation in sounds. Physiol. Rev., 62:187, 1983.

Kiang, N. Y. S.: Peripheral neural processing of auditory information. In Darian-Smith, I. (ed.): Handbook of Physiology. Sec. 1, Vol. III. Bethesda, Md., American Physiological Society, 1984, p. 639.

Lee, K. J.: Textbook of Otolaryngology and Head and Neck Surgery. New York, Elsevier Science Publishing Co., 1989.

Lucente, F. E., and Sobol, S. M.: Essentials of Otolaryngology, 2nd Ed. New York, Raven Press, 1988.

Masterton, R. B., and Imig, T. J.: Neural mechanisms of sound localization. Annu. Rev. Physiol., 46:275, 1984.

Patuzzi, R., and Robertson, D.: Tuning in the mammalian cochlea. Physiol. Rev., 68:1009, 1988.

Rhode, W. S.: Cochlear mechanisms. Annu. Rev. Physiol., 46:231, 1984.

Rubel, E. W.: Ontogeny of auditory system function. Annu. Rev. Physiol., 46:213, 1984.

Sachs, M. B.: Neural coding of complex sounds: Speech. Annu. Rev. Physiol., 46:261, 1984.

Sataloff, J., et al.: Hearing Loss, 2nd Ed. Philadelphia, J. B. Lippincott Co., 1980.

Schneiderman, C. R.: Basic Anatomy and Physiology in Speech and Hearing. San Diego, College-Hill Press, 1984.

Simmons, J. A., and Kick, S. A.: Physiological mechanisms for spatial filtering and image enhancement in the sonar of bats. Annu. Rev. Physiol., 46:599, 1984.

Singh, R. P.: Anatomy of Hearing and Speech. New York, Oxford University Press, 1980.

Sterkers, O., et al.: How are inner ear fluids formed? News Physiol. Sci., 2:176, 1987.

Stevens, S. S.: Hearing. Its Psychology and Physiology. New York, Acoustical Society of America, 1983.

Syka, J., and Masterton, R. B. (eds.): Auditory Pathway. Structure and Function. New York, Plenum Publishing Corp., 1988.

Weiss, T. F.: Relation of receptor potentials of cochlear hair cells to spike discharges of cochlear neurons. Annu. Rev. Physiol., 46:247, 1984.

Wever, E. G., and Lawrence, M.: Physiological Acoustics. Princeton, Princeton University Press, 1954.

Woolsey, D. N. (ed.): Cortical Sensory Organization. Multiple Auditory Areas. Clifton, N. J., Humana Press, 1982.

15

The Chemical Senses— Taste and Smell

The senses of taste and smell allow us to separate undesirable or even lethal foods from those that are nutritious. And the sense of smell allows animals to recognize the proximity of other animals, or even individuals among animals. Finally, both senses are strongly tied to primitive emotional and behavioral functions of our nervous systems.

■ THE SENSE OF TASTE

Taste is mainly a function of the *taste buds* in the mouth, but it is common experience that one's sense of smell also contributes strongly to taste perception. In addition, the texture of food, as detected by tactual senses of the mouth, and the presence of such substances in the food as pepper, which stimulate pain endings, greatly condition the taste experience. The importance of taste lies in the fact that it allows a person to select food in accord with desires and perhaps also in accord with the needs of the tissues for specific nutritive substances.

THE PRIMARY SENSATIONS OF TASTE

The identities of the specific chemicals that excite different taste receptors are still very incomplete. Even so, psychophysiological and neurophysiological studies have identified at least 13 possible or probable chemical receptors in the taste cells as follows: 2 sodium receptors, 2 potassium receptors, 1 chloride receptor, 1 adenosine receptor, 1 inosine receptor, 2 sweet receptors, 2 bitter receptors, 1 glutamate receptor, and 1 hydrogen ion receptor.

For practical analysis of taste, however, the above receptor capabilities have been collected into four general categories called the *primary sensations of taste*. These are *sour, salty, sweet,* and *bitter*.

We know, of course, that a person can perceive literally hundreds of different tastes. These are all sup-

posed to be combinations of the elementary sensations in the same manner that all the colors that we can see are combinations of the three primary colors as described in Chapter 12.

The Sour Taste. The sour taste is caused by acids, and the intensity of the taste sensation is approximately proportional to the logarithm of the *hydrogen ion concentration*. That is, the more acidic the acid, the stronger becomes the sensation.

The Salty Taste. The salty taste is elicited by ionized salts. The quality of the taste varies somewhat from one salt to another because the salts also elicit other taste sensations besides saltiness. The cations of the salts are mainly responsible for the salty taste, but the anions also contribute to a lesser extent.

The Sweet Taste. The sweet taste is not caused by any single class of chemicals. A list of some of the types of chemicals that cause this taste includes sugars, glycols, alcohols, aldehydes, ketones, amides, esters, amino acids, sulfonic acids, halogenated acids, and inorganic salts of lead and beryllium. Note specifically that most of the substances that cause a sweet taste are organic chemicals. It is especially interesting that very slight changes in the chemical structure, such as addition of a simple radical, can often change the substance from sweet to bitter.

The Bitter Taste. The bitter taste, like the sweet taste, is not caused by any single type of chemical agent; but, here again, the substances that give the bitter taste are almost entirely organic substances. Two particular classes of substances are especially likely to cause bitter taste sensations: (1) long chain organic substances containing nitrogen and (2) alkaloids. The alkaloids include many of the drugs used in medicines such as quinine, caffeine, strychnine, and nicotine.

Some substances that at first taste sweet have a bitter aftertaste. This is true of saccharin, which makes this substance objectionable to some people.

The bitter taste, when it occurs in high intensity, usually causes the person or animal to reject the food. This is undoubtedly an important purposive function

TABLE 15–1 Relative Taste Indices of Different Substances

Sour Substances	Index	Bitter Substances	Index	Sweet Substances	Index	Salty Substances	Index
Hydrochloric acid	1	Quinine	1	Sucrose	1	NaCl	1
Formic acid	1.1	Brucine	11	1-propoxy-2-amino-		NaF	2
Chloracetic acid	0.9	Strychnine	3.1	4-nitrobenzene	5000	CaCl$_2$	1
Acetyllactic acid	0.85	Nicotine	1.3	Saccharin	675	NaBr	0.4
Lactic acid	0.85	Phenylthiourea	0.9	Chloroform	40	NaI	0.35
Tartaric acid	0.7	Caffeine	0.4	Fructose	1.7	LiCl	0.4
Malic acid	0.6	Veratrine	0.2	Alanine	1.3	NH$_4$Cl	2.5
Potassium H tartrate	0.58	Pilocarpine	0.16	Glucose	0.8	KCl	0.6
Acetic acid	0.55	Atropine	0.13	Maltose	0.45		
Citric acid	0.46	Cocaine	0.02	Galactose	0.32		
Carbonic acid	0.06	Morphine	0.02	Lactose	0.3		

(From Derma: Proc. Oklahoma Acad. Sci., 27:9, 1947; and Pfaffman: Handbook of Physiology. Sec. I, Vol. I. Baltimore, Williams & Wilkins, 1959, p. 507.)

of the bitter taste sensation, because many of the deadly toxins found in poisonous plants are alkaloids, which all cause intensely bitter taste.

Threshold for Taste

The threshold for stimulation of the sour taste by hydrochloric acid averages 0.0009 M; for stimulation of the salty taste by sodium chloride, 0.01 M; for the sweet taste by sucrose, 0.01 M; and for the bitter taste by quinine, 0.000008 M. Note especially how much more sensitive is the bitter taste sense than all the others, which would be expected, because this sensation provides an important protective function.

Table 15–1 gives the relative taste indices (the reciprocals of the taste thresholds) of different substances. In this table, the intensities of the four different primary sensations of taste are referred, respectively, to the intensities of taste of hydrochloric acid, quinine, sucrose, and sodium chloride, each of which is considered to have a taste index of 1.

Taste Blindness. Many persons are taste blind for certain substances, especially for different types of thiourea compounds. A substance used frequently by psychologists for demonstrating taste blindness is *phenylthiocarbamide,* for which approximately 15 to 30 per cent of all people exhibit taste blindness, the exact percentage depending on the method of testing and the concentration of the substance.

THE TASTE BUD AND ITS FUNCTION

Figure 15–1 illustrates a taste bud, which has a diameter of about $\frac{1}{30}$ mm and a length of about $\frac{1}{16}$ mm. The taste bud is composed of about 40 modified epithelial cells, some of which are supporting cells called *sustentacular cells* and others are *taste cells.* The taste cells are continually being replaced by mitotic division from the surrounding epithelial cells so that some are young cells and others are mature cells that lie toward the center of the bud and soon break up and dissolve. The life span of each taste cell is about 10 days in lower mammals but is unknown for the human being.

The outer tips of the taste cells are arranged around a minute *taste pore,* shown in Figure 15–1. From the tip of each cell, several *microvilli,* or *taste hairs,* protrude outward into the taste pore to approach the cavity of the mouth. These microvilli provide the receptor surface for taste.

Interwoven among the taste cells is a branching terminal network of several *taste nerve fibers* that are stimulated by the taste receptor cells. Some of these fibers invaginate into folds of the taste cell membranes. Many vesicles form beneath the membrane near the fibers, suggesting that these might secrete a neurotransmitter to excite the nerve fibers in response to taste stimulation.

Location of the Taste Buds. The taste buds are found on three different types of papillae of the tongue, as follows: (1) A large number of taste buds are on the walls of the troughs that surround the circumvallate papillae, which form a V line toward the posterior of the tongue. (2) Moderate numbers of taste buds are on the fungiform papillae over the flat anterior surface of the tongue. (3) Moderate numbers are on the foliate papillae located in the folds along the lateral surfaces of the tongue. Additional taste buds are located on the palate and a few on the tonsillar pillars, the epiglottis, and even the proximal esophagus. Adults have approximately 10,000 taste buds, and children a few more. Beyond the age of 45 years many taste buds rapidly degenerate, causing the taste sensation to become progressively less critical.

Especially important in relation to taste is the tendency for taste buds subserving particular primary sensations of taste to be located in special areas. The sweet and salty tastes are located *principally* on the tip of the tongue, the sour taste on

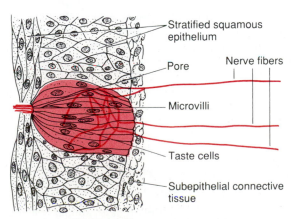

Stratified squamous epithelium

Pore

Nerve fibers

Microvilli

Taste cells

Subepithelial connective tissue

Figure 15–1. The taste bud.

the two lateral sides of the tongue, and the bitter taste on the posterior tongue and soft palate.

Specificity of Taste Buds for the Primary Taste Stimuli. Microelectrode studies from single taste buds while they are stimulated successively by the four different primary taste stimuli have shown that most of them can be excited by two, three, or even four of the primary taste stimuli as well as by a few other taste stimuli that do not fit into the "primary" categories. Usually, though, one or two of the taste categories will predominate.

Mechanism of Stimulation of Taste Buds. *The Receptor Potential.* The membrane of the taste cell, like that of other sensory receptor cells, is negatively charged on the inside with respect to the outside. Application of a taste substance to the taste hairs causes partial loss of this negative potential — that is, the taste cell is *depolarized*. The decrease in potential, within a wide range, is approximately proportional to the logarithm of concentration of the stimulating substance. This change in potential in the taste cell is the *receptor potential* for taste.

The mechanism by which the stimulating substance reacts with the taste villi to initiate the receptor potential is believed to be by binding of the taste chemicals to protein receptor molecules that protrude through the villus membrane. This in turn opens ion channels, which allow sodium ions to enter and depolarize the cell. Then the taste chemical is gradually washed away from the taste hair by the saliva, which removes the stimulus. Supposedly, the types of receptors in each taste hair determine the types of taste that will elicit responses.

Generation of Nerve Impulses by the Taste Bud. On first application of the taste stimulus, the rate of discharge of the nerve fibers rises to a peak in a small fraction of a second, but then it adapts within the next 2 sec back to a lower steady level. Thus, a strong immediate signal is transmitted by the taste nerve, and a weaker continuous signal is transmitted as long as the taste bud is exposed to the taste stimulus.

TRANSMISSION OF TASTE SIGNALS INTO THE CENTRAL NERVOUS SYSTEM

Figure 15–2 illustrates the neuronal pathways for transmission of taste signals from the tongue and pharyngeal region into the central nervous system. Taste impulses from the anterior two thirds of the tongue pass first into the *fifth nerve* and then through the *chorda tympani* into the *facial nerve*, thence into the *tractus solitarius* in the brain stem. Taste sensations from the circumvallate papillae on the back of the tongue and from other posterior regions of the mouth are transmitted through the *glossopharyngeal nerve* also into the *tractus solitarius* but at a slightly lower level. Finally, a few taste signals are transmitted into the *tractus solitarius* from the base of the tongue and other parts of the pharyngeal region by way of the *vagus nerve*.

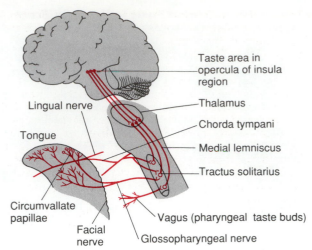

Figure 15–2. Transmission of taste impulses into the central nervous system.

All taste fibers synapse in the *nuclei of the tractus solitarius* and send second-order neurons to a small area of the *ventral posterior medial nucleus of the thalamus* located slightly medial to the thalamic terminations of the facial regions of the dorsal column-medial lemniscal system. From the thalamus, third-order neurons are transmitted to the *lower tip of the postcentral gyrus in the parietal cortex,* where it curls deep into the sylvian fissure and also into the adjacent *operculorinsular area,* also in the sylvian fissure. This lies slightly lateral, ventral, and rostral to the tongue area of somatic area I.

From this description of the taste pathways, it immediately becomes evident that they parallel closely the somatic pathways from the tongue.

Taste Reflexes Integrated in the Brain Stem. From the tractus solitarius a large number of impulses are transmitted within the brain stem itself directly into the *superior* and *inferior salivatory nuclei,* and these in turn transmit impulses to the submandibular, sublingual, and parotid glands to help control the secretion of saliva during the ingestion of food.

Adaptation of Taste. Everyone is familiar with the fact that taste sensations adapt rapidly, often with almost complete adaptation within a minute or so of continuous stimulation. Yet, from electrophysiological studies of taste nerve fibers, it is clear that the taste buds themselves adapt enough to account for no more than about one half of this. Therefore, the extreme degree of adaptation that occurs in the sensation of taste almost certainly occurs in the central nervous system itself, though the mechanism and site of this are not known. At any rate, it is a mechanism different from that of most other sensory systems, which adapt mainly at the receptors.

TASTE PREFERENCE AND CONTROL OF THE DIET

Taste preferences mean simply that an animal will choose certain types of food in preference to others, and it automatically uses this to help control the type of diet it eats. Furthermore, its taste preferences often change in accord with the needs of the body for certain specific substances. The following experiments illustrate this ability of animals to choose food in accord with the needs of its body: First, adrenalectomized animals automatically select drinking water with a high concentration of sodium chloride in preference to pure water, and this in many instances is sufficient to supply the needs of the body and prevent death as a result of salt depletion. Second, an animal given injections of excessive amounts of insulin develops a depleted blood sugar, and it automatically chooses the sweetest food from among many samples. Third, parathyroidectomized animals automatically choose drinking water with a high concentration of calcium chloride.

These same phenomena are also observed in many instances of everyday life. For instance, the salt licks of the desert region are known to attract animals from far and wide, and even the human being rejects any food that has an unpleasant affective sensation, which certainly in many instances protects our bodies from undesirable substances.

The phenomenon of taste preference almost certainly results from some mechanism located in the central nervous system and not from a mechanism in the taste receptors themselves, although it is true that the receptors often do become sensitized to the needed nutrient. An important reason for believing taste preference to be mainly a central phenomenon is that previous experience with unpleasant or pleasant tastes plays a major role in determining one's different taste preferences. For instance, if a person becomes sick soon after eating a particular type of food, the person generally develops a negative taste preference, or *taste aversion,* for that particular food thereafter; the same effect can be demonstrated in animals.

■ THE SENSE OF SMELL

Smell is the least understood of our senses. This results partly from the fact that the sense of smell is a subjective phenomenon that cannot be studied with ease in lower animals. Still another complicating problem is that the sense of smell is almost rudimentary in the human being in comparison with that of some lower animals.

THE OLFACTORY MEMBRANE

The olfactory membrane lies in the superior part of each nostril, as illustrated in Figure 15–3. Medially it folds downward over the surface of the septum, and laterally it folds over the superior turbinate and even over a small portion of the upper surface of the middle turbinate. In each nostril the olfactory membrane has a surface area of approximately 2.4 cm².

The Olfactory Cells. The receptor cells for the smell sensation are the *olfactory cells,* which are actually bipolar nerve cells derived originally from the central nervous system itself. There are about 100 million of these cells in the olfactory epithelium interspersed among *sustentacular cells,* as shown in Figure 15–4. The mucosal end of the olfactory cell forms a knob from which 6 to 12 *olfactory hairs,* or *cilia,* 0.3 μm in diameter and up to 200 μm in length, project into the

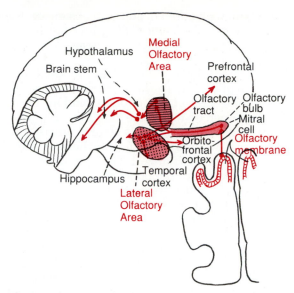

Figure 15–3. Neural connections of the olfactory system.

mucus that coats the inner surface of the nasal cavity. These projecting olfactory cilia form a dense mat in the mucus, and it is these cilia that react to odors in the air and then stimulate the olfactory cells, as discussed later. Spaced among the olfactory cells in the olfactory membrane are many small *glands of Bowman* that secrete mucus onto the surface of the olfactory membrane.

STIMULATION OF THE OLFACTORY CELLS

Mechanism of Excitation of the Olfactory Cells. The portion of the olfactory cells that responds to the olfactory chemical stimuli is the *cilia.* The membranes of the cilia contain large numbers of protein molecules that protrude all the way through the membrane and

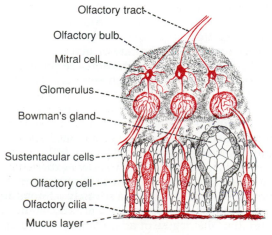

Figure 15–4. Organization of the olfactory membrane.

that can bind with different odorant substances. These proteins are called *odorant binding proteins*. It is presumed that this binding is the necessary stimulus for exciting the olfactory cells.

Two different theories have been proposed for the mechanism of excitation: The simplest theory suggests that the molecules of the odorant binding proteins themselves open up to become ion channels when the odorant binds, allowing mainly large numbers of positively charged sodium ions to flow to the interior of the olfactory cell and depolarize it. The second theory proposes that binding of the odorant causes the odorant binding protein to become an activated adenylate cyclase at its end that protrudes to the interior of the cell. The cyclase in turn catalyzes the formation of cyclic adenosine monophosphate (cAMP), and the cAMP acts on many other membrane proteins to open ion channels through them. This second mechanism would provide an extremely sensitive receptor because of a cascade effect that would occur and allow even the most minute of stimulation to cause a reaction.

Regardless of the basic chemical mechanism by which the olfactory cells are stimulated, several physical factors also affect the degree of stimulation. First, only volatile substances that can be sniffed into the nostrils can be smelled. Second, the stimulating substance must be at least slightly water soluble, so that it can pass through the mucus to reach the olfactory cells. And, third, it must be at least slightly lipid soluble, presumably because the lipid constituents of the cell membrane repel odorants from the membrane receptor proteins.

Membrane Potentials and Action Potentials in Olfactory Cells. The membrane potential of unstimulated olfactory cells, as measured by microelectrodes, averages about -55 mV. At this potential, most of the cells generate continuous action potentials at a rate varying from once every 20 sec up to 2 to 3/sec.

Most odorants cause depolarization of the olfactory membrane, decreasing the negative potential in the olfactory cell from -55 down to as low as -30 mV or even less. Along with this, the number of action potentials increases to about 20/sec, which is a very high rate for the very minute, fraction-of-a-micrometer olfactory nerve fibers.

A few odorants hyperpolarize the olfactory cell membrane, thus decreasing, instead of increasing, the nerve firing rate.

Over a wide range, the rate of olfactory nerve impulses is approximately proportional to the logarithm of the stimulus strength, which illustrates that the olfactory receptors tend to obey principles of transduction similar to those of other sensory receptors.

Adaptation. The olfactory receptors adapt approximately 50 per cent in the first second or so after stimulation. Thereafter, they adapt very little and very slowly. Yet we all know from our own experience that smell sensations adapt almost to extinction within a minute or so after one enters a strongly odorous atmosphere. Because this psychological adaptation is far greater than the degree of adaptation of the receptors themselves, it is almost certain that most of the adaptation occurs in the central nervous system, which seems also to be true for the adaptation of taste sensations. A postulated neuronal mechanism for this adaptation is the following: Large numbers of centrifugal nerve fibers pass from the olfactory regions of the brain backward along the olfactory tract and terminate on special inhibitory cells in the olfactory bulb, the *granule cells*. It is postulated that after the onset of an olfactory stimulus the central nervous system gradually develops a strong feedback inhibition to suppress relaying of the smell signals through the olfactory bulb.

SEARCH FOR THE PRIMARY SENSATIONS OF SMELL

Most physiologists are convinced that the many smell sensations are subserved by a few rather discrete primary sensations, in the same way that vision and taste are subserved by a select few sensations. But, thus far, only minor success has been achieved in classifying the primary sensations of smell. Yet, on the basis of psychological tests and action potential studies from various points in the olfactory nerve pathways, it has been postulated that about seven different primary classes of olfactory stimulants preferentially excite separate olfactory cells. These classes of olfactory stimulants are characterized as follows:

1. Camphoraceous
2. Musky
3. Floral
4. Pepperminty
5. Ethereal
6. Pungent
7. Putrid

However, it is unlikely that this list actually represents the true primary sensations of smell, even though it does illustrate the results of one of the many attempts to classify them. Indeed, several clues in recent years have indicated that there may be as many as *50 or more* primary sensations of smell—a marked contrast to only 3 primary sensations of color detected by the eyes and only a few primary sensations of taste detected by the tongue. For instance, persons have been found who have *odor blindness* for single substances; and such discrete odor blindness has been identified for more than 50 different substances. Because it is presumed that odor blindness for each substance represents a lack of the appropriate receptor protein in olfactory cells for that substance, it is possible that the sense of smell might be subserved by 50 or more primary smell sensations.

Affective Nature of Smell. Smell, equally as much as taste, has the affective qualities of either pleasantness or unpleasantness. Because of this, smell is as important as, if not more important than, taste in the selection of food. Indeed, a person who has previously eaten food that has disagreed with him is often nauseated by even the smell of that same type of food on a second occasion. Other types of odors that have proved to be unpleasant in the past may also provoke a disagreeable feeling; on the other hand, perfume of the right quality can wreak havoc with masculine emotions.

In addition, in some lower animals odors are the primary excitant of sexual drive.

Threshold for Smell. One of the principal characteristics of smell is the minute quantity of the stimulating agent in the air often required to effect a smell sensation. For instance, the substance *methyl mercaptan* can be smelled when only one 25 billionth of a milligram is present in each milliliter of air. Because of this low threshold, this substance is mixed with natural gas to give the gas an odor that can be detected when it leaks from a gas pipe.

Measurement of Smell Threshold. One of the problems in studying smell has been difficulty in obtaining accurate measurements of the threshold stimulus required to induce smell. The simplest technique is simply to allow a person to sniff different substances in the usual manner of smelling. Indeed, some investigators feel that this is as satisfactory as almost any other procedure. However, to eliminate variations from person to person, more objective methods have been developed: One of these has been to place a box containing the volatilized agent over the subject's head. Appropriate precautions are taken to exclude odors from the person's own body. The person is allowed to breathe naturally, but the volatilized agent is distributed evenly in the air that is breathed.

Gradations of Smell Intensities. Though the threshold concentrations of substances that evoke smell are extremely slight, concentrations only 10 to 50 times above the threshold values for certain substances sometimes evoke maximum intensity of smell. This is in contrast to most other sensory systems of the body, in which the ranges of detection are tremendous—for instance, 500,000 to 1 in the case of the eyes and 1 trillion to 1 in the case of the ears. This perhaps can be explained by the fact that smell is concerned more with detecting the presence or absence of odors than with quantitative detection of their intensities.

TRANSMISSION OF SMELL SIGNALS INTO THE CENTRAL NERVOUS SYSTEM

The olfactory portions of the brain are among its oldest structures, and much of the remainder of the brain developed around these olfactory beginnings. In fact, part of the brain that originally subserved olfaction later evolved into the basal brain structures that in the human being control emotions and other aspects of behavior; this is the system we call the *limbic system,* discussed in Chapter 20.

Transmission of Olfactory Signals Into the Olfactory Bulb

The olfactory bulb, which is also called cranial nerve I, is illustrated in Figure 15–3. Although it looks like a nerve, in reality it is an anterior outgrowth of brain tissue from the base of the brain having a bulbous enlargement, the *olfactory bulb,* at its end that lies over the *cribriform plate* separating the brain cavity from the upper reaches of the nasal cavity. The cribriform plate has multiple small perforations through which an equal number of small nerves enter the olfactory bulb from the olfactory membrane. Figure 15–4 illustrates the close relationship between the *olfactory cells* in the olfactory membrane and the olfactory bulb, showing

very short axons terminating in multiple globular structures of the olfactory bulb called *glomeruli.* Each bulb has several thousand such glomeruli, each of which is the terminus for about 25,000 axons from olfactory cells. Each glomerulus also is the terminus for dendrites from about 25 large *mitral cells* and about 60 smaller *tufted cells* the cell bodies of which lie also in the olfactory bulb superior to the glomeruli. These cells in turn send axons through the olfactory tract into the central nervous system.

Recent research work suggests that different glomeruli respond to different odors. Therefore, it is possible that the specific glomeruli that are stimulated are the real clue to the analysis of different odor signals transmitted into the central nervous system.

The Very Old, the Old, and the Newer Olfactory Pathways Into the Central Nervous System

The olfactory tract enters the brain at the junction between the mesencephalon and cerebrum; there the tract divides into two pathways, one passing medially into the *medial olfactory area* and the other laterally into the *lateral olfactory area.* The medial olfactory area represents a very old olfactory system, while the lateral olfactory area is the input to both a less old olfactory system and a newer system.

The Very Old Olfactory System—The Medial Olfactory Area. The medial olfactory area consists of a group of nuclei located in the midbasal portions of the brain anterior and superior to the hypothalamus. Most conspicuous are the *septal nuclei,* which are midline nuclei that feed into the hypothalamus and other portions of the brain's limbic system, the system that is concerned with basic behavior as described in Chapter 20.

The importance of this medial olfactory area is best understood by considering what happens in animals when the lateral olfactory areas on both sides of the brain are removed and only the medial system remains. The answer is that this hardly affects the more primitive responses to olfaction such as licking the lips, salivation, and other feeding responses caused by the smell of food, or such as primitive emotional drives associated with smell. On the other hand, removal of the lateral areas does abolish the more complicated olfactory conditioned reflexes.

The Old Olfactory System—The Lateral Olfactory Area. The lateral olfactory area is composed mainly of the *prepyriform* and *pyriform cortex* plus the *cortical portion of the amygdaloid nuclei.* From these areas, signal pathways pass into almost all portions of the limbic system, especially into the hippocampus, which is most important in the learning process—in this case presumably learning of likes and dislikes of certain foods depending upon experiences with the foods. For instance, it is this lateral olfactory area and its many connections with the limbic behavioral system that cause a person to develop absolute aversion to foods that have previously caused nausea and vomiting.

An important feature of the lateral olfactory area is that many signal pathways from this area feed directly into an older type of cerebral cortex called the paleocortex in the anteromedial portion of the temporal lobe. This is the only area of the entire cerebral cortex where sensory signals pass directly to the cortex without passing through the thalamus.

The Newer Pathway. Still a newer olfactory pathway has now been found that does indeed pass through the thalamus, passing to the dorsomedial thalamic nucleus and thence to the lateroposterior quadrant of the orbitofrontal cortex. Based on studies in monkeys, this newer system probably helps especially in the conscious analysis of odor.

Thus, there appears to be a very old olfactory system that subserves the basic olfactory reflexes, an old system that provides automatic but learned control of food intake and aversion to toxic and unhealthy foods, and finally a newer system that is comparable to most of the other cortical sensory systems and is used for conscious perception of olfaction.

Centrifugal Control of Activity in the Olfactory Bulb by the Central Nervous System. Many nerve fibers originating in the olfactory portions of the brain pass in the backward direction in the olfactory tract to the olfactory bulb, that is, "centrifugally" from the brain to the periphery. These terminate on a very large number of small *granule cells* located in the center of the bulb. These in turn send short, inhibitory *dendrites* to the mitral and tufted cells. It is believed that this inhibitory feedback to the olfactory bulb might be a means of helping to sharpen one's specific capability of distinguishing one odor from another.

Electrical Activity in the Olfactory Nerves and Tracts. Electrophysiological studies show that the mitral and tufted cells are continually active, the same as is true for the olfactory receptors as discussed earlier. Superimposed on this background are increases or decreases in impulse traffic caused by different odors. Thus, the olfactory stimuli *modulate* the frequency of impulses in the olfactory system and in this way transmit the olfactory information.

REFERENCES

Alberts, J. R.: Producing and interpreting experimental olfactory deficits. Physiol. Behav., 12:657, 1974.

Chanel, J.: The olfactory system as a molecular descriptor. News Physiol. Sci., 2:203, 1987.

Dastoli, F. R.: Taste receptor proteins. Life Sci., 14:1417, 1974.

Denton, D. A.: Salt appetite. In Code, C. F., and Heidel, W. (eds.): Handbook of Physiology. Sec. 6, Vol. 1. Baltimore, Md., Williams & Wilkins, 1967, p. 433.

Douek, E.: The Sense of Smell and Its Abnormalities. New York, Churchill Livingstone, 1974.

Getchell, T. V.: Functional properties of vertebrate olfactory receptor neurons. Physiol. Rev., 66:772, 1986.

Kashara, Y. (ed.): Proceedings of the Seventeenth Japanese Symposium on Taste and Smell. Arlington, Va, IRL Press, 1984.

Lat, J.: Self-selection of dietary components. In Code, C. F., and Heidel, W. (eds.): Handbook of Physiology. Sec. 6, Vol. 1, Baltimore, Md., Williams and Wilkins, 1967, p. 367.

Margolis, F. L., and Getchell, T. V. (eds.): Molecular Neurobiology of the Olfactory System. New York, Plenum Publishing Corp., 1988.

McBurney, D. H.: Taste and olfaction: Sensory discrimination. In Darian-Smith, I. (ed.): Handbook of Physiology. Sec. 1, Vol. III. Bethesda, Md., American Physiological Society, 1984, p. 1067.

Monmaney, T.: Are we led by the nose? Discover, September, 1987, p. 48.

Moulton, D. G., and Beidler, L. M.: Structure and function in the peripheral olfactory system. Physiol. Rev., 47:1, 1967.

Norgren, R.: Central neural mechanisms of taste. In Darian-Smith, I. (ed.): Handbook of Physiology. Sec. 1, Vol. III. Bethesda, Md., American Physiological Society, 1984, p. 1087.

Oakley, B., and Benjamin, R. M.: Neural mechanisms of taste. Physiol. Rev., 46:173, 1966.

Roper, S. D.: The cell biology of vertebrate taste receptors. Annu. Rev. Neurosci., 12:329, 1989.

Schiffman, S. S.: Taste transduction and modulation. News Physiol. Sci., 3:109, 1988.

Shepherd, G. M.: The olfactory bulb: A simple system in the mammalian brain. In Brookhart, J. M., and Mountcastle, V. B. (eds.): Handbook of Physiology. Sec. 1, Vol. I. Baltimore, Md., Williams & Wilkins, 1977, p. 945.

Takagi, S. F.: The olfactory nervous system of the Old World monkey. Jpn. J. Physiol., 34:51, 1984.

Zotterman, Y.: Olfaction and Taste. New York, Macmillan Co., 1963.

VI

THE CENTRAL NERVOUS SYSTEM:

C. Motor and Integrative Neurophysiology

16

Motor Functions of the Spinal Cord; the Cord Reflexes

In the discussion of the nervous system thus far, we have considered principally the input of sensory information. In the following chapters we discuss the origin and output of motor signals, the signals that cause muscle contraction, secretory function, and other motor effects throughout the body.

Sensory information is integrated at all levels of the nervous system and causes appropriate motor responses, beginning in the spinal cord with relatively simple reflexes, extending into the brain stem with still more complicated responses, and finally extending to the cerebrum, where the most complicated responses are controlled.

In the present chapter we discuss the control of muscle function by the spinal cord. The spinal cord is not merely a conduit for sensory signals to the brain or for motor signals from the brain back to the periphery. In fact, without the special neuronal circuits of the cord, even the most fundamental motor control systems in the brain cannot cause any purposeful muscle movement. To give an example, there is no neuronal circuit anywhere in the brain that causes the specific to-and-fro movement of the legs that is required in walking. Instead, the circuits for these movements are in the cord, and the brain simply sends *command* signals to set into motion the walking process. Thus, under appropriate conditions, a cat or a dog with its cord transected in the neck can be made to walk.

Yet let us not belittle the role of the brain as well, for the brain gives the sequential directions to the cord activities, to promote turning movements when they are required, to lean the body forward during acceleration, to change the movements from walking to jumping as needed, and to monitor continuously and control equilibrium. All of this is done through "command" signals from above. But it also requires the many neuronal circuits of the spinal cord that are themselves the objects of the commands. These circuits in turn provide all but a small fraction of the direct control of the muscles.

Experimental Preparations for Studying Cord Reflexes — The Spinal Animal and the Decerebrate Animal. Two different types of experimental preparations have been especially useful in studying spinal cord function: (1) the *spinal animal,* in which the spinal cord is transected, frequently in the neck so that most of the cord still remains functional; and (2) the *decerebrate animal,* in which the brain stem is transected in the lower part of the mesencephalon.

Immediately after preparing a *spinal animal,* most spinal cord function is severely depressed below the level of the transection. However, after a few hours in lower animals, and after a few days to weeks in monkeys, most of the intrinsic spinal cord functions return nearly to normal and provide a suitable experimental preparation for study.

In the *decerebrate animal* the brain stem is transected at the lower mesencephalic level, which blocks the normal inhibitory signals from the higher control centers of the brain to the pontile reticular and vestibular nuclei. This causes these nuclei to become tonically active, transmitting facilitatory signals to most of the spinal cord motor control circuits. The result is that these become easy to activate by even the slightest sensory input signals to the cord. Using this preparation, one can study very easily the intrinsic motor functions of the cord itself.

■ ORGANIZATION OF THE SPINAL CORD FOR MOTOR FUNCTIONS

The cord gray matter is the integrative area for the cord reflexes and other motor functions. Figure 16–1 shows the typical organization of the cord gray matter in a single cord segment. Sensory signals enter the cord almost entirely through the sensory (posterior) roots. After entering the cord, every sensory signal travels to two separate destinations. First, one branch of the sensory nerve terminates in the gray matter of the cord and elicits local segmental reflexes and other effects. Second, another branch transmits signals to higher levels of the nervous system — to higher levels in the

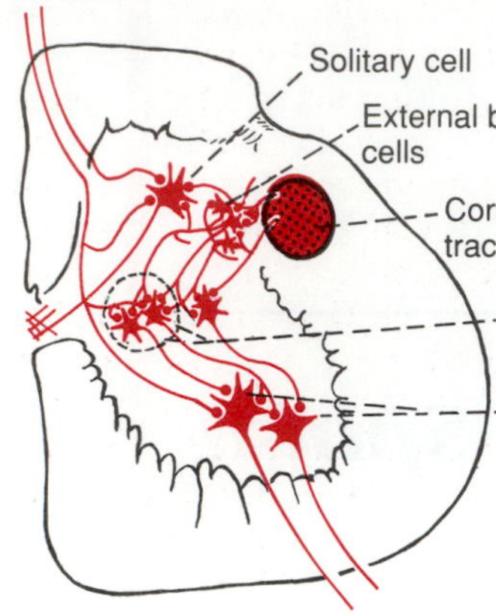

Figure 16–1. Connections of the sensory fibers and corticospinal fibers with the interneurons and anterior motor neurons of the spinal cord.

cord itself, to the brain stem, or even to the cerebral cortex, as described in earlier chapters.

Each segment of the spinal cord has several million neurons in its gray matter. Aside from the sensory relay neurons discussed in Chapters 9 and 10, the remainder of these neurons are of two separate types, the *anterior motor neurons* and the *interneurons.*

The Anterior Motor Neurons. Located in each segment of the anterior horns of the cord gray matter are several thousand neurons that are 50 to 100 per cent larger than most of the others and called *anterior motor neurons.* These give rise to the nerve fibers that leave the cord via the anterior roots and innervate the skeletal muscle fibers. The neurons are of two types, the *alpha motor neurons* and the *gamma motor neurons.*

The Alpha Motor Neurons. The alpha motor neurons give rise to large type A alpha (Aα) nerve fibers ranging from 9 to 20 μm in diameter that innervate the large skeletal muscle fibers. Stimulation of a single nerve fiber excites from as few as three to as many as several hundred skeletal muscle fibers, which are collectively called the *motor unit.* Transmission of nerve impulses into skeletal muscles and their stimulation of the muscles are discussed in Chapters 24 and 25.

The Gamma Motor Neurons. In addition to the alpha motor neurons that excite contraction of the skeletal muscle fibers, about one half as many much smaller gamma motor neurons are located along with the alpha motor neurons in the anterior horns. These transmit impulses through type A gamma (Aγ) fibers, averaging 5 μm in diameter, to very small, special skeletal muscle fibers called *intrafusal fibers.* These are part of the *muscle spindle,* which is discussed later in the chapter.

The Interneurons. Interneurons are present in all areas of the cord gray matter—in the dorsal horns, in the anterior horns, and in the intermediate areas between these two. These cells are numerous—approximately 30 times as numerous as the anterior

motor neurons. They are small and highly excitable, often exhibiting spontaneous activity and capable of firing as rapidly as 1500 times per second. They have many interconnections, and many of them directly innervate the anterior motor neurons as illustrated in Figure 16–1. The interconnections among the interneurons and anterior motor neurons are responsible for many of the integrative functions of the spinal cord that are discussed in the remainder of this chapter.

Essentially all the different types of neuronal circuits described in Chapter 8 are found in the interneuron pool of cells of the spinal cord, including the *diverging, converging,* and *repetitive-discharge* circuits. In this chapter we see many applications of these different circuits to the performance of specific reflex acts by the spinal cord.

Only a few incoming sensory signals from the spinal nerves or signals from the brain terminate directly on the anterior motor neurons. Instead, most of them are transmitted first through interneurons, where they are appropriately processed. Thus, in Figure 16–1, it is shown that the corticospinal tract terminates almost entirely on interneurons, and it is only after the signals from this tract have been integrated in the interneuron pool with signals from other spinal tracts or from the spinal nerves that they finally impinge on the anterior motor neurons to control muscular function.

The Renshaw Cell Inhibitory System. Located also in the ventral horns of the spinal cord in close association with the motor neurons are a large number of small interneurons called *Renshaw cells.* Almost immediately after the axon leaves the body of the anterior motor neuron, collateral branches from the axon pass to the adjacent Renshaw cells. These in turn are inhibitory cells that transmit inhibitory signals to the nearby motor neurons. Thus, stimulation of each motor neuron tends to inhibit the surrounding motor neurons, an effect called *recurrent inhibition.* This effect is probably important for the following major reason:

It shows that the motor system utilizes the principle of lateral inhibition to focus, or sharpen, its signals in the same way that the sensory system utilizes this principle—that is, to allow unabated transmission of the primary signal while suppressing the tendency for signals to spread to adjacent neurons.

MULTISEGMENTAL CONNECTIONS IN THE SPINAL CORD—THE PROPRIOSPINAL FIBERS

More than half of all the nerve fibers ascending and descending in the spinal cord are *propriospinal fibers.* These are fibers that run from one segment of the cord to another. In addition, the sensory fibers as they enter the cord branch both up and down the spinal cord, some of the branches transmitting signals only a segment or two in each direction whereas others transmit signals many segments. These ascending and descending fibers of the cord provide pathways for the multisegmental reflexes that are described later in this chapter, including reflexes that coordinate simultaneous movements in the forelimbs and hindlimbs.

■ THE MUSCLE RECEPTORS — MUSCLE SPINDLES AND GOLGI TENDON ORGANS — AND THEIR ROLES IN MUSCLE CONTROL

Proper control of muscle function requires not only excitation of the muscle by the anterior motor neurons but also continuous feedback of information from each muscle to the nervous system, giving the status of the muscle at each instant. That is, what is the length of the muscle, what is its instantaneous tension, and how rapidly is its length or tension changing? To provide this information, the muscles and their tendons are supplied abundantly with two special types of sensory receptors: (1) *muscle spindles*, which are distributed throughout the belly of the muscle and send information to the nervous system about either the muscle length or the rate of change of its length; and (2) *Golgi tendon organs*, which are located in the muscle tendons and transmit information about tension or the rate of change of tension.

The signals from these two receptors are either entirely or almost entirely for the purpose of muscle control itself, because they operate almost entirely at a subconscious level. Even so, they transmit tremendous amounts of information into the spinal cord, the cerebellum, and even the cerebral cortex, helping each of these portions of the nervous system in its function for controlling muscle contraction.

RECEPTOR FUNCTION OF THE MUSCLE SPINDLE

Structure and Innervation of the Muscle Spindle. The physiologic organization of the muscle spindle is illustrated in Figure 16–2. Each spindle is built around 3 to 12 small *intrafusal muscle fibers* that are pointed at their ends and that are attached to the glycocalyx of the surrounding *extrafusal* skeletal muscle fibers. Each in-

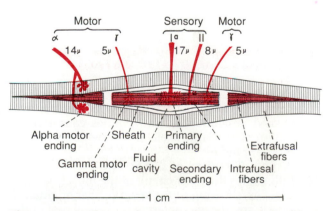

Figure 16–2. The muscle spindle, showing its relationship to the large extrafusal skeletal muscle fibers. Note also both the motor and the sensory innervation of the muscle spindle.

trafusal fiber is a very small skeletal muscle fiber. However, the central region of each of these fibers — that is, the area midway between its two ends — has either no or few actin and myosin filaments. Therefore, this central portion does not contract when the ends do. Instead, it functions as a sensory receptor, as we describe later. The end portions are excited by the small *gamma motor nerve fibers* originating from the gamma motor neurons that were described earlier. These fibers are frequently also called *gamma efferent fibers*, in contradistinction to the *alpha efferent fibers* that innervate the extrafusal skeletal muscle.

Excitation of the Spindle Receptors. The receptor portion of the muscle spindle is its central portion, where the intrafusal muscle fibers have no contractile elements. As illustrated in Figure 16–2, and also in more detail in Figure 16–3, sensory fibers originate in this area, and these are stimulated by stretching of this midportion of the spindle. One can readily see that the muscle spindle receptor can be excited in two different ways:

1. Lengthening the whole muscle will obviously stretch the midportion of the spindle and therefore excite the receptor.

2. Even if the length of the entire muscle does not change, contraction of the end portions of the intrafusal fibers will also stretch the midportions of the spindle fibers and therefore excite the receptor.

Two types of sensory endings are found in the receptor area of the muscle spindle. These are:

The Primary Ending. In the very center of the receptor area a large sensory fiber encircles the central portion of each intrafusal fiber, forming the so-called *primary ending* or *annulospiral ending*. This nerve fiber is a type Ia fiber averaging 17 μm in diameter, and it transmits sensory signals to the spinal cord at a velocity of 70 to 120 m/sec, as rapidly as any type of sensory nerve fiber in the entire body.

The Secondary Ending. Usually one but sometimes two smaller sensory nerve fibers, type II fibers with an average diameter of 8 μm, innervate the receptor region on one side of the primary ending, as illustrated in Figures 16–2 and 16–3. This sensory ending is called the *secondary ending*, or sometimes the *flower spray ending*, because in some preparations it looks like a flower spray even though it mainly encircles the intrafusal fibers in the same way that the type Ia fiber does.

Division of the Intrafusal Fibers Into Nuclear Bag and Nuclear Chain Fibers — Dynamic and Static Responses of the Muscle Spindle. There are also two different types of intrafusal fibers: (1) *nuclear bag fibers* (one to three in each spindle), in which a large number of nuclei are congregated into an expanded bag in the central portion of the receptor area as shown by the top fiber in Figure 16–3; and (2) *nuclear chain fibers* (three to nine), which are about half as large in diameter and half as long as the nuclear bag fibers and have nuclei aligned in a chain throughout the receptor area, as illustrated by the bottom fiber in the figure. The primary ending innervates both the nuclear bag intrafusal

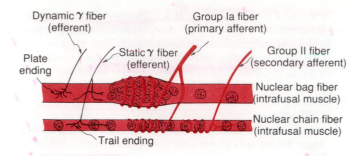

Figure 16–3. Details of nerve connections to the nuclear bag and nuclear chain muscle spindle fibers. (Modified from Stein: Physiol. Rev., 54:225, 1974, and Boyd: Philios. Trans. R. Soc. Lond. [Biol Sci.], 245:81, 1962.)

fibers *and* the nuclear chain fibers. On the other hand, the secondary ending usually innervates only the nuclear chain fibers. These relationships are all illustrated in Figure 16–3.

Response of Both the Primary and the Secondary Endings to the Length of the Receptor — The "Static" Response. When the receptor portion of the muscle spindle is stretched slowly, the number of impulses transmitted from both the primary and the secondary endings increases almost directly in proportion to the degree of stretching, and the endings continue to transmit these impulses for many minutes. This effect is called the *static response* of the spindle receptor, meaning simply that both the primary and the secondary endings continue to transmit their signals for as long as the receptor itself remains stretched. Because only the *nuclear chain* type of intrafusal fiber is innervated by both the primary and the secondary endings, it is believed that these nuclear chain fibers are responsible for the static response.

Response of the Primary Ending (but Not the Secondary Ending) to the Rate of Change of Receptor Length — The "Dynamic" Response. When the length of the spindle receptor increases suddenly, the primary ending (but not the secondary ending) is stimulated especially powerfully, much more powerfully than the stimulus caused by the static response. This excess stimulus of the primary ending is called the *dynamic response*, which means that the primary ending responds extremely actively to a rapid *rate of change* in length. When the length of a spindle receptor increases only a fraction of a micrometer, if this increase occurs in a fraction of a second, the primary receptor transmits tremendous numbers of excess impulses into the Ia fiber, but only *while the length is actually increasing.* As soon as the length has stopped increasing, the rate of impulse discharge returns back to the level of the much smaller static response that is still present in the signal.

Conversely, when the spindle receptor shortens, this change momentarily decreases the impulse output from the primary ending; then, as soon as the receptor area has reached its new shortened length, the impulses reappear in the Ia fiber within a fraction of a second. Thus, the primary ending sends extremely strong signals to the central nervous system to apprise it of any change in length of the spindle receptor area.

Because only the primary endings transmit the dynamic response and the nuclear bag intrafusal fibers have only primary endings, it is assumed that the nuclear bag fibers are responsible for the powerful dynamic response.

Control of the Static and Dynamic Responses by the Gamma Motor Nerves. The gamma motor nerves to the muscle spindle can be divided into two different types: gamma-dynamic (gamma-d) and gamma-static (gamma-s). The first of these excites mainly the nuclear bag intrafusal fibers, and the second mainly the nuclear chain intrafusal fibers. When the gamma-d fibers excite the nuclear bag fibers, the dynamic response of the muscle spindle becomes tremendously enhanced, whereas the static response is hardly affected. On the other hand, stimulation of the gamma-s fibers, which excite the nuclear chain fibers, enhances the static response while having little influence on the dynamic response. We shall see in subsequent paragraphs that these two different types of responses of the muscle spindle are exceedingly important in different types of muscle control.

Continuous Discharge of the Muscle Spindles Under Normal Conditions. Normally, particularly when there is a slight amount of gamma motor excitation, the muscle spindles emit sensory nerve impulses continuously. Stretching the muscle spindles increases the rate of firing, while shortening the spindle decreases this rate of firing. Thus, either the spindles can send *positive signals* to the spinal cord—that is, increased numbers of impulses to indicate increased stretch of a muscle—or they can send *negative signals*—decreased numbers of impulses below the normal level—to indicate that the muscle is actually being unstretched.

THE MUSCLE STRETCH REFLEX (ALSO CALLED MYOTATIC REFLEX)

The simplest manifestation of muscle spindle function is the *muscle stretch reflex*—that is, whenever a muscle is stretched, excitation of the spindles causes reflex contraction of the large skeletal muscle fibers that lie around the spindles.

Neuronal Circuitry of the Stretch Reflex. Figure 16–4 illustrates the basic circuit of the muscle spindle stretch reflex, showing a type Ia nerve fiber originating in a muscle spindle and entering the dorsal root of the spinal cord. Then, in contrast to most other nerve fibers entering the cord, one branch of it passes directly to the

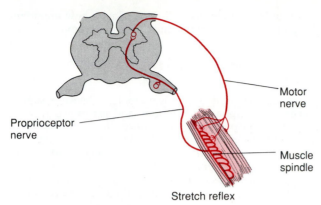

Figure 16–4. Neuronal circuit of the stretch reflex.

anterior horn of the cord gray matter and synapses directly with anterior motor neurons that send nerve fibers back to the same muscle whence the muscle spindle fiber originated. Thus, this is a *monosynaptic pathway* that allows a reflex signal to return with the shortest possible delay back to the same muscle after excitation of the spindle.

Some of the type II fibers from the secondary spindle endings also terminate monosynaptically with the anterior motor neurons. However, most of the type II fibers (as well as many collaterals from the Ia fibers from the primary endings) terminate on multiple interneurons in the cord gray matter, and these in turn transmit more delayed signals to the anterior motor neurons and also serve other functions.

The Dynamic Stretch Reflex Versus the Static Stretch Reflex. The stretch reflex can be divided into two separate components called respectively the dynamic stretch reflex and the static stretch reflex. The *dynamic stretch reflex* is elicited by the potent dynamic signal transmitted from the primary endings of the muscle spindles. That is, when a muscle is suddenly stretched, a strong signal is transmitted to the spinal cord, and this causes an instantaneous, very strong reflex contraction of the same muscle from which the signal originated. Thus, the reflex functions to oppose sudden changes in the length of the muscle, because the muscle contraction opposes the stretch.

The dynamic stretch reflex is over within a fraction of a second after the muscle has been stretched to its new length, but then a weaker *static stretch reflex* continues for a prolonged period of time thereafter. This reflex is elicited by the continuous static receptor signals transmitted by both the primary and secondary endings. The importance of the static stretch reflex is that it continues to cause muscle contraction as long as the muscle is maintained at excessive length. The muscle contraction in turn opposes the force that is causing the excess length.

The Negative Stretch Reflex. When a muscle is suddenly shortened, exactly opposite effects occur because of negative signals from the spindles. If the muscle is already taut, any sudden release of the load on the muscle that allows it to shorten will elicit both dynamic and static reflex *muscle inhibition* rather than reflex excitation. Thus, this *negative stretch reflex* opposes the shortening of the muscle in the same way that the positive stretch reflex opposes lengthening of the muscle. Therefore, one can begin to see that the stretch reflex tends to maintain the status quo for the length of a muscle.

The Damping Function of the Dynamic and Static Stretch Reflexes. An especially important function of the stretch reflex is its ability to prevent some types of oscillation and jerkiness of the body movements. This is a *damping*, or smoothing, function. An example is the following:

Use of the Damping Mechanism in Smoothing Muscle Contraction. Occasionally, signals from other parts of the nervous system are transmitted to a muscle in a very unsmooth form, increasing in intensity for a few milliseconds, then decreasing in intensity, then changing to another intensity level, and so forth. When the muscle spindle apparatus is not functioning satisfactorily, the muscle contraction is very jerky during the course of such a signal. This effect is illustrated in Figure 16–5, which shows an experiment in which a sensory nerve signal entering one side of the cord is transmitted to a motor nerve on the other side of the cord to excite a muscle. In curve A the muscle spindle reflex of the excited muscle is intact. Note that the contraction is relatively smooth even though the sensory nerve is excited at a very slow frequency of 8/sec. Curve B, on the other hand, is the same experiment in an animal whose muscle spindle sensory nerves from the muscle had been sectioned 3 months earlier. Note the very unsmooth muscle contraction. Thus, curve A illustrates very graphically the ability of the damping mechanism of the muscle spindle to smooth muscle contractions even though the input signals to the muscle motor system may themselves be very jerky. This effect can also be called a *signal averaging* function of the muscle spindle reflex.

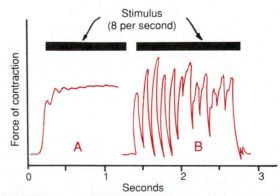

Figure 16–5. Muscle contraction caused by a spinal cord signal under two different conditions: *A*, in a normal muscle, and *B*, in a muscle whose muscle spindles had been denervated by section of the posterior roots of the cord 82 days previously. Note the smoothing effect of the muscle spindle reflex in *A*. (Modified from Creed, et al.: Reflex Activity of the Spinal Cord. New York, Oxford University Press, 1932.)

ROLE OF THE MUSCLE SPINDLE IN VOLUNTARY MOTOR ACTIVITY

To emphasize the importance of the gamma efferent system, one needs to recognize that 31 per cent of all the motor nerve fibers to the muscle are gamma efferent fibers rather than large, type A alpha motor fibers. Whenever signals are transmitted from the motor cortex or from any other area of the brain to the alpha motor neurons, almost always the gamma motor neurons are stimulated simultaneously, an effect called *coactivation* of the alpha and gamma motor neurons. This causes both the extrafusal and the intrafusal muscle fibers to contract at the same time.

The purpose of contracting the muscle spindle fibers at the same time that the large skeletal muscle fibers contract is probably twofold: First, it keeps the length of the receptor portion of the muscle spindle from changing and therefore keeps the muscle spindle from opposing the muscle contraction. Second, it maintains proper damping function of the muscle spindle regardless of change in muscle length. For instance, if the muscle spindle should not contract and relax along with the large muscle fibers, the receptor portion of the spindle would sometimes be flail and at other times be overstretched, in neither instance operating under optimal conditions for spindle function.

Possible "Servo-Assist" Function of the Muscle Spindle Reflex

It is possible that the muscle spindle reflex also functions as a "servo-assist" mechanism during muscle contraction. But, first, let us explain what is meant by "servo-assist mechanism."

When both the alpha and gamma motor neurons are stimulated simultaneously, if the intra- and extrafusal fibers contract equal amounts, the degree of stimulation of the muscle spindles will not change at all—neither increase nor decrease. However, in case the extrafusal muscle fibers should contract less than the intrafusal fibers (as might occur when the muscle is contracting against a great load), this mismatch would stretch the receptor portions of the spindles and, therefore, elicit a stretch reflex that would provide extra excitation of the extrafusal fibers. This is exactly the same mechanism employed by power steering in an automobile. That is, if the front wheels are resistant to following the movement of the steering wheel, a servo-assist device becomes activated that applies additional force to turn the wheels.

A servo-assist function of the muscle spindle reflex could have several important advantages, as follows:

1. It could allow the brain to cause a muscle contraction against a load without the brain having to expend much extra nervous energy—instead, the spindle reflex would provide most of the nervous energy.

2. It could make the muscle contract almost the desired length even when the load is increased or decreased between successive contractions. In other words, it would make the length of contraction less load-sensitive.

3. It could compensate for fatigue or other abnormalities of the muscle itself, because any failure of the muscle to provide the proper contraction would elicit an additional muscle spindle reflex stimulus to make the contraction occur.

But, unfortunately, we still do not know how important this possible function of the muscle spindle reflex actually is.

Brain Areas for Control of the Gamma Efferent System

The gamma efferent system is excited by the same signals that excite the alpha motor neurons and also by signals from the *bulboreticular facilitatory* region of the brain stem, and secondarily by impulses transmitted into the bulboreticular area from (a) the *cerebellum,* (b) the *basal ganglia,* and even (c) the *cerebral cortex.* Unfortunately, though, little is known about the precise mechanisms of control of the gamma efferent system. However, since the bulboreticular facilitatory area is particularly concerned with antigravity contractions, and also because the antigravity muscles have an especially high density of muscle spindles, emphasis is given to the possible or probable importance of the gamma efferent mechanism in controlling muscle contraction for positioning the different parts of the body and for damping the movements of the different parts.

CLINICAL APPLICATIONS OF THE STRETCH REFLEX

The stretch reflex is tested by the clinician almost every time he performs a physical examination. His purpose is to determine how much background excitation, or "tone," the brain is sending to the spinal cord. This reflex is elicited as follows:

The Knee Jerk and Other Muscle Jerks. Clinically, a method used to determine the sensitivity of the stretch reflexes is to elicit the knee jerk and other muscle jerks. The knee jerk can be elicited by simply striking the patellar tendon with a reflex hammer; this stretches the quadriceps muscle and initiates a *dynamic stretch reflex* that in turn causes the lower leg to jerk forward. The upper part of Figure 16–6 illustrates a myogram from the quadriceps muscle recorded during a knee jerk.

Similar reflexes can be obtained from almost any muscle of the body either by striking the tendon of the muscle or by striking the belly of the muscle itself. In other words, sudden stretch of muscle spindles is all that is required to elicit a stretch reflex.

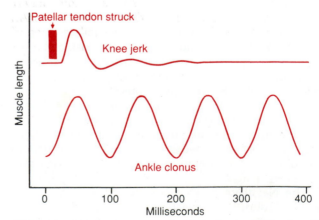

Figure 16–6. Myograms recorded from the quadriceps muscle during elicitation of the knee jerk and from the gastrocnemius muscle during ankle clonus.

The muscle jerks are used by neurologists to assess the degree of facilitation of spinal cord centers. When large numbers of facilitatory impulses are being transmitted from the upper regions of the central nervous system into the cord, the muscle jerks are greatly exacerbated. On the other hand, if the facilitatory impulses are depressed or abrogated, the muscle jerks are considerably weakened or completely absent. These reflexes are used most frequently in determining the presence or absence of muscle spasticity following lesions in the motor areas of the brain or muscle spasticity in diseases that excite the bulboreticular facilitatory area of the brain stem. Ordinarily, large lesions in the contralateral motor areas of the cerebral cortex, especially those caused by strokes or brain tumors, cause greatly exacerbated muscle jerks.

Clonus. Under appropriate conditions, the muscle jerks can oscillate, a phenomenon called *clonus* (see lower myogram, Fig. 16–6). Oscillation can be explained particularly well in relation to ankle clonus, as follows:

If a man standing on his tiptoes suddenly drops his body downward to stretch one of the gastrocnemius muscles, impulses are transmitted from the muscle spindles into the spinal cord. These reflexly excite the stretched muscle, which lifts the body back up again. After a fraction of a second, the reflex contraction of the muscle dies out and the body falls again, thus stretching the spindles a second time. Again a dynamic stretch reflex lifts the body, but this too dies out after a fraction of a second, and the body falls once more to elicit still a new cycle. In this way, the stretch reflex of the gastrocnemius muscle continues to oscillate, often for long periods of time; this is clonus.

Clonus ordinarily occurs only if the stretch reflex is highly sensitized by facilitatory impulses from the brain. For instance, in the decerebrate animal, in which the stretch reflexes are highly facilitated, clonus develops readily. Therefore, to determine the degree of facilitation of the spinal cord, neurologists test patients for clonus by suddenly stretching a muscle and keeping a steady stretching force applied to the muscle. If clonus occurs, the degree of facilitation is certain to be very high.

THE GOLGI TENDON REFLEX

The Golgi Tendon Organ and Its Excitation. The Golgi tendon organ, illustrated in Figure 16–7, is an encapsulated sensory receptor through which a small bundle of muscle tendon fibers pass immediately beyond their point of fusion with the muscle fibers. An average of 10 to 15 muscle fibers is usually connected in series with each Golgi tendon organ, and the organ is stimulated by the tension produced by this small bundle of muscle fibers. Thus, the major difference between the function of the Golgi tendon organ and the muscle spindle is that the spindle detects muscle length and changes in muscle length, while the tendon organ detects muscle *tension.*

The tendon organ, like the primary receptor of the muscle spindle, has both a *dynamic response* and a *static response*, responding very intensely when the muscle tension suddenly increases (the dynamic response) but within a small fraction of a second settling down to a lower level of steady-state firing that is almost directly proportional to the muscle tension (the static response). Thus, the Golgi tendon organs provide the nervous system with instantaneous information on the degree of tension in each small segment of each muscle.

Transmission of Impulses from the Tendon Organ Into the Central Nervous System. Signals from the tendon organ are transmitted through large, rapidly conducting type Ib nerve fibers averaging 16 μm in diameter, only slightly smaller than those from the primary ending of the muscle spindle. These fibers, like those from the primary endings, transmit signals both into local areas of the cord and through long fiber pathways such as the spinocerebellar tracts into the cerebellum and through still other tracts to the cerebral cortex. The local cord signal excites a single *inhibitory* interneuron that in turn inhibits the anterior motor neuron. This local circuit directly inhibits the individual muscle without affecting adjacent muscles. The signals to the brain are discussed in Chapter 18.

Inhibitory Nature of the Tendon Reflex and Its Importance

When the Golgi tendon organs of a muscle are stimulated by increased muscle tension, signals are transmitted into the spinal cord to cause reflex effects in the respective muscle. However, this reflex is entirely inhibitory, the exact opposite of the muscle spindle reflex. Thus, this reflex provides a negative feedback mechanism that prevents the development of too much tension on the muscle.

When tension on the muscle and, therefore, on the tendon becomes extreme, the inhibitory effect from the tendon organ can be so great that it leads to a sudden reaction in the spinal cord and instantaneous relaxation of the entire muscle. This effect is called the *lengthening reaction*; it is possibly or even probably a protective mechanism to prevent tearing of the muscle or avulsion of the tendon from its attachments to the bone. We know, for instance, that direct electrical stimulation of muscles in the laboratory, which cannot be opposed by this negative reflex, can frequently cause such destructive effects.

Possible Role of the Tendon Reflex to Equalize Contractile Force Among the Muscle Fibers. Another likely function of the Golgi tendon reflex is to equalize the contractile forces of the separate muscle

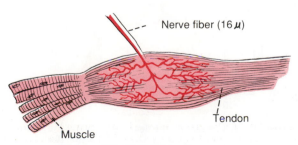

Figure 16–7. Golgi tendon organ.

Nerve fiber (16 μ)

Tendon

Muscle

fibers. That is, those fibers that exert excess tension become inhibited by the reflex while those that exert too little tension become more excited because of absence of reflex inhibition. Obviously this would spread the muscle load over all the fibers and especially would prevent local muscle damage where small numbers of fibers would be overloaded.

FUNCTION OF THE MUSCLE SPINDLES AND GOLGI TENDON ORGANS IN CONJUNCTION WITH MOTOR CONTROL FROM HIGHER LEVELS OF THE BRAIN

Although we have emphasized the function of the muscle spindles and Golgi tendon organs in spinal cord control of motor function, these two sensory organs also apprise the higher motor control centers of instantaneous changes taking place in the muscles. For instance, the dorsal spinocerebellar tracts carry instantaneous information from both the muscle spindles and the Golgi tendon organs directly to the cerebellum at conduction velocities approaching 120 m/sec. Additional pathways transmit similar information into the reticular regions of the brain stem and also all the way to the motor areas of the cerebral cortex. We shall learn in the following two chapters that information from these receptors is crucial for feedback control of motor signals originating in all of these areas.

■ THE FLEXOR REFLEX (THE WITHDRAWAL REFLEXES)

In the spinal or decerebrate animal, almost any type of cutaneous sensory stimulus on a limb is likely to cause the flexor muscles of the limb to contract, thereby withdrawing the limb from the stimulus. This is called the *flexor reflex.*

In its classic form the flexor reflex is elicited most powerfully by stimulation of pain endings, such as by a pinprick, heat, or some other painful stimulus, for which reason it is also frequently called a *nociceptive reflex,* or simply *pain reflex.* However, stimulation of the touch receptors can also elicit a weaker and less prolonged flexor reflex.

If some part of the body besides one of the limbs is painfully stimulated, this part, in a similar manner, will be *withdrawn from the stimulus,* but the reflex may not be confined entirely to flexor muscles even though it is basically the same type of reflex. Therefore, the many patterns of reflexes of this type in the different areas of the body are called the *withdrawal reflexes.*

Neuronal Mechanism of the Flexor Reflex. The left-hand portion of Figure 16–8 illustrates the neuronal pathways for the flexor reflex. In this instance, a painful stimulus is applied to the hand; as a result, the flexor muscles of the upper arm become reflexly excited, thus withdrawing the hand from the painful stimulus.

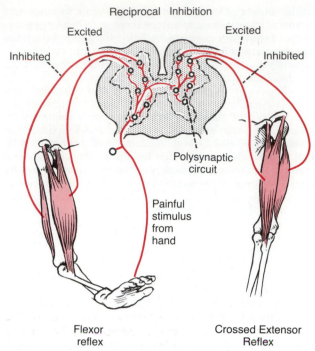

Figure 16–8. The flexor reflex, the crossed extensor reflex, and reciprocal inhibition.

The pathways for eliciting the flexor reflex do not pass directly to the anterior motor neurons but, instead, pass first into the interneuron pool of neurons and then to the motor neurons. The shortest possible circuit is a three- or four-neuron arc; however, most of the signals of the reflex traverse many more neurons than this and involve the following basic types of circuits: (1) diverging circuits to spread the reflex to the necessary muscles for withdrawal, (2) circuits to inhibit the antagonist muscles, called *reciprocal inhibition circuits,* and (3) circuits to cause a prolonged repetitive afterdischarge even after the stimulus is over.

Figure 16–9 illustrates a typical myogram from a flexor muscle during a flexor reflex. Within a few milliseconds after a pain nerve begins to be stimulated, the flexor response appears. Then, in the next few seconds the reflex begins to *fatigue,* which is characteristic of essentially all of the more complex integrative reflexes of the spinal cord. Then, soon after the stimulus is over, the contraction of the muscle begins to return toward

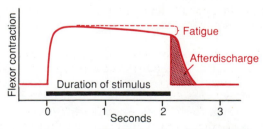

Figure 16–9. Myogram of the flexor reflex, showing rapid onset of the reflex, an interval of fatigue, and finally afterdischarge after the stimulus is over.

the base line, but, because of *afterdischarge,* will not return all the way for many milliseconds. The duration of the afterdischarge depends on the intensity of the sensory stimulus that had elicited the reflex; a weak tactile stimulus causes almost no afterdischarge in contrast to an afterdischarge lasting for a second or more following a very strong pain stimulus.

The afterdischarge that occurs in the flexor reflex almost certainly results from both types of repetitive-discharge circuits that were discussed in Chapter 8. Electrophysiological studies indicate that the immediate afterdischarge, lasting for about 6 to 8 msec, results from repetitive firing of the excited interneurons themselves. However, the prolonged afterdischarge that occurs following strong pain stimuli almost certainly results from recurrent pathways that excite reverberating interneuron circuits, these transmitting impulses to the anterior motor neurons sometimes for several seconds after the incoming sensory signal is completely over.

Thus, the flexor reflex is appropriately organized to withdraw a pained or otherwise irritated part of the body away from the stimulus. Furthermore, because of the afterdischarge, the reflex can still hold the irritated part away from the stimulus for as long as 1 to 3 sec after the irritation is over. During this time, other reflexes and actions of the central nervous system can move the entire body away from the painful stimulus.

The Pattern of Withdrawal. The pattern of withdrawal that results when the flexor reflex (or the many other types of withdrawal reflexes) is elicited depends on the sensory nerve that is stimulated. Thus, a painful stimulus on the inside of the arm not only elicits a flexor reflex in the arm but also contracts the abductor muscles to pull the arm outward. In other words, the integrative centers of the cord cause those muscles to contract that can most effectively remove the pained part of the body from the object that causes pain. This same principle, which is called the principle of "local sign," applies to any part of the body but especially to the limbs, because they have highly developed flexor reflexes.

■ THE CROSSED EXTENSOR REFLEX

Approximately 0.2 to 0.5 sec after a stimulus elicits a flexor reflex in one limb, the opposite limb begins to extend. This is called the *crossed extensor reflex.* Extension of the opposite limb obviously can push the entire body away from the object causing the painful stimulus.

Neuronal Mechanism of the Cross Extensor Reflex. The right-hand portion of Figure 16–8 illustrates the neuronal circuit responsible for the crossed extensor reflex, showing that signals from the sensory nerves cross to the opposite side of the cord to cause reactions exactly opposite those that cause the flexor reflex. Because the crossed extensor reflex usually does not begin until 200 to 500 msec following the initial

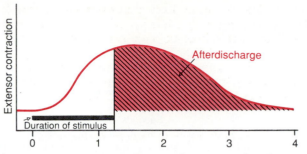

Figure 16–10. Myogram of a crossed extensor reflex, showing slow onset but prolonged afterdischarge.

pain stimulus, it is certain that many interneurons are involved in the circuit between the incoming sensory neuron and the motor neurons of the opposite side of the cord responsible for the crossed extension. Furthermore, after the painful stimulus is removed, the crossed extensor reflex continues for an even longer period of afterdischarge than that for the flexor reflex. Therefore, again, it is presumed that this prolonged afterdischarge results from reverberatory circuits among the internuncial cells.

Figure 16–10 illustrates a typical myogram recorded from a muscle involved in a crossed extensor reflex. This shows the relatively long latency before the reflex begins and also the long afterdischarge following the end of the stimulus. The prolonged afterdischarge obviously would be of benefit in holding the body away from a painful object until other nervous reactions could cause the body to move away.

■ RECIPROCAL INHIBITION AND RECIPROCAL INNERVATION

In the foregoing paragraphs we have pointed out several times that excitation of one group of muscles is usually associated with inhibition of another group. For instance, when a stretch reflex excites one muscle, it simultaneously inhibits the antagonist muscles. This is the phenomenon of *reciprocal inhibition,* and the neuronal circuit that causes this reciprocal relationship is called *reciprocal innervation.* Likewise, reciprocal relationships exist between the two sides of the cord, as exemplified by the flexor and extensor reflexes described above.

Figure 16–11 illustrates a typical example of reciprocal inhibition. In this instance, a moderate but prolonged flexor reflex is elicited from one limb of the body; and while this reflex is still being elicited, a still stronger flexor reflex is elicited in the opposite limb. This reflex then sends reciprocal inhibitory signals to the first limb and depresses its degree of flexion. Finally, removal of the stronger reflex allows the original reflex to reassume its previous intensity.

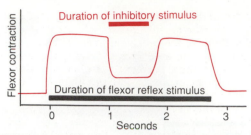

Figure 16–11. Myogram of a flexor reflex, illustrating reciprocal inhibition caused by a stronger flexor reflex in the opposite limb.

■ THE REFLEXES OF POSTURE AND LOCOMOTION

THE POSTURAL AND LOCOMOTIVE REFLEXES OF THE CORD

The Positive Supportive Reaction. Pressure on the footpad of a decerebrate animal causes the limb to extend against the pressure being applied to the foot. Indeed, this reflex is so strong that an animal whose spinal cord has been transected for several months — that is, after the reflexes have become exacerbated — can often be placed on its feet, and the reflex will stiffen the limbs sufficiently to support the weight of the body — the animal will stand in a rigid position. This reflex is called the *positive supportive reaction.*

The positive supportive reaction involves a complex circuit in the interneurons similar to those responsible for the flexor and the cross-extensor reflexes. The locus of the pressure on the pad of the foot determines the position to which the limb will extend; pressure on one side causes extension in that direction, an effect called the *magnet reaction.* This obviously helps keep an animal from falling to that side.

The Cord "Righting" Reflexes. When a spinal cat or even a well-recovered young spinal dog is laid on its side, it will make incoordinate movements that indicate that it is trying to raise itself to the standing position. This is called a *cord righting reflex.* Such a reflex illustrates that relatively complicated reflexes associated with posture are integrated in the spinal cord. Indeed, a puppy with a well-healed transected thoracic cord between the level for the forelimbs and the hindlimbs can completely right itself from the lying position and can even walk on its hindlimbs. And, in the case of the opossum with a similar transection of the thoracic cord, the walking movements of the hindlimbs are hardly different from those in the normal opossum — except that the hindlimb movements are not synchronized with those of the forelimbs as is normally the case.

Stepping and Walking Movements

Rhythmic Stepping Movements of a Single Limb. Rhythmic stepping movements are frequently observed in the limbs of spinal animals. Indeed, even when the lumbar portion of the spinal cord is separated from the remainder of the cord and a longitudinal section is made down the center of the cord to block neuronal connections between the two limbs, each hindlimb can still perform individual stepping functions. Forward flexion of the limb is followed a second or so later by backward extension. Then flexion occurs again, and the cycle is repeated over and over.

This oscillation back and forth between the flexor and extensor muscles can occur even after the sensory nerves have been cut, and it seems to result mainly from mutually reciprocal inhibition circuits that oscillate between the agonist and antagonist muscles within the matrix of the cord itself. That is, the forward flexion of the limb is coincident with reciprocal inhibition of the cord center controlling the extensor muscles. Then, as the flexion begins to die out, *rebound* excitation of the extensors causes the leg to move downward and backward, with simultaneous reciprocal inhibition of the flexor muscles. And the oscillating cycle continues again and again.

The sensory signals from the footpads and from the position sensors around the joints play a strong role in controlling foot pressure and rate of stepping when the foot is allowed to walk along a surface. In fact, the cord mechanism for control of stepping can be still more complex. For instance, if during the forward thrust of the foot the top of the foot encounters an obstruction, the forward thrust will stop temporarily, the foot will be lifted higher, and then the foot will proceed forward to be placed over the obstruction. Thus, the cord is an intelligent walking controller.

Reciprocal Stepping of Opposite Limbs. If the lumbar spinal cord is not split down its center as noted above, every time stepping occurs in the forward direction in one limb, the opposite limb ordinarily steps backward. This effect results from reciprocal innervation between the two limbs.

Diagonal Stepping of All Four Limbs — The "Mark Time" Reflex. If a well-healed spinal animal, with the spinal transection above the forelimb area of the cord, is held up from the floor and its legs are allowed to fall downward as illustrated in Figure 16–12, the stretch on the limbs occasionally elicits stepping reflexes that involve all four limbs. In general, stepping occurs diagonally between the forelimbs and hindlimbs. This diagonal response is another manifestation of reciprocal innervation, this time occurring the entire distance up and down the cord between the forelimbs and hindlimbs. Such a walking pattern is called a *mark time reflex.*

The Galloping Reflex. Another type of reflex that occasionally develops in the spinal animal is the galloping reflex, in which both forelimbs move backward in unison while both hindlimbs move forward. This often occurs when stretch or pressure stimuli are applied almost exactly equally to opposite limbs at the same time, whereas unequal stimulation of one side versus the other elicits the diagonal walking reflex. This is in keeping with the normal patterns of walking and of galloping, for, in walking, only one limb at a time is stimulated, and this would predispose to con-

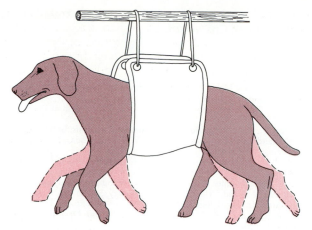

Figure 16–12. Diagonal stepping movements exhibited by a spinal animal.

tinued walking. Conversely, when the animal strikes the ground during galloping, the limbs on both sides are stimulated approximately equally; this obviously would predispose to further galloping and, therefore, would continue this pattern of motion in contradistinction to the walking pattern.

■ THE SCRATCH REFLEX

An especially important cord reflex in some animals is the scratch reflex, which is initiated by the *itch and tickle sensation.* It actually involves two different functions: (1) a *position sense* that allows the paw to find the exact point of irritation on the surface of the body, and (2) a *to-and-fro scratching movement.*

Obviously, the *to-and-fro movement,* like the stepping movements of locomotion, involves reciprocal innervation circuits that cause oscillation, which can still function even when all the sensory roots from the oscillating limb are sectioned, as is true for basic walking movements.

The *position sense* of the scratch reflex is a very highly developed function, for even though a flea might be crawling as far forward as the shoulder of a spinal animal, the hind paw can often find its position even though 19 different muscles in the limb must be contracted simultaneously in a precise pattern to bring the paw to the position of the crawling flea. To make the reflex even more complicated, when the flea crosses the midline, the first paw stops scratching, and the opposite paw begins the to-and-fro motion and eventually finds the flea.

■ THE SPINAL CORD REFLEXES THAT CAUSE MUSCLE SPASM

In human beings, local muscle spasm is often observed. The mechanism of this has not been elucidated to complete satisfaction even in experimental animals, but it is known that pain stimuli can cause reflex spasm of local muscles, which presumably is the cause of much if not most of the muscle spasm observed in localized regions of the human body.

Muscle Spasm Resulting from a Broken Bone. One type of clinically important spasm occurs in muscles surrounding a broken bone. This seems to result from the pain impulses

initiated from the broken edges of the bone, which cause the muscles surrounding the area to contract powerfully and tonically. Relief of the pain by injection of a local anesthetic relieves the spasm; a general anesthetic also relieves the spasm. One of these procedures is often necessary before the spasm can be overcome sufficiently for the two ends of the bone to be set back into appropriate positions.

Abdominal Muscle Spasm in Peritonitis. Another type of local spasm caused by cord reflexes is the abdominal spasm resulting from irritation of the parietal peritoneum by peritonitis. Here, again, relief of the pain caused by the peritonitis allows the spastic muscle to relax. Almost the same type of spasm often occurs during surgical operations; pain impulses from the parietal peritoneum cause the abdominal muscles to contract extensively and sometimes actually to extrude the intestines through the surgical wound. For this reason deep surgical anesthesia is usually required for intra-abdominal operations.

Muscle Cramps. Still another type of local spasm is the typical muscle cramp. Electromyographic studies indicate that the cause of at least some muscle cramps is the following:

Any local irritating factor or metabolic abnormality of a muscle—such as severe cold, lack of blood flow to the muscle, or overexercise of the muscle—can elicit pain or other types of sensory impulses that are transmitted from the muscle to the spinal cord, thus causing reflex muscle contraction. The contraction in turn stimulates the same sensory receptors still more, which causes the spinal cord to increase the intensity of contraction still further. Thus, a positive feedback develops so that a small amount of initial irritation causes more and more contraction until a full-blown muscle cramp ensues. Reciprocal inhibition of the muscle can sometimes relieve the cramp. That is, if a person purposefully contracts the muscle on the side of the joint opposite the cramped muscle while at the same time using the other hand or foot to prevent movement of the joint, the reciprocal inhibition that occurs in the cramped muscle can at times relieve the cramp.

■ THE AUTONOMIC REFLEXES IN THE SPINAL CORD

Many different types of segmental autonomic reflexes occur in the spinal cord, most of which are discussed in Chapter 28. Briefly, these include (1) changes in vascular tone, resulting from local skin heat and cold; (2) sweating, which results from localized heat on the surface of the body; (3) intestinointestinal reflexes that control some motor functions of the gut; (4) peritoneointestinal reflexes that inhibit gastric motility in response to peritoneal irritation; and (5) evacuation reflexes for emptying the bladder and the colon. In addition, all the segmental reflexes can at times be elicited simultaneously in the form of the so-called mass reflex as follows:

The Mass Reflex. In a spinal animal or human being, the spinal cord sometimes suddenly becomes excessively active, causing massive discharge of large portions of the cord. The usual stimulus that causes this is a strong nociceptive stimulus to the skin or excessive filling of a viscus, such as overdistension of the bladder or of the gut. Regardless of the type of stimulus, the resulting reflex, called the *mass reflex,* involves large portions or even all of the cord, and its pattern of reaction is the same. The effects are (1) a major portion of the body goes into strong flexor spasm, (2) the colon and bladder are likely to evacuate, (3) the arterial pressure often rises to maximal values—sometimes to a mean pressure well over 200 mm Hg, and (4) large areas of the body break out into profuse sweating. The mass reflex might be likened to epileptic seizures that involve the central nervous system in which large portions of the brain become massively activated.

The precise neuronal mechanism of the mass reflex is unknown. However, because it lasts for minutes, it presumably results from activation of great masses of reverberating circuits that excite large areas of the cord at once.

■ SPINAL CORD TRANSECTION AND SPINAL SHOCK

When the spinal cord is suddenly transected, essentially all cord functions, including the cord reflexes, immediately become depressed to the point of total silence, a reaction called *spinal shock*. The reason for this is that normal activity of the cord neurons depends to a great extent on continual tonic discharges of nerve fibers entering the cord from higher centers, particularly discharges transmitted through the reticulospinal tracts, vestibulospinal tracts, and corticospinal tracts.

After a few hours to a few weeks, the spinal neurons gradually regain their excitability. This seems to be a natural characteristic of neurons everywhere in the nervous system —that is, after they lose their source of facilitatory impulses, they increase their own natural degree of excitability to make up for the loss. But there is also some possibility of sprouting of multiple new nerve endings in the cord, which could also increase excitability. In most nonprimates, the excitability of the cord centers returns essentially to normal within a few hours to a day or so, but in human beings the return is often delayed for several weeks and occasionally is never complete; or, on the other hand, recovery is sometimes excessive, with resultant hyperexcitability of some or all cord functions.

Some of the spinal functions specifically affected during or following spinal shock are the following: (1) The arterial blood pressure falls immediately—sometimes to as low as 40 mm Hg—thus illustrating that sympathetic activity becomes blocked almost to extinction. However, the pressure ordinarily returns to normal within a few days even in the human being. (2) All skeletal muscle reflexes integrated in the spinal cord are completely blocked during the initial stages of shock. In lower animals, a few hours to a few days are required for these reflexes to return to normal, and in human beings 2 weeks to several months are usually required. Sometimes, both in animals and in people, some reflexes eventually become hyperexcitable, particularly if a few facilitatory pathways remain intact between the brain and the cord while the remainder of the spinal cord is transected. The first reflexes to return are the stretch reflexes, followed in order by the progressively more complex reflexes: the flexor reflexes, the postural antigravity reflexes, and remnants of stepping reflexes. (3) The sacral reflexes for control of bladder and colon evacuation are completely suppressed in human beings for the first few weeks following cord transection, but they eventually return. These effects are discussed in Chapter 28.

REFERENCES

Adams, R. D., and Victor, M. (eds.): Principles of Neurology, 4th Ed. New York, McGraw-Hill Book Co., 1989.

Austin, G.: The Spinal Cord. New York, Igaku Shoin Medical Publishers, 1981.

Baldissera, F., et al.: Integration in spinal neuronal systems. In Brooks, V. B. (ed.): Handbook of Physiology. Sec. 1, Vol. II. Bethesda, Md., American Physiological Society, 1981, p. 509.

Brooks, V. B.: The Neural Basis of Motor Control. New York, Oxford University Press, 1986.

Burke, R. E.: Motor units: Anatomy, physiology, and functional organization. In Brooks, V. B. (ed.): Handbook of Physiology. Sec. 1, Vol. II. Bethesda, Md., American Physiological Society, 1981, p. 345.

Creed, R. S., et al.: Reflex Activity of the Spinal Cord. New York, Oxford University Press, 1932.

Emonet-Denand, F., et al.: How muscle spindles signal changes of muscle length. News Physiol. Sci., 3:105, 1988.

Evarts, E. V., et al. (eds.): Motor System in Neurobiology. New York, Elsevier Science Publishing Co., 1986.

Hammond, D. L.: New insights regarding organization of spinal cord pain pathways. News Physiol. Sci., 4:98, 1989.

Hasan, A., and Stuart, D. G.: Animal solutions to problems of movement control: The role of proprioceptors. Annu. Rev. Neurosci., 11:199, 1988.

Henneman, E., and Mendell, L. M.: Functional organization of motoneuron pool and its inputs. In Brooks, V. B. (ed.): Handbook of Physiology. Sec. 1, Vol. II. Bethesda, Md., American Physiological Society, 1981, p. 423.

Hnik, P., et al. (eds.): Mechanoreceptors. Development, Structure, and Function. New York, Plenum Publishing Corp., 1988.

Houk, J. C., and Rymer, W. Z.: Neural control of muscle length and tension. In Brooks, V. B. (ed.): Handbook of Physiology. Sec. 1, Vol. II. Bethesda, Md., American Physiological Society, 1981, p. 257.

Houk, J. C.: Control strategies in physiological systems. FASEB J., 2:97, 1988.

Hunt, C. C., and Perl, E. R.: Spinal reflex mechanisms concerned with skeletal muscle. Physiol. Rev., 40:538, 1960.

Illis, L. S.: Spinal Cord Dysfunction. New York, Oxford University Press, 1988.

Janig, W., and McLachlan, E. M.: Organization of lumbar spinal outflow to distal colon and pelvic organs. Physiol. Rev., 67:1332, 1987.

Kao, C. C. (ed.): Spinal Cord Reconstruction. New York, Raven Press, 1983.

Le Douarin, N. M., et al.: From the neural crest to the ganglia of the peripheral nervous system. Annu. Rev. Physiol., 43:653, 1981.

Matthews, P. B. C.: Muscle spindles and their fusimotor supply. In Brooks, V. B. (ed.): Handbook of Physiology. Sec. 1, Vol. II. Bethesda, Md., American Physiological Society, 1981, p. 189.

Mendell, L. M.: Modifiability of spinal synapses. Physiol. Rev., 64:260, 1984.

Rack, P. M. H.: Limitations of somatosensory feedback in control of posture and movement. In Brooks, V. B. (ed.): Handbook of Physiology. Sec. 1, Vol. II. Bethesda, Md., American Physiological Society, 1981, p. 229.

Redman, S. J.: Monosynaptic transmission in the spinal cord. News Physiol. Sci., 1:171, 1986.

Rowell, L. B.: Reflex control of regional circulation in humans. J. Auton. Nerv. Syst., 11:101, 1984.

Sachs, F.: Mechanical transduction: Unification? News Physiol. Sci., 1:98, 1986.

Sherrington, C. S.: The Integrative Action of the Nervous System. New Haven, Conn., Yale University Press, 1911.

Stein, D. G., and Sabel, B. A. (eds.): Pharmacological Approaches to the Treatment of Brain and Spinal Cord Injury. New York, Plenum Publishing Corp., 1988.

Stein, R. B.: Peripheral control of movement. Physiol. Rev., 54:215, 1974.

Stein, R. B., and Lee, R. G.: Tremor and clonus. In Brooks, V. B. (ed.): Handbook of Physiology. Sec. 1, Vol. II. Bethesda, Md., American Physiological Society, 1981, p. 325.

Wiesendanger, M., and Miles, T. S.: Ascending pathway of low-threshold muscle afferents to the cerebral cortex and its possible role in motor control. Physiol. Rev., 62:1234, 1982.

Youmans, J. R. (ed.): Neurological Surgery. Philadelphia, W. B. Saunders Co., 1989.

17

Cortical and Brain Stem Control of Motor Function

In this chapter we will discuss the control of the body's movements by the cerebral cortex and brain stem. These two neural areas, along with the basal ganglia and cerebellum, which are discussed in the following chapter, control the very complex movements that the human being and other higher animals have developed for their special purposes.

Virtually all "voluntary" movements involve conscious activity in the cerebral cortex. Yet this does not mean that each contraction of each muscle is willed by the cortex itself. Instead, most control used by the cortex involves the patterns of function in lower brain areas — in the cord, in the brain stem, in the basal ganglia, in the cerebellum — and these lower centers in turn send most of the specific activating signals to the muscles. However, for a few types of movements, the cortex does have almost a direct pathway to the anterior motor neurons of the cord, bypassing other motor centers on the way, especially for control of the very fine dexterous movements of our fingers and hands. It will be the goal of this and the following chapter to explain the interplay among the different motor areas of the brain and spinal cord that provides this overall synthesis of motor function.

■ THE MOTOR CORTEX AND THE CORTICOSPINAL TRACT

Figure 17–1 illustrates the functional areas of the cerebral cortex. Anterior to the central sulcus, occupying approximately the posterior third of the frontal lobes, is the *motor cortex*. Posterior to the central sulcus is the *somatic sensory cortex,* an area discussed in detail in earlier chapters that feeds many signals into the motor cortex for control of motor activities.

The motor cortex itself is further divided into three separate subareas, each of which has its own topographical representation of all the muscle groups of the body: (1) the *primary motor cortex*, (2) the *premotor area*, and (3) the *supplemental motor area*.

THE PRIMARY MOTOR CORTEX

The primary motor cortex, shown in Figure 17–1, lies in the first convolution of the frontal lobes anterior to the central sulcus. It begins laterally in the sylvian fissure and spreads superiorly to the uppermost portion of the brain, then dips over into the longitudinal fissure. This area is the same as area 4 in Brodmann's classification of the brain cortical areas shown in Figure 9–5 of Chapter 9.

Figure 17–1 lists the topographical representations of the different muscle areas of the body in the primary motor cortex, beginning with the face and mouth region, near the sylvian fissure; the arm and hand area, in the midportions of the primary motor cortex; the trunk, near the apex of the brain; and the leg and foot areas, in that part of the primary motor cortex that dips into the longitudinal fissure. This topographical organization is illustrated even more graphically in Figure 17–2, which shows the degrees of representation of the different muscle areas as mapped by Penfield and Rasmussen. This mapping was done by electrically stimulating the different areas of the motor cortex in human beings who were undergoing neurosurgical operations. Note that more than half of the entire primary motor cortex is concerned with controlling the hands and the muscles of speech. Point stimulations in these motor areas will cause contraction of a single muscle or even a portion of a single muscle. But in those areas with a lesser degree of representation, such as the trunk area, electrical stimulation usually contracts a group of muscles instead.

THE PREMOTOR AREA

The premotor area, also shown in Figure 17–1, lies immediately anterior to the primary motor cortex, projecting 1 to 3 cm anteriorly and extending inferiorly into the sylvian fissure and superiorly into the longitudinal fissure, where it abuts the supplemental motor

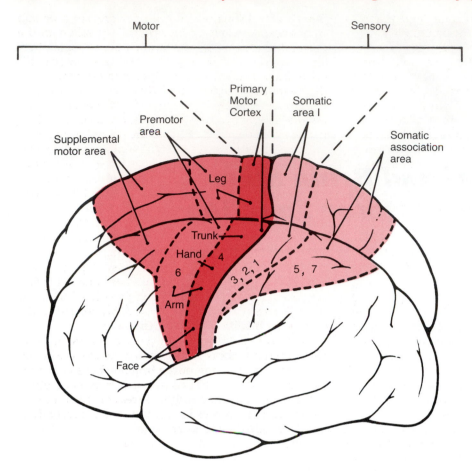

Figure 17–1. The motor and somatic sensory functional areas of the cerebral cortex.

area. Note that the topographical organization of the premotor cortex is roughly the same as that of the primary motor cortex, with the face area located most laterally and then in the upward direction the arm, trunk, and leg areas. The premotor area is frequently called simply motor area 6 because it occupies a large share of area 6 in the Brodmann classification of brain topology.

Most nerve signals generated in the premotor area cause patterns of movement involving groups of muscles that perform specific tasks. For instance, the task may be to position the shoulders and arms so that the hands become properly oriented to perform specific tasks. To achieve these results, the premotor area sends its signals either directly into the primary motor cortex to excite multiple groups of muscles or, more likely, by way of the basal ganglia and then back through the thalamus to the primary motor cortex. Thus, the premotor cortex, the basal ganglia, the thalamus, and the primary motor cortex constitute a complex overall system for control of many of the body's more complex patterns of coordinated muscle activity.

THE SUPPLEMENTAL MOTOR AREA

The supplemental motor area has still another topographical organization for control of motor function. It lies immediately superior and anterior to the premotor area, lying mainly in the longitudinal fissure but extending a centimeter or so over the edge onto the superiormost portion of the exposed cortex. Note in Figure 17–1 that the leg area lies most posteriorly and the face most anteriorly.

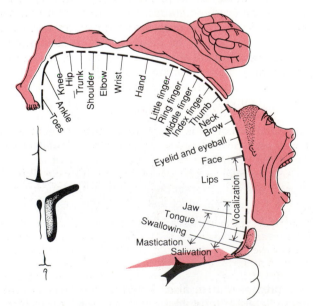

Figure 17–2. Degree of representation of the different muscles of the body in the motor cortex. (From Penfield and Rasmussen: The Cerebral Cortex of Man: A Clinical Study of Localization of Function. New York, Macmillan Co., 1968.)

Considerably stronger electrical stimuli are required in the supplemental motor area to cause muscle contraction than in the other motor areas. However, when contractions are elicited, they are often bilateral rather than unilateral. And stimulation frequently leads to movements such as unilateral grasping of a hand or at other times bilateral grasping of both hands simultaneously; these movements are perhaps rudiments of the hand functions required for climbing. Also, there may be rotation of the hands, movement of the eyes, vocalization, or yawning. But other than these subtle clues to the function of the supplemental motor area, very little else is known. In general, this area probably functions in concert with the premotor area to provide attitudinal movements, fixation movements of the different segments of the body, positional movements of the head and eyes, and so forth, as background for the finer motor control of the hands and feet by the premotor and primary motor cortex.

SOME SPECIALIZED AREAS OF MOTOR CONTROL FOUND IN THE HUMAN MOTOR CORTEX

Neurosurgeons have found a few highly specialized motor regions of the human cerebral cortex, located mainly in the premotor areas as illustrated in Figure 17–3, that control very specific motor functions. These have been localized either by electrical stimulation or by noting the loss of motor function when destructive lesions have occurred in specific cortical areas. Some of the more important of these are the following:

Broca's Area and Speech. Figure 17–3 illustrates a premotor area lying immediately anterior to the primary motor cortex and immediately above the sylvian fissure labeled "word formation." This region is called *Broca's area.* Damage to it does not prevent a person from vocalizing, but it does make it impossible for the person to speak whole words other than simple utterances such as "no" or "yes." A closely associated cortical area also causes appropriate respiratory function so that respiratory activation of vocal cords can occur simultaneously with the movements of the mouth and tongue during speech. Thus, the premotor activities that are related to Broca's area are highly complex.

The "Voluntary" Eye Movement Field. Immediately above Broca's area is a locus for controlling eye movements. Damage to this area prevents a person from voluntarily moving the eyes toward different objects. Instead, the eyes tend to lock on specific objects, an effect controlled by signals from the occipital cortex, as explained in Chapter 13. This frontal area also controls eyelid movements such as blinking.

Head Rotation Area. Still slightly higher in the motor association area, electrical stimulation will elicit head rotation. This area is closely associated with the eye movement field and is presumably related to directing the head toward different objects.

Area for Hand Skills. In the premotor area immediately anterior to the primary motor cortex for the hands and fingers is a region neurosurgeons have called an area for hand skills. That is, when tumors or other lesions cause destruction in this area, the hand movements become incoordinate and nonpurposeful, a condition called *motor apraxia.*

TRANSMISSION OF SIGNALS FROM THE MOTOR CORTEX TO THE MUSCLES

Motor signals are transmitted directly from the cortex to the spinal cord through the *corticospinal tract* and indirectly through multiple accessory pathways that involve the *basal ganglia,* the *cerebellum,* and various *nuclei of the brain stem.* In general, the direct pathways are concerned more with discrete and detailed movements, especially of the distal segments of the limbs, particularly the hands and the fingers.

The Corticospinal Tract (Pyramidal Tract)

The most important output pathway from the motor cortex is the *corticospinal tract,* also called the *pyramidal tract,* which is illustrated in Figure 17–4.

The corticospinal tract originates about 30 per cent from the primary motor cortex, 30 per cent from the premotor and supplementary motor areas, and 40 per cent from the somatic sensory areas posterior to the central sulcus. After leaving the cortex it passes through the posterior limb of the internal capsule (between the caudate nucleus and the putamen of the basal ganglia) and then downward through the brain stem, forming the *pyramids of the medulla.* By far the majority of the pyramidal fibers then cross to the opposite side and descend in the *lateral corticospinal tracts* of the cord; most terminate on the interneurons in the intermediate regions of the cord gray matter, but a few end on sensory relay neurons in the dorsal horn, and some end directly on the anterior motor neurons.

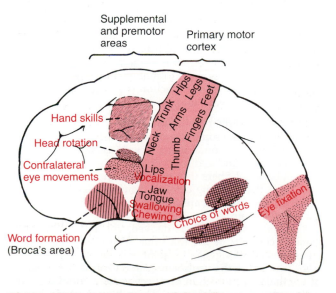

Figure 17–3. Representation of the different muscles of the body in the motor cortex and location of other cortical areas responsible for certain types of motor movements.

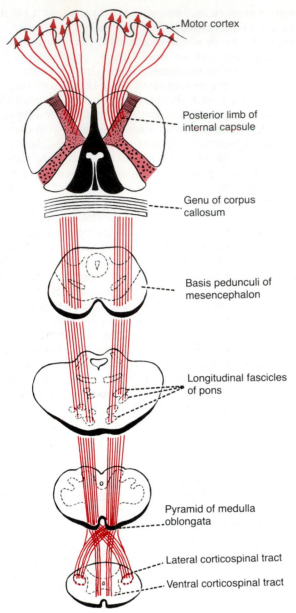

Figure 17–4. The pyramidal tract. (Modified from Ranson and Clark: Anatomy of the Nervous System. Philadelphia, W. B. Saunders Co., 1959.)

A few of the fibers do not cross to the opposite side in the medulla but pass ipsilaterally down the cord in the *ventral corticospinal tracts,* but many of these fibers also cross to the opposite side of the cord either in the neck or in the upper thoracic region. These fibers are perhaps concerned with control by the supplementary motor area of bilateral postural movements.

The most impressive fibers in the pyramidal tract are a population of large myelinated fibers with mean diameter of 16 μm. These originate from the *giant pyramidal cells,* also called *Betz cells,* that are found only in the primary motor cortex. These cells are about 60 μm in diameter, and their fibers transmit nerve impulses to the spinal cord at a velocity of about 70 m/sec, the most rapid rate of transmission of any signals from the brain to the cord. There are approximately 34,000 of these large Betz cell fibers in each corticospinal tract. However, the total number of fibers in each corticospinal tract is more than 1,000,000; so these large fibers rep-

resent only 3 per cent of all of them. The other 97 per cent are mainly fibers smaller than 4 μm in diameter.

Other Fiber Pathways From the Motor Cortex

The motor cortex gives rise to very large numbers of fibers from the cortex or collaterals from the pyramidal tract that go to deeper regions of the cerebrum and also into the brain stem, including the following:

1. The axons from the giant Betz cells send short collaterals back to the cortex itself. It is believed that these collaterals mainly inhibit adjacent regions of the cortex when the Betz cells discharge, thereby "sharpening" the boundaries of the excitatory signal.

2. A large body of fibers passes into the *caudate nucleus* and *putamen.* From here additional pathways extend through several neurons into the brain stem, as discussed in the following chapter.

3. A moderate number of fibers pass to the *red nuclei.* From these, additional fibers pass down the cord through the *rubrospinal tract.*

4. A moderate number of fibers deviate into the *reticular substance* and *vestibular nuclei* of the brain stem; from here signals go to the cord via *reticulospinal* and *vestibulospinal tracts,* and others go to the cerebellum via *reticulocerebellar* and *vestibulocerebellar tracts.*

5. A tremendous number of fibers synapse in the pontile nuclei, which give rise to the *pontocerebellar fibers,* carrying signals into the cerebellar hemispheres.

6. Collaterals also terminate in the *inferior olivary nuclei,* and from here secondary *olivocerebellar fibers* transmit signals to many areas of the cerebellum.

Thus, the basal ganglia, the brain stem, and the cerebellum all receive strong signals from the corticospinal system every time a signal is transmitted down the spinal cord to cause a motor activity.

INCOMING FIBER PATHWAYS TO THE MOTOR CORTEX

The functions of the motor cortex are controlled mainly by the somatic sensory system but also to a lesser extent by the other sensory systems such as hearing and vision. Once the sensory information is derived from these sources, the motor cortex operates in association with the basal ganglia and cerebellum to process the information and to determine the appropriate course of motor action. The more important incoming fiber pathways to the motor cortex are as follows:

1. Subcortical fibers from adjacent regions of the cortex, especially from the somatic sensory areas of the parietal cortex and from the frontal areas as well as subcortical fibers from the visual and auditory cortices.

2. Subcortical fibers that pass through the corpus callosum from the opposite cerebral hemisphere. These fibers connect corresponding areas of the motor cortices in the two sides of the brain.

3. Somatic sensory fibers derived directly from the ventrobasal complex of the thalamus. These transmit mainly cutaneous tactile signals and joint and muscle signals.

4. Tracts from the ventrolateral and ventroanterior nuclei of the thalamus, which in turn receive tracts from the cerebellum and the basal ganglia. These tracts provide signals that are necessary for coordination between the functions of the motor cortex, the basal ganglia, and the cerebellum.

5. Fibers from the intralaminar nuclei of the thalamus. These fibers probably control the general level of excitability of the motor cortex in the same manner that they also control the general level of excitability of most other regions of the cerebral cortex.

THE RED NUCLEUS SERVES AS AN ALTERNATE PATHWAY FOR TRANSMITTING CORTICAL SIGNALS TO THE SPINAL CORD

The *red nucleus,* located in the mesencephalon, functions in close association with the corticospinal tract. As illustrated in Figure 17–5, it receives a large number of direct fibers from the primary motor cortex through the *corticorubral tract* as well as branching fibers from the corticospinal tract as it passes through the mesencephalon. These fibers synapse in the lower portion of the red nucleus, the *magnocellular portion,* that contains large neurons similar to the Betz cells in the motor cortex. These large neurons give rise to the *rubrospinal tract,* which crosses to the opposite side in the lower brain stem and follows a course parallel to the corticospinal tract into the lateral columns of the spinal cord. This tract partially overlaps the corticospinal tract but on the average lies slightly anterior to it. The rubrospinal fibers terminate mainly on the interneurons of the intermediate areas of the cord gray matter along with the corticospinal fibers, but a few of the rubrospinal fibers also terminate on the anterior motor neurons, along with some corticospinal fibers.

The red nucleus also has close connections with the cerebellum, similar to the connections between the motor cortex and the cerebellum.

Function of the Corticorubrospinal System. The magnocellular portion of the red nucleus has a somatographic representation of all the muscles of the body, as is true of the motor cortex. Therefore, stimulation of a single point in this portion of the red nucleus will cause contraction of either a single muscle or a small group of muscles. However, the fineness of representation of the different muscles is far less developed than in the motor cortex. This is especially true in human beings who have a relatively small red nucleus.

The corticorubrospinal pathway serves as an accessory route for the transmission of relatively discrete signals from the motor cortex to the spinal cord. When the corticospinal fibers are destroyed without destroying this other pathway, discrete movements can still occur, except that the movements of the fingers and hands are considerably impaired. Wrist movements are still well developed, which is not true when the corticorubrospinal pathway is also blocked. Therefore, the pathway through the red nucleus to the spinal cord is associated far more with the corticospinal system than with the other major brain stem motor pathway, the vestibuloreticulospinal system that controls mainly the axial and girdle muscles of the body, as we shall discuss later in the chapter. Furthermore, the rubrospinal tract lies in the lateral columns of the spinal cord, along with the corticospinal tract, and terminates more on the interneurons and motor neurons that control the distal muscles of the limbs. Therefore, the corticospinal and rubrospinal tracts together are frequently called the *lateral motor system of the cord,* in contradistinction to the vestibuloreticulospinal system that lies mainly medially in the cord and is called the *medial motor system of the cord.*

THE EXTRAPYRAMIDAL SYSTEM

The term *"extrapyramidal motor system"* is widely used in clinical circles to denote all those portions of the brain and brain stem that contribute to motor control that are not part of the direct corticospinal-pyramidal system. This includes pathways through the basal ganglia, the reticular formation of the brain stem, the vestibular nuclei, and often the red nuclei as well. However, this is such an all-inclusive and diverse group of motor control areas that it is difficult to ascribe specific neurophysiological functions to the extrapy-

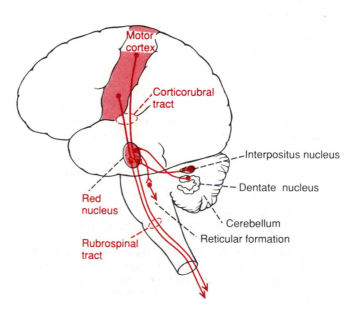

Figure 17–5. The corticorubrospinal pathway for motor control, showing also the relationship of this pathway to the cerebellum.

ramidal system as a whole. For this reason, the term "extra-pyramidal" is beginning to have less usage clinically as well as physiologically.

EXCITATION OF THE SPINAL CORD BY THE PRIMARY MOTOR CORTEX AND THE RED NUCLEUS

Vertical Columnar Arrangement of the Neurons in the Motor Cortex. In Chapters 9 and 13 it is pointed out that the cells in the somatic sensory cortex and visual cortex—and in all other parts of the brain as well—are organized in vertical columns of cells. In a like manner, the cells of the motor cortex are also organized in vertical columns a fraction of a millimeter in diameter and having thousands of neurons in each column.

Each column of cells functions as a unit, stimulating either a single muscle or a group of synergistic muscles. Also, each column is arranged in six distinct layers of cells, like the arrangement throughout almost all the cerebral cortex. The pyramidal cells that give rise to the corticospinal fibers all lie in the fifth layer of cells from the cortical surface, whereas the input signals to the column of cells all enter layers 2 through 4. The sixth layer gives rise mainly to fibers that communicate with other regions of the cerebral cortex itself.

Function of Each Column of Neurons. The neurons of each column operate as an integrative processing system, utilizing information from multiple input sources to determine the output response from the column. In addition, each column can function as an amplifying system to stimulate large numbers of pyramidal fibers to the same muscle or to synergistic muscles simultaneously. This is important because stimulation of a single pyramidal cell can rarely excite a muscle. Instead, as many as 50 to 100 pyramidal cells usually need to be excited simultaneously or in rapid succession to achieve muscle contraction.

Dynamic and Static Signals Transmitted by the Pyramidal Neurons. If a strong signal is sent at first to a muscle to cause initial rapid contraction, then a much weaker signal can maintain the contraction for long periods thereafter. This is the manner in which excitation for causing muscle contractions is usually provided. To do this, each column of cells excites two separate populations of pyramidal cell neurons, one called *dynamic neurons* and the other *static neurons.* The dynamic neurons are excessively excited for a short period of time at the beginning of the contraction, causing the initial *development of force.* Then the static neurons fire at a much slower rate, but they continue at this slow rate indefinitely to *maintain the force* of contraction as long as the contraction is required.

The neurons of the red nucleus have similar dynamic and static characteristics, except that more dynamic neurons are in the red nucleus and more static neurons in the primary motor cortex. This perhaps relates to the fact that the red nucleus is closely allied with the cerebellum, and we shall learn later in this chapter that the cerebellum also plays an important role in the rapid initiation of muscle contraction.

Somatic Sensory Feedback to the Motor Cortex

When nerve signals from the motor cortex cause a muscle to contract, somatic sensory signals return from the activated region of the body to the neurons in the motor cortex that are causing the action. Most of these somatic sensory signals arise in the muscle spindles or in the tactile receptors of the skin overlying the muscles. In general, the somatic signals cause a positive feedback enhancement of the muscle contraction in the following ways: In the case of the muscle spindles, if the fusimotor muscle fibers in the spindles contract more than the large skeletal muscle itself does, then the spindles become excited, and the signals from these spindles stimulate the pyramidal cells in the motor cortex, which further excites the muscle, helping its contraction catch up with the contraction of the spindles. In the case of the tactile receptors, if the muscle contraction causes compression of the skin against an object, such as compression of the fingers around an object that is being grasped, the signals from these receptors cause further excitement of the muscles and therefore increase the muscle contraction—such as increasing the tightness of the grasp of the hand.

Stimulation of the Spinal Motor Neurons

Figure 17–6 shows a segment of the spinal cord, illustrating multiple motor tracts entering the cord from the brain and also showing a representative anterior motor neuron. The corticospinal tract and the rubrospinal tract lie in the dorsal portions of the lateral columns. At most levels of the cord, their fibers terminate mainly on interneurons in the intermediate area of the cord gray matter. However, in the cervical enlargement of the cord where the hands and fingers are represented,

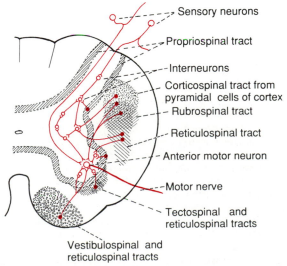

Sensory neurons

Propriospinal tract

Interneurons

Corticospinal tract from pyramidal cells of cortex

Rubrospinal tract

Reticulospinal tract

Anterior motor neuron

Motor nerve

Tectospinal and reticulospinal tracts

Vestibulospinal and reticulospinal tracts

Figure 17–6. Convergence of all the different motor pathways on the anterior motor neurons.

moderate numbers of both corticospinal and rubrospinal fibers terminate directly on the anterior motor neurons, thus allowing a direct route from the brain for activating muscle contraction. This is in keeping with the fact that the primary motor cortex has an extremely high degree of representation for fine control of hand, finger, and thumb actions.

Patterns of Movement Elicited by Spinal Cord Centers. From the previous chapter, recall that the spinal cord can provide specific reflex patterns of movement in response to sensory nerve stimulation. Many of these patterns are also important when the anterior motor neurons are excited by signals from the brain. For instance, the stretch reflex is functional at all times, helping to damp the motor movements initiated from the brain and probably providing at least part of the motive power required to cause the muscle contractions, employing the servo-assist mechanism that was described in Chapter 16.

Also, when a brain signal excites an agonist muscle, it is not necessary to transmit an inverse signal to the antagonist at the same time; this transmission will be achieved by the reciprocal innervation circuit that is always present in the cord for coordinating the functions of antagonistic pairs of muscles.

Finally, parts of the other reflex mechanisms, such as withdrawal, stepping and walking, scratching, postural mechanisms, and so forth, can be activated by "command" signals from the brain. Thus, very simple signals from the brain can initiate many of our normal motor activities, particularly for such functions as walking and the attainment of different postural attitudes of the body.

EFFECT OF LESIONS IN THE MOTOR CORTEX OR CORTICOSPINAL PATHWAY— THE "STROKE"

The motor cortex or corticospinal pathway is frequently damaged, especially by the common abnormality called a "stroke." This is caused either by a ruptured blood vessel that allows hemorrhage into the brain or by thrombosis of one of the major arteries supplying the brain, in either case causing loss of blood supply to the cortex, or very frequently to the corticospinal tract, where it passes through the internal capsule between the caudate nucleus and the putamen. Also, experiments have been performed in animals to remove selectively different parts of the motor cortex.

Removal of the Primary Motor Cortex (the Area Pyramidalis). Removal of a portion of the primary motor cortex —the area that contains the giant Betz pyramidal cells—in a monkey causes varying degrees of paralysis of the represented muscles. If the sublying caudate nucleus and the adjacent premotor area are not damaged, gross postural and limb "fixation" movements can still be performed, but the animal *loses voluntary control of discrete movements of the distal segments of the limbs—especially of the hands and fingers.* This does not mean that the muscles themselves cannot contract, but that the animal's ability to control the fine movements is gone.

From these results one can conclude that the area pyramidalis is essential for voluntary initiation of finely controlled movements, especially of the hands and fingers.

Muscle Spasticity Caused by Lesions That Damage Large Areas Adjacent to the Motor Cortex. Ablation of the primary motor cortex alone causes *hypotonia,* not spasticity, because the primary motor cortex normally exerts a continual tonic stimulatory effect on the motor neurons of the spinal cord; when this is removed, hypotonia results.

On the other hand, most lesions of the motor cortex, especially those caused by a stroke, involve not only the primary motor cortex but also adjacent cortical areas and deeper structures of the cerebrum as well, especially the basal ganglia. In these instances, muscle spasm almost invariably occurs in the afflicted muscle areas on the opposite side of the body (because all the motor pathways cross to the opposite side). Obviously, this spasm is not caused by loss of either the primary motor cortex or blockage of the corticospinal fibers to the cord.

Instead, it is believed to result mainly from damage to accessory pathways from the cortex that normally inhibit the vestibular and reticular brain stem nuclei. When these nuclei lose this inhibition (that is, when they are said to be "disinhibited"), they become spontaneously active and cause excessive spastic tone in the involved areas of the body, as we discuss more fully later in the chapter. This is the spasticity that normally accompanies a "stroke" in the human being.

The Babinski Sign Used As a Clinical Tool in Testing Corticospinal Integrity. Destruction of the foot region of the area pyramidalis or transection of the foot portion of the corticospinal tract causes a peculiar response of the foot called the *Babinski sign.* This response is demonstrated when a firm tactile stimulus is applied to the lateral sole of the foot: The great toe extends upward, and the other toes fan outward. This is in contradistinction to the normal effect in which all the toes bend downward. The Babinski sign does not occur when damage occurs in the noncorticospinal portions of the motor control system without involving the corticospinal tract. Therefore, the sign is used clinically to detect damage specifically in the corticospinal portion of the motor control system.

The cause of the Babinski sign is believed to be the following: The corticospinal tract is a major controller of muscle activity for performance of voluntary, purposeful activity. On the other hand, the noncorticospinal pathways constitute a much older motor control system and are concerned to a great extent with protection of the body from damage. Therefore, when only the noncorticospinal system is functional, stimuli to the bottom of the feet cause a typical withdrawal protective type of reflex, expressed by the upturned great toe and fanning of the other toes. But when the corticospinal system is also fully functional, it suppresses the protective reflex and instead excites a higher order of motor function, including the normal effect of causing downward bending of the toes and foot in response to sensory stimuli from the bottom of the feet, a response that helps us to walk.

■ ROLE OF THE BRAIN STEM IN CONTROLLING MOTOR FUNCTION

The brain stem consists of the *medulla, pons,* and *mesencephalon.* In one sense it is an extension of the spinal cord upward into the cranial cavity, because it contains motor and sensory nuclei that perform motor and sensory functions for the face and head regions in the same way that the anterior and posterior gray horns of the spinal cord perform these same functions from the neck down. But in another sense, it is its own master, because it provides many special control functions, such as the following:

1. Control of respiration.
2. Control of the cardiovascular system.
3. Control of gastrointestinal function.
4. Control of many stereotyped movements of the body.
5. Control of equilibrium.
6. Control of eye movement.

Finally, the brain stem serves also as an instrument of higher neural centers that transmit many "command" signals into the brain stem to initiate or modify the brain stem's specific control functions.

In the following sections of this chapter we discuss the role of the brain stem in controlling whole body movement and equilibrium. Especially important for this purpose are the brain stem's *reticular nuclei* and *vestibular nuclei* plus the *vestibular apparatus* that sends most of the equilibrium control signals to the vestibular nuclei and to a lesser extent to the reticular nuclei as well.

SUPPORT OF THE BODY AGAINST GRAVITY—ROLES OF THE RETICULAR AND VESTIBULAR NUCLEI

Excitatory-Inhibitory Antagonism Between Pontine and Medullary Reticular Nuclei

Figure 17–7 illustrates the locations of the reticular and vestibular nuclei. The reticular nuclei are divided into two major groups, (1) the *pontine reticular nuclei,* located mainly in the pons but extending into the mesencephalon as well, lying more laterally in the brain stem, and (2) the *medullary reticular nuclei,* which extend the entire extent of the medulla, lying ventrally and medially near the midline. These two sets of nuclei function mainly antagonistically to each other, the pontine exciting the antigravity muscles and the medullary inhibiting them. The pontine reticular nuclei transmit excitatory signals downward into the cord through the *pontine* (or *medial*) *reticulospinal tract,* illustrated in Figure 17–8. The fibers of this pathway terminate on the medial anterior motor neurons that excite the muscles that support the body against gravity, that is, the muscles of the spinal column and the extensor muscles of the limbs.

The pontine reticular nuclei have a high degree of natural excitability. In addition, they receive excitatory signals from local circuits within the brain stem and especially strong excitatory signals from the vestibular nuclei and also the deep nuclei of the cerebellum. Therefore, when the pontine reticular excitatory system is unopposed by the medullary reticular system, it causes powerful excitation of the antigravity muscles throughout the body, so much so that animals can then stand up against gravity without any signals from the higher levels of the brain.

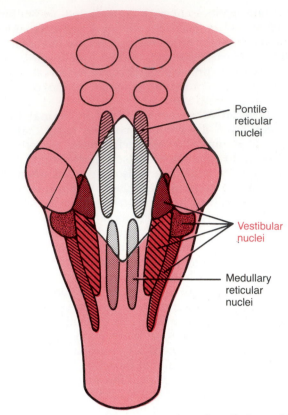

Figure 17–7. Locations of the reticular and vestibular nuclei in the brain stem.

The Medullary Reticular System. The medullary nuclei, on the other hand, transmit inhibitory signals to the same antigravity anterior motor neurons by way of a different tract, the *medullary* (or *lateral*) *reticulospinal tract,* also illustrated in Figure 17–8. The medullary reticular nuclei receive strong input collaterals from (1) the corticospinal tract, (2) the rubrospinal tract, and (3) other motor pathways. These normally activate the medullary reticular inhibitory system to counterbalance the excitatory signals from the pontine reticular system. Yet other signals from the cerebral, red nu-

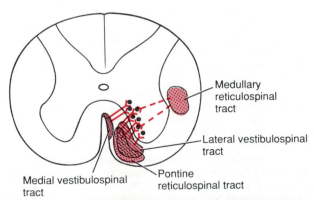

Figure 17–8. The vestibulospinal and reticulospinal tracts descending in the spinal cord to excite (solid lines) or inhibit (dashed lines) the anterior motor neurons that control the body's axial musculature.

cleus, and cerebellar pathways "disinhibit" the medullary system when the brain wishes for excitation by the pontine system to cause standing. Or at other times, excitation of the medullary reticular system can inhibit the antigravity muscles in certain portions of the body to allow those portions to perform other motor activities, which would be impossible if the antigravity muscles opposed the necessary movements.

Therefore, the excitatory and inhibitory reticular nuclei constitute a controllable system that is manipulated by motor signals from the cortex and elsewhere to provide the necessary muscle contractions for standing against gravity and yet to inhibit appropriate groups of muscles as needed so that other functions can be performed as required.

Role of the Vestibular Nuclei in Exciting the Antigravity Muscles

The vestibular nuclei, illustrated in Figure 17–7, function also in association with the pontine reticular nuclei to excite the antigravity muscles. The *lateral vestibular nuclei* (indicated by the heavy dots in the figure), especially, transmit strong excitatory signals by way of both the *lateral* and *medial vestibulospinal tracts* in the anterior column of the spinal cord, as illustrated in Figure 17–8. In fact, without the support of the vestibular nuclei, the pontine reticular system loses much of its force. The specific role of the vestibular nuclei, however, is to control selectively the excitatory signals to the different antigravity muscles to maintain equilibrium in response to signals from the vestibular apparatus. We shall discuss this more fully later in the chapter.

The Decerebrate Animal Develops Spastic Rigidity

When the brain stem is sectioned between the pons and the mesencephalon, leaving both the pontine and medullary reticular systems as well as the vestibular system intact, the animal develops a condition called *decerebrate rigidity*. This rigidity does not occur in all muscles of the body, but instead occurs in the antigravity muscles — the muscles of the neck and the trunk and the extensors of the legs.

The cause of decerebrate *rigidity* is blockage of the normally strong excitatory input to the medullary reticular nuclei from the cerebral cortex, red nuclei, and basal ganglia. As a result, the medullary vestibular inhibitor system becomes nonfunctional because of loss of its usual excitatory drive, thus allowing full overactivity of the pontine excitatory system.

A specific characteristic of decerebrate rigidity is that the antigravity muscles exhibit the phenomenon called *spasticity* as well as rigidity. This means that any attempt to change the position of a limb or other part of the body, especially attempts to stretch the muscles suddenly, is resisted by very powerful stretch reflexes described in the previous chapter. This occurs because the pontine and vestibular antigravity signals to the cord selectively excite the gamma motor neurons in the spinal cord much more than they excite the alpha motor neurons. This tightens the intrafused muscle fibers of the muscle spindles, which in turn strongly sensitizes the stretch reflex feedback loop.

We see later that other types of rigidity occur in other neuromotor diseases, especially in lesions of the basal ganglia. In many of these the rigidity involves all muscles equally without the excessive spastic stretch reflex component.

■ VESTIBULAR SENSATIONS AND THE MAINTENANCE OF EQUILIBRIUM

THE VESTIBULAR APPARATUS

The vestibular apparatus is the organ that detects sensations of equilibrium. It is composed of a system of bony tubes and chambers in the petrous portion of the temporal bone called the *bony labyrinth* and within this a system of membranous tubes and chambers called the *membranous labyrinth,* which is the functional part of the apparatus. The top of Figure 17–9 illustrates the membranous labyrinth; it is composed mainly of the *cochlea,* three *semicircular ducts,* and two large

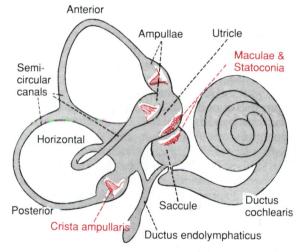

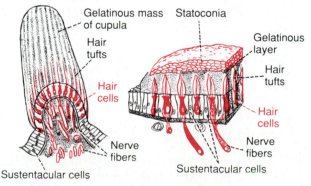

Figure 17–9. The membranous labyrinth and organization of the crista ampullaris and the macula. (Modified from Goss: Gray's Anatomy of the Human Body. 25th ed. Philadelphia, Lea & Febiger, 1948; modified from Kolmer by Buchanan: Functional Neuroanatomy. Philadelphia, Lea & Febiger.)

chambers known as the *utricle* and the *saccule*. The cochlea is the major sensory area for hearing that was discussed in Chapter 14 and has nothing to do with equilibrium. However, the *utricle*, the *semicircular ducts*, and the *saccule* are all integral parts of the equilibrium mechanism.

The Maculae — The Sensory Organs of the Utricle and the Saccule for Detecting the Orientation of the Head With Respect to Gravity. Located on the inside surface of each utricle and saccule is a small sensory area slightly over 2 mm in diameter called a *macula*. The macula of the utricle lies mainly in the horizontal plane on the inferior surface of the utricle and plays an important role in determining the normal orientation of the head with respect to the direction of gravitational or acceleratory forces when a person is upright. On the other hand, the macula of the saccule is located mainly in a vertical plane and therefore is important in equilibrium when the person is lying down.

Each macula is covered by a gelatinous layer in which many small calcium carbonate crystals called *statoconia* (or *otoliths*) are imbedded. Also, in the macula are thousands of *hair cells*, one of which is illustrated in Figure 17–10; these project *cilia* up into the gelatinous layer. The bases and sides of the hair cells synapse with sensory endings of the vestibular nerve.

Directional Sensitivity of the Hair Cells — The Kinocilium. Each hair cell has an average of 50 to 70 small cilia called *stereocilia*, plus one very large cilium, the *kinocilium*, as illustrated in the figure. The kinocilium is located always to one side, and the stereocilia become progressively shorter toward the other side of the cell. Very minute filamentous attachments, almost invisible even to the electron microscope, connect the tip of each stereocilium to the next longer stereocilium and finally to the kinocilium. Because of these attachments, when the brush pile of stereocilia and kinocilium is bent in the direction of the kinocilium, the filamentous attachments tug one after the other on the stereocilia, pulling them away from the cell body. This opens several hundred channels in each cilium membrane for conducting positive sodium ions, and large quantities of these positive ions pour into the cell from the surrounding fluids, causing *depolarization*. Conversely, bending the pile of cilia in the opposite direction (away from the kinocilium) reduces the tension on the attachments, and this closes the ion channels, thus causing *hyperpolarization*.

Under normal resting conditions, the nerve fibers leading from the hair cells transmit continuous nerve impulses at rates of about 100/sec. When the cilia are bent toward the kinocilium, the impulse traffic can increase to several hundred per second; conversely, bending the cilia in the opposite direction decreases the impulse traffic, often turning it off completely. Therefore, as the orientation of the head in space changes and the weight of the statoconia (whose specific gravity is about three times that of the surrounding tissues) bends the cilia, appropriate signals are transmitted to the brain to control equilibrium.

In each macula the different hair cells are oriented in different directions so that some of them are stimu-

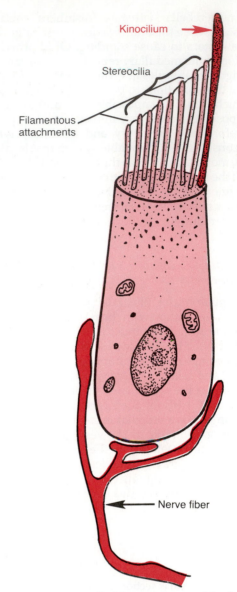

Figure 17–10. A hair cell of the membranous labyrinth of the equilibrium apparatus.

lated when the head bends forward, some when it bends backward, others when it bends to one side, and so forth. Therefore, a different pattern of excitation occurs in the nerve fibers from the macula for each position of the head; it is this "pattern" that apprises the brain of the head's orientation.

The Semicircular Ducts. The three semicircular ducts in each vestibular apparatus, known respectively as the *anterior, posterior,* and *horizontal semicircular ducts*, are arranged at right angles to each other so that they represent all three planes in space. When the head is bent forward approximately 30 degrees, the horizontal semicircular ducts are located approximately horizontal with respect to the surface of the earth. The anterior ducts are then located in vertical planes that project *forward and 45 degrees outward,* and the posterior ducts are also in vertical planes but project *backward and 45 degrees outward.* Thus, the anterior duct on each side of the head is in a plane parallel to that of the posterior duct on the opposite side of the head, whereas the horizontal ducts on the two sides are located in approximately the same plane.

Each semicircular duct has an enlargement at one of its ends called the *ampulla,* and the ducts are filled with a viscous fluid called *endolymph.* Flow of this fluid from one of the ducts into the ampulla excites the sensory organ of the ampulla in the following manner: Figure 17–11 illustrates in each ampulla a small crest called a *crista ampullaris.* On top of this crista is a gelatinous mass, the *cupula.* When the head begins to rotate in any direction, the inertia of the fluid in one or more of the semicircular ducts will cause the fluid to remain stationary while the semicircular duct rotates with the head. This causes fluid flow from the duct into the ampulla, bending the cupula to one side as illustrated by the position of the shaded cupula in Figure 17–11. Rotation of the head in the opposite direction causes the cupula to bend to the opposite side.

Into the cupula are projected hundreds of cilia from hair cells located along the ampullary crest. The *kinocilium* of each of these hair cells is always directed toward the same side of the cupula as the others, and bending the cupula in that direction causes depolarization of the hair cells, whereas bending it in the opposite direction will hyperpolarize the cells. From the hair cells, appropriate signals are sent by way of the *vestibular nerve* to apprise the central nervous system of changes in the rate and direction of rotation of the head in the three different planes of space.

FUNCTION OF THE UTRICLE AND SACCULE IN THE MAINTENANCE OF STATIC EQUILIBRIUM

It is especially important that the different hair cells are oriented in all different directions in the maculae of the

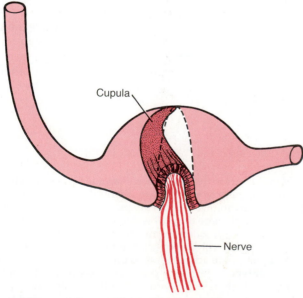

Figure 17–11. Movement of the cupula and its embedded hairs at the onset of rotation.

utricles and saccules so that at different positions of the head, different hair cells become stimulated. The "patterns" of stimulation of the different hair cells apprise the nervous system of the position of the head with respect to the pull of gravity. In turn, the vestibular, cerebellar, and reticular motor systems excite by reflex the appropriate muscles to maintain proper equilibrium.

The maculae, especially those in the utricles, function extremely effectively for maintaining equilibrium when the head is in the near-vertical position. Indeed, a person can determine as little as a half-degree of malequilibrium when the head leans from the precise upright position. On the other hand, as the head is leaned farther and farther from the upright, the determination of head orientation by the vestibular sense becomes poorer and poorer. Obviously, extreme sensitivity in the upright position is of major importance for maintenance of precise vertical static equilibrium, which is the most essential function of the vestibular apparatus.

Detection of Linear Acceleration by the Maculae. When the body is suddenly thrust forward—that is, when the body accelerates—the statoconia, which have greater inertia than the surrounding fluids, fall backward on the hair cell cilia, and information of malequilibrium is sent into the nervous centers, causing the individual to feel as though he were falling backward. This automatically causes him to lean his body forward until the anterior shift of the statoconia caused by leaning exactly equals the tendency for the statoconia to fall backward. At this point, the nervous system detects a state of proper equilibrium and therefore leans the body no farther forward. Thus, the maculae operate to maintain equilibrium during linear acceleration in exactly the same manner as they operate in static equilibrium.

The maculae *do not* operate for the detection of linear *velocity.* When runners first begin to run, they must lean far forward to keep from falling over backward because of *acceleration,* but once they have achieved running speed, they would not have to lean forward at all if they were running in a vacuum. When running in air they lean forward to maintain equilibrium only because of the air resistance against their bodies; in this instance it is not the maculae that make them lean but the pressure of the air acting on pressure end-organs in the skin, which initiate the appropriate equilibrium adjustments to prevent falling.

DETECTION OF HEAD ROTATION BY THE SEMICIRCULAR DUCTS

When the head suddenly *begins* to rotate in any direction (this is called angular acceleration), the endolymph in the semicircular ducts, because of its inertia, tends to remain stationary while the semicircular ducts themselves turn. This causes relative fluid flow in the ducts in the direction opposite to the rotation of the head.

Figure 17–12 illustrates a typical discharge signal from a single hair cell in the crista ampullaris when an animal is rotated for 40 sec, showing that (1) even when the cupula is in its resting position the hair cell emits a tonic discharge of approximately 100 impulses/sec; (2) when the animal is rotated, the hairs bend to one side and the rate of discharge increases greatly; and (3) with continued rotation, the excess discharge of the hair cell gradually subsides back to the resting level in about 20 sec.

The reason for this adaptation of the receptor is that within a second or more of rotation, back-pressure from the bent cupula causes the endolymph to rotate as rapidly as the semicircular canal itself; then in an additional 15 to 20 sec the cupula slowly returns to its resting position in the middle of the ampulla because of its own elastic recoil.

When the rotation suddenly stops, exactly the opposite effects take place: the endolymph continues to rotate while the semicircular duct stops. This time the cupula is bent in the opposite direction, causing the hair cell to stop discharging entirely. After another few seconds, the endolymph stops moving, and the cupula returns gradually to its resting position in about 20 sec, thus allowing the discharge of the hair cell to return to its normal tonic level as shown to the right in Figure 17–12.

Thus, the semicircular duct transmits a signal of one polarity when the head *begins* to rotate and of opposite polarity when it *stops* rotating. Furthermore, at least some hair cells will always respond to rotation in any plane—horizontal, sagittal, or coronal—for fluid movement always occurs in at least one semicircular duct.

Rate of Angular Acceleration Required to Stimulate the Semicircular Ducts. The angular acceleration required to stimulate the semicircular ducts in the human being averages about 1 degree/sec per second. In other words, when one begins to rotate, the velocity of rotation must be as much as 1 degree/sec by the end of the first second, 2 degrees/sec by the end of the second second, 3 degrees/sec by the end of the third second, and so forth, in order for the person barely to detect that the rate of rotation is increasing.

"Predictive" Function of the Semicircular Ducts in the Maintenance of Equilibrium. Because the semicircular ducts do not detect that the body is off balance in the forward direction, in the side direction, or in the backward direction, one might at first ask: What is the function of the semicircular ducts in the maintenance of equilibrium? All they detect is that the person's head is beginning to rotate or stopping rotation in one direction or another. Therefore, the function of the semicircular ducts is not likely to be to maintain static equilibrium or to maintain equilibrium during linear acceleration or when the person is exposed to steady centrifugal forces. Yet loss of function of the semicircular ducts causes a person to have very poor equilibrium when attempting to perform *rapid* and *intricate* body movements.

We can explain the function of the semicircular ducts best by the following illustration. If a person is running forward rapidly, and then suddenly begins to turn to one side, he falls off balance a fraction of a second later unless appropriate corrections are made *ahead of time*. But, unfortunately, the macula of the utricle cannot detect that he is off balance until *after* this has occurred. On the other hand, the semicircular ducts will have already detected that the person is turning, and this information can easily apprise the central nervous system of the fact that the person *will* fall off balance within the next fraction of a second or so unless some correction is made. In other words, the semicircular duct mechanism *predicts ahead of time* that malequilibrium is going to occur even before it does occur and thereby causes the equilibrium centers to make appropriate preventive adjustments. In this way, the person need not fall off balance before he begins to correct the situation.

Removal of the flocculonodular lobes of the cerebellum prevents normal function of the semicircular ducts but has less effect on the function of the macular receptors. It is especially interesting in this connection that the cerebellum serves as a "predictive" organ for most of the other rapid movements of the body as well as those having to do with equilibrium. These other functions of the cerebellum are discussed in the following chapter.

VESTIBULAR POSTURAL REFLEXES

Sudden changes in the orientation of an animal in space elicit reflexes that help to maintain equilibrium and posture. For instance, if an animal is suddenly pushed to the right, even before it can fall more than a few degrees its right legs extend instantaneously. In other words, this mechanism *anticipates* that the animal will be off balance in a few seconds and makes appropriate adjustments to prevent this.

Another type of vestibular postural reflex occurs when the animal suddenly falls forward. When this occurs, the forepaws extend forward, the extensor muscles tighten, and the muscles in the back of the neck stiffen to prevent the ani-

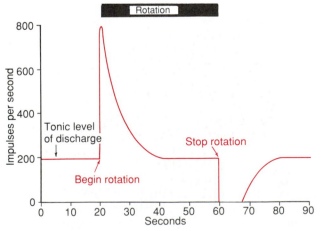

Figure 17–12. Response of a hair cell when a semicircular canal is stimulated first by rotation and then by stopping rotation.

mal's head from striking the ground. This reflex is probably also of importance in locomotion, for, in the case of the galloping horse, the downward thrust of the head can automatically provide reflex thrust of the forelimbs to move the animal forward for the next gallop.

VESTIBULAR MECHANISM FOR STABILIZING THE EYES

When a person changes his direction of movement rapidly, or even leans his head sideways, forward, or backward, it would be impossible for him to maintain a stable image on the retinae of his eyes unless he had some automatic control mechanism to stabilize the direction of gaze of the eyes. In addition, the eyes would be of little use in detecting an image unless they remained "fixed" on each object long enough to gain a clear image. Fortunately, each time the head is suddenly rotated, signals from the semicircular ducts cause the eyes to rotate in a direction equal and opposite to the rotation of the head. This results from reflexes transmitted from the canals through the *vestibular nuclei* and the *medial longitudinal fasciculus* to the *ocular nuclei* that were described in Chapter 13.

OTHER FACTORS CONCERNED WITH EQUILIBRIUM

The Neck Proprioceptors. The vestibular apparatus detects the orientation and movements *only of the head.* Therefore, it is essential that the nervous centers also receive appropriate information depicting the orientation of the head with respect to the body. This information is transmitted from the proprioceptors of the neck and body directly into the vestibular and reticular nuclei of the brain stem and also indirectly by way of the cerebellum.

By far the most important proprioceptive information needed for the maintenance of equilibrium is that derived from the *joint receptors of the neck.* When the head is leaned in one direction by bending the neck, impulses from the neck proprioceptors keep the vestibular apparatus from giving the person a sense of malequilibrium. They do this by transmitting signals that exactly oppose the signals transmitted from the vestibular apparatuses. However, *when the entire body* leans in one direction, the impulses from the vestibular apparatuses *are not opposed* by the neck proprioceptors; therefore, the person in this instance does perceive a change in equilibrium status of the entire body.

The Neck Reflexes. In an animal *whose vestibular apparatuses have been destroyed,* bending the neck causes immediate muscular reflexes called *neck reflexes* occurring especially in the forelimbs. For instance, bending the head forward causes both forelimbs to relax. However, when the vestibular apparatuses are intact, this effect does *not* occur because the vestibular reflexes function in a manner exactly opposite that of the neck reflexes. Because the equilibrium of the entire body and not of the head alone must be maintained, it is easy to understand that the vestibular and neck reflexes must function oppositely.

Proprioceptive and Exteroceptive Information from Other Parts of the Body. Proprioceptive information from other parts of the body besides the neck is also important in the maintenance of equilibrium. For instance, pressure sensations from the footpads can tell one (1) whether weight is distributed equally between the two feet and (2) whether weight is more forward or backward on the feet.

An instance in which exteroceptive information is necessary for maintenance of equilibrium occurs when a person is running. The air pressure against the front of the body signals that a force is opposing the body in a direction different from that caused by gravitational pull; as a result, the person leans forward to oppose this.

Importance of Visual Information in the Maintenance of Equilibrium. After complete destruction of the vestibular apparatus, and even after loss of most proprioceptive information from the body, a person can still use the visual mechanisms effectively for maintaining equilibrium. Even slight linear or rotational movement of the body instantaneously shifts the visual images on the retina, and this information is relayed to the equilibrium centers. Many persons with complete destruction of the vestibular apparatus have almost normal equilibrium as long as their eyes are open and as long as they perform all motions slowly. But when one is moving rapidly or when the eyes are closed, equilibrium is immediately lost.

Neuronal Connections of the Vestibular Apparatus with the Central Nervous System

Figure 17–13 illustrates the central connections of the vestibular nerve. Most of the vestibular nerve fibers end in the vestibular nuclei, which are located approximately at the junction of the medulla and the pons, but some fibers pass without synapsing to the brain stem reticular nuclei and the fastigial nuclei, uvula, and flocculonodular lobes of the cerebellum. The fibers that end in the vestibular nuclei synapse with second-order neurons that also send fibers into these areas of the cerebellum as well as to the cortex of other portions of the cerebellum, into the vestibulospinal tracts, into the medial longitudinal fasciculus, and to other areas of the brain stem, particularly the reticular nuclei.

The primary pathway for the reflexes of equilibrium begins in the vestibular nerves and passes next to both the vestibular nuclei and the cerebellum. Then, along with two-way traffic of impulses between these two, signals are also sent into the reticular nuclei of the brain stem as well as down the spinal cord via the vestibulospinal and reticulospinal tracts. In turn, the signals to the cord control the interplay between facilitation and inhibition of the antigravity muscles, thus automatically controlling equilibrium.

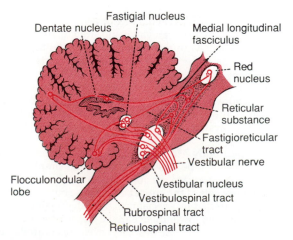

Figure 17–13. Connections of vestibular nerves in the central nervous system.

The *flocculonodular lobes* of the cerebellum seem to be especially concerned with equilibrium functions of the semicircular ducts, because destruction of these lobes gives almost exactly the same clinical symptoms as destruction of the semicircular ducts themselves. That is, severe injury to either of these structures causes loss of equilibrium during *rapid changes in direction of motion* but does not seriously disturb equilibrium under static conditions, as discussed earlier. It is also believed that the *uvula* of the cerebellum plays a similar important role in static equilibrium.

Signals transmitted upward in the brain stem from both the vestibular nuclei and the cerebellum via the *medial longitudinal fasciculus* cause corrective movements of the eyes every time the head rotates, so that the eyes can remain fixed on a specific visual object. Signals also pass upward (either through this same tract or through reticular tracts) to the cerebral cortex, probably terminating in a primary cortical center for equilibrium located in the parietal lobe deep in the sylvian fissure, on the opposite side of the fissure from the auditory area of the superior temporal gyrus. These signals apprise the psyche of the equilibrium status of the body.

The vestibular nuclei on either side of the brain stem are divided into four separate subdivisions: (1 and 2) the *superior* and *medial vestibular nuclei,* which receive signals mainly from the semicircular ducts and in turn send large numbers of nerve signals into the *medial longitudinal fasciculus* to cause corrective movements of the eyes as well as signals through the *medial vestibulospinal tract* to cause appropriate movements of the neck and head; (3) the *lateral vestibular nucleus,* which receives its innervation primarily from the utricle and saccule and in turn transmits outflow signals to the spinal cord through the *lateral vestibulospinal tract* to control body movement; (4) the *inferior vestibular nucleus,* which receives signals from both the semicircular ducts and the utricle and in turn sends signals into both the cerebellum and the reticular formation of the brain stem.

■ FUNCTIONS OF SPECIFIC BRAIN STEM NUCLEI IN CONTROLLING SUBCONSCIOUS, STEREOTYPED MOVEMENTS

Rarely, a child called an *anencephalic monster* is born without brain structures above the mesencephalic region, and some of these children have been kept alive for many months. They are able to perform essentially all of the functions of feeding, such as suckling, extrusion of unpleasant food from the mouth, and moving the hands to the mouth to suck the fingers. In addition, they can yawn and stretch. They can cry and follow objects with movements of the eyes and head. Also, placing pressure on the upper anterior parts of their legs will cause them to pull to the sitting position.

Therefore, it is obvious that many of the stereotyped motor functions of the human being are integrated in the brain stem. Unfortunately, the loci of most of the different motor control systems have not been found except for the following:

Stereotyped Body Movements. Most movements of the trunk and head can be classified into several simple movements, such as forward flexion, extension, rotation, and turning movements of the entire body. These types of movements are controlled by special nuclei located mainly in the mesencephalic and lower diencephalic region. For instance, *rotational* movements of the head and eyes are controlled by the *interstitial nucleus.* This nucleus lies in the mesencephalon in close approximation to the *medial longitudinal fasciculus,* through which it transmits a major portion of its control impulses. The *raising movements* of the head and body are controlled by the *prestitial nucleus,* which is located approximately at the juncture of the diencephalon and mesencephalon. On the other hand, the *flexing movements* of the head and body are controlled by the *nucleus precommissuralis* located at the level of the posterior commissure. Finally, the *turning movements* of the entire body, which are much more complicated, involve both the pontile and mesencephalic reticular nuclei.

REFERENCES

Adams, R. D., and Victor, M. (eds.): Principles of Neurology, 4th Ed. New York, McGraw Hill Book Co., 1989.

Asanuma, H.: The pyramidal tract. In Brooks, V. B. (ed.): Handbook of Physiology. Sec. 1, Vol. II. Bethesda, Md., American Physiological Society, 1981, p. 703.

Bahill, A. T., and Hamm, T. M.: Using open-loop experiments to study physiological systems, with examples from the human eye-movement systems. News Physiol. Sci., 4:104, 1989.

Brooks, V. B.: The Neural Basis of Motor Control. New York, Oxford University Press, 1986.

Carpenter, M. B.: Anatomy of the corpus striatum and brain stem integrating systems. In Brooks, V. B. (ed.): Handbook of Physiology. Sec. 1, Vol. II. Bethesda, Md., American Physiological Society, 1981, p. 947.

Dampney, R. A., et al.: Identification of cardiovascular cell groups in the brain stem. Clin. Exp. Hypertens., 6:205, 1984.

Desmedt, J. E. (ed.): Cerebral Motor Control in Man: Long Loop Mechanisms. New York, S. Karger, 1978.

Dublin, W. B.: Fundamentals of Vestibular Pathology. St. Louis, Warren H. Green, Inc., 1985.

Dutia, M. B.: Mechanisms of head stabilization. News Physiol. Sci., 4:101, 1989.

Elder, H. Y., and Trueman, E. R. (eds.): Aspects of Animal Movement. New York, Cambridge University Press, 1980.

Evarts, E. V., et al. (eds.): Motor System in Neurobiology. New York, Elsevier Science Publishing Co., 1986.

Evarts, E. V.: Role of motor cortex in voluntary movements in primates. In Brooks, V. B. (ed.): Handbook of Physiology. Sec. 1, Vol. II. Bethesda, Md., American Physiological Society, 1981, p. 1083.

Fernstrom, J. D.: Role of precursor availability on control of monoamine biosynthesis in brain. Physiol. Rev., 63:484, 1983.

Fournier, E., and Pierrot-Deseilligny, E.: Changes in transmission in some reflex pathways during movement in humans. News Physiol. Sci., 4:29, 1989.

Goldberg, J. M., and Fernandez, C.: The vestibular system. In Darian-Smith, I. (ed.): Handbook of Physiology, Sec. 1, Vol. III. Bethesda, Md., American Physiological Society, 1984, p. 977.

Graham, M. D., and House, W. F.: New York, Raven Press, 1987.

Grillner, S.: Control of locomotion in bipeds, tetrapods, and fish. In Brooks, V. B. (ed.): Handbook of Physiology. Sec. 1, Vol. II. Bethesda, Md., American Physiological Society, 1981, p. 1179.

Grillner, S.: Locomotion in vertebrates: Central mechanisms and reflex interaction. Physiol. Rev., 55:247, 1975.

Hasan, Z., and Stuart, D. G.: Animal solutions to problems of movement control: The role of proprioceptors. Annu. Rev. Neurosci., 11:199, 1988.

Hobson, J. A., and Brazier, M. A. B. (eds.): The Reticular Formation Revisited: Specifying Function for a Nonspecific System. New York, Raven Press, 1980.

Keele, S. W.: Behavioral analysis of movement. In Brooks, V. B. (ed.): Handbook of Physiology. Sec. 1, Vol. II. Bethesda, Md., American Physiological Society, 1981, p. 1391.

Kuypers, H. B. J. M.: Anatomy of the descending pathways. In Brooks, V. B. (ed.): Handbook of Physiology. Sec. 1, Vol. II. Bethesda, Md., American Physiological Society, 1981, p. 597.

Llinas, R.: Electrophysiology of the cerebellar networks. In Brooks, V. B. (ed.): Handbook of Physiology. Sec. 1, Vol. II. Bethesda, Md., American Physiological Society, 1981, p. 831.

Luschei, E. S., and Goldberg, L. J.: Neural mechanisms of mandibular control: Mastication and voluntary biting. In Brooks, V. B. (ed.): Handbook of Physiology. Sec. 1, Vol. II. Bethesda, Md., American Physiological Society, 1981, p. 1237.

McCloskey, D. I.: Corollary discharges: Motor commands and perception. In Brooks, V. B. (ed.): Handbook of Physiology. Sec. 1, Vol. II. Bethesda, Md., American Physiological Society, 1981, p. 1415.

Oosterveld, W. J. (ed.): Audio-Vestibular System and Facial Nerve. New York, S. Karger, 1977.

Pearson, K.: The control of walking. Sci. Am., 235(6):72, 1976.

Penfield, W., and Rasmussen, T.: The Cerebral Cortex of Man. New York, Macmillan Co., 1950.

Peterson, B. W., and Richmond, F. J. (eds.): Control of Head Movement. New York, Oxford University Press, 1988.

Peterson, B. W.: Reticulospinal projections of spinal motor nuclei. Annu. Rev. Physiol., 41:127, 1979.

Porter, R.: Influences of movement detectors on pyramidal tract neurons in primates. Annu. Rev. Physiol., 38:121, 1976.

Porter, R.: Internal organization of the motor cortex for input-output arrangements. In Brooks, V. B. (ed.): Handbook of Physiology. Sec. 1, Vol. II. Bethesda, Md., American Physiological Society, 1981, p. 1063.

Poulton, E. C.: Human manual control. In Brooks, V. B. (ed.): Handbook of Physiology. Sec. 1, Vol. II. Bethesda, Md., American Physiological Society, 1981, p. 1337.

Precht, W.: Vestibular mechanisms. Annu. Rev. Neurosci., 2:265, 1979.

Rack, P. M. H.: Limitations of somatosensory feedback in control of posture and movement. In Brooks, V. B. (ed.): Handbook of Physiology. Sec. 1, Vol. II. Bethesda, Md., American Physiological Society, 1981, p. 229.

Scheibel, A. B.: The brain stem reticular core and sensory function. In Darian-Smith, I. (ed.): Handbook of Physiology. Sec. 1, Vol. III. Bethesda, Md., American Physiological Society, 1984, p. 213.

Sherrington, C. S.: Decerebrate rigidity and reflex coordination of movements. J. Physiol. (Lond.), 22:319, 1898.

Shik, M. L., and Orlovsky, B. N.: Neurophysiology of locomotor automatism. Physiol. Rev., 56:465, 1976.

Silverman, A. J.: Magnocellular neurosecretory system. Annu. Rev. Neurosci., 6:357, 1983.

Stein, P. S. B.: Motor systems with reference to the control of locomotion. Annu. Rev. Neurosci., 1:61, 1978.

Valentinuzzi, M.: The Organs of Equilibrium and Orientation as a Control System. New York, Harwood Academic Publishers, 1980.

Wiesendanger, M., and Miles, T. S.: Ascending pathway of low-threshold muscle afferents to the cerebral cortex and its possible role in motor control. Physiol. Rev., 62:1234, 1982.

Wilson, V. J., and Peterson, B. W.: Peripheral and central substrates of vestibulospinal reflexes. Physiol. Rev., 58:80, 1978.

18

The Cerebellum, the Basal Ganglia, and Overall Motor Control

Aside from the cerebral cortical areas for control of muscle activity, two other brain structures are also essential for normal motor function. These are the *cerebellum* and the *basal ganglia*. Yet neither of these two can initiate muscle function by themselves. Instead, *they always function in association with other systems of motor control.*

Basically, the cerebellum plays major roles in the timing of motor activities and in rapid progression from one movement to the next; it also helps to control the instantaneous interplay between agonist and antagonist muscle groups.

The basal ganglia, on the other hand, help to control complex patterns of muscle movement, controlling the relative intensities of movements, directions of movement, and sequencing of multiple successive and parallel movements for achieving specific motor goals.

It is the purpose of this chapter to explain the basic mechanisms of function of the cerebellum and basal ganglia and also to discuss what we know about the overall brain mechanisms for achieving the intricate coordination of total motor activity.

■ THE CEREBELLUM AND ITS MOTOR FUNCTIONS

The cerebellum has long been called a *silent area* of the brain, principally because electrical excitation of this structure does not cause any sensation and rarely any motor movement. However, as we shall see, removal of the cerebellum does cause movement to become highly abnormal. The cerebellum is especially vital to the control of rapid muscular activities such as running, typing, playing the piano, and even talking. Loss of this area of the brain can cause almost total incoordination of these activities even though its loss causes paralysis of no muscles.

But how is it that the cerebellum can be so important when it has no direct ability to cause muscle contraction? The answer to this is that it both helps *sequence the motor activities* and also *monitors and makes corrective adjustments in the motor activities elicited by other parts of the brain.* It receives continuously updated information on the desired program of muscle contractions from the motor control areas of the other parts of the brain; and it receives continuous sensory information from the peripheral parts of the body to determine the sequential changes in the status of each part of the body — its position, its rate of movement, forces acting on it, and so forth. The cerebellum *compares* the actual movements as depicted by the peripheral sensory feedback information with the movements intended by the motor system. If the two do not compare favorably, then appropriate corrective signals are transmitted instantaneously back into the motor system to increase or decrease the levels of activation of the specific muscles.

In addition, the cerebellum aids the cerebral cortex in planning the next sequential movement a fraction of a second in advance while the present movement is still being executed, thus helping one to progress smoothly from one movement to the next. Also, it learns by its mistakes — that is, if a movement does not occur exactly as intended, the cerebellar circuit learns to make a stronger or weaker movement the next time. To do this, long-lasting changes occur in the excitability of the appropriate cerebellar neurons, thus bringing the subsequent contractions into better correspondence with the intended movements.

THE ANATOMICAL FUNCTIONAL AREAS OF THE CEREBELLUM

Anatomically, the cerebellum is divided into three separate lobes by two deep fissures, as shown in Figures 18–1 and

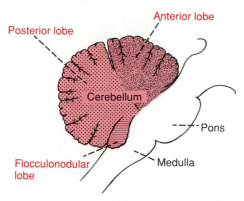

Figure 18–1. The anatomical lobes of the cerebellum as seen from the lateral side.

18–2: (1) the *anterior lobe,* (2) the *posterior lobe,* and (3) the *flocculonodular lobe.* The flocculonodular lobe is the oldest of all portions of the cerebellum; it developed along with (and functions with) the vestibular system in controlling equilibrium, as discussed in the previous chapter.

The Longitudinal Functional Divisions of the Anterior and Posterior Lobes. From a functional point of view, the anterior and posterior lobes are organized not by lobes, but instead along the longitudinal axis, as illustrated in Figure 18–2, which shows the human cerebellum after the lower end of the posterior cerebellum has been rolled downward from its normally hidden position. Note down the center of the cerebellum a narrow band separated from the remainder of the cerebellum by shallow grooves. This is called the *vermis.* In this area most cerebellar control functions for the muscle movements of the axial body, the neck, and the shoulders and hips are located.

To each side of the vermis is a large, laterally protruding *cerebellar hemisphere,* and each of these hemispheres is divided into an *intermediate zone* and a *lateral zone.*

The intermediate zone of the hemisphere is concerned with the control of muscular contractions in the distal portions of the upper and lower limbs, especially of the hands and fingers and feet and toes.

The lateral zone of the hemisphere operates at a much more remote level, for this area joins in the overall planning of sequential motor movements. Without this lateral zone, most discrete motor activities of the body lose their appropriate timing and therefore become incoordinate, as we discuss more fully later.

Topographical Representation of the Body in the Vermis and Intermediate Zones. In the same manner that the sensory cortex, the motor cortex, the basal ganglia, the red nuclei, and the reticular formation all have topographical representations of the different parts of the body, so also is this true for the vermis and intermediate zones of the cerebellum. Figure 18–3 illustrates two separate such representations. Note that the axial portions of the body lie in the vermal part of the cerebellum, whereas the limbs and facial regions lie in the intermediate zones. These topographical representations receive afferent nerve signals from all the respective parts of the body as well as from the corresponding topographical areas of the motor cortex and brain stem motor areas. In turn, they send motor signals into the same respective topographical areas of the motor cortex, the red nucleus, and the reticular formation.

However, note that the large lateral portions of the cerebellar hemispheres *do not* have topographical representations of the body. These areas of the cerebellum connect mainly with corresponding association areas of the cerebral cortex, especially the premotor area of the frontal cortex and the somatic sensory and sensory association areas of the parietal cortex. Presumably this connectivity with the association areas allows the lateral portions of the cerebellar hemispheres to play important roles in planning and coordinating the sequential muscular activities.

The Input Pathways to the Cerebellum

Afferent Pathways From the Brain. The basic input pathways to the cerebellum are illustrated in Figure 18–4. An extensive and important afferent pathway is the *corticopontocerebellar pathway,* which originates mainly in the *motor* and *premotor cortices* but to a lesser extent in the sensory cortex as well and then passes by way of the *pontile nuclei* and *pontocerebellar tracts* to the contralateral hemisphere of the cerebellum.

In addition, important afferent tracts originate in the brain stem; they include (1) an extensive *olivocerebellar* tract,

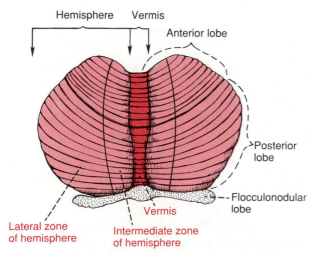

Figure 18–2. The functional parts of the cerebellum as seen from the posteroinferior view, with the inferiormost portion of the cerebellum rolled outward to flatten the surface.

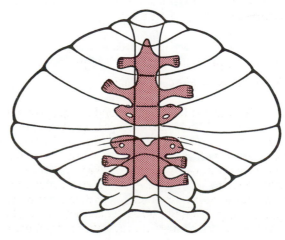

Figure 18–3. The somatic sensory projection areas in the cerebellar cortex.

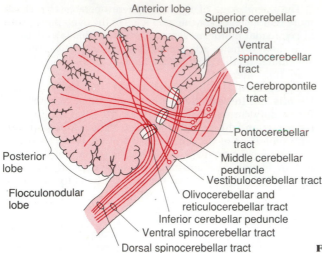

Figure 18–4. The principal afferent tracts to the cerebellum.

which passes from the *inferior olive* to all parts of the cerebellum and is excited by fibers from the *motor cortex,* the *basal ganglia,* widespread areas of the *reticular formation,* and the *spinal cord;* (2) *vestibulocerebellar fibers,* some of which originate in the vestibular apparatus itself and others from the vestibular nuclei and most of which terminate in the *flocculonodular lobe* and *fastigial nucleus* of the cerebellum; and (3) *reticulocerebellar fibers,* which originate in different portions of the reticular formation and terminate mainly in the midline cerebellar areas (the vermis).

Afferent Pathways From the Periphery. The cerebellum also receives important sensory signals directly from the peripheral parts of the body through four separate tracts, two of which are located dorsally in the cord and two ventrally. The two most important of these tracts are illustrated in Figure 18–5: the *dorsal spinocerebellar tract* and the *ventral spinocerebellar tract* (plus similar tracts from the neck and facial regions). The dorsal tracts enter the cerebellum through the inferior cerebellar peduncle and terminate in the vermis and intermediate zones of the cerebellum on the same side as their origin. The two ventral tracts enter the same areas of the cerebellum through the superior cerebellar peduncle, but they terminate in both sides of the cerebellum.

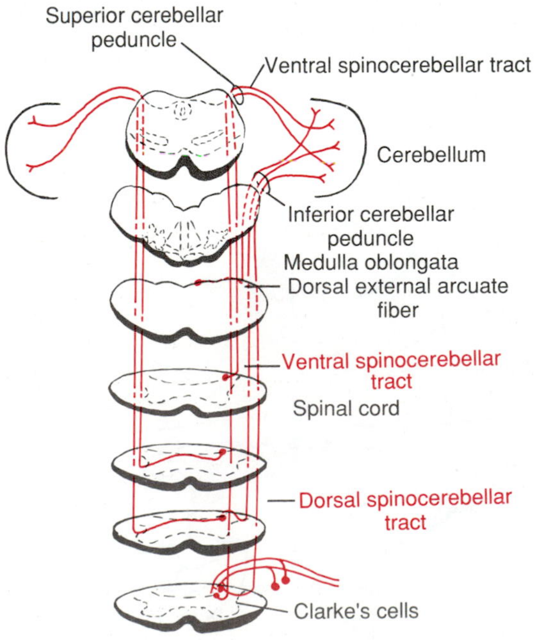

The signals transmitted in the dorsal spinocerebellar tracts come mainly from the muscle spindles and to a lesser extent from other somatic receptors throughout the body, such as from the Golgi tendon organs, the large tactile receptors of the skin, and the joint receptors. All these signals apprise the cerebellum of the momentary status of muscle contraction, degree of tension on the muscle tendons, positions and rates of movement of the parts of the body, and forces acting on the surfaces of the body.

On the other hand, the ventral spinocerebellar tracts receive less information from the peripheral receptors. Instead, they are excited mainly by the motor signals arriving in the anterior horns of the spinal cord from the brain through the corticospinal and rubrospinal tracts as well as from the internal motor pattern generators in the cord itself. Thus, this ventral fiber pathway tells the cerebellum what motor signals have arrived at the anterior horns; this feedback is called the *efference copy* of the anterior horn motor drive.

The spinocerebellar pathways can transmit impulses at velocities of up to 120 m/sec, which is the most rapid conduction of any pathway in the entire central nervous system. This extremely rapid conduction is important for the instantaneous apprisal of the cerebellum of the changes that take place in peripheral motor actions.

In addition to the signals in the spinocerebellar tracts, other signals are transmitted through the dorsal columns to the dorsal column nuclei of the medulla and then relayed from there to the cerebellum. Likewise, signals are transmitted through the *spinoreticular pathway* to the reticular formation of the brain stem and through the *spino-olivary pathway* to the inferior olivary nucleus and then relayed from both these areas to the cerebellum. Thus, the cerebellum continually collects information about all parts of the body even though it is operating at a subconscious level.

Output Signals From the Cerebellum

The Deep Cerebellar Nuclei and the Efferent Pathways. Located deep in the cerebellar mass are three *deep cerebellar nuclei*—the *dentate, interpositus,* and *fastigial.* The *vestibular*

Figure 18–5. The spinocerebellar tracts.

nuclei in the medulla also function in some respects as if they were deep cerebellar nuclei because of their direct connections with the cortex of the flocculonodular lobe. All the deep cerebellar nuclei receive signals from two different sources: (1) the cerebellar cortex and (2) the sensory afferent tracts to the cerebellum. Each time an input signal arrives in the cerebellum, it divides and goes in two directions: (1) directly to one of the deep nuclei and (2) to a corresponding area of the cerebellar cortex overlying the deep nucleus. Then, a short time later, the cerebellar cortex relays its output signals back to the same deep nucleus. Thus, all the input signals that enter the cerebellum eventually end in the deep nuclei, from which output signals are then distributed to other parts of the brain.

Three major efferent pathways lead out of the cerebellum, as illustrated in Figure 18–6:

1. A pathway that originates in the *midline structures of the cerebellum* (the *vermis*) and then passes through the *fastigial nuclei* into the *medullary* and *pontile regions of the brain stem*. This circuit functions in close association with the equilibrium apparatus to help control equilibrium and also, in association with the reticular formation of the brain stem, to help control the postural attitudes of the body. It was discussed in detail in the previous chapter in relation to equilibrium.

2. A pathway that originates in the *intermediate zone of the cerebellar hemisphere*, then passes (a) through the *nucleus interpositus* to the *ventrolateral and ventroanterior nuclei of the thalamus*, and thence to the *cerebral cortex*, (b) to several *midline structures* of the *thalamus* and thence to the *basal ganglia*, and (c) to the *red nucleus* and *reticular formation* of the upper portion of the brain stem. This circuit is believed to coordinate mainly the reciprocal contractions of agonist and antagonist muscles in the peripheral portions of the limbs — especially in the hands, fingers, and thumbs.

3. A pathway that begins in the *cortex of the lateral zone of the cerebellar hemisphere*, then passes to the *dentate nucleus*, next to the *ventrolateral and ventroanterior nuclei of the thalamus*, and finally to the *cerebral cortex*. This pathway plays an important role in helping coordinate sequential motor activities initiated by the cerebral cortex.

THE NEURONAL CIRCUIT OF THE CEREBELLUM

The human cerebellar cortex is actually a large folded sheet, approximately 17 cm wide by 120 cm long, with the folds lying crosswise, as illustrated in Figures 18–2 and 18–3. Each fold is called a *folium*. Lying deep in the folded mass of cortex are the deep nuclei.

The Functional Unit of the Cerebellar Cortex — The Purkinje Cell and the Deep Nuclear Cell. The cerebellum has approximately 30 million nearly identical functional units, one of which is illustrated to the left in Figure 18–7. This functional unit centers on a single very large *Purkinje cell*, 30 million of which are in the cerebellar cortex.

To the right in Figure 18–7, the three major layers of the cerebellar cortex are illustrated, the *molecular layer*, the *Purkinje cell layer*, and the *granular cell layer*. And far beneath these cortical layers, in the center of the cerebellar mass, are the deep nuclei.

The Neuronal Circuit of the Functional Unit. As illustrated in the left half of Figure 18–7, the output from the functional unit is from a deep nuclear cell. However, this cell is continually under the influence of both excitatory and inhibitory influences. The excitatory influences arise from direct connections with the afferent fibers that enter the cerebellum from the brain or the periphery. The inhibitory influences arise entirely from the Purkinje cell in the cortex of the cerebellum.

The afferent inputs to the cerebellum are mainly of two types, one called the *climbing fiber type* and the other called the *mossy fiber type*.

The climbing fibers *all originate from the inferior olivary complex of the medulla*. There is one climbing fiber for about 10 Purkinje cells. After sending branches to several deep nuclear cells, the climbing fiber projects all the way to the molecular layer of the cerebellar cortex, where it makes about 300 synapses with the soma and dendrites of each Purkinje cell. This climbing fiber is distinguished by the fact that a single impulse in it will always cause a single, very prolonged (up to 1 sec), and peculiar oscillatory type of action potential in each Purkinje cell with which it connects. This action potential is called the *complex spike*.

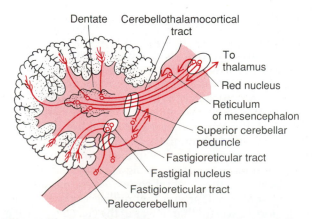

Figure 18–6. Principal efferent tracts from the cerebellum.

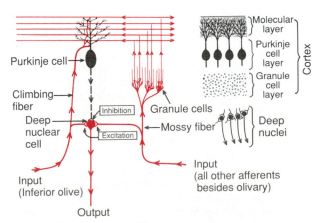

Figure 18–7. The left side of this figure shows the basic neuronal circuit of the cerebellum, with excitatory neurons shown in red. To the right is illustrated the physical relationship of the deep cerebellar nuclei to the cerebellar cortex with its three layers.

The mossy fibers are all the other fibers that enter the cerebellum from multiple sources: the higher brain, the brain stem, and the spinal cord. These fibers also send collaterals to excite deep nuclear cells. Then they proceed to the granular layer of the cortex, where they synapse with hundreds of *granule cells*. In turn, the granule cells send very small axons, less than 1 μm in diameter, up to the outer surface of the cerebellar cortex to enter the molecular layer. Here the axons divide into two branches that extend 1 to 2 mm in each direction parallel to the folia. There are literally billions of these *parallel nerve fibers,* for there are some 500 to 1000 granule cells for every Purkinje cell. It is into this molecular layer that the dendrites of the Purkinje cells project, and 80,000 to 200,000 of these parallel fibers synapse with each Purkinje cell. As these fibers pass along their 1 to 2 mm course, each of them contacts about 250 to 500 Purkinje cells.

Yet the mossy fiber input to the Purkinje cell is quite different from the climbing fiber input because their synaptic connections are very weak, so that large numbers of mossy fibers must be stimulated simultaneously to alter the activation of the Purkinje cell. Furthermore, this activation usually takes the form of facilitation or excitation that causes repetitive Purkinje cell firing of short-duration action potentials called *simple spikes,* rather than the prolonged complex action potential occurring in response to the climbing fiber input.

Continual Firing of the Cerebellum Purkinje Cells and Deep Nuclear Cells Under Normal Resting Conditions. One of the characteristics of both the Purkinje cells and the deep nuclear cells is that normally they fire continually, the Purkinje cell fires at about 50 to 100 action potentials/sec and the deep nuclear cells at still much higher rates. Therefore, the output activity of both these cells can be modulated either upward or downward. For instance, a decrease in the firing rate of the deep nuclear cells below the normal level would actually provide an *inhibitory output signal* to the motor system. On the other hand, any factor that should increase the firing rate above normal would provide an *excitatory output signal.* In this way, the cerebellum can provide either excitation or inhibition as the need arises.

Balance Between Excitation and Inhibition of the Deep Cerebellar Nuclei. Referring again to the circuit of Figure 18–7, *one should note that direct stimulation of the deep nuclear cells by both the climbing and the mossy fibers excites them. By contrast, the signals arriving from the Purkinje cells inhibit them.* Normally, the balance between these two effects is slightly in favor of excitation, so that the output from the deep nuclear cell remains relatively constant at a moderate level of continuous stimulation. On the other hand, in the execution of rapid motor movements, the *timing* of the two effects on the deep nuclei is such that the excitation appears before the inhibition. Then a few milliseconds later inhibition occurs. In this way, there is first a very rapid excitatory signal fed back into the motor pathway to modify the motor movement, but this is followed within a few milliseconds by an inhibitory signal. This inhibitory signal resembles a "delay-line" negative feedback signal of the type that is very effective in providing *damping*. That is, when the motor system is excited, a negative feedback signal presumably occurs after a short delay to stop the muscle movement from overshooting its mark, which is the usual cause of oscillation.

Other Inhibitory Cells in the Cerebellar Cortex. In addition to the granule cells and Purkinje cells, three other types of neurons are also located in the cerebellar cortex: *basket cells, stellate cells,* and *Golgi cells.* All of these are inhibitory cells with very short axons. Both the basket cells and the stellate cells are located in the molecular layer of the cortex, lying among and stimulated by the parallel fibers. These cells in turn send their axons at right angles across the parallel fibers and cause *lateral inhibition* of the adjacent Purkinje cells, thus sharpening the signal in the same manner that lateral inhibition sharpens the contrast of signals in many other areas of the nervous system. The Golgi cells, on the other hand, lie beneath the parallel fibers, though their dendrites are also stimulated by the parallel fibers. Their axons then feed back to inhibit the granule cells. The function of this feedback is to limit the duration of the signal transmitted into the cerebellar cortex from the granule cells. That is, within a short fraction of a second after the granule cells are stimulated, their initial burst of excitation is reduced back to a lower level of excitation that is sustained only as long as the input signal lasts.

The Turn-On/Turn-Off and Turn-Off/Turn-On Output Signals From the Cerebellum

The typical function of the cerebellum is to help provide rapid turn-on signals for agonist muscles and simultaneous reciprocal turn-off signals for the antagonist muscles at the onset of a movement. Then, at the termination of the movement, the cerebellum is mainly responsible for timing and executing the turn-off signals to the agonists and turn-on signals to the antagonists. Although the exact means by which the cerebellum achieves these turn-on and turn-off signals are not fully known, one can speculate from the basic cerebellar circuit of Figure 18–7 how this might work as follows:

First, let us suppose that the turn-on/turn-off pattern of agonist/antagonist contraction at the onset of movement begins with signals from the cerebral cortex that pass directly to the agonist muscle to begin the initial contraction. At the same time, parallel signals are also sent by way of the pontile mossy fibers into the cerebellum. One branch of each mossy fiber goes directly to deep nuclear cells in the dentate or other deep nucleus; this instantly sends an excitatory signal back into the corticospinal motor system, either by way of the return signals through the thalamus to the cortex or by way of neuronal circuitry in the brain stem, to support the muscle contraction signal that had already been begun by the cerebral cortex. As a consequence, the turn-on signal, after a few milliseconds, becomes even more powerful than it was at the start because it is now the sum of both the cortical and the cerebellar signals. This is the normal effect when the cerebellum is intact, but in the absence of the cerebellum the secondary extra supportive signal is missing. Obviously, this cerebellar support makes the turn-on muscle contraction much stronger than it otherwise would be.

Now, what causes the turn-off signal for the agonist muscles at the termination of the movement? Remember that all mossy fibers have a second branch that transmits signals by way of the granule cells to the cerebellar cortex and eventually to the Purkinje cells, and the Purkinje cells in turn *inhibit* the deep nuclear cells. This pathway passes through some of the smallest nerve fibers known in the entire nervous system, the parallel fibers of the cerebellar cortical molecular layer that have diameters of only a fraction of a millimeter. Also, the signals from these fibers are weak, so that they require a finite period of time to build up enough excitation in the dendrites of the Purkinje cell to excite it. But once the Purkinje cell is excited, it sends *inhibitory* signals to the same deep nuclear cells that had originally turned on the movement. Therefore, theoretically, this could turn off the cerebellar excitation of the agonist muscles.

Thus, one can see how this circuit could cause a rapid turn-on of agonist contraction at the beginning of a movement and yet cause also a precisely timed turn-off of the same agonist contraction after a given period of time.

Now let us speculate on a circuit for the antagonist muscles. Most importantly, remember that throughout the spinal cord, there are reciprocal agonist/antagonist circuits for virtually every movement that the cord can initiate. Therefore, these circuits are probably the major basis for the antagonist turn-off at the onset of movement and turn-on at its termination, always mirroring whatever occurs in the agonist muscles. But we must remember, too, that the cerebellum contains several other types of inhibitory cells besides the Purkinje cells. The functions of some of these are still to be determined; these, too, could play roles in the initial inhibition of the antagonist muscles and then subsequent excitation.

Obviously, these theoretical mechanisms are still mainly speculation. They are presented here only to illustrate possible ways by which the cerebellum could indeed cause reciprocal turn-on and turn-off signals in the agonist and antagonist muscles, and with controlled timing as well.

The Purkinje Cells Can "Learn" to Correct Motor Errors — The Role of the Climbing Fibers

The degree to which the cerebellum supports the onset and offset of muscle contractions, as well as the timing of the contractions, can be learned by the cerebellum itself. Typically, when a person first performs a new motor act, the degree of motor enhancement provided by the cerebellum to the onset agonist contraction, the degree of onset inhibition of the antagonist, the timing of the offset, the extent of inhibition of the agonist at the offset, and the extent of contraction of the antagonist at the offset, all are almost always incorrect for precise performance of the movement. But after the act has been performed many times, these individual events become progressively more precise in performing the movement exactly as desired, sometimes requiring only a few movements before the desired result is achieved but at other times requiring hundreds of movements.

Yet how do these adjustments come about? The exact answer is not known, although it is known that sensitivity levels of cerebellar circuits themselves progressively adapt during the training process. For instance, the sensitivity of the Purkinje cells to respond to the parallel fibers from the granule cells becomes altered. Furthermore, research studies suggest that this sensitivity change is brought about by signals from the climbing fibers entering the cerebellum from the inferior olivary complex. These signals adjust the long-term sensitivity of the Purkinje cells to stimulation by the parallel fibers.

Under resting conditions, the climbing fibers fire about once per second. But each time they do fire, they cause extreme depolarization of the entire dendritic tree of the Purkinje cell, lasting for up to a second. During this time, the Purkinje cell fires with one initial very strong output spike followed by a series of oscillatory waves in the membrane potential. When a person performs a new movement for the first time and the achieved movement does not match the intended movement, the firing by the climbing fibers changes markedly, either greatly increased or decreased as needed, up to a maximum of about 4/sec or all the way down to zero. These changes in stimulatory rate are believed to alter the long-term sensitivity of the Purkinje cells to the subsequent signals from the mossy fiber circuit. That is, the greater or lesser the climbing fiber input, the greater becomes the accumulative change in long-term sensitivity to the mossy fiber input. Over a period of time, this change in sensitivity, along with other possible "learning" functions of the cerebellum, is believed to make the timing and other aspects of cerebellar control of movements approach perfection. When this has been achieved, the climbing fibers no longer send their "error" signals to the cerebellum to cause further change.

Finally, we need to answer how the climbing fibers themselves know to alter their own rate of firing when a performed movement is imperfect. What is known about this is that the inferior olivary complex receives full information from the corticospinal tracts as well as from the motor centers of the brain stem detailing the *intent* of each motor movement; and it also receives full information from the sensory nerve endings in the muscles and surrounding tissues detailing the movement that actually occurs. Therefore, it is presumed that the inferior olivary complex then functions as a *comparator* to test how well the actual performance matches the intended performance. If there is a match, no change in firing of the climbing fibers occurs. But if there is a mismatch, then the climbing fibers are stimulated or inhibited as needed in proportion to the degree of mismatch, thus leading to progressive changes in

Purkinje cell sensitivity until no further mismatch occurs—or so the theory goes.

OVERALL FUNCTION OF THE CEREBELLUM IN CONTROLLING MOVEMENTS

It is already clear that the cerebellum functions in motor control only in association with motor activities initiated elsewhere in the nervous system. These activities may originate in the spinal cord, in the brain stem reticular nuclei, or in the cerebral cortex. We discuss first the operation of the cerebellum in association with the spinal cord and brain stem for control of postural movements and equilibrium and then discuss its function in association with the motor cortex for control of voluntary movements.

FUNCTION OF THE CEREBELLUM WITH THE SPINAL CORD AND BRAIN STEM TO CONTROL POSTURAL AND EQUILIBRIUM MOVEMENTS

The cerebellum originated phylogenetically at about the same time that the vestibular apparatus developed. Furthermore, as discussed in the previous chapter, loss of the flocculonodular lobes and portions of the vermis of the cerebellum causes extreme disturbance of equilibrium.

Yet we still must ask the question, what role does the cerebellum play in equilibrium that cannot be provided by the other neuronal machinery of the brain stem? A clue is the fact that in persons with cerebellar dysfunction, equilibrium is far more disturbed during performance of rapid motions than during stasis—especially so when the movements involve changes in direction that stimulate the semicircular ducts. This suggests that the cerebellum is especially important in controlling the balance between agonist and antagonist muscle contractions during *rapid changes* in body positions as dictated by the vestibular apparatus.

One of the major problems in controlling balance is the time required to transmit position signals and velocity of movement signals from the different parts of the body to the brain. Even when the most rapidly conducting sensory pathways, up to 120 m/sec, are used, as by the spinocerebellar system, the delay for transmission from the feet to the brain is still 15 to 20 msec. The feet of a person running rapidly can move as much as 10 in. during this time. Therefore, it is never possible for the return signals from the peripheral parts of the body to reach the brain at the same time that the movements actually occur. How, then, is it possible for the brain to know when to stop a movement in order to perform the next sequential act, especially when the movements are performed very rapidly? The answer is that the signals from the periphery tell the brain not only positions of the different parts of the body but also how rapidly and in what directions they are moving. It is the function of the cerebellum then to *calculate* from these rates and directions where the different parts of the body will be during the next few milliseconds. The results of these calculations are the key to the brain's progression to the next sequential movement.

Thus, during the control of equilibrium, it is presumed that the information from the vestibular apparatus is used in a typical feedback control circuit to provide almost instantaneous correction of postural motor signals as necessary for maintaining equilibrium even during extremely rapid motion, including rapidly changing directions of motion. The feedback signals from the peripheral areas of the body help in this process. Their help is mediated mainly through the *cerebellar vermis* that functions in association with the axial and girdle muscles of the body; it is the role of the cerebellum to compute actual positions of the respective parts of the body at any given time, despite the long delay time from the periphery to the cerebellum.

FUNCTION OF THE CEREBELLUM IN VOLUNTARY MUSCLE CONTROL

In addition to the feedback circuitry between the body periphery and the cerebellum, an almost entirely independent feedback circuitry exists between the motor cortex of the cerebrum and the cerebellum. This circuitry affects only slightly if at all the control of equilibrium and other postural movements of the axial and girdle muscles of the body. Instead, it serves two other principal functions: (1) It helps the cerebral cortex to coordinate patterns of movement involving mostly the distal parts of the limbs—especially the hands, fingers, and feet. The part of the cerebellum involved in this function is mainly the *intermediate zone of the cerebellar cortex and its associated nucleus interpositus.* (2) It helps the cerebral cortex to plan the timing and sequencing of the next successive movement that will be performed after the present movement is completed. The part of the cerebellum involved in this is the large *lateral zone of the cerebellar hemisphere,* along with its associated *dentate nucleus.* Let us discuss each of these two functions separately.

Cerebellar Feedback Control of Distal Limb Movements by Way of the Intermediate Cerebellar Cortex and the Nucleus Interpositus

As illustrated in Figure 18–8, the intermediate zone of each cerebellar hemisphere receives two types of information when a movement is performed: (1) direct information from the motor cortex and red nucleus, telling the cerebellum the sequential *intended plan of movement* for the next few fractions of a second; and (2) feedback information from the peripheral parts of the body, especially from the distal parts of the limbs, telling the cerebellum what *actual movements* result. After the intermediate zone of the cerebellum has compared the intended movements with the actual movement, the nucleus interpositus sends *corrective*

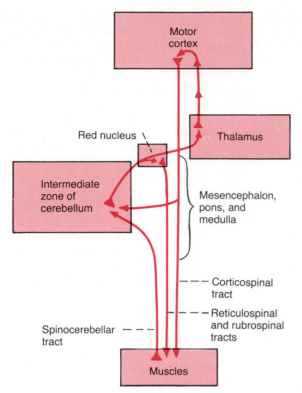

Figure 18–8. Cerebral and cerebellar control of voluntary movements, involving especially the intermediate zone of the cerebellar cortex and its associated nucleus interpositus.

output signals (a) back to the *motor cortex* through relay nuclei in the *thalamus* and (b) to the *magnocellular portion* (the lower portion) *of the red nucleus,* which gives rise to the *rubrospinal tract.* The rubrospinal tract, in turn, joins the corticospinal tract in innervating the lateralmost motor neurons in the anterior horns of the spinal cord gray matter, the neurons that control the distal parts of the limbs, particularly the hands and fingers.

This part of the cerebellar motor control system provides smooth, coordinate movements of the agonist and antagonist muscles of the distal limbs for the performance of acute purposeful patterned movements. The cerebellum seems to compare the "intentions" of the higher levels of the motor control system, as transmitted to the intermediate cerebellar zone through the corticopontocerebellar tract, with the "performance" by the respective parts of the body as transmitted back to the cerebellum from the periphery. In fact, the ventral spinocerebellar tract even transmits back to the cerebellum an "efference" copy of the actual motor control signals that reach the anterior motor neurons, and this, too, is integrated with the signals arriving from the muscle spindles and other proprioceptor sensory organs. We learned earlier that similar comparator signals also go to the inferior olivary complex; if the signals do not compare favorably, the olivary-Purkinje cell system, along with possible other cerebellar learning mechanisms, will eventually correct the motions until they perform the desired function.

Once the cerebellum has learned its role in each pattern of movement, it provides rapid turn-on of agonist muscle activity at the onset of each movement while inhibiting the antagonist muscles. Then it continues agonist contraction until near the end of the movement, when the cerebellar circuit again plays the major role in rapid turn-off of the agonist muscles and turn-on of the antagonist muscles. The point at which the reversal of excitation between agonist and antagonist muscles occurs depends on (1) the rate of movement and (2) the previously learned knowledge of the inertia of the system. The faster the movement and the greater the inertia, the earlier the reversal point must occur in the course of movement to stop the movement at the proper point.

Function of the Cerebellum to Prevent Overshoot of Movements and to "Damp" Movements. Almost all movements of the body are "pendular." For instance, when an arm is moved, momentum develops, and the momentum must be overcome before the movement can be stopped. Because of the momentum, all pendular movements have a tendency to *overshoot.* If overshooting does occur in a person whose cerebellum has been destroyed, the conscious centers of the cerebrum eventually recognize this and initiate a movement in the opposite direction to bring the arm to its intended position. But again the arm, by virtue of its momentum, overshoots, and appropriate corrective signals must again be instituted. Thus, the arm oscillates back and forth past its intended point for several cycles before it finally fixes on its mark. This effect is called an *action tremor,* or *intention tremor.*

However, if the cerebellum is intact, appropriate learned, subconscious signals stop the movement precisely at the intended point, thereby preventing the overshoot and also the tremor. This is the basic characteristic of a damping system. All control systems regulating pendular elements that have inertia must have damping circuits built into the mechanisms. In the motor control system of our central nervous system, the cerebellum provides most of this damping function.

Cerebellar Control of Ballistic Movements. Many rapid movements of the body, such as the movements of the fingers in typing, occur so rapidly that it is not possible to receive feedback information either from the periphery to the cerebellum or from the cerebellum back to the motor cortex before the movements are over. These movements are called *ballistic movements,* meaning that the entire movement is preplanned and is set into motion to go a specific distance and then to stop. Another important example is the saccadic movements of the eyes, in which the eyes jump from one position to the next when reading or when looking at successive points along a road as a person is moving in a car.

Much can be understood about the function of the cerebellum by studying the changes that occur in the ballistic movements when the cerebellum is removed. Three major changes occur: (1) the movements are slow to develop and do not have the extra onset surge that the cerebellum usually gives to an agonist movement, (2) the force development is weak, and (3) the movements are slow to turn off, usually allowing the movement to go well beyond the intended mark. Therefore, in the absence of the cerebellar circuit, the motor cortex has to think extra hard to turn ballistic movements on and again has to think hard and take extra time to turn the movement off. Thus, the automatism of ballistic movements is lost.

If one will consider once again the circuitry of the cerebellum as described earlier in the chapter, one will see that it is beautifully organized to perform this biphasic, first excitatory and then delayed inhibitory, function that is required for ballistic movements. One will also see that the time delay circuits of the cerebellar cortex are fundamental to this particular ability of the cerebellum.

Function of the Large Lateral Zone of the Cerebellar Hemisphere — The "Sequencing" and "Timing" Functions

In human beings, the lateral zones of the two cerebellar hemispheres have become very highly developed and greatly enlarged, along with the human ability to perform intricate sequential patterns of movement, especially with the hands and fingers, and along with the ability to speak. Yet, strangely enough, these large lateral portions of the cerebellar hemispheres have no direct input of information from the peripheral parts of the body. Also, almost all the communication between these lateral cerebellar areas and the cortex is not with the primary motor cortex itself but instead with the premotor area and primary and association somatic sensory areas. Even so, destruction of the lateral portions of the cerebellar hemispheres along with their deep nuclei, the dentate nuclei, can lead to extreme incoordination of the purposeful movements of the hands, fingers, feet, and speech apparatus. This has been difficult to understand because of lack of direct communication between this part of the cerebellum and the primary motor cortex. However, recent experimental studies suggest that these portions of the cerebellum are concerned with two other important aspects of motor control: (1) the planning of sequential movements and (2) the "timing" of the sequential movements.

The Planning of Sequential Movements. The planning of sequential movements seems to be related to the fact that the lateral hemispheres communicate with the premotor and sensory portions of the cerebral cortex and that there is also two-way communication between these same areas and corresponding areas of the basal ganglia. It seems that the "plan" of the sequential movements is transmitted from the sensory and premotor areas of the cortex to the lateral zones of the cerebellar hemispheres, and two-way traffic between the cerebellum and the cortex is necessary to provide appropriate transition from one movement to the next. An exceedingly interesting observation that supports this view is that many of the neurons in the dentate nuclei display the activity pattern of the movement that is yet to follow at the same time that the

present movement is occurring. Thus, the lateral hemispheres appear to be involved not with what is happening at a given moment, but instead with *what will be happening during the next sequential movement*.

To summarize, one of the most important features of normal motor function is one's ability to progress smoothly from one movement to the next in orderly succession. In the absence of the cerebellar hemispheres, this capability is seriously disturbed, especially for rapid movements.

The Timing Function. Another important function of the lateral cerebellar hemispheres is to provide appropriate timing for each movement. In the absence of these lateral areas, one loses the subconscious ability to predict ahead of time how far the different parts of the body will move in a given time. And without this timing capability, the person becomes unable to determine when the next movement should begin. As a result, the succeeding movement may begin too early or, more likely, too late. Therefore, cerebellar lesions cause complex movements, such as those required for writing, running, or even talking, to become totally incoordinate, lacking completely in the ability to progress in an orderly sequence from one movement to the next. Such cerebellar lesions are said to cause *failure of smooth progression of movements*.

Extramotor Predictive Functions of the Cerebellum. The cerebellum also plays a role in predicting events other than movements of the body. For instance, the rates of progression of both auditory and visual phenomena can be predicted, and both of these require cerebellar participation. As an example, a person can predict from the changing visual scene how rapidly he or she is approaching an object. A striking experiment that demonstrates the importance of the cerebellum in this ability is the removal of the "head" portion of the cerebellum in monkeys. Such a monkey occasionally charges the wall of a corridor and literally bashes its brains out because it is unable to predict when it will reach the wall.

Unfortunately, we are only now beginning to learn about these extramotor predictive functions of the cerebellum. It is quite possible that the cerebellum provides a "time base," perhaps utilizing time-delay circuits, against which signals from other parts of the central nervous system can be compared. It is often stated that the cerebellum is especially important in interpreting *spatiotemporal relationships* in sensory information.

CLINICAL ABNORMALITIES OF THE CEREBELLUM

An important feature of clinical cerebellar abnormalities is that destruction of small portions of the cerebellar *cortex* rarely causes detectable abnormalities in motor function. In fact, several months after as much as half the cerebellar cortex has been removed, if the deep cerebellar nuclei are not removed along with the cortex, the motor functions of an animal appear to be almost entirely normal as long as the animal performs all movements slowly. Thus, the remaining portions of the motor control system are capable of compensating tremendously for loss of parts of the cerebellum.

Therefore, to cause serious and continuing dysfunction of the cerebellum, the cerebellar lesion must usually involve one or more of the deep cerebellar nuclei — the *dentate, interpositus,* and *fastigial nuclei* — as well as the cerebellar cortex.

Dysmetria and Ataxia. Two of the most important symptoms of cerebellar disease are dysmetria and ataxia. It was pointed out earlier that in the absence of the cerebellum the subconscious motor control system cannot predict ahead of time how far movements will go. Therefore, the movements ordinarily overshoot their intended mark, and then the conscious portion of the brain overcompensates in the opposite direction for the succeeding movements. This effect is called *dysmetria,* and it results in incoordinate movements that are called *ataxia.*

Dysmetria and ataxia can also result from lesions in the spinocerebellar tracts, for the feedback information from the moving parts of the body is essential for accurate control of the movements.

Past Pointing. Past pointing means that in the absence of the cerebellum a person ordinarily moves the hand or some other moving part of the body considerably beyond the point of intention. This probably results from the fact that normally the cerebellum provides most of the motor signal that turns off a movement after it has begun; and if the cerebellum is not available to do this, the movement ordinarily goes beyond the intended point. Therefore, past pointing is actually a manifestation of dysmetria.

Failure of Progression. *Dysdiadochokinesia.* When the motor control system fails to predict ahead of time where the different parts of the body will be at a given time, it temporarily "loses" the parts during rapid motor movements. As a result, the succeeding movement may begin much too early or much too late, so that no orderly "progression of movement" can occur. One can demonstrate this readily by having a patient with cerebellar damage turn one hand upward and downward at a rapid rate. The patient rapidly "loses" all perception of the instantaneous position of the hand during any portion of the movement. As a result, a series of jumbled movements occurs instead of the normal coordinate upward and downward motions. This is called *dysdiadochokinesia.*

Dysarthria. Another instance in which failure of progression occurs is in talking, for the formation of words depends on rapid and orderly succession of individual muscular movements in the larynx, mouth, and respiratory system. Lack of coordination between these and inability to predict either the intensity of the sound or the duration of each successive sound cause jumbled vocalization, with some syllables loud, some weak, some held long, some held for short intervals, and resultant speech that is almost completely unintelligible. This is called *dysarthria.*

Intention Tremor. When a person who has lost the cerebellum performs a voluntary act, the movements tend to oscillate, especially when they approach the intended mark, first overshooting the mark and then vibrating back and forth several times before settling on the mark. This reaction is called an *intention tremor* or an *action tremor,* and it results from cerebellar overshooting and failure of the cerebellar system to damp the motor movements.

Cerebellar Nystagmus. Cerebellar nystagmus is a tremor of the eyeballs that occurs usually when one attempts to fixate the eyes on a scene to one side of the head. This off-center type of fixation results in rapid, tremulous movements of the eyes rather than a steady fixation, and it is another manifestation of the failure of damping by the cerebellum. It occurs especially when the flocculonodular lobes are damaged; in this instance it is associated with loss of equilibrium, presumably because of dysfunction of the pathways through the cerebellum from the semicircular ducts.

Rebound. If a person with cerebellar disease is asked to pull upward strongly with his or her arm while the physician holds it back at first and then lets go, the arm will fly back until it strikes the face instead of being automatically stopped. This is called *rebound,* and it results from *loss of the cerebellar component of the stretch reflex.* That is, the normal cerebellum ordinarily instantaneously adds a large amount of additional feedback support to the spinal cord stretch reflex mechanism whenever a portion of the body begins to move unexpectedly in an unwilled direction. Without the cerebellum, strong activation of the muscles fails to occur, thus allowing overmovement of the limb in the unwanted direction.

Hypotonia. Loss of the deep cerebellar nuclei, particularly the dentate and interpositus, causes decreased tone of the peripheral musculature on the side of the lesion, though after several months the cerebral motor cortex usually compensates for this by an increase in its intrinsic activity. The hypotonia results from loss of cerebellar facilitation of the motor cortex and brain stem motor nuclei by the tonic discharge of the deep cerebellar nuclei.

■ THE BASAL GANGLIA — THEIR MOTOR FUNCTIONS

The basal ganglia, like the cerebellum, are another accessory motor system that functions not by itself but always in close association with the cerebral cortex and corticospinal system. In fact, the basal ganglia receive almost all their input signals from the cortex itself and in turn return almost all of their output signals back to the cortex.

Figure 18–9 illustrates the anatomical relationships of the basal ganglia to the other structures of the brain. Note that they are located mainly lateral to the thalamus, occupying a large portion of the deeper regions of both cerebral hemispheres. Note also that almost all of the motor and sensory nerve fibers connecting the cerebral cortex and spinal cord pass between the two major masses of the basal ganglia, the *caudate nucleus* and the *putamen.* This mass of nerve fibers is called the *internal capsule* of the brain. It is important to our present discussion because of the intimate association between the basal ganglia and the corticospinal system for motor control.

The Neuronal Circuitry of the Basal Ganglia. The anatomical connections between the basal ganglia and the other elements of motor control are very complex, as illustrated in Figure 18–10. To the left is shown the motor cortex, the thalamus, the corticospinal pathways, and associated brain stem and cerebellar circuitry. To the right is the major circuitry of the basal ganglia system, showing the tremendous number of interconnections among the basal ganglia themselves — plus extensive input and output pathways between the motor regions of the cerebral cortex and the basal ganglia.

Anatomists consider the motor portions of the basal ganglia to be the *caudate nucleus,* the *putamen,* and the *globus pallidus.* But, physiologically, two other struc-

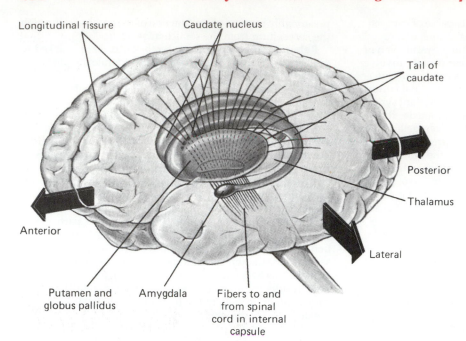

Figure 18–9. Anatomical relationships of the basal ganglia to the cerebral cortex and thalamus, shown in three-dimensional view.

tures that are not normally classified as basal ganglia are intimately involved as well, the *subthalamus* and *substantia nigra,* which are located inferior and posterior to the thalamus in the lower diencephalon and upper mesencephalon. Several specific reentrant circuits interconnect the subthalamus and substantia nigra with all three of the basal ganglia. Also, both the subthalamus and substantia nigra feed back to the thalamus and thence to the cortical areas for motor control.

From Figure 18–10, it is already clear that the circuitry of the basal ganglial system is intensely complex. However, we try in the next few sections to dissect out the major pathways of action and attempt to describe their functional attributes. We concentrate especially on two major circuits called the *putamen circuit* and the *caudate circuit.*

FUNCTION OF THE BASAL GANGLIA IN EXECUTING PATTERNS OF MOTOR ACTIVITY—THE PUTAMEN CIRCUIT

One of the principal roles of the basal ganglia in motor control is to function in association with the corticospinal system to control complex patterns of motor activity. An example is the writing of letters of the alphabet. When there is serious damage to the basal ganglia, the cortical system of motor control can no longer provide these patterns. Instead, one's writing becomes crude, as if one were learning for the first time how to write.

Other patterns requiring the basal ganglia are cutting paper with scissors, hammering nails, shooting basketballs through a hoop, passing a football, throwing a baseball, the movements of shoveling dirt, some aspects of vocalization, and virtually any other of our skilled movements.

The Neural Circuit Through the Putamen for Executing Patterns of Movements. Figure 18–11 illustrates the principal pathways through the basal ganglia for executing learned patterns of movement.

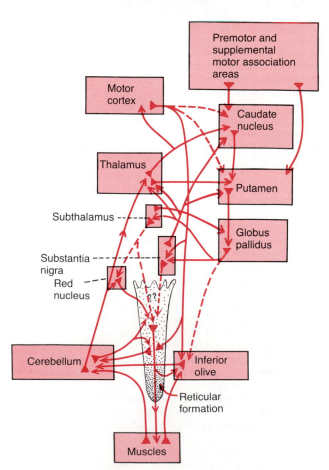

Figure 18–10. Relation of the basal ganglial circuitry to the corticospinal-cerebellar system for movement control.

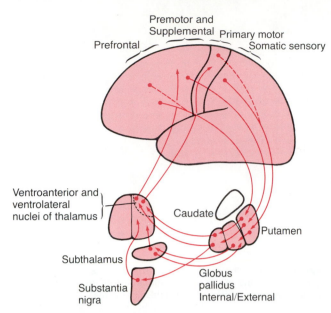

Figure 18–11. The *putamen circuit* through the basal ganglia for subconscious execution of learned patterns of movement.

These begin mainly in the premotor and supplemental motor areas of the motor cortex and also the primary somatic sensory area of the sensory cortex. Next they pass, as shown in the bright red in the figure, to the putamen (mainly bypassing the caudate nucleus), then to the internal portion of the globus pallidus, next to the ventroanterior and ventrolateral nuclei of the thalamus, and then finally return to the primary motor cortex and portions of the premotor and supplemental areas closely associated with the primary motor cortex. Thus, this putamen circuit has its inputs mainly from those parts of the brain adjacent to the primary motor cortex, but not much from the primary motor cortex itself. Then its outputs do go mainly back to the *primary* motor cortex.

Functioning in close association with this primary putamen circuit are three ancillary circuits: (1) from the putamen to the external globus pallidus, to the subthalamus, to the relay nuclei of the thalamus, and back to the motor cortex; (2) from the putamen to the internal globus pallidus, to the substantia nigra, to the relay nuclei of the thalamus, and also returning to the motor cortex; and (3) a local feedback circuit from the external globus pallidus to the subthalamus and returning again to the external globus pallidus.

Athetosis, Hemiballismus, and Chorea. How does the above putamen circuit function in the execution of patterns of movement? The answer is only poorly known. However, when any portion of the circuit is damaged or blocked, certain patterns of movement become severely abnormal. For instance, lesions in the *globus pallidus* frequently lead to spontaneous *writhing movements* of a hand, an arm, the neck, or the face, movements called *athetosis*.

A lesion in the *subthalamus* often leads to sudden *flailing movements* of an entire limb, a condition called *hemiballismus.*

Multiple small lesions in the *putamen* lead to *flicking movements* in the hands, face, and other parts of the body, which is called *chorea.*

And lesions of the *substantia nigra* lead to the common and extremely severe disease of rigidity and tremors known as *Parkinson's disease,* which we shall discuss in more detail later.

ROLE OF THE BASAL GANGLIA FOR COGNITIVE CONTROL OF SEQUENCES OF MOTOR PATTERNS — THE CAUDATE CIRCUIT

The term cognition means the thinking processes of the brain, utilizing both the sensory input to the brain as well as information already stored in memory. Obviously, most of our motor actions occur as a consequence of thoughts generated in the mind, a process called *cognitive control of motor activity.* The caudate nucleus plays a major role in this cognitive control of motor activity.

The neural connections between the corticospinal motor control system and the caudate nucleus, illustrated in Figure 18–12, are somewhat different from those of the putamen circuit. Part of the reason for this is that the caudate nucleus extends into all lobes of the cerebrum, beginning anteriorly in the frontal lobes, then passing posteriorly through the parietal and occipital lobes, and finally curving forward again like a letter "C" into the temporal lobes. Furthermore, the caudate nucleus receives large amounts of its input from the *association areas* of the cerebral cortex, the areas that integrate the different types of sensory and motor information into usable thought patterns.

After the signals pass from the cerebral cortex to the caudate nucleus, they are transmitted next to the internal globus pallidus, then to the relay nuclei of the ventroanterior and ventrolateral thalamus, and finally back to the prefrontal, premotor, and supplemental motor areas of the cerebral cortex, but with almost none of the returning signals passing directly to the primary motor cortex. Instead, the returning signals go to those accessory motor regions that are concerned with patterns of movement instead of individual muscle movements.

A good example of this would be for a person to see a lion approach and then respond instantaneously and automatically by (1) turning away from the lion, (2) beginning to run, and (3) even attempting to climb a tree. Without the cognitive functions, the person might not have the instinctive knowledge, without thinking for too long a time, to respond quickly and appropriately. Thus, cognitive control of motor activity determines which patterns of movement will be used together and in what sequence to achieve a complex goal.

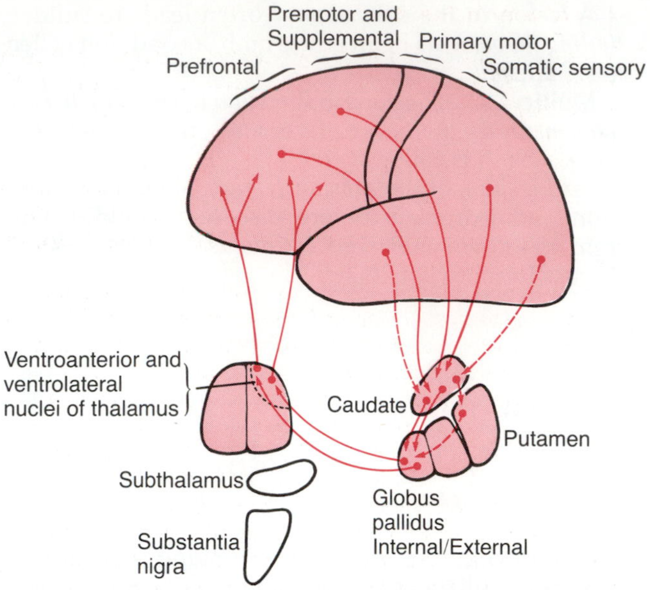

Premotor and
Supplemental   Primary motor
Prefrontal   Somatic sensory

Ventroanterior and
ventrolateral
nuclei of thalamus

Caudate

Putamen

Subthalamus

Globus
pallidus
Internal/External

Substantia
nigra

Figure 18–12. The *caudate circuit* through the basal ganglia for cognitive planning of the combinations of sequential and parallel motor patterns to achieve specific conscious goals.

FUNCTION OF THE BASAL GANGLIA TO CHANGE THE TIMING AND TO SCALE THE INTENSITY OF MOVEMENTS

Two important capabilities of the brain in controlling movement are (1) to determine how rapidly it is to be performed and (2) to control how large the movement will be. For instance, one may write the letter "a" slowly or rapidly. Also, he may write a small "a" or a very large letter "a" on a chalk board. Regardless of his choices, the proportional characteristics of the letter will remain the same. This is also true even though the person might use the fingers for writing the letter in one instance or the whole arm at another time.

In the absence of the basal ganglia, these timing and scaling functions are very poor, in fact almost nonexistent. Of course, here again, the basal ganglia do not function alone; they function in close association with the cerebral cortex as well. One especially important cortical area is the posterior parietal cortex, which is the locus of the spatial coordinates for all parts of the body as well as for the relationship of the body and its parts to all surroundings. Figure 18–13 illustrates the way in which a person lacking a left posterior parietal cortex might draw the face of another human being, providing proper proportions for the right side of the face, but almost ignoring the left side (which is in his right field of vision). Also, such a person will try always to avoid using his right arm, right hand, or other portions of his right body for the performance of tasks, almost not knowing that these parts of his body even exist.

Because it is the caudate circuit of the basal ganglial system that functions mainly with the association areas of the cortex, such as the posterior parietal cortex, presumably the timing and scaling of movements are functions of this caudate cognitive motor control circuit.

FUNCTIONS OF SPECIFIC NEUROTRANSMITTERS IN THE BASAL GANGLIAL SYSTEM

Figure 18–14 illustrates the interplay of some specific neurotransmitters that are known to function within

Figure 18–13. A typical drawing made by a person who has severe damage in his or her left parietal cortex, where the spatial coordinates of the right side of the body and right field of vision are calculated.

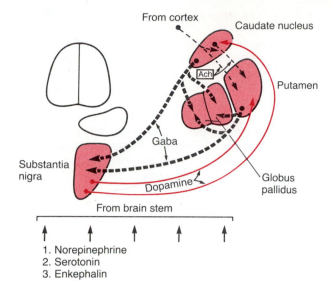

Figure 18–14. Neuronal pathways that secrete different types of neurotransmitter substances in the basal ganglia.

the basal ganglia, showing (1) a *dopamine* pathway from the substantia nigra to the caudate nucleus and putamen; (2) a *gamma-aminobutyric acid (GABA)* pathway from the caudate nucleus and putamen to the globus pallidus and substantia nigra; (3) *acetylcholine* pathways from the cortex to the caudate nucleus and putamen; and (4) multiple general pathways from the brain stem that secrete *norepinephrine, serotonin, enkephalin,* and several other neurotransmitters in the basal ganglia as well as in other parts of the cerebrum. We will have more to say about some of these hormonal systems in the following sections when we discuss diseases of the basal ganglia as well as in subsequent chapters when we discuss behavior, sleep, wakefulness, and functions of the autonomic nervous system.

For the present, it should be remembered that the neurotransmitter GABA always functions as an inhibitory agent. Therefore, the GABA neurons in the feedback loops from the cortex through the basal ganglia and then back to the cortex make virtually all these loops *negative feedback loops,* rather than positive feedback loops, thus lending stability to the motor control systems. Dopamine also functions as an inhibitory neurotransmitter in most parts of the brain, so that it too may function as a stabilizer. Acetylcholine, on the other hand, usually functions as an excitatory transmitter and therefore probably provides many of the positive features of motor action.

CLINICAL SYNDROMES RESULTING FROM DAMAGE TO THE BASAL GANGLIA

Aside from athetosis and hemiballismus, which have already been mentioned in relation to lesions in the globus pallidus and the subthalamus, two other major diseases result from damage in the basal ganglia. These are Parkinson's disease and Huntington's chorea.

Parkinson's Disease

Parkinson's disease, also known as *paralysis agitans,* results from *widespread destruction of that portion of the substantia nigra, the pars compacta, that sends dopamine-secreting nerve fibers to the caudate nucleus and putamen.* The disease is characterized by (1) *rigidity* of much if not most of the musculature of the body, (2) *involuntary tremor* of the involved areas even when the person is resting and always at a fixed rate of 3 to 6 cycles/sec, and (3) a serious inability to initiate movement called *akinesia.*

The causes of these abnormal motor effects are almost entirely unknown. However, if the dopamine secreted in the caudate nucleus and putamen functions as an inhibitory transmitter, then destruction of the substantia nigra theoretically would allow these structures to become overly active and possibly cause continuous output of excitatory signals to the corticospinal motor control system. These signals could certainly overly excite many or all muscles of the body, thus leading to *rigidity.* And some of the feedback circuits might easily *oscillate* because of high feedback gains after loss of their inhibition, leading to the *tremor* of Parkinson's disease. This tremor is quite different from that of cerebellar disease, for it occurs during all waking hours and is therefore called an *involuntary tremor,* in contradistinction to cerebellar tremor, which occurs only when the person performs intentionally initiated movements and therefore is called *intention tremor.*

The *akinesia* that occurs in Parkinson's disease is often much more distressing to the patient than are the symptoms of muscle rigidity and tremor, for to perform even the simplest movement in severe Parkinsonism the person must exert the highest degree of concentration. The mental effort, even mental anguish, that is necessary to make the movement "go" is often at the limit of the patient's willpower. Then, when the movement does occur, it is stiff and staccato in character instead of occurring smoothly. Unfortunately, the cause of this akinesia is still entirely speculative. It is presumed that loss of dopamine secretion in the caudate nucleus and putamen might lead to loss of balance between the excitatory and inhibitory systems. Since *patterns of movement* require sequential changes between excitation and inhibition, any effect that would lock basal ganglia activity always in one direction would obviously prevent the initiation of and progression through sequential patterns, which is exactly what happens in akinesia.

Treatment With L-Dopa. Administration of the drug L-dopa to patients with Parkinson's disease ameliorates many of the symptoms, especially the rigidity and akinesia, in most patients. The reason for this is believed to be that L-dopa is converted in the brain into dopamine, and the dopamine then restores the normal balance between inhibition and excitation in the caudate nucleus and putamen. Unfortunately, administration of dopamine itself does not have the same effect because dopamine has a chemical structure that will not allow it to pass through the blood-brain barrier, even though the slightly different structure of L-dopa does allow it to pass.

Coagulation of the Ventrolateral and Ventroanterior Nuclei of the Thalamus for Treatment of Parkinson's Disease. Many researchers have also treated Parkinson's disease, with varying degrees of success, by surgically destroying portions of the basal ganglia, the thalamus, or even the motor cortex to block basal ganglial feedback to the cortex. The most widely employed of these has been destruction of the *ventrolateral* and *ventroanterior nuclei of the thalamus,* usually by electrocoagulation. Almost all of the feedback pathways from the basal ganglia to the cerebral cortex pass through these nuclei. It is presumed that blockage of these feedbacks prevents function of the neuronal loops that cause the tremor and some other symptoms of Parkinson's disease.

Huntington's Chorea

Huntington's chorea is a hereditary disorder that usually begins to cause symptoms in the third or fourth decade of life. It is characterized at first by flicking movements at individual joints and then progressive severe distortional movements of the entire body. In addition, severe dementia also develops along with the motor dysfunctions.

The abnormal movements of Huntington's chorea are *believed to be caused by loss of most of the cell bodies of the GABA-secreting neurons in the caudate nucleus and putamen.* The axon terminals of these neurons normally cause inhibition in the globus pallidus and substantia nigra. This loss of inhibition is believed to allow spontaneous outbursts of globus pallidus and substantia nigra activity that cause the distortional movements.

The dementia in Huntington's chorea probably does not result from the loss of GABA neurons but instead from loss of many acetylcholine-secreting neurons at the same time. This loss occurs not only in the basal ganglia but also in much of the cerebral cortex, which could easily block much of the thinking process.

■ INTEGRATION OF ALL PARTS OF THE TOTAL MOTOR CONTROL SYSTEM

Finally, we need to summarize as best we can what is known about overall control of movement. To do this, let us first give a synopsis of the different levels of control:

THE SPINAL LEVEL

Programmed in the spinal cord are local patterns of movement for all muscle areas of the body—for instance, programmed withdrawal reflexes that pull any part of the body away from a source of pain. And the cord is the locus even of complex patterns of rhythmical motions such as to-and-fro movement of the limbs for walking, plus reciprocal activity of opposite sides of the body, or hind limbs versus forelimbs.

All these programs of the cord can be commanded into action by the higher levels of motor control, or they can be inhibited while the higher levels take over control.

THE HINDBRAIN LEVEL

The hindbrain provides two major functions for general motor control of the body: (1) maintenance of axial tone of the body for the purpose of standing and (2) continuous modification of the different directions of this tone in response to continuous information from the vestibular apparatuses for the purpose of maintaining equilibrium.

THE CORTICOSPINAL LEVEL

The corticospinal system transmits most of the motor signals from the motor cortex to the spinal cord. It functions partly by issuing commands to set into motion the various cord patterns of motor control. It can also change the intensity of the different patterns or modify their timing or other characteristics. When needed, the corticospinal system can bypass the cord patterns by issuing inhibitory commands and replacing them with higher-level patterns from the brain stem or from the cerebral cortex. Usually the cortical patterns are more complex; also, they can be learned by practice, while the cord patterns are mainly set by heredity and are said to be "hard wired."

The Associated Function of the Cerebellum. The cerebellum functions with all levels of muscle control. It functions with the spinal cord especially to enhance the stretch reflex, so that when a contracting muscle meets an unexpectedly heavy load, a long stretch reflex arc through the cerebellum and back again to the cord strongly facilitates the load-resisting effect of the basic stretch reflex.

At the brain stem level, the cerebellum functions to make the postural movements of the body, especially the rapid movements required by the equilibrium system, smooth and continuous and without abnormal oscillations.

At the cerebral cortex level, the cerebellum functions to provide many accessory motor commands, especially to provide extra motor force to turn on muscle contraction very rapidly and forcefully at the start of movements. And near the end of each movement, the cerebellum turns on antagonist muscles at exactly the right time and with proper force to stop the movement at the intended point. Furthermore, there is good physiological evidence that all aspects of this turn-on/turn-off patterning by the cerebellum can be learned with experience.

In addition, the cerebellum functions with the cerebral cortex at still another level of motor planning: it helps to program in advance the muscle contractions that are required for smooth progression from the present movement in one direction to the next movement in another direction. The neural circuit for this passes from the cerebral cortex to the large lateral hemispheres of the cerebellum and then back to the cortex.

It should be noted especially that the cerebellum functions mainly with very rapid movements. Without the cerebellum, slow and calculated movements can still occur, but it is difficult for the corticospinal system to achieve well-controlled rapid intended movements to a particular goal, or especially to progress smoothly from one movement to the next.

The Associated Functions of the Basal Ganglia. The basal ganglia are essential to motor control in ways entirely different from those of the cerebellum. Their two most important functions are (1) to help the cortex execute subconscious but *learned* patterns of movement and (2) to help plan multiple parallel and sequential patterns of movement that the mind must put together to accomplish a purposeful task.

The types of motor patterns that require the basal ganglia include those for writing all the different letters of the alphabet, for throwing a ball, for typing, and so forth. Also, the basal ganglia are required to modify these patterns for slow execution, for rapid execution, to write small, or to write very large — thus controlling both timing and dimensions of the patterns.

At still a higher level of control is another cerebral cortex-basal ganglia circuit, beginning in the thinking processes of the brain and providing the overall sequence of action for responding to each new situation — such as planning one's immediate response to an assailant who hits the person in the face or one's sequential response to an unexpectedly fond embrace.

An important part of all these basal ganglial planning processes is not only the motor cortex and basal ganglia but also the somatic sensory cortex of the parietal lobe, especially the posterior portion where the instantaneous spatial coordinates of all parts of one's body are continuously calculated, and even the spatial coordinates of the relationships of the body parts to the physical surroundings. If one of the two parietal cortices is severely damaged, then the person simply ignores the opposite side of his or her body and even ignores objects on the opposite side; then the movements are planned around use of only the consciously recognized side of the body.

WHAT DRIVES US TO ACTION?

Finally, what is it that arouses us from inactivity and sets into play our trains of movement? Fortunately, we are beginning to learn about the motivational systems of the brain. Basically, the brain has an older core centered beneath, anterior, and lateral to the thalamus — including the hypothalamus, the amygdala, the hippocampus, the septal region anterior to the hypothalamus and thalamus, and even older regions of the thalamus and cerebral cortex themselves — all of which function together to motivate most of the motor and other functional activities of the brain. These areas are collectively called the *limbic system* of the brain. We discuss this system in detail in Chapter 20.

REFERENCES

Atkeson, C. G.: Learning arm kinematics and dynamics. Annu. Rev. Neurosci., 12:157, 1989.

Baldessarini, R. J., and Tarsey, D.: Dopamine and the pathophysiology of dyskinesias induced by antipsychotic drugs. Annu. Rev. Pharmacol. Toxicol., 20:533, 1980.

Bloedel, J. R., and Courville, J.: Cerebellar afferent systems. In Brooks, V. B. (ed.): Handbook of Physiology. Sec. 1, Vol. II. Bethesda, Md., American Physiological Society, 1981, p. 735.

Brooks, V. B.: The Neural Basis of Motor Control. New York, Oxford University Press, 1986.

Brooks, V. B., and Thach, W. T.: Cerebellar control of posture and movement. In Handbook of Physiology. Sec. 1, Vol. II. Bethesda, Md., American Physiological Society, 1981, p. 877.

Carpenter, M. B.: Anatomy of the corpus striatum and brain stem integrating system. In Handbook of Physiology. Sec. 1, Vol. II. Bethesda, Md., American Physiological Society, 1981, p. 947.

Collier, T. J., and Sladek, J. R., Jr.: Neural transplantation in animal models of neurodegenerative disease. News Physiol. Sci., 3:204, 1988.

Courville, J., et al. (eds.): The Inferior Olivary Nucleus. New York, Raven Press, 1980.

DeLong, M., and Georgopoulos, A. P.: Motor functions of the basal ganglia. In Handbook of Physiology. Sec. 1, Vol. II. Bethesda, Md., American Physiological Society, 1981, p. 1017.

Di Chiara, G. (ed.): GABA and the Basal Ganglia. New York, Raven Press, 1981.

Duvoisin, R. C.: Parkinson's Disease. New York, Raven Press, 1978.

Eckmiller, R.: Neural control of pursuit eye movements. Physiol. Rev., 67:797, 1987.

Evarts, E. V.: Role of motor cortex involuntary movements in primates. In Handbook of Physiology. Sec. 1, Vol. II. Bethesda, Md., American Physiological Society, 1981, p. 1083.

Evarts, E. V., et al. (eds.): Motor System in Neurobiology. New York, Elsevier Science Publishing Co., 1986.

Fernstrom, J. D.: Role of precursor availability on control of monoamine biosynthesis in the brain. Physiol. Rev., 63:484, 1983.

Fuster, J. M.: Prefrontal cortex in motor control. In Handbook of Physiology. Sec. 1, Vol. II. Bethesda, Md., American Physiological Society, 1981, p. 1149.

Georgopoulos A. P.: Neural integration of movement: role of motor cortex in reaching. FASEB J., 1:2849, 1988.

Glickstein M., and Yeo, C. (eds.): Cerebellum and Neuronal Plasticity. New York, Plenum Publishing Corp., 1987.

Goldstein, M., et al. (eds.): Central D_1 Dopamine Receptors. New York, Plenum Publishing Corp., 1988.

Grillner, S.: Control of locomotion in bipeds, tetrapods, and fish. In Handbook of Physiology. Sec. 1, Vol. II. Bethesda, Md., American Physiological Society, 1981, p. 1179.

Ito, M.: The Cerebellum and Neural Control. New York, Raven Press, 1984.

Ito, M.: Where are neurophysiologists going? News Physiol. Sci., 1:30, 1986.

Jones, E. G., and Peters, A. (eds.): Sensory-Motor Areas and Aspects of Cortical Connectivity. New York, Plenum Publishing Corp., 1986.

Keele, S. W.: Behavioral analysis of movement. In Handbook of Physiology. Sec. 1, Vol. II. Bethesda, Md., American Physiological Society, 1981, p. 1391.

Kitai, S. T.: Electrophysiology of the corpus striatum and brain stem integrating systems. In Handbook of Physiology. Sec. 1, Vol. II. Bethesda, Md., American Physiological Society, 1981, p. 997.

Kuypers, H. G. J. M.: Anatomy of the descending pathways. In Handbook of Physiology. Sec. 1, Vol. II. Bethesda, Md., American Physiological Society, 1981, p. 597.

Lewin, R.: Brain grafts benefit Parkinson's patients. Science, 236:149, 1987.

Llinas, R.: Eighteenth Bowditch lecture. Motor aspects of cerebellar control. Physiologist, 17:19, 1974.

Llinas, R.: Electrophysiology of the cerebellar networks. In Handbook of Physiology. Sec. 1, Vol. II. Bethesda, Md., American Physiological Society, 1981, p. 831.

McCloskey, D. I., et al.: Sensing position and movements of the fingers. News Physiol. Sci., 2:226, 1987.

Olsen, R. W.: Drug interactions at the GABA receptor-ionophore complex. Annu. Rev. Pharmacol. Toxicol., 22:245, 1982.

Palacios, J. M.: Neurotransmitters, their receptors and the degenerative diseases of the aging brain. Triangle, 25:85, 1986.

Palay, S. L., and Chan-Palay, V.: The Cerebellum — New Vistas. New York, Springer-Verlag, 1982.

Penney, J. B., Jr., and Young, A. B.: Speculations on the functional anatomy of basal ganglia disorders. Annu. Rev. Neurosci., 6:73, 1983.

Peterson, B. W., and Richmond, F. J. (eds.): Control of Head Movement. New York, Oxford University Press, 1988.

Porter, R.: Internal organization of the motor cortex for input-output arrangements. In Handbook of Physiology. Sec. 1, Vol. II. Bethesda, Md., American Physiological Society, 1981, p. 1063.

Poulton, E. C.: Human manual control. In Handbook of Physiology. Sec. 1, Vol. II. Bethesda, American Physiological Society, 1981, p. 1337.

Riklan, M.: L-Dopa and Parkinsonism. Springfield, Ill., Charles C Thomas, 1973.

Robinson, D. A.: The windfalls of technology in the oculomotor system. Inv. Ophthal. Vis. Sci., 28:1912, 1987.

Sandler, M., et al. (eds.): Neurotransmitter Interactions in the Basal Ganglia. New York, Raven Press, 1987.

Scheibel, A. B.: The brain stem reticular core and sensory function. In Handbook of Physiology. Sec. 1, Vol. II. Bethesda, Md., American Physiological Society, 1981, p. 213.

Shik, M. L., and Orlovsky, G. N.: Neurophysiology of locomotor automatism. Physiol. Rev., 56:465, 1976.

Stein, R. B., and Lee, R. G.: Tremor and clonus. In Handbook of Physiology. Sec. 1, Vol. II. Bethesda, Md., American Physiological Society, 1981, p. 325.

Wiesendanger, M.: Organization of secondary motor areas of cerebral cortex. In Handbook of Physiology. Sec. 1, Vol. II. Bethesda, Md., American Physiological Society, 1981, p. 1121.

Wiesendanger, M., and Miles, T. S.: Ascending pathway of low-threshold muscle afferents to the cerebral cortex and its possible role in motor control. Physiol. Rev., 62:1234, 1982.

19

The Cerebral Cortex; Intellectual Functions of the Brain and Learning and Memory

It is ironic that of all the parts of the brain, we know least about the mechanisms of the cerebral cortex, even though it is by far the largest portion of the nervous system. Yet we do know the effects of destruction or of specific stimulation of various portions of the cortex. In the early part of the present chapter the facts known about cortical functions are discussed; then some basic theories of the neuronal mechanisms involved in thought processes, memory, analysis of sensory information, and so forth, are presented briefly.

■ PHYSIOLOGIC ANATOMY OF THE CEREBRAL CORTEX

The functional part of the cerebral cortex is composed mainly of a thin layer of neurons 2 to 5 mm in thickness, covering the surface of all the convolutions of the cerebrum and having a total area of about 0.25 m². The total cerebral cortex probably contains 100 billion or more neurons.

Figure 19–1 illustrates the typical structure of the cerebral cortex, showing successive layers of different types of cells. Most of the cells are of three types: *granular* (also called *stellate*), *fusiform,* and *pyramidal,* the latter named for their characteristic pyramidal shape. The *granule cells,* in general, have short axons and therefore function mainly as intracortical interneurons. Some are excitatory, probably releasing the excitatory neurotransmitter *glutamate;* others are inhibitory and release the inhibitory neurotransmitter *gamma-aminobutyric acid (GABA).* The sensory areas of the cortex, as well as the association areas between sensory and motor, have large concentrations of these granule cells, suggesting a high degree of intracortical processing of the incoming sensory signals in the sensory areas and of the cognitive analytical signals in the association areas.

The *pyramidal* and *fusiform cells,* on the other hand, give rise to almost all of the output fibers from the cortex. The pyramidal cells are the larger of the two and are more numer-ous than the fusiform cells. They are the source of the long, large nerve fibers that go all the way to the spinal cord. They also give rise to most of the large subcortical association fiber bundles that pass from one major part of the brain to the other.

To the right in Figure 19–1 is illustrated the typical organization of nerve fibers within the different layers of the cortex. Note particularly the large number of *horizontal fibers* extending between adjacent areas of the cortex, but note also the *vertical fibers* that extend to and from the cortex to lower areas of the brain and to the spinal cord or to distant regions of the cerebral cortex through the long association bundles.

The functions of the specific layers of the cerebral cortex have been discussed briefly in Chapters 9 and 13. By way of review, let us recall that most incoming specific sensory signals terminate in cortical layer IV. Most of the output signals leave the cortex from neurons located in layers V and VI, the very large fibers to the brain stem and cord arise generally in layer V, and the tremendous numbers of fibers to the thalamus arise in layer VI. Layers I, II, and III perform most of the intracortical association functions, with especially large numbers of neurons in layers II and III making short horizontal connections with adjacent cortical areas.

Anatomical and Functional Relationships of the Cerebral Cortex to the Thalamus and Other Lower Centers. All areas of the cerebral cortex have extensive to-and-fro efferent and afferent connections with the deeper structures of the brain. It is especially important to emphasize the relationship between the cerebral cortex and the thalamus. When the thalamus is damaged along with the cortex, the loss of cerebral function is far greater than when the cortex alone is damaged, for thalamic excitation of the cortex is necessary for almost all cortical activity.

Figure 19–2 shows the areas of the cerebral cortex connected with specific parts of the thalamus. These connections act in *two* directions, both from the thala-

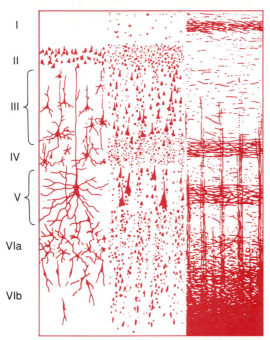

Figure 19–1. Structure of the cerebral cortex, illustrating *I*, molecular layer; *II*, external granular layer; *III*, layer of pyramidal cells; *IV*, internal granular layer; *V*, large pyramidal cell layer; and *VI*, layer of fusiform or polymorphic cells. (From Ranson and Clark [after Brodmann]: Anatomy of the Nervous System. Philadelphia, W. B. Saunders Company, 1959.)

mus to the cortex and then from the cortex back to essentially the same area of the thalamus. Furthermore, when the thalamic connections are cut, the functions of the corresponding cortical area become entirely abrogated. Therefore, the cortex operates in close association with the thalamus and can almost be considered both anatomically and functionally to be a unit with the thalamus; for this reason the thalamus and the cortex together are sometimes called the *thalamo-cortical system.* Also, all pathways from the sensory organs to the cortex pass through the thalamus, with the single exception of most sensory pathways of the olfactory tract.

■ FUNCTIONS OF SPECIFIC CORTICAL AREAS

Studies in human beings by neurosurgeons, neurologists, and neuropathologists have shown that different cortical areas have their own separate functions. Figure 19–3 is a map of some of these functions as determined by Penfield and Rasmussen from electrical stimulation of the cortex in awake patients or during neurological examination of patients after portions of the cortex had been removed. The electrically stimulated patients either told the surgeons their thoughts evoked by the stimulation or at times experienced a movement or a spontaneously emitted sound or even a word or some other evidence of the stimulation. In the patients in whom portions of the cortex had been removed, the subsequent neurological examinations demonstrated different deficits of brain function.

Information of the type illustrated in Figure 19–3 from many different sources gives a more general map, as illustrated in Figure 19–4. This figure shows the major primary and secondary motor areas of the cortex, as well as the major primary and secondary sensory areas for somatic sensation, vision, and hearing, all of which have been discussed in previous chapters. The primary areas have direct connections with specific muscles or specific sensory receptors, for causing discrete muscle movements or experiencing a sensation—visual, auditory, or somatic—from a minute receptor area. The secondary areas, on the other hand, make sense out of the functions of the primary areas. For instance, the supplemental and premotor areas function along with the primary motor cortex and basal ganglia to provide highly specific patterns of motor activity. On the sensory side, the secondary sensory areas, which are located within a few centimeters of the primary areas, begin to make sense out of the specific sensory signals, such as interpreting the shape or texture of an object in one's hand; the color, the light intensity, the directions of lines and angles, and other

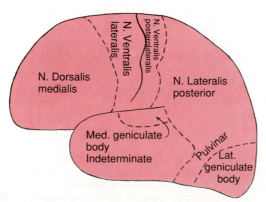

Figure 19–2. Areas of the cerebral cortex that connect with specific portions of the thalamus. (Modified from Elliott: Textbook of the Nervous System. Philadelphia, J. B. Lippincott Company.)

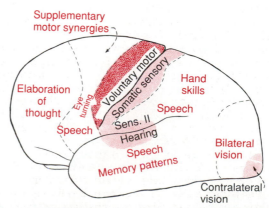

Figure 19–3. Functional areas of the human cerebral cortex as determined by electrical stimulation of the cortex during neurosurgical operations and by neurological examinations of patients with destroyed cortical regions. (From Penfield and Rasmussen: The Cerebral Cortex of Man: A Clinical Study of Localization of Function. New York, Macmillan Company, 1968.)

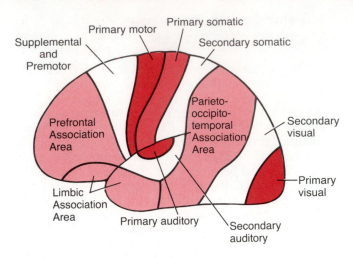

Figure 19–4. Locations of the major association areas of the cerebral cortex, shown in relation to the primary and secondary motor and sensory areas.

aspects of vision; and the combination of tones, sequence of tones, and beginning interpretation of the meanings of auditory signals.

THE ASSOCIATION AREAS

Figure 19–4 also shows several large areas of the cerebral cortex that do not fit into the rigid categories of primary or secondary motor and sensory areas. These are called *association areas* because they receive and analyze signals from multiple regions of the cortex and even subcortical structures. Yet even the association areas have their own specializations, as we shall see. The three most important association areas are (1) the *parieto-occipitotemporal association area,* (2) the *prefrontal association area,* and (3) the *limbic association area.* The functions of these are the following:

The Parieto-occipitotemporal Association Area. This association area lies in the large cortical space between the somatic sensory cortex anteriorly, the vi-

sual cortex posteriorly, and the auditory cortex laterally. As would be expected, it provides a high level of interpretive meaning for the signals from all the surrounding sensory areas. However, even the parieto-occipitotemporal association area has its own functional subareas, which are illustrated in Figure 19–5:

1. An area beginning in the *posterior parietal cortex and extending into the superior occipital cortex provides continuous analysis of the spatial coordinates of all parts of the body as well as of the surroundings of the body.* This area receives visual information from the posterior occipital cortex and simultaneous somatic information from the anterior parietal cortex; from this it computes the coordinates. But why does a person need to know these spatial coordinates? The answer is that to control the body movements, the brain must know at all times where each part of the body is located and also the relation to the surroundings. The person also needs this information to analyze incoming somatic sensory signals. In fact, as was illustrated in Figure 18–13 of

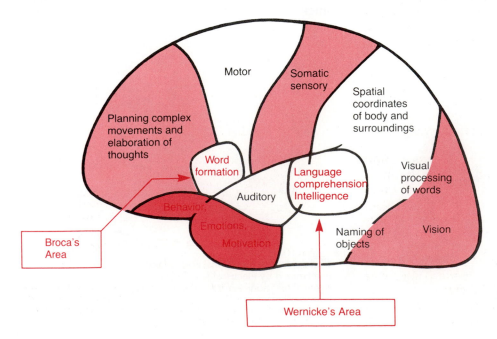

Figure 19–5. Map of specific functional areas in the cerebral cortex, showing especially Wernicke's and Broca's areas for language comprehension and speech production, which in 95 per cent of all persons are located in the left hemisphere.

Chapter 18, a person missing this area of the brain actually loses recognition of the fact that he or she has an opposite side of the body or surroundings and, as a consequence, will fail to consider the existence of the opposite side either for sensation or for planning voluntary movements.

2. The major area for language comprehension, called *Wernicke's area,* lies behind *the primary auditory cortex in the posterior part of the superior temporal lobe.* We discuss this area much more fully later; it is the most important region of the entire brain for higher intellectual functions because almost all intellectual functions are language-based.

3. *Posterior to the language comprehension area, lying mainly in the angular gyrus region of the occipital lobe, is a secondary visual processing area that feeds the visual signals of words read from a page into Wernicke's area, the language comprehension area.* This angular gyrus area is needed to make meaning out of the visually perceived words. In its absence, a person can still have excellent language comprehension through hearing but not through reading.

4. *In the most lateral portions of both the anterior occipital lobe and posterior temporal lobe is an area for naming objects.* The names presumably originate mainly through auditory input, whereas the nature of the objects originates mainly through visual input. In turn, the names are essential for language comprehension and intelligence, functions performed in Wernicke's area, located immediately superior to the "names" region.

The Prefrontal Association Area. In the previous chapter we learned that the prefrontal association area functions in close association with the motor cortex to plan complex patterns and sequences of motor movements. To aid in this function, it receives very strong input through a massive subcortical bundle of fibers connecting the parieto-occipitotemporal association area with the prefrontal association area. Through this bundle the prefrontal cortex receives much preanalyzed sensory information, especially information on the spatial coordinates of the body that is absolutely necessary in the planning of effective movements. Much of the output from the prefrontal area into the motor control system passes through the caudate portion of the basal ganglia-thalamic feedback circuit for motor planning, which provides many of the sequential and parallel components of the movement complex.

The prefrontal association area is also essential to carrying out prolonged thought processes in the mind. This presumably results from some of the same capabilities of the prefrontal cortex that allow it to plan motor activities. That is, it seems to be capable of combining nonmotor information from widespread areas of the brain and therefore to achieve nonmotor types of thinking as well as motor types. In fact, the prefrontal association area is frequently described simply as important for the *elaboration of thoughts.*

A special region in the frontal cortex, called *Broca's area, provides the neural circuitry for word formation.*

This area, illustrated in Figure 19–5, is located partly in the posterior lateral prefrontal cortex and partly in the premotor area. It is here that the plans and motor patterns for the expression of individual words or even short phrases are initiated and executed. This area also works in close association with Wernicke's language comprehension center in the temporal association cortex, as we discuss more fully later in the chapter.

The Limbic Association Area. Figure 19–4 illustrates still another association area called the *limbic area.* This is found in the anterior pole of the temporal lobe, in the ventral portions of the frontal lobes, and in the cingulate gyri on the midsurfaces of the cerebral hemispheres. This region is concerned primarily with *behavior, emotions,* and *motivation,* as illustrated in Figure 19–5. We will learn in the following chapter that the limbic cortex is part of a much more extensive system, the *limbic system,* that includes a complex set of neuronal structures in the midbasal regions of the brain. It is this limbic system that provides most of the drives for setting the other areas of the brain into action and even provides the motivational drive for the process of learning itself.

An Area for Recognition of Faces

An interesting type of brain abnormality called *prosophenosia* is the inability to recognize faces. This occurs in persons who have extensive damage on the medial undersides of both occipital lobes and along the medioventral surfaces of the temporal lobes, as illustrated in Figure 19–6. Loss of these face recognition areas, strangely enough, results in very little other abnormality of brain function.

One wonders why so much of the cerebral cortex should be reserved for the simple task of face recognition. However, when it is remembered that most of our daily tasks involve associations with other people, one can see the importance of this intellectual function.

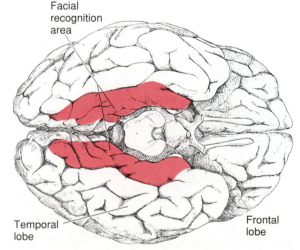

Figure 19–6. Facial recognition areas located on the underside of the brain in the medial occipital and temporal lobes. (From Geschwind: Sci. Am., 241:180, 1979. © 1979 by Scientific American, Inc. All rights reserved.)

The occipital portion of this area is contiguous with the visual cortex, and the temporal portion is closely associated with the limbic system that has to do with emotions, brain activation, and control of one's behavioral response to the environment, as we see later in the following chapter.

INTERPRETATIVE FUNCTION OF THE POSTERIOR SUPERIOR TEMPORAL LOBE — WERNICKE'S AREA (A GENERAL INTERPRETATIVE AREA)

The somatic, visual, and auditory secondary and association areas, which can actually be called sensory interpretative areas, all meet one another in the posterior part of the superior temporal lobe, as illustrated in Figure 19–7, where the temporal, parietal, and occipital lobes all come together. This area of confluence of the different sensory interpretative areas is especially highly developed in the *dominant* side of the brain—the *left side* in almost all right-handed persons—and it plays the greatest single role of any part of the cerebral cortex in the higher levels of brain function that we call *intelligence*. Therefore, this region has frequently been called by different names suggestive of the area having almost global importance: the *general interpretative area,* the *gnostic area,* the *knowing area,* the *tertiary association area,* and so forth. However, it is best known as *Wernicke's area* in honor of the neurologist who first described its special significance in intellectual processes.

Following severe damage in Wernicke's area, a person might hear perfectly well and even recognize different words but still be unable to arrange these words into a coherent thought. Likewise, the person may be able to read words from the printed page but be unable to recognize the thought that is conveyed.

Electrical stimulation in Wernicke's area of the conscious patient occasionally causes a highly complex thought. This is particularly true when the stimulatory electrode is passed deep enough into the brain to approach the corresponding connecting areas of the thalamus. The types of thoughts that might be experienced include complicated visual scenes that one might remember from childhood, auditory hallucinations such as a specific musical piece, or even a discourse by a specific person. For this reason it is believed that activation of Wernicke's area can call forth complicated memory patterns involving more than one sensory modality even though many of the memory patterns may be stored elsewhere. This belief is in accord with the importance of Wernicke's area in interpretation of the complicated meanings of different sensory experiences.

The Angular Gyrus—Interpretation of Visual Information. The angular gyrus is the most inferior portion of the posterior parietal lobe, lying immediately behind Wernicke's area and fusing posteriorly into the visual areas of the occipital lobe as well. If this region is destroyed while Wernicke's area in the temporal lobe is still intact, the person can still interpret auditory experiences as usual, but the stream of visual experiences passing into Wernicke's area from the visual cortex is mainly blocked. Therefore, the person may be able to see words and even know they are words but, nevertheless, not be able to interpret their meanings. This is the condition called *dyslexia,* or *word blindness.*

Let us again emphasize the global importance of Wernicke's area for most intellectual functions of the brain. Loss of this area in an adult usually leads thereafter to a lifetime of almost demented existence.

The Concept of the Dominant Hemisphere

The general interpretative functions of Wernicke's area and of the angular gyrus, and also the functions of the speech and motor control areas, are usually much more highly developed in one cerebral hemisphere than in the other. Therefore, this hemisphere is called the *dominant hemisphere.* In about 95 per cent of all persons the left hemisphere is the dominant one. Even at birth, the area of the cortex that will eventually become Wernicke's area is as much as 50 per cent larger in the left hemisphere than in the right in more than one half of newborn babies. Therefore, it is easy to understand why the left side of the brain might become dominant over the right side. However, if for some reason this left side area is damaged or removed in early childhood, the opposite side of the brain can develop full dominant characteristics.

A theory that can explain the capability of one hemisphere to dominate the other is the following:

The attention of the "mind" seems to be directed to one portion of the brain at a time. Presumably, because its size is usually larger at birth, the left temporal lobe normally begins to be used to a greater extent than the right, and, thenceforth, because of the tendency to

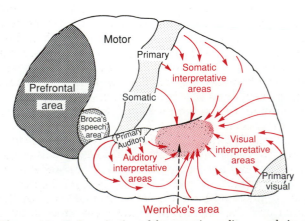

Figure 19–7. Organization of the somatic, auditory, and visual association areas into a general mechanism for interpretation of sensory experience. All these feed also into *Wernicke's area,* located in the posterosuperior portion of the temporal lobe. Note also the prefrontal area and Broca's speech area.

direct one's attention to the better developed region, the rate of learning in the cerebral hemisphere that gains the first start increases rapidly while that in the opposite side remains slight. Therefore, in the normal human being, one side becomes dominant over the other.

In about 95 per cent of all persons the left temporal lobe and angular gyrus become dominant, and in the remaining 5 per cent either both sides develop simultaneously to have dual dominance, or, more rarely, the right side alone becomes highly developed.

Usually associated with the dominant temporal lobe and angular gyrus is dominance of certain portions of the somatic sensory cortex and motor cortex for control of voluntary motor functions. For instance, as is discussed later in the chapter, the prefrontal and premotor speech area (Broca's area), located far laterally in the intermediate frontal lobe, also is almost always dominant on the left side of the brain. This speech area causes the formation of words by exciting simultaneously the laryngeal muscles, the respiratory muscles, and the muscles of the mouth.

Also, the motor areas for controlling the hands are dominant on the left side of the brain in about nine of ten persons, thus causing "right-handedness" in most people.

Although the interpretative areas of the temporal lobe and angular gyrus, as well as many of the motor areas, are highly developed in only a single hemisphere, they are capable of receiving sensory information from both hemispheres and are also capable of controlling motor activities in both hemispheres, utilizing mainly fiber pathways in the *corpus callosum* for communication between the two hemispheres. This unitary, cross-feeding organization prevents interference between the functions of the two sides of the brain; such interference, obviously, could create havoc with both thoughts and motor responses.

Role of Language in the Function of Wernicke's Area and in Intellectual Functions

A major share of our sensory experience is converted into its language equivalent before being stored in the memory areas of the brain and before being processed for other intellectual purposes. For instance, when we read a book, we do not store the visual images of the printed words but, instead, store the words themselves in language form. Also, the information conveyed by the words is usually converted to language form before its meaning is discerned.

The sensory area of the dominant hemisphere for interpretation of language is Wernicke's area, and this is very closely associated with both the primary hearing area and the secondary auditory areas of the temporal lobe. This very close relationship probably results from the fact that the first introduction to language is by way of hearing. Later in life, when visual perception of language through the medium of reading develops, the visual information is then pre-

sumably channeled into the already developed language regions of the dominant temporal lobe.

FUNCTIONS OF THE PARIETO-OCCIPITOTEMPORAL CORTEX IN THE NONDOMINANT HEMISPHERE

When Wernicke's area in the dominant hemisphere is destroyed, the person normally loses almost all intellectual functions associated with language or verbal symbolism, such as ability to read, ability to perform mathematical operations, and even the ability to think through logical problems. However, many other types of interpretative capabilities, some of which utilize the temporal lobe and angular gyrus regions of the opposite hemisphere, are retained. Psychological studies in patients with damage to their nondominant hemispheres have suggested that this hemisphere may be especially important for understanding and interpreting music, nonverbal visual experiences (especially visual patterns), spatial relationships between the person and the surroundings, the significance of "body language" and intonations of persons' voices, and probably also many somatic experiences related to use of the limbs and hands.

Thus, even though we speak of the "dominant" hemisphere, this dominance is primarily for language – or verbal symbolism – related intellectual functions; the opposite hemisphere is actually dominant for some other types of intelligence.

THE HIGHER INTELLECTUAL FUNCTIONS OF THE PREFRONTAL ASSOCIATION AREA

For years it has been taught that the prefrontal cortex is the locus of the higher intellect in the human being, principally because the main difference between the brain of monkeys and of human beings is the great prominence of the human prefrontal areas. Yet efforts to show that the prefrontal cortex is more important in higher intellectual functions than other portions of the brain have not been entirely successful. Indeed, destruction of the language comprehension area in the posterior superior temporal lobe (Wernicke's area) and the angular gyrus region in the dominant hemisphere causes infinitely more harm to the intellect than does destruction of the prefrontal area. The prefrontal areas do, however, have less definable but nevertheless very important intellectual functions of their own. These can be explained best by describing what happens to patients in whom the prefrontal lobes have become nonfunctional as follows:

Several decades ago, before the advent of modern drugs for treating psychiatric conditions, it was found that some patients could receive significant relief from severe psychotic depression by severing of the neuronal connections between the prefrontal areas of the brain and the remainder of the brain, that is, by a procedure called *prefrontal lobotomy*. This was done by inserting a blunt, thin-bladed knife through small openings in the lateral frontal skull on both sides and slicing the brain from top to bottom. Subsequent studies in these patients showed the following mental changes:

1. The patients lost their ability to solve complex problems.

2. They became unable to string together sequential tasks to reach specific goals, and in general lost all ambition.

3. They became unable to learn to do several parallel tasks at the same time.

4. Their level of aggressiveness was decreased, sometimes markedly.

5. Their social responses were often inappropriate for the occasion, including loss of morals and little embarrassment in relation to sex and excretion.

6. The patients could still talk and comprehend language, but they were unable to carry through any long trains of thought, and their moods changed rapidly from sweetness to rage to exhilaration to madness.

7. The patients could also still perform most of the usual patterns of motor function that they had performed throughout life, but often without purpose.

From this information, let us try to piece together a coherent understanding of the function of the prefrontal association areas.

Decreased Aggressiveness and Inappropriate Social Responses. These two characteristics probably result from loss of the ventral parts of the frontal lobes on the underside of the brain. As explained earlier and illustrated in Figure 19–4, this area is considered to be part of the limbic association cortex, rather than the prefrontal association cortex. This limbic area helps to control behavior.

Inability to Progress Toward Goals or to Carry Through With Sequential Thoughts. We learned earlier in the chapter that the prefrontal association areas appear to have the capability of calling forth information from widespread areas of the brain and then using it in deeper thought patterns for attaining goals. If these goals include motor action, so be it. If they do not, then the thought processes attain intellectual analytical goals. Although persons without prefrontal cortices can still think, they show little concerted thinking in logical sequence for longer than a few seconds or a few minutes at most. One of the results is that persons without prefrontal cortices are *easily distracted from the central theme of the thought*, whereas persons with functioning prefrontal cortices can drive themselves to completion of their thought goals irrespective of distractions.

Elaboration of Thought, Prognostication, and Performance of Higher Intellectual Functions by the Prefrontal Areas. Another function that has been ascribed to the prefrontal areas by psychologists and neurologists is *elaboration of thought*. This means simply an increase in depth and abstractness of the different thoughts. Psychological tests have shown that prefrontal lobectomized lower animals presented with successive bits of sensory information fail to keep track of these bits even in temporary memory—probably because they are distracted so easily that they cannot hold thoughts long enough for storage to take place.

This ability of the prefrontal areas to keep track of many bits of information simultaneously, and then to cause recall of this information bit by bit as it is needed for subsequent thoughts, could well explain the many functions of the brain that we associate with higher intelligence, such as the abilities to (1) prognosticate, (2) plan for the future, (3) delay action in response to incoming sensory signals so that the sensory information can be weighed until the best course of response is decided, (4) consider the consequences of motor actions even before these are performed, (5) solve complicated mathematical, legal, or philosophical problems, (6) correlate all avenues of information in diagnosing rare diseases, and (7) control one's activities in accord with moral laws.

■ FUNCTION OF THE BRAIN IN COMMUNICATION

One of the most important differences between the human being and lower animals is the facility with which human beings can communicate with one another. Furthermore, because neurological tests can easily assess the ability of a person to communicate with others, we know more about the sensory and motor systems related to communication than about any other segment of cortical function. Therefore, we will review, with the help of the anatomical maps of neural pathways in Figure 19–8, the function of the cortex in communication, and from this one can see immediately how the principles of sensory analysis and motor control apply to this art.

There are two aspects to communication: first, the *sensory aspect*, involving the ears and eyes, and, second, the *motor aspect*, involving vocalization and its control.

Sensory Aspects of Communication

We noted earlier in the chapter that destruction of portions of the *auditory* and *visual association areas* of the cortex can result in inability to understand the spoken word or the written word. These effects are called, respectively, *auditory receptive aphasia* and *visual receptive aphasia* or, more commonly, *word deafness* and *word blindness* (also called *dyslexia*).

Wernicke's Aphasia and Global Aphasia. Some persons are perfectly capable of understanding either the spoken word or the written word but are *unable to interpret the thought* that is expressed. This results most frequently when *Wernicke's area* in the *posterior portion of the dominant hemisphere superior temporal gyrus* is damaged or destroyed. Therefore, this type of aphasia is generally called *Wernicke's aphasia*.

When the lesion in Wernicke's area is widespread and extends (1) backward into the angular gyrus region, (2) inferiorly into the lower areas of the temporal lobe, and (3) superiorly into the superior border of the sylvian fissure, the person is likely to be almost totally demented and therefore is said to have *global aphasia*.

Motor Aspects of Communication

The process of speech involves two principal stages of mentation: (1) formation in the mind of thoughts to be expressed and choice of words to be used, then (2) motor control of

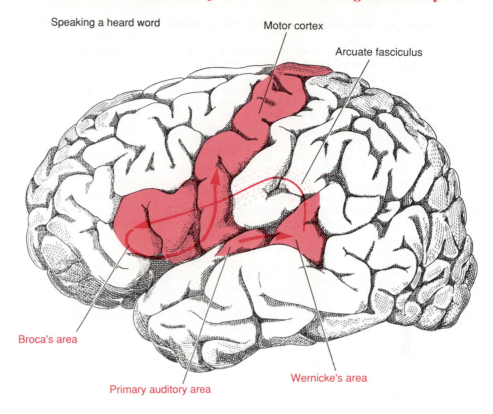

Speaking a heard word

Motor cortex

Arcuate fasciculus

Broca's area

Primary auditory area

Wernicke's area

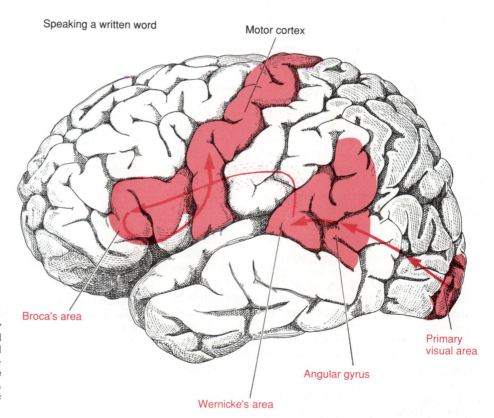

Speaking a written word

Motor cortex

Broca's area

Primary visual area

Angular gyrus

Wernicke's area

Figure 19–8. Brain pathways for *(top)* perception of the heard word and then speaking the same word and *(bottom)* perception of the written word and then speaking the same word. (From Geschwind: Sci. Am., 241:180, 1979. © 1979 by Scientific American, Inc. All rights reserved.)

vocalization and the actual act of vocalization itself. The formation of thoughts and even most choices of words are the function of the sensory areas of the brain. Again, it is Wernicke's area in the posterior part of the superior temporal gyrus that is most important for this ability. Therefore, per-

sons with either Wernicke's aphasia or global aphasia are unable to formulate the thoughts that are to be communicated. Or, if the lesion is less severe, the person may be able to formulate thoughts but yet be unable to put together the appropriate sequence of words to express the thought.

Often, the person is very fluent in words but the words are jumbled.

Motor Aphasia. Often a person is perfectly capable of deciding what he wishes to say, and he is capable of vocalizing, but he simply cannot make his vocal system emit words instead of noises. This effect, called *motor aphasia*, results from damage to *Broca's speech area*, which lies in the *prefrontal* and *premotor* facial region of the cortex—about 95 per cent of the time in the left hemisphere, as illustrated in Figures 19–5 and 19–8. Therefore, we assume that the *skilled motor patterns* for control of the larynx, lips, mouth, respiratory system, and other accessory muscles of articulation are all initiated from this area.

Articulation. Finally, we have the act of articulation itself, which means the muscular movements of the mouth, tongue, larynx, and so forth, that are responsible for the actual emission of sound. The *facial and laryngeal regions of the motor cortex* activate these muscles, and the *cerebellum, basal ganglia,* and *sensory cortex* all help control the muscle contractions by feedback mechanisms described in Chapters 17 and 18. Destruction of these regions can cause either total or partial inability to speak distinctly.

Summary

Figure 19–8 illustrates two principal pathways for communication. The upper half of the figure shows the pathway involved in hearing and speaking. This sequence is the following: (1) reception in the primary auditory area of the sound signals that encode the words; (2) interpretation of the words in Wernicke's area; (3) determination, also in Wernicke's area, of the thoughts and the words to be spoken; (4) transmission of signals from Wernicke's area to Broca's area via the *arcuate fasciculus;* (5) activation of the skilled motor programs in Broca's area for control of word formation; and (6) transmission of appropriate signals into the motor cortex to control the speech muscles.

The lower figure illustrates the comparable steps in reading and then speaking in response. The initial receptive area for the words is in the primary visual area rather than in the primary auditory area. Then the information passes through early stages of interpretation in the *angular gyrus region* and finally reaches its full level of recognition in Wernicke's area. From here, the sequence is the same as for speaking in response to the spoken word.

▪ FUNCTION OF THE CORPUS CALLOSUM AND ANTERIOR COMMISSURE TO TRANSFER THOUGHTS, MEMORIES, TRAINING, AND OTHER INFORMATION TO THE OPPOSITE HEMISPHERE

Fibers in the *corpus callosum* connect most of the respective cortical areas of the two hemispheres with each other except for the anterior portions of the temporal lobes; these temporal areas, including especially the *amygdala,* are interconnected by fibers that pass through the *anterior commissure.* Because of the tremendous number of fibers in the corpus callosum, it was assumed from the beginning that this massive structure must have some important function to correlate activities of the two cerebral hemispheres. However, after cutting the corpus callosum in experimental animals, it was difficult to discern deficits in brain function. Therefore, for a long time the function of the corpus callosum was a mystery.

Yet properly designed psychological experiments have now demonstrated the extremely important functions of the corpus callosum and anterior commissure. These can be explained best by recounting one of the experiments. A monkey is first prepared by cutting the corpus callosum and splitting the optic chiasm longitudinally so that signals from each eye can go only to the cerebral hemisphere on the side of the eye. Then the monkey is taught to recognize different types of objects with its right eye while its left eye is covered. Next, the right eye is covered and the monkey is tested to determine whether or not its left eye can recognize the same object. The answer to this is that the left eye *cannot* recognize the object. Yet, on repeating the same experiment in another monkey with the optic chiasm split but the corpus callosum intact, it is found invariably that recognition in one hemisphere of the brain creates recognition also in the opposite hemisphere.

Thus, one of the functions of the corpus callosum and the anterior commissure is to make information stored in the cortex of one hemisphere available to cortical areas of the opposite hemisphere. Three important examples of such cooperation between the two hemispheres are the following:

1. Cutting of the corpus callosum blocks transfer of information from Wernicke's area of the dominant hemisphere to the motor cortex on the opposite side of the brain. Therefore, the intellectual functions of the brain, located primarily in the dominant hemisphere, lose their control over the right motor cortex and therefore also of the voluntary motor functions of the left hand and arm even though the usual subconscious movements of the left hand and arm are completely normal.

2. Cutting of the corpus callosum prevents transfer of somatic and visual information from the right hemisphere into Wernicke's area of the dominant hemisphere. Therefore, somatic and visual information from the left side of the body frequently fails to reach this general interpretative area of the brain and therefore cannot be used for decision-making.

3. Finally, persons whose corpus callosum is completely sectioned are found to have two entirely separate conscious portions of the brain. For example, in a recently studied teenage boy with a sectioned corpus callosum, only the left half of his brain could understand the spoken word, because it was the dominant hemisphere. On the other hand, the right side of the brain could understand the written word and could elicit a motor response to it without the left side of the brain ever knowing why the response was performed. Yet the effect was quite different when an emotional response was evoked in the right side of the brain: in this case a subconscious emotional response occurred in the left side of the brain as well. This undoubtedly occurred because the areas of the two sides of the brain for emotions, the anterior temporal cortices and adjacent areas, were still communicating with each other through the anterior commissure that was not sectioned. For instance, when the command "kiss" was written for the right half of his brain to see, the boy immediately and with full emotion, said "No way!" This response obviously required function of Wernicke's area and the motor areas for speech in the left hemispheres. But, when questioned why he said this, the boy could not explain. Thus, the two halves of the brain have independent capabilities for consciousness, memory storage, communication, and control of motor activities. The corpus callosum is required for the two sides to operate cooperatively, and the anterior commissure plays an important additional role in unifying the emotional responses of the two sides of the brain.

■ THOUGHTS, CONSCIOUSNESS, AND MEMORY

Our most difficult problem in discussing consciousness, thoughts, memory, and learning is that we do not know the neural mechanism of a thought. We know that destruction of large portions of the cerebral cortex does not prevent a person from having thoughts, but it usually does reduce the *degree* of awareness of the surroundings.

Each thought almost certainly involves simultaneous signals in many portions of the cerebral cortex, thalamus, limbic system, and reticular formation of the brain stem. Some crude thoughts probably depend almost entirely on lower centers; the thought of pain is probably a good example, for electrical stimulation of the human cortex rarely elicits anything more than the mildest degree of pain, whereas stimulation of certain areas of the hypothalamus and mesencephalon often causes excruciating pain. On the other hand, a type of thought pattern that requires mainly the cerebral cortex is that of vision, because loss of the visual cortex causes complete inability to perceive visual form or color.

Therefore, we might formulate a definition of a thought in terms of neural activity as follows: A thought results from the "pattern" of stimulation of many different parts of the nervous system at the same time and in definite sequence, probably involving most importantly the cerebral cortex, the thalamus, the limbic system, and the upper reticular formation of the brain stem. This is called the *holistic theory* of thoughts. The stimulated areas of the limbic system, thalamus, and reticular formation are believed to determine the general nature of the thought, giving it such qualities as pleasure, displeasure, pain, comfort, crude modalities of sensation, localization to gross areas of the body, and other general characteristics. On the other hand, the stimulated areas of the cerebral cortex determine the discrete characteristics of the thought such as specific localization of sensations on the body and of objects in the fields of vision, discrete patterns of sensation such as the rectangular pattern of a concrete block wall or the texture of a rug, and other individual characteristics that enter into the overall awareness of a particular instant.

Consciousness can perhaps be described as our continuing stream of awareness of either our surroundings or our sequential thoughts.

MEMORY — ROLES OF SYNAPTIC FACILITATION AND SYNAPTIC INHIBITION

Physiologically, memories are caused by changes in the capability of synaptic transmission from one neuron to the next as a result of previous neural activity. These changes in turn cause new pathways to develop for transmission of signals through the neural circuits of the brain. The new pathways are called *memory traces.* They are important because, once established, they can be activated by the thinking mind to reproduce the memories.

Experiments in lower animals have demonstrated that memory traces can occur at all levels of the nervous system. Even spinal cord reflexes can change at least slightly in response to repetitive cord activation, which is part of the memory process. Also, even some long-term memories result from changed synaptic conduction in the lower brain centers. To give an example, the blinking reflex is a learned function involving neuronal circuits in the cerebellum.

Yet there is much reason to believe that most of the memory that we associate with intellectual processes is based on memory traces mainly in the cerebral cortex.

Positive and Negative Memory — "Sensitization" or "Habituation" of Synaptic Transmission. Although we often think of memories as being positive recollections of previous thoughts or experiences, probably the greater share of our memories is negative memories, not positive. That is, our brain is inundated with sensory information from all of our senses. If our minds attempted to remember all of this information, the memory capacity of the brain would be exceeded within minutes. Fortunately, though, the brain has the peculiar capacity to learn to ignore information that is of no consequence. This results from *inhibition* of the synaptic pathways for this type of information, and the resulting effect is called *habituation.* This is, in a sense, a type of negative memory.

On the other hand, for those types of incoming information that cause important consequences, such as pain or pleasure, the brain also has the automatic capability of enhancing and storing the memory traces. Obviously, this is positive memory. It results from *facilitation* of the synaptic pathways, and the process is called *memory sensitization.* We learn later that special areas in the basal limbic regions of the brain determine whether information is important or unimportant and make the subconscious decision whether to store the thought as an enhanced memory trace or to suppress it.

Classification of Memories. We all know that some memories last only a few seconds, and others hours, days, months, or years. For the purpose of discussing these, let us use a common classification of memories that divides memories into (1) *immediate memory,* which includes memories that last for seconds or at most minutes unless they are converted into short-term memories; (2) *short-term memories,* which last for days to weeks but eventually are lost; and (3) *long-term memory,* which, once stored, can be recalled up to years or even a lifetime later.

IMMEDIATE MEMORY

Immediate memory is typified by one's memory of up to 7 to 10 telephone numbers at a time (or other discrete facts) for a few seconds to a few minutes at a time, but lasting only so long as the person continues to think about the numbers or facts.

Many physiologists have suggested that immediate memory is caused by continual neural activity resulting from nerve signals that travel around and around in a temporary memory trace through a *circuit of reverberating neurons.* Unfortunately, it has not yet been possible to prove this theory.

Another possible explanation of immediate memory is *presynaptic facilitation or inhibition.* This occurs at synapses that lie on presynaptic terminals, not on the subsequent neuron. The neurotransmitters secreted at such terminals frequently cause prolonged facilitation or inhibition (depending upon the type of transmitter secreted) for as long as seconds or even several minutes. Obviously, circuits of this type could lead to immediate memory.

A final possibility for explaining immediate memory is synaptic potentiation, which can enhance synaptic conduction. It can result from the accumulation of large amounts of calcium ions in the presynaptic terminals. That is, when a train of impulses passes through a presynaptic terminal, the amount of calcium ions increases with each impulse. When the amount of calcium ions becomes greater than the mitochondria and endoplasmic reticulum can absorb, the excess calcium then causes prolonged release of transmitter substance at the synapse. Thus, this, too, could be a mechanism for immediate memory.

SHORT-TERM MEMORY

Now we come to short-term memories that may last for many minutes or even weeks. Yet these will eventually be lost unless the memory traces become more permanent; they are then classified as long-term memories. Recent experiments in primitive animals have demonstrated that memories of this type can result from temporary chemical or physical changes, or both, either in the presynaptic terminals or in the postsynaptic membrane, changes that can persist for up to several weeks. These mechanisms are so important that they deserve special description.

Memory Based on Chemical and Physical Changes in the Presynaptic Terminal or Postsynaptic Neuronal Membrane

Figure 19–9 illustrates a mechanism of memory studied especially by Kandel and his colleagues that can cause memories lasting for up to 3 weeks in the large snail *Aplysia.* In this figure there are two separate presynaptic terminals. One terminal is from a primary input sensory neuron and terminates on the surface of the neuron that is to be stimulated; this is called the *sensory terminal.* The other terminal lies on the surface of the sensory terminal and is called the *facilitator terminal.* When the sensory terminal is stimulated repeatedly but without stimulating the facilitator terminal, signal transmission at first is very great, but this becomes less and less intense with repeated stimulation

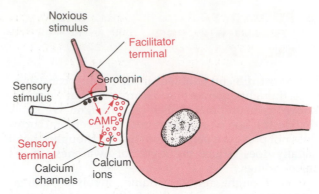

Figure 19–9. A memory system that has been discovered in the snail *Aplysia.*

until transmission almost ceases. This phenomenon is called *habituation.* It is a type of memory that causes the neuronal circuit to lose its response to repeated events that are insignificant.

On the other hand, if a noxious stimulus excites the facilitator terminal at the same time that the sensory terminal is stimulated, then, instead of the transmitted signal becoming progressively weaker, the ease of transmission becomes much stronger and will remain strong for hours, days, or, with more intense training, up to about 3 weeks even without further stimulation of the facilitator terminal. Thus, the noxious stimulus causes the memory pathway to become facilitated for days or weeks thereafter. It is especially interesting that once habituation has occurred, the pathway can be converted to a facilitated pathway with only a few noxious stimuli.

At the molecular level, the habituation effect in the sensory terminal results from progressive closure of calcium channels of the terminal membrane, although the cause of this is not fully known. Nevertheless, much smaller than normal amounts of calcium can then diffuse into this terminal when action potentials occur, and much less transmitter is therefore released because calcium entry is the stimulus for transmitter release (as was discussed in Chapter 7).

In the case of facilitation, the molecular mechanism is believed to be the following:

1. Stimulation of the facilitator neuron at the same time that the sensory neuron is stimulated causes serotonin release at the facilitator synapse on the sensory presynaptic terminal.

2. The serotonin acts on *serotonin receptors* in the sensory terminal membrane, and these activate the enzyme *adenylate cyclase* inside the membrane. This causes the formation of *cyclic adenosine monophosphate (cAMP)* inside the sensory presynaptic terminal.

3. The cAMP activates a *protein kinase* that causes phosphorylation of a protein that is part of the potassium channels in the sensory terminal membrane. This blocks these channels for potassium conductance. This blockage of the potassium channels can last for minutes up to several weeks.

4. Lack of potassium conductance causes a greatly

prolonged action potential in the presynaptic terminal because the flow of potassium ions out of the terminal is necessary for recovery from the action potential.

5. The prolonged action potential causes prolonged activation of the calcium pores, allowing tremendous quantities of calcium ions to enter the sensory terminal. These calcium ions then cause greatly increased transmitter release, thereby greatly facilitating synaptic transmission.

Thus, in a very indirect way the associative effect of stimulating the facilitator neuron at the same time that the sensory neuron is stimulated causes a prolonged change in the sensory terminal that produces the memory trace.

In addition, recent studies by Byrne and his colleagues, also in the snail *Aplysia,* have suggested still another mechanism of cellular memory; their studies have shown that stimuli from two separate sources acting on a single neuron can, under appropriate conditions, cause long-term changes in the membrane properties of the entire postsynaptic neuron. Thus, this is another possible mechanism of short-term memory.

LONG-TERM MEMORY

There is no real demarcation between the more prolonged types of short-term memory and long-term memory. The distinction is one of degree. However, long-term memory is generally believed to result from actual *structural changes* at the synapses that enhance or suppress signal conduction. Again, let us recall experiments in primitive animals (where the nervous systems are much easier to study) that have aided immensely in understanding possible mechanisms of long-term memory.

Structural and Other Physical Changes in Synapses During the Development of Long-Term Memory

If the reader will refer to Figure 25 – 2 in Chapter 25, he or she will see that the vesicles in a presynaptic terminal release their transmitter substance into the synaptic cleft through a special release site. When extra amounts of calcium enter the terminal, those vesicles near the release site attach to receptors at the site; then it is this attachment that causes vesicular exocytosis of the transmitter substance into the synaptic cleft.

Electron-microscopic pictures in invertebrate animals have now demonstrated that the total area of this vesicular release site increases in the presynaptic terminal during the development of long-term memory traces. Conversely, during long periods of synaptic inactivity, the release site diminishes and may actually disappear. Furthermore, the growth of the site depends upon activation of specific genetic control mechanisms for synthesizing proteins that are required for assembling the release structures.

An intriguing characteristic of this mechanism for learning is that increased areas of vesicular release sites

can be seen experimentally within hours after initiating the training sessions. Thus, it is entirely possible that much of what we now consider to be the longer types of short-term memory are actually early stages of this purely anatomically based long-term memory.

Therefore, at least in these primitive animals, at last we are beginning to understand a physical, structural basis for the development of long-term memory.

Other Possible Physical, Anatomical Mechanisms for Long-Term Memory

Aside from increasing the physical capability for release of neurotransmitter from the presynaptic terminals, memory development is also associated with an increase in the number of transmitter vesicles in the presynaptic terminals. And in some instances, the number of terminals themselves increases. Both of these effects could account for enhanced signal transmission. In fact, as a child grows and learns, the number of synapses in the brain increases greatly. Conversely, if a newborn animal is prevented from seeing with one of its eyes, those stripes of the visual cortex that are normally connected to the blinded eye fail to develop the same profusion of synapses as occurs for the other eye. Furthermore, taking the cover off the blinded eye reveals that the eye has not learned to see and has even lost much of its future capacity for learning to see.

Finally, in addition to changed synaptic conduction as a basis for learning, there is also the possibility of changing numbers of neurons in the used circuits. However, in this case, the process seems to be one of negative selection, for the brain has its greatest peak number of neurons at the time of birth or shortly after birth. Then, during the subsequent period of most rapid learning, those neurons that are excited appear to flourish; whereas those that remain unexcited may disappear altogether.

Therefore, it is likely that the human brain utilizes several different methods for enhancing or suppressing neuronal transmission when establishing memories.

CONSOLIDATION OF MEMORY

For an immediate memory to be converted into either a more prolonged short-term memory or a long-term memory that can be recalled weeks or years later, it must become "consolidated." That is, the memory must in some way initiate the chemical, physical, and anatomical changes in the synapses that are responsible for the long-term type of memory. This process requires 5 to 10 min for minimal consolidation and an hour or more for maximal consolidation. For instance, if a strong sensory impression is made on the brain but is then followed within a minute or so by an electrically induced brain convulsion, the sensory experience will not be remembered at all. Likewise, brain concussion, sudden application of deep general anesthesia, or any other effect that temporarily blocks the dynamic function of the brain can prevent consolidation.

However, if the strong electrical shock is delayed for more than 5 to 10 min, at least part of the memory trace will have become established. If the shock is delayed for an hour, the memory will have become even much more fully consolidated.

The process of consolidation and the time required for consolidation can probably be explained by the phenomenon of *rehearsal* of the immediate memory as follows:

Role of Rehearsal in Transference of Immediate Memory into Longer Memory. Psychological studies have shown that rehearsal of the same information again and again accelerates and potentiates the degree of transfer of immediate memory into longer-term memory, and therefore also accelerates and potentiates the process of consolidation. The brain has a natural tendency to rehearse newfound information, and especially to rehearse newfound information that catches the mind's attention. Therefore, over a period of time the important features of sensory experiences become progressively more and more fixed in the secondary memory stores. This explains why a person can remember small amounts of information studied in depth far better than large amounts of information studied only superficially. It also explains why a person who is wide awake will consolidate memories far better than a person who is in a state of mental fatigue.

Codifying of Memories During the Process of Consolidation. One of the most important features of the process of consolidation is that memories placed permanently into the longer-term memory storehouse are codified into different classes of information. During this process similar information is recalled from the memory storage bins and is used to help process the new information. The new and old are compared for similarities and for differences, and part of the storage process is to store the information about these similarities and differences, rather than simply to store the information unprocessed. Thus, during the process of consolidation, the new memories are not stored randomly in the brain, but instead are stored in direct association with other memories of the same type. This is obviously necessary if one is to be able to "search" the memory store at a later date to find the required information.

Role of Specific Parts of the Brain in the Memory Process

Role of the Hippocampus for Storage of Memories — Anterograde Amnesia Following Hippocampal Lesions. The hippocampus is the most medial portion of the temporal lobe cortex where it folds underneath the brain and then upward into the lower surface of the lateral ventricle. The two hippocampi have been removed for the treatment of epilepsy in a number of patients. This procedure does not seriously affect the person's memory for information stored in the brain prior to removal of the hippocampi. However, after removal, these persons have very little capacity for storing *verbal and symbolic types* of memories in long-term memory, or even in short-term memory lasting longer than a few minutes. Therefore, these persons are unable to establish new long-term memories of those types of information that are the basis of intelligence. This is called *anterograde amnesia.*

But why is the hippocampus so important in helping the brain to store new memories? The probable answer is that the hippocampus is one of the important output pathways from the "reward" and "punishment" areas of the limbic system. These areas are found in many of the basal regions of the brain, and they feed into the hippocampus. Those sensory stimuli, or even thoughts, that cause pain or aversion excite the *punishment centers*, whereas those stimuli that cause pleasure, happiness, or a sense of reward excite the *reward centers*. All of these together provide the background mood and motivations of the person. Among these motivations is the drive in the brain to remember those experiences and thoughts that are either pleasant or unpleasant. The hippocampus especially and to a lesser degree the dorsal medial nuclei of the thalamus, another limbic structure, have proved especially important in making a decision about which of our thoughts are important enough on a basis of reward or punishment to be worthy of memory.

Lesions in other parts of the temporal lobes besides the hippocampi, especially of the amygdalas, are also frequently associated with reduced ability to store new memories. This probably results from two factors: (1) the association of the other parts of the temporal lobes with the hippocampi and therefore failure of the usual consolidation process for memories, and (2) the fact that Wernicke's area, which is the major locus of intellectual operations of the brain, is located in the temporal lobe. The reason that lesions affecting Wernicke's area might diminish memory storage is probably that consolidation of memories requires analysis of the memory so that it can be stored in association with other memories of like kind.

Retrograde Amnesia. *Retrograde amnesia* means inability to recall memories from the past — that is, from the long-term memory storage bins — even though the memories are known to be still there. When retrograde amnesia occurs, the degree of amnesia for recent events is likely to be much greater than for events of the distant past. The reason for this difference is probably that the distant memories have been rehearsed so many times that the memory traces are deeply engrained so that elements of these memories are stored in widespread areas of the brain.

In some persons who have hippocampal lesions, some degree of retrograde amnesia occurs along with anterograde amnesia just discussed, which suggests that these two types of amnesia are at least partially related and that hippocampal lesions can cause both. However, it has also been claimed that damage in some thalamic areas can lead specifically to retrograde amnesia without causing significant anterograde amnesia. A possible explanation of this is that the thalamus might play a role in helping the person "search" the memory storehouses and thus "read out" the memories. That is, the memory process requires not only the storing of memories but also the ability to search and find the memory at a later date. The possible function of the thalamus in this process is discussed in the following chapter.

Lack of Importance of the Hippocampi in Reflexive Learning. It should be noted, however, that persons with either temporal lobe or hippocampal lesions usually do not have difficulty in learning physical skills that do not involve verbalization or symbolic types of intelligence. For instance, these persons can still learn hand and physical skills such as those required in many types of sports. This type of learning is called *reflexive learning;* it depends on physically repeating the required tasks over and over again, rather than on symbolic rehearsing in the mind.

REFERENCES

Avoli, M., et al. (eds.): Neurotransmitters and Cortical Function. New York, Plenum Publishing Corp., 1988.

Baddeley, A.: Working Memory. New York, Oxford University Press, 1987.

Bear, M. F., et al.: A physiological basis for a theory of synapse modification. Science, 237:42, 1987.

Benson, D. F.: Aphasia, Alexia, and Agraphia. New York, Churchill Livingstone, 1979.

Brown, T. H., et al.: Long-term synaptic potentiation. Science, 242:724, 1988.

Byrne, J. H.: Can learning and memory be understood? News Physiol. Sci., 1:182, 1986.

Byrne, J. H.: Cellular analysis of associative learning. Physiol. Rev., 67:329, 1987.

Cotman, C. W., et al.: The role of the NMDA receptor in central nervous system plasticity and pathology. J. NIH Res., 1:65, 1989.

Damasio, A. R., and Geschwind, N.: The neural basis of language. Annu. Rev. Neurosci., 7:127, 1984.

De Valois, R. L., and De Valois, K. K.: Spatial Vision. New York, Oxford University Press, 1988.

De Wied, D.: Neuroendocrine aspects of learning and memory processes. News Physiol. Sci., 4:32, 1989.

DeFelipe, J., and Jones, E. G. (eds.): Cajal on the Cerebral Cortex. New York, Oxford University Press, 1988.

Geschwind, N.: Specializations of the human brain. Sci. Am., 241(3):180, 1979.

Gold, P. E.: Sweet memories. Am. Sci., 75:151, 1987.

Goldman-Rakic, P. S.: Topography of cognition: Parallel distributed networks in primate association cortex. Annu. Rev. Neurosci., 11:137, 1988.

Gregory, R. L. (ed.): The Oxford Companion to the Mind. Oxford University Press, 1987.

Heppenheimer, T. A.: Nerves of silicon. Discover, February 1988, p. 70.

Hixon, T. J., et al. (eds.): Introduction to Communicative Disorders. Englewood Cliffs, N. J., Prentice-Hall, 1980.

Hubel, D. H.: The brain. Sci. Am., 241(3):44, 1979.

Hundert, E. M.: Philosophy, Psychiatry and Neuroscience—Three Approaches to the Mind. New York, Oxford University Press, 1989.

Hyvarinen, J.: Posterior parietal lobe of the primate brain. Physiol. Rev., 62:1060, 1982.

Ito, M.: Long-term depression. Annu. Rev. Neurosci., 12:85, 1989.

John, E. R., et al.: Double-labeled metabolic maps of memory. Science, 233:1167, 1986.

Kandel, E. R.: A Cell-Biological Approach to Learning. Bethesda, Md., Society for Neuroscience, 1977, p. 1137.

Kandel, E. R.: Neuronal plasticity and the modification of behavior. In Brookhart, J. M., and Mountcastle, V. B. (eds.): Handbook of Physiology. Sec. 1, Vol. I. Baltimore, Williams & Wilkins, 1977, p. 1137.

Kolata, G.: Associations or rules in learning language? Science, 237:133, 1987.

Kosslyn, S. M.: Aspects of a cognitive neuroscience of mental imagery. Science, 240:1621, 1988.

Laduron, P. M.: Presynaptic heteroreceptors in regulation of neuronal transmission. Biochem. Pharmacol., 34:467, 1985.

Lieke, E. E., et al.: Optical imaging of cortical activity. Annu. Rev. Physiol., 51:543, 1989.

McCloskey, D. I.: Corollary discharges: Motor commands and perception. In Brooks, V. B. (ed.): Handbook of Physiology. Sec. 1, Vol. II. Bethesda, Md., American Physiological Society, 1981, p. 1415.

McNeil, M. R. (ed.): The Dysarthrias. Physiology, Acoustics, and Perception Management. San Diego, College-Hill Press, 1984.

Mitzdorf, U.: Current source-density method and application in cat cerebral cortex: Investigation of evoked potentials and EEG phenomena. Physiol. Rev., 65:37, 1985.

Peters, A., and Jones, E. G. (eds.): Development and Maturation of Cerebral Cortex. New York, Plenum Publishing Corp., 1988.

Plum, F. (ed.): Language, Communication, and the Brain. New York, Raven Press, 1988.

Quinn, W. G., and Greenspan, R. J.: Learning and courtship in Drosophila: Two stories with mutants. Annu. Rev. Neurosci., 7:67, 1984.

Rakic, P.: Specification of cerebral cortical areas. Science, 241:170, 1988.

Rosenbek, J. C. (ed.): Apraxia of Speech. Physiology, Acoustics, Linguistics, Management. San Diego, College-Hill Press, 1984.

Squire, L. R.: Mechanisms of memory. Science, 232:1612, 1986.

Squire, L. R.: Memory and Brain. New York, Oxford University Press, 1987.

Sutcliffe, J. G.: mRNA in the mammalian central nervous system. Annu. Rev. Neurosci., 11:157, 1988.

Thompson, R. F., et al.: Cellular processes of learning and memory in the mammalian CNS. Annu. Rev. Neurosci., 6:447, 1983.

Trevarthen, C.: Hemispheric specialization. In Darian-Smith, I. (ed.): Handbook of Physiology. Sec. 1, Vol. III. Bethesda, Md., American Physiological Society, 1984, p. 1129.

Truman, J. W.: Cell death in invertebrate nervous systems. Annu. Rev. Neurosci., 7:171, 1984.

Waldrop, M. M.: Soar: A unified theory of cognition? Science, 241:296, 1988.

Walters, E. T., and Byrne, J. H.: Associative conditioning of single sensory neurons suggest a cellular mechanism for learning. Science, 219:405, 1983.

Weiskrantz, L. (ed.): Thought Without Language. New York, Oxford University Press, 1988.

Wong, R. K., et al.: Local circuit interactions in synchronization of cortical neurones. J. Exp. Biol., 112:169, 1984.

Woody, C. D., et al. (eds.): Cellular Mechanisms of Conditioning and Behavioral Plasticity. New York, Plenum Publishing Corp., 1988.

Zucker, R. S.: Short-term synaptic plasticity. Annu. Rev. Neurosci., 12:13, 1989.

20

Behavioral and Motivational Mechanisms of the Brain— The Limbic System and the Hypothalamus

The control of behavior is a function of the entire nervous system. Even the discrete cord reflexes are an element of behavior, and the wakefulness and sleep cycle discussed in the following chapter is certainly one of the most important of our behavioral patterns. However, in this chapter we deal first with those mechanisms that control the levels of activity in the different parts of the brain. Then we discuss the bases of motivational drives, especially the motivational control of the learning process and the feelings of pleasure or punishment. These functions of the nervous system are performed mainly by the basal regions of the brain, which together are loosely called the *limbic*, meaning "border," *system*.

■ THE ACTIVATING-DRIVING SYSTEMS OF THE BRAIN

In the absence of continuous transmission of nerve signals from the brain stem into the cerebrum, the brain becomes useless. In fact, severe compression of the brain stem at the juncture between the mesencephalon and cerebrum, often resulting from a pineal tumor, usually causes the person to go into unremitting coma lasting for the remainder of the person's life.

Nerve signals in the brain stem activate the cerebral part of the brain in two different ways: (1) by directly stimulating the background level of activity in wide areas of the brain and (2) by activating neurohormonal systems that release specific facilitory or inhibitory hormonal substances into selected areas of the brain. These two activating systems always function together and cannot be distinguished entirely from each other; nevertheless, let us discuss each as a separate entity.

CONTROL OF CEREBRAL ACTIVITY BY CONTINUOUS EXCITATORY SIGNALS FROM THE BRAIN STEM

The Reticular Excitatory Area of the Brain Stem

Figure 20–1 illustrates a general system for controlling the level of activity of the brain. The central driving component of this system is an excitatory area called the *bulboreticular facilitory area.* This lies in the reticular substance of the middle and lateral pons and mesencephalon. Actually, we have already discussed this area in Chapter 17, for it is the same brain stem reticular area that transmits facilitory signals downward to the spinal cord to maintain tone in the antigravity muscles and also to control the level of activity of the spinal cord reflexes. In addition to these downward signals, this area sends a profusion of signals in the upward direction as well. Most of these synapse in the thalamus and are distributed from there to all regions of the cerebral cortex, although others go to most of the other subcortical structures besides the thalamus as well.

The signals passing through the thalamus are of two types. One type is rapidly transmitted action potentials that excite the cerebrum for only a few milliseconds. These originate from very large neuronal cell bodies that lie throughout the reticular area. Their nerve endings release the neurotransmitter substance *acetylcholine,* which serves as the excitatory agent, lasting for only a few milliseconds before it is destroyed.

The second type of excitatory signal originates from large numbers of very small neurons spread throughout the reticular excitatory area. Again, most of these pass to the thalamus, but this time through small, very slowly conducting fibers, and synapsing mainly in the intralaminar nuclei of the thalamus and reticular nu-

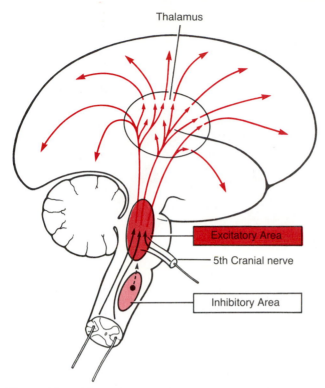

Thalamus

Excitatory Area

5th Cranial nerve

Inhibitory Area

Figure 20 – 1. The excitatory-activating system of the brain. Also shown is an inhibitory area in the medulla that can inhibit or depress the activating system.

clei over the surface of the thalamus. From here, additional very small fibers are distributed everywhere in the cerebral cortex. The excitatory effect caused by this system of fibers can build up progressively for many seconds to a minute or more, which suggests that its signals are especially important for controlling the longer-term background excitability level of the brain.

Excitation of the Brain Stem Excitatory Area by Peripheral Sensory Signals. The level of activity of the brain stem excitatory area, and therefore the level of activity of the entire brain, is determined to a great extent by the sensory signals that enter the excitatory area from the periphery. Pain signals, in particular, increase the activity in this area and therefore strongly excite the brain to attention.

The importance of sensory signals in activating the excitatory area is illustrated by the effect of cutting the brain stem above the point where the bilateral fifth nerves enter the pons. These are the highest nerves that transmit significant numbers of somatosensory signals into the brain. When all of these signals are gone, the level of activity in the excitatory area diminishes abruptly, and the brain proceeds instantly to greatly reduced activity, actually approaching a permanent state of coma. Yet when the brain stem is transected below the fifth nerves, which leaves much input of the sensory signals from the facial and oral regions, the coma is averted.

Increased Activity of the Brain Stem Excitatory Area Caused by Feedback Signals From the Cerebrum. Not only do excitatory signals pass to the cere-

brum from the bulboreticular excitatory area of the brain stem, but signals in turn return from the cerebrum back to the bulbar regions. Therefore, any time the cerebral cortex becomes activated by either thinking or motor processes, reverse signals are sent back to the brain stem excitatory areas; this obviously helps to maintain the level of excitation of the cerebral cortex or even to enhance it. Thus, this is a general mechanism of *positive feedback* that allows any beginning activity in the cerebrum to support still more activity, thus leading to an awake mind.

The Thalamus Is a Distribution Center That Controls Activity in Specific Regions of the Cortex. It was pointed out in the previous chapter, and illustrated in Figure 19 – 2, that almost every area of the cerebral cortex connects with its own highly specific area in the thalamus. Therefore, electrical stimulation of a specific point in the thalamus will activate a specific small region of the cortex. Furthermore, signals regularly reverberate back and forth between the thalamus and the cerebral cortex, the thalamus exciting the cortex and the cortex then re-exciting the thalamus by way of return fibers. It has been suggested that the part of the thinking process that helps us to establish long-term memories might result from just such back-and-forth reverberation of signals.

However, can the thalamus also function internally within the brain to call forth specific memories or to activate specific thought processes? The answer to this is not known. Yet, certainly, the thalamus does have the appropriate neuronal circuitry to do this.

An Inhibitory Reticular Area Located in the Lower Brain Stem

Figure 20 – 1 illustrates still another area that is important in controlling brain activity. This is the *reticular inhibitory area* located medially and ventrally in the medulla. In Chapter 17 we saw that this area can inhibit the reticular facilitory area of the upper brain stem and thereby reduce the tonic nerve signals transmitted through the spinal cord to the antigravity muscles. Likewise, this same inhibitory area, when excited, will decrease activity in the superior portions of the brain as well. One of the mechanisms that it uses is to excite serotonergic neurons; these, in turn, secrete the inhibitory neurohormone serotonin at crucial points in the brain; we discuss this in more detail later.

NEUROHORMONAL CONTROL OF BRAIN ACTIVITY

Aside from direct control of brain activity by specific transmission of nerve signals from the lower brain areas to the cortical regions of the brain, still another method is also used to control brain activity. This is the release of excitatory or inhibitory neurotransmitter hormonal agents into the substance of the brain. These neurohormones often persist for minutes or even hours and thereby provide long periods of control rather than instantaneous activation or inhibition.

Figure 20–2 illustrates three neurohormonal systems that have been mapped in detail in the rat brain, a *norepinephrine system*, a *dopamine system*, and a *serotonin system*. Usually norepinephrine functions as an excitatory hormone, while serotonin is usually inhibitory, and dopamine is excitatory in some areas but inhibitory in others. Therefore, as would be expected, these three different systems have different effects on the levels of excitability in different parts of the brain. The norepinephrine system spreads to virtually every area of the brain, whereas the serotonin and dopamine systems are directed to much more specific brain regions, the dopamine system mainly into the basal ganglial regions and the serotonin system more into the midline structures.

Neurohormonal Systems in the Human Brain. Figure 20–3 illustrates the brain stem areas in the human brain for activating four different neurohormonal systems, the same three discussed above for the

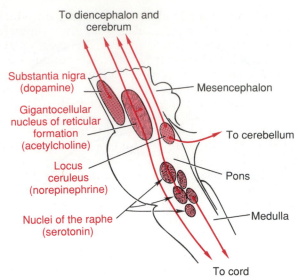

Figure 20–3. Multiple centers in the brain stem, the neurons of which secrete different transmitter substances. These neurons send control signals upward into the diencephalon and cerebrum and downward into the spinal cord.

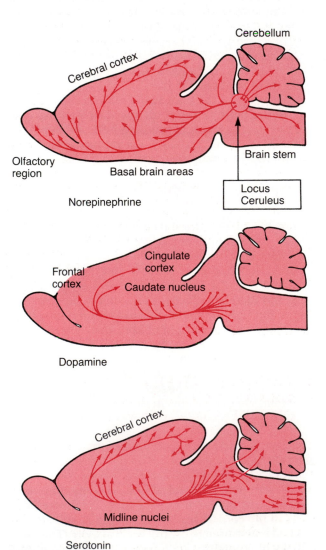

Figure 20–2. Three neurohormonal systems that have been mapped in the rat brain: a *norepinephrine system*, a *dopamine system*, and a *serotonin system*. (Adapted from Kelly, J. P. (after Cooper, Bloom, and Roth) in Kandel and Schwartz: Principles of Neural Science, 2nd Ed. New York, Elsevier, 1985.)

rat and one other, the *acetylcholine system*. Some of the specific functions of these are as follows:

1. *The locus ceruleus and the norepinephrine system.* The *locus ceruleus* is a small area located bilaterally and posteriorly at the juncture between the pons and the mesencephalon. Nerve fibers from this area spread throughout the brain, the same as illustrated for the rat in the top frame of Figure 20–2, and they secrete *norepinephrine*. The norepinephrine excites the brain to generalized increased activity. However, it has inhibitory effects in a few areas because of inhibitory receptors at certain neuronal synapses. In the following chapter we see that this system probably plays a very important role in causing a dreaming type of sleep called REM sleep.

2. *The substantia nigra and the dopamine system.* The *substantia nigra* is discussed in Chapter 18 in relation to the basal ganglia. It lies anteriorly in the superior mesencephalon, and its neurons send nerve endings mainly to the caudate nucleus and putamen, where they secrete *dopamine*. Other neurons located in adjacent regions also secrete dopamine, but these send their endings into the ventral areas of the cerebrum, especially to the hypothalamus and the limbic system. The dopamine is believed to act as an inhibitory transmitter in the basal ganglia, but in some of the other areas of the brain it is possibly excitatory. Also, remember from Chapter 18 that destruction of the dopaminergic neurons in the substantia nigra is the basic cause of Parkinson's disease.

3. *The raphe nuclei and the serotonin system.* In the midline of the lower pons and medulla are several very thin nuclei called the *raphe nuclei*. Many of the neurons in these nuclei secrete *serotonin*. They send many fibers into the diencephalon and fewer fibers to the cerebral cortex; still many others descend to the spinal cord.

The cord fibers have the ability to suppress pain, which is discussed in Chapter 10. The serotonin released in the diencephalon and cerebrum almost certainly plays an essential inhibitory role to help cause normal sleep, as we discuss in the following chapter.

4. *The gigantocellular neurons of the reticular excitatory area and the acetylcholine system.* Earlier, we discussed the gigantocellular neurons (the *giant cells*) in the reticular excitatory area of the pons and mesencephalon. The fibers from these large cells divide immediately into two branches, one passing upward to the higher levels of the brain and the other passing downward through the reticulospinal tracts into the spinal cord. The neurohormone secreted at their terminals is *acetylcholine.* In most places, the acetylcholine functions as an excitatory neurotransmitter at specific synapses.

Still other acetylcholine-secreting neurons are present in some regions of the diencephalon; some psychiatric disorders of the brain have been found to be associated with decreased function or even destruction of some of these neurons.

Other Neurotransmitters and Neurohormonal Substances Secreted in the Brain. Without describing their function, the following is a list of still other neurohormonal substances that, among others, function either at synapses or by release into the fluids of the brain: enkephalins, gamma-aminobutyric acid (GABA), glutamate, vasopressin, adrenocorticotropic hormone, epinephrine, endorphins, angiotensin II, neurotensin.

Thus, there are multiple neurohormonal systems in the brain, the activation of each of which plays its own role in controlling a different quality of brain function.

■ THE LIMBIC SYSTEM

The word "limbic" means "border." Originally, the term "limbic" was used to describe the border structures around the basal regions of the cerebrum, but as we have learned more about the functions of the limbic system, the term *"limbic system"* has been expanded to mean the entire neuronal circuitry that controls emotional behavior and motivational drives.

A major part of the limbic system is the *hypothalamus,* with its related structures. In addition to their roles in behavioral control, these areas also control many internal conditions of the body as well, such as body temperature, osmolality of the body fluids, the drive to eat and drink and control body weight, and so forth. These internal functions are collectively called *vegetative functions* of the brain, and their control is obviously closely related to behavior.

■ FUNCTIONAL ANATOMY OF THE LIMBIC SYSTEM; ITS RELATION ALSO TO THE HYPOTHALAMUS

Figure 20–4 illustrates the anatomical structures of the limbic system, showing these to be an interconnected complex of basal brain elements. Located in the midst of all these is the *hypothalamus,* which is considered by some anatomists to be a structure separate from the remainder of the limbic system but which, from a physiological point of view, is one of the central elements of the system. Figure 20–5 illustrates schematically this key position of the hypothalamus in the limbic system and shows that surrounding it are the other subcortical structures of the limbic system, including the *septum,* the *paraolfactory area,* the *epithalamus,* the *anterior nucleus of the thalamus,* portions of the *basal ganglia,* the *hippocampus,* and the *amygdala.*

Surrounding the subcortical limbic areas is the *limbic cortex,* composed of a ring of cerebral cortex (1) beginning in the *orbitofrontal area* on the ventral surface of the frontal lobes, (2) extending upward in the *subcallosal gyrus* beneath the anterior limb of the corpus callosum, (3) over the top of the corpus callosum onto the medial aspect of the cerebral hemisphere in the *cingulate gyrus,* and finally (4) passing behind the corpus callosum and downward onto the ventromedial surface of the temporal lobe to the *parahippocampal gyrus* and *uncus.* Thus, on the medial and ventral surfaces of each cerebral hemisphere is a ring mostly of *paleocortex* that surrounds a group of deep structures intimately associated with overall behavior and with emotions. In turn, this ring of limbic cortex functions as a two-way communication and association linkage between the *neocortex* and the lower limbic structures.

It is also important to recognize that many of the behavioral functions elicited from the hypothalamus and other limbic structures are mediated through the reticular nuclei and their associated nuclei in the brain stem. It is pointed out in Chapter 17 and earlier in this chapter that stimulation of the excitatory portion of the reticular formation can cause high degrees of somatic and cerebral cortical excitability; in Chapter 22 we see that most of the hypothalamic signals for control of the autonomic nervous system also are transmitted through nuclei located in the brain stem.

A very important route of communication between the limbic system and the brain stem is the *medial forebrain bundle* that extends from the septal and orbitofrontal cortical regions downward through the middle of the hypothalamus to the brain stem reticular formation. This bundle carries fibers in both directions, forming a trunk line communication system. A second route of communication is through short pathways among the reticular formation of the brain stem, the thalamus, the hypothalamus, and most of the other contiguous areas of the basal brain.

■ THE HYPOTHALAMUS, A MAJOR OUTPUT PATHWAY OF THE LIMBIC SYSTEM

The hypothalamus has communicating pathways with all levels of the limbic system. In turn, it and its closely allied structures send output signals in three directions: (1) downward to the brain stem, mainly into the reticular areas of the mesencephalon, pons, and medulla; (2) upward toward many higher areas of the diencephalon and cerebrum, especially to the anterior thalamus and the limbic cortex; and (3) into the infundibulum to control most of the secretory functions of both the posterior and anterior pituitary glands.

Thus, the hypothalamus, which represents less than 1 per cent of the brain mass, nevertheless is one of the

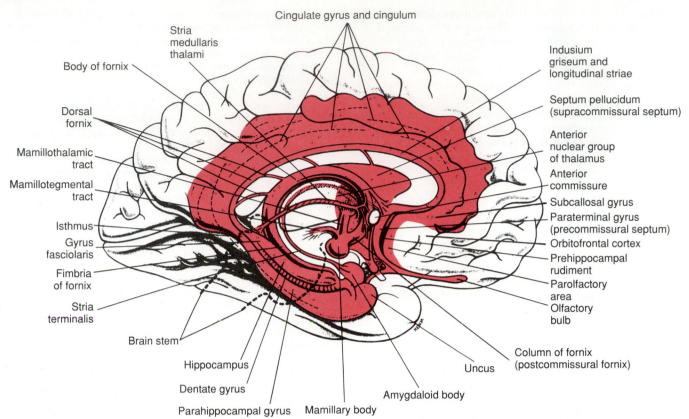

Figure 20 – 4. Anatomy of the limbic system illustrated by the shaded areas of the figure. (From Warwick and Williams: Gray's Anatomy. 35th Br. Ed. London, Longman Group, Ltd., 1973.)

most important of the motor output pathways of the limbic system. It controls most of the vegetative and endocrine functions of the body as well as many aspects of emotional behavior. Let us discuss first the vegetative and endocrine control functions and then return to the behavioral functions of the hypothalamus to see how all these operate together.

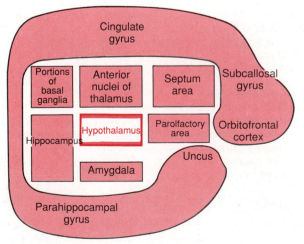

Figure 20 – 5. The limbic system.

VEGETATIVE AND ENDOCRINE CONTROL FUNCTIONS OF THE HYPOTHALAMUS

The different hypothalamic mechanisms for controlling the vegetative and endocrine functions of the body are discussed in many different chapters throughout this text. For instance, the role of the hypothalamus in arterial pressure regulation is discussed in Chapter 27, water conservation in Chapter 28, temperature regulation in Chapter 28, and endocrine control in Chapter 29. However, to illustrate the organization of the hypothalamus as a functional unit, let us summarize the more important of its vegetative and endocrine functions here as well.

Figures 20 – 6 and 20 – 7 show enlarged coronal and sagittal views of the hypothalamus, which represents only a small area in Figure 20 – 4. Please take a few minutes to study these diagrams, especially to read in Figure 20 – 6 the multiple activities that are excited or inhibited when respective hypothalamic nuclei are stimulated. In addition to those centers that are illustrated in Figure 20 – 6, a large *lateral hypothalamic* area overlies the illustrated areas on each side of the hypothalamus. The lateral areas are especially important in controlling thirst, hunger, and many of the emotional drives.

A word of caution must be issued for studying these diagrams, however, for the areas that cause specific

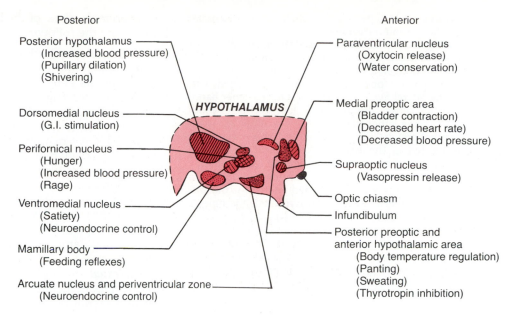

Posterior   Anterior

Posterior hypothalamus
(Increased blood pressure)
(Pupillary dilation)
(Shivering)

Paraventricular nucleus
(Oxytocin release)
(Water conservation)

HYPOTHALAMUS

Dorsomedial nucleus
(G.I. stimulation)

Medial preoptic area
(Bladder contraction)
(Decreased heart rate)
(Decreased blood pressure)

Perifornical nucleus
(Hunger)
(Increased blood pressure)
(Rage)

Supraoptic nucleus
(Vasopressin release)

Ventromedial nucleus
(Satiety)
(Neuroendocrine control)

Optic chiasm

Infundibulum

Mamillary body
(Feeding reflexes)

Posterior preoptic and
anterior hypothalamic area
(Body temperature regulation)
(Panting)
(Sweating)
(Thyrotropin inhibition)

Arcuate nucleus and periventricular zone
(Neuroendocrine control)

Figure 20–6. Control centers of the hypothalamus.

Lateral hypothalamic area (not shown)
(Thirst and hunger)

activities are not nearly so accurately localized as suggested in the figure. Also, it is not known whether the effects noted in the figure result from stimulation of specific control nuclei or whether they result merely from activation of fiber tracts leading from control nuclei located elsewhere. With this caution in mind, we can give the following general description of the vegetative and control functions of the hypothalamus.

Cardiovascular Regulation. Stimulation of different areas throughout the hypothalamus can cause every known type of neurogenic effect on the cardiovascular system, including increased arterial pressure, decreased arterial pressure, increased heart rate, and decreased heart rate. In general, stimulation in the *posterior* and *lateral hypothalamus* increases the arterial pressure and heart rate, whereas stimulation in the *preoptic area* often has opposite effects, causing a decrease in both heart rate and arterial pressure. These effects are transmitted mainly through the cardiovascular control centers in the reticular regions of the medulla and pons.

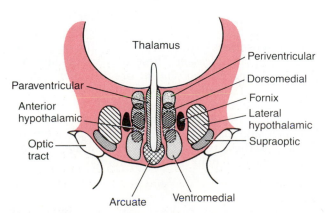

Thalamus

Periventricular

Paraventricular

Dorsomedial

Anterior
hypothalamic

Fornix

Lateral
hypothalamic

Optic
tract

Supraoptic

Arcuate   Ventromedial

Figure 20–7. Coronal view of the hypothalamus, showing the mediolateral positions of the respective hypothalamic nuclei.

Regulation of Body Temperature. The anterior portion of the hypothalamus, especially the *preoptic area*, is concerned with regulation of body temperature. An increase in the temperature of the blood flowing through this area increases the activity of temperature-sensitive neurons, whereas a decrease in temperature decreases their activity. In turn, these neurons control the mechanisms for increasing or decreasing body temperature, as discussed in Chapter 28.

Regulation of Body Water. The hypothalamus regulates body water in two separate ways: (1) by creating the sensation of thirst, which makes an animal drink water; and (2) by controlling the excretion of water into the urine. An area called the *thirst center* is located in the lateral hypothalamus. When the electrolytes inside the neurons either of this center or allied areas of the hypothalamus become too concentrated, the animal develops an intense desire to drink water; it will search out the nearest source of water and drink enough to return the electrolyte concentration of the thirst center neurons to normal.

Control of renal excretion of water is vested mainly in the *supraoptic* nucleus. When the body fluids become too concentrated, the neurons of this area become stimulated. The nerve fibers from these neurons project downward through the infundibulum into the posterior pituitary gland, where they secrete a hormone called *antidiuretic hormone* (also called *vasopressin*). This hormone is then absorbed into the blood and acts on the collecting ducts of the kidneys to cause massive reabsorption of water, thereby decreasing the loss of water into the urine.

Regulation of Uterine Contractility and of Milk Ejection by the Breasts. Stimulation of the *paraventricular nucleus* causes its neuronal cells to secrete the hormone *oxytocin*. This in turn causes increased contractility of the uterus and also contraction of the myoepithelial cells that surround the alveoli of the breasts, which then causes the alveoli to empty the milk through the nipples. At the end of pregnancy, especially large quantities of oxytocin are secreted, and this secretion helps to promote labor contractions that expel the baby. Also, when a baby suckles the mother's breast, a reflex signal from the nipple to the hypothalamus causes oxytocin release, and the oxytocin then performs the necessary function of expelling the milk through the nipples so that the baby can nourish itself.

Gastrointestinal and Feeding Regulation. Stimulation of several areas of the hypothalamus causes an animal to experience extreme hunger, a voracious appetite, and an intense desire to search for food. The area most associated with hunger is the *lateral hypothalamic area.* On the other hand, damage to this causes the animal to lose desire for food, sometimes causing lethal starvation.

A center that opposes the desire for food, called the *satiety center,* is located in the *ventromedial nucleus.* When this center is stimulated, an animal that is eating food suddenly stops eating and shows complete indifference to food. On the other hand, if this area is destroyed bilaterally, the animal cannot be satiated; instead, its hypothalamic hunger centers become overactive, so that it has a voracious appetite, resulting in tremendous obesity.

Another area closely associated with the hypothalamus that enters into the overall control of gastrointestinal activity is the *mamillary bodies;* these control the patterns of many feeding reflexes, such as licking the lips and swallowing.

Hypothalamic Control of the Anterior Pituitary Gland

Stimulation of certain areas of the hypothalamus also causes the *anterior* pituitary gland to secrete its hormones. This subject is discussed in detail in Chapter 29 in relation to the neural control of the endocrine glands. Briefly, the basic mechanisms are the following:

The anterior pituitary gland receives its blood supply mainly from venous blood that flows into the anterior pituitary sinuses after having passed first through the lower part of the hypothalamus. As the blood courses through the hypothalamus before reaching the anterior pituitary, *releasing* and *inhibitory hormones* are secreted into the blood by various hypothalamic nuclei. These hormones are then transported in the blood to the anterior pituitary, where they act on the glandular cells to control the release of the anterior pituitary hormones.

The cell bodies of the neurons that secrete these releasing hormones and inhibitory hormones are located mainly in the medial basal nuclei of the hypothalamus, especially in the *paraventricular zone,* the *arcuate nucleus,* and part of the *ventromedial nucleus.* However, the axons from these nuclei then project to the *median eminence,* which is an enlarged area of the infundibulum where it arises from the inferior border of the hypothalamus. It is here that the nerve terminals actually secrete their releasing and inhibitory hormones. These hormones are then absorbed into the blood capillaries in the median eminence and carried in the venous blood down along the infundibulum to the anterior pituitary gland.

Summary. A number of areas of the hypothalamus control specific vegetative functions. However, these areas are still poorly delimited, so much so that the above specification of different areas for different hypothalamic functions is still tentative.

BEHAVIORAL FUNCTIONS OF THE HYPOTHALAMUS AND ASSOCIATED LIMBIC STRUCTURES

Aside from the vegetative and endocrine functions of the hypothalamus, stimulation of or lesions in the hypothalamus often have profound effects on the emotional behavior of animals or human beings.

In animals, some of the behavioral effects of stimulation are the following:

1. Stimulation in the *lateral hypothalamus* not only causes thirst and eating as discussed above but also increases the general level of activity of the animal, sometimes leading to overt rage and fighting, as is discussed subsequently.

2. Stimulation in the *ventromedial nucleus* and surrounding areas mainly causes effects opposite to those caused by lateral hypothalamic stimulation — that is, a sense of *satiety, decreased eating,* and *tranquility.*

3. Stimulation of a *thin zone of the periventricular nucleus,* located immediately adjacent to the third ventricle (or also stimulation of the central gray area of the mesencephalon that is continuous with this portion of the hypothalamus) usually leads to *fear* and *punishment reactions.*

4. *Sexual drive* can be stimulated from several areas of the hypothalamus, especially the most anterior and most posterior portions of the hypothalamus.

Lesions in the hypothalamus, in general, cause the opposite effects. For instance:

1. Bilateral lesions in the lateral hypothalamus will decrease drinking and eating almost to zero, often leading to lethal starvation. These lesions cause extreme *passivity* of the animal as well, with loss of most of its overt drives.

2. Bilateral lesions of the ventromedial areas of the hypothalamus cause effects that are mainly opposite to those caused by lesions of the lateral hypothalamus: excessive drinking and eating as well as hyperactivity and often continuous savagery along with frequent bouts of extreme rage on the slightest provocation.

Stimulation or lesions in other regions of the limbic system, especially the amygdala, the septal area, and areas in the mesencephalon, often cause effects similar to those elicited from the hypothalamus. We discuss some of these in more detail later.

THE REWARD AND PUNISHMENT FUNCTION OF THE LIMBIC SYSTEM

From the preceding discussion, it is already clear that several limbic structures, including the hypothalamus, are particularly concerned with the *affective* nature of sensory sensations — that is, whether the sensations are *pleasant* or *unpleasant.* These affective qualities are also called *reward* or *punishment,* or *satisfaction* or *aversion.* Electrical stimulation of certain regions pleases or satisfies the animal, whereas electrical stimulation of other regions causes terror, pain, fear, defense, escape reactions, and all the other elements of punishment. Obviously, these two oppositely responding systems greatly affect the behavior of the animal.

Reward Centers

Figure 20–8 illustrates a technique that has been used for localizing specific reward and punishment areas of

Figure 20–8. Technique for localizing reward and punishment centers in the brain of a monkey.

the brain. In this figure a lever is placed at the side of the cage and is arranged so that depressing the lever makes electrical contact with a stimulator. Electrodes are placed successively at different areas in the brain so that the animal can stimulate the area by pressing the lever. If stimulating the particular area gives the animal a sense of reward, then it will press the lever again and again, sometimes as much as thousands of times per hour. Furthermore, when offered the choice of eating some delectable food as opposed to the opportunity to stimulate the reward center, it often chooses the electrical stimulation.

By using this procedure, the major reward centers have been found to be located *along the course of the medial forebrain bundle,* especially in the *lateral* and *ventromedial nucleus of the hypothalamus.* It is strange that the lateral nucleus should be included among the reward areas—indeed, they are one of the most potent of all—because even stronger stimuli in this area can cause rage. But this is true in many areas, with weaker stimuli giving a sense of reward and stronger ones a sense of punishment.

Less potent reward centers, which are perhaps secondary to the major ones in the hypothalamus, are found in the septum, the amygdala, certain areas of the thalamus and basal ganglia, and finally extending downward into the basal tegmentum of the mesencephalon.

Punishment Centers

The apparatus illustrated in Figure 20–8 can also be connected so that pressing the lever turns off, rather than turning on, an electrical stimulus. In this case, the animal will not turn the stimulus off when the electrode is in one of the reward areas; but when it is in certain other areas, it immediately learns to turn it off. Stimulation in these areas causes the animal to show all the signs of displeasure, fear, terror, and punishment. Furthermore, prolonged stimulation for 24 hours or more can cause the animal to become severely sick and actually lead to death.

By means of this technique, the most potent areas for punishment and escape tendencies have been found in the *central gray area surrounding the aqueduct of Sylvius in the mesencephalon* and extending upward into the *periventricular zones of the hypothalamus and thalamus.* Also, less potent punishment areas are found in the *amygdala* and the *hippocampus.*

It is particularly interesting that stimulation in the punishment centers can frequently inhibit the reward and pleasure centers completely, illustrating that punishment and fear can take precedence over pleasure and reward.

Importance of Reward and Punishment in Behavior

Almost everything that we do is related in some way to reward and punishment. If we are doing something that is rewarding, we continue to do it; if it is punishing, we cease to do it. Therefore, the reward and punishment centers undoubtedly constitute one of the most important of all the controllers of our bodily activities, our drives, our aversions, our motivations.

Effect of Tranquilizers on the Reward and Punishment Centers. Administration of a tranquilizer, such as chlorpromazine, inhibits both the reward and punishment centers, thereby greatly decreasing the affective reactivity of the animal. Therefore, it is presumed that tranquilizers function in psychotic states by suppressing many of the important behavioral areas of the hypothalamus and its associated regions of the brain, a subject that we discuss more fully later.

Importance of Reward and Punishment in Learning and Memory — Habituation or Reinforcement

Animal experiments have shown that a sensory experience causing neither reward nor punishment is remembered hardly at all. Electrical recordings show that new and novel sensory stimuli always excite the cerebral cortex. But repetition of the stimulus over and over leads to almost complete extinction of the cortical response if the sensory experience does not elicit either a sense of reward or punishment. Thus, the animal becomes *habituated* to the sensory stimulus and thereafter ignores it.

However, if the stimulus causes either reward or punishment rather than indifference, the cortical response becomes progressively more and more intense with repeated stimulation instead of fading away, and the response is said to be *reinforced.* Thus, an animal builds up strong memory traces for sensations that are either rewarding or punishing but, on the other hand, develops complete habituation to indifferent sensory stimuli. Therefore, it is evident that the reward and punishment centers of the limbic system have much to do with selecting the information that we learn.

RAGE

An emotional pattern that involves the hypothalamus and many other limbic structures, and has also been well characterized, is the *rage pattern*. This can be described as follows:

Strong stimulation of the punishment centers of the brain, especially in the *periventricular zone of the hypothalamus* or in the *lateral hypothalamus*, causes the animal to (1) develop a defensive posture, (2) extend its claws, (3) lift its tail, (4) hiss, (5) spit, (6) growl, and (7) develop piloerection, wide-open eyes, and dilated pupils. Furthermore, even the slightest provocation causes an immediate savage attack. This is approximately the behavior that one would expect from an animal being severely punished, and it is a pattern of behavior that is called *rage*.

Stimulation of the more rostral areas of the punishment areas — in the midline preoptic areas — causes mainly fear and anxiety, associated with a tendency for the animal to run away.

In the normal animal the rage phenomenon is held in check mainly by counterbalancing activity of the ventromedial nucleus of the hypothalamus. In addition, the hippocampus, the amygdala, and the anterior portions of the limbic cortex, especially the limbic cortex of the anterior cingulate gyrus and the subcallosal gyrus, help suppress the rage phenomenon. Conversely, if these portions of the limbic system are damaged or destroyed, the animal (also the human being) becomes far more susceptible to bouts of rage.

Placidity and Tameness. Exactly the opposite emotional behavior patterns occur when the reward centers are stimulated: placidity and tameness.

■ SPECIFIC FUNCTIONS OF OTHER PARTS OF THE LIMBIC SYSTEM

FUNCTIONS OF THE AMYGDALA

The amygdala is a complex of nuclei located immediately beneath the cortex of the medial anterior pole of each temporal lobe. It has abundant bidirectional connections with the hypothalamus.

In lower animals, the amygdala is concerned to a great extent with association of olfactory stimuli with stimuli from other parts of the brain. Indeed, it is pointed out in Chapter 15 that one of the major divisions of the olfactory tract leads directly to a portion of the amygdala called the *corticomedial nuclei* that lies immediately beneath the cortex in the pyriform area of the temporal lobe. However, in the human being, another portion of the amygdala, the *basolateral nuclei,* has become much more highly developed than this olfactory portion and plays exceedingly important roles in many behavioral activities not generally associated with olfactory stimuli.

The amygdala receives neuronal signals from all portions of the limbic cortex as well as from the neocortex of the temporal, parietal, and occipital lobes, especially from the auditory and visual association areas. Because of these multiple connections, the amygdala has been called the "window" through which the limbic system sees the place of the person in the world. In turn, the amygdala transmits signals (1) back into these same cortical areas, (2) into the hippocampus, (3) into the septum, (4) into the thalamus, and (5) especially into the hypothalamus.

Effects of Stimulating the Amygdala. In general, stimulation in the amygdala can cause almost all the same effects as those elicited by stimulation of the hypothalamus, plus still other effects. The effects that are mediated through the hypothalamus include (1) increases or decreases in arterial pressure, (2) increases or decreases in heart rate, (3) increases or decreases in gastrointestinal motility and secretion, (4) defecation and micturition, (5) pupillary dilatation or, rarely, constriction, (6) piloerection, (7) secretion of the various anterior pituitary hormones, especially the gonadotropins and adrenocorticotropic hormone.

Aside from these effects mediated through the hypothalamus, amygdala stimulation can also cause different types of involuntary movement. These include (1) tonic movements, such as raising the head or bending the body, (2) circling movements, (3) occasionally clonic, rhythmic movements, and (4) different types of movements associated with olfaction and eating, such as licking, chewing, and swallowing.

In addition, stimulation of certain amygdaloid nuclei can, rarely, cause a pattern of rage, escape, punishment, and fear similar to the rage pattern elicited from the hypothalamus as described earlier. And stimulation of other nuclei can give reactions of reward and pleasure.

Finally, excitation of still other portions of the amygdala can cause sexual activities that include erection, copulatory movements, ejaculation, ovulation, uterine activity, and premature labor.

Effects of Bilateral Ablation of the Amygdala — The Klüver-Bucy Syndrome. When the anterior portions of both temporal lobes in a monkey are destroyed, this removes not only the temporal cortex but also the amygdalas that lie deep in these parts of the temporal lobes. This causes a combination of changes in behavior called the Klüver-Bucy syndrome, which includes (1) excessive tendency to examine objects orally, (2) loss of fear, (3) decreased aggressiveness, (4) tameness, (5) changes in dietary habits, even to the extent that a herbivorous animal frequently becomes carnivorous, (6) sometimes psychic blindness, and (7) often excessive sex drive. The characteristic picture is of an animal that is not afraid of anything, has extreme curiosity about everything, forgets very rapidly, has a tendency to place everything in its mouth and sometimes even tries to eat solid objects, and, finally, often has a sex drive so strong that it attempts to copulate with immature animals, animals of the wrong sex, or animals of a different species.

Though similar lesions in human beings are rare, afflicted persons respond in a manner not too different from that of the monkey.

Overall Function of the Amygdala. The amygdala seems to be a behavioral awareness area that operates at a semiconscious level. It also seems to project into the limbic system one's present status in relation both to surroundings and thoughts. On the basis of this information, the amygdala is believed to help pattern the person's behavioral response so that it is appropriate for each occasion.

FUNCTIONS OF THE HIPPOCAMPUS

The hippocampus is the elongated, medial portion of the temporal cortex that folds upward and inward to form the ventral surface of the inferior horn of the lateral ventricle.

One end of the hippocampus abuts the amygdaloid nuclei, and it also fuses along one of its borders with the parahippocampal gyrus, which is the cortex of the ventromedial surface of the temporal lobe.

The hippocampus has numerous but mainly indirect connections with many portions of the cerebral cortex as well as with the basic structures of the limbic system—the amygdala, the hypothalamus, the septum, and the mamillary bodies. Almost any type of sensory experience causes activation of at least some part of the hippocampus, and the hippocampus in turn distributes many outgoing signals to the anterior thalamus, the hypothalamus, and other parts of the limbic system, especially through the *fornix*, its major output pathway. Thus, the hippocampus, like the amygdala, is an additional channel through which incoming sensory signals can lead to appropriate behavioral reactions, but perhaps for different purposes, as can be seen later.

As in other limbic structures, stimulation of different areas in the hippocampus can cause almost any one of different behavioral patterns, such as rage, passivity, excess sex drive, and so forth.

Another feature of the hippocampus is that very weak electrical stimuli can cause local epileptic seizures that persist for many seconds after the stimulation is over, suggesting that the hippocampus can perhaps give off prolonged output signals even under normal functioning conditions. During hippocampal seizures, the person experiences various psychomotor effects, including olfactory, visual, auditory, tactile, and other types of hallucinations that cannot be suppressed even though the person has not lost consciousness and knows these hallucinations to be unreal. Probably one of the reasons for this hyperexcitability of the hippocampus is that it is composed of a different type of cortex from that elsewhere in the cerebrum, having only three nerve cell layers instead of the six layers found elsewhere.

Role of the Hippocampus in Learning

Effect of Bilateral Removal of the Hippocampi—Inability to Learn. The hippocampi have been surgically removed bilaterally in a few human beings for the treatment of epilepsy. These persons can recall most previously learned memories satisfactorily. However, they can learn essentially no new information that is based on verbal symbolism. In fact, they cannot even learn the names of persons with whom they come in contact every day. Yet they can remember for a moment or so what transpires during the course of their activities. Thus, they are capable of the type of very short-term memory called "immediate memory" even though their ability to establish secondary memories lasting longer than a few minutes is either completely or almost completely abolished, which is the phenomenon called *anterograde amnesia* discussed in the previous chapter.

Destruction of the hippocampi also causes some deficit in previously learned memories (retrograde amnesia), a little more so for memories up to the past year or so than for memories of the distant past.

Theoretical Function of the Hippocampus in Learning. The hippocampus originated as part of the olfactory cortex. In the very lowest animals it plays essential roles in determining whether the animal will eat a particular food, whether the smell of a particular object suggests danger, and whether the odor is sexually inviting and in making other decisions that are of life-or-death importance. Thus, very early in the development of the brain, the hippocampus presumably became a critical decision-making neuronal mechanism, determining the importance and type of the incoming sensory signals. Once this critical decision-making capability had been established, presumably the remainder of the brain began to call on it for the same decision-making. If the hippocampus says that a neuronal signal is important, it is likely to be committed to memory.

Earlier in this chapter (and also in the previous chapter), it was pointed out that reward and punishment play a major role in determining the importance of information and especially whether or not the information will be stored in memory. A person rapidly becomes habituated to indifferent stimuli but learns assiduously any sensory experience that causes either pleasure or punishment. Yet, what is the mechanism by which this occurs? It has been suggested that the hippocampus provides the drive that causes translation of immediate memory into secondary memory—that is, it transmits some type of signal or signals that seem to make the mind rehearse over and over the new information until permanent storage takes place.

Whatever the mechanism, without the hippocampi *consolidation* of long-term memories of verbal or symbolic type does not take place.

FUNCTION OF THE LIMBIC CORTEX

Probably the most poorly understood portion of the entire limbic system is the ring of cerebral cortex called the *limbic cortex* that surrounds the subcortical limbic structures. This cortex functions as a transitional zone through which signals are transmitted from the remainder of the cortex into the limbic system. Therefore, it is presumed that the limbic cortex functions as a cerebral *association area for control of behavior.*

Stimulation of the different regions of the limbic cortex has failed to give any real idea of their functions. However, as is true of so many other portions of the limbic system, essentially all the behavioral patterns that have already been described can also be elicited by stimulation in different portions of the limbic cortex. Likewise, ablation of a few limbic cortical areas can cause persistent changes in an animal's behavior, as follows:

Ablation of the Temporal Cortex. When the anterior temporal cortex is ablated bilaterally, the amygdala is almost invariably damaged as well. This was discussed earlier, and it was pointed out that the Klüver-Bucy syndrome occurs. The animal especially develops consummatory behavior, investigates any and all objects, has intense sex drives toward inappropriate animals or even inanimate objects, and loses all fear—thus develops tameness as well.

Ablation of the Posterior Orbital Frontal Cortex. Bilateral removal of the posterior portion of the orbital frontal cortex often causes an animal to develop insomnia and an intense degree of motor restlessness, becoming unable to sit still but moving about continually.

Ablation of the Anterior Cingulate Gyri and Subcallosal Gyri. The anterior cingulate gyri and the subcallosal gyri are the portions of the limbic cortex that communicate between the prefrontal cerebral cortex and the subcortical limbic structures. Destruction of these gyri bilaterally releases the rage centers of the septum and hypothalamus from any prefrontal inhibitory influence. Therefore, the animal can become vicious and much more subject to fits of rage than normally.

Summary. Until further information is available, it is perhaps best to state that the cortical regions of the limbic system occupy intermediate associative positions between the functions of the remainder of the cerebral cortex and the functions of the subcortical limbic structures for control of behavioral patterns. Thus, in the anterior temporal cortex one especially finds gustatory and olfactory associations. In the parahippocampal gyri there is a tendency for complex auditory associations and also complex thought associations derived from Wernicke's area of the posterior temporal lobe. In the middle and posterior cingulate cortex, there is reason to believe that sensorimotor associations occur.

REFERENCES

Aoki, C., and Siekevitz, P.: Plasticity in brain development. Sci. Am., December, 1988, p. 56.

Ashton, H.: Brain Systems, Disorders and Psychotropic Drugs. New York, Oxford University Press, 1987.

Avoli, M., et al. (eds.): Neurotransmitters and Cortical Function. New York, Plenum Publishing Corp., 1988.

Barnes, D. M.: The biological tangle of drug addiction. Science, 241:415, 1988.

Barnes, D. M.: Neural models yield data on learning. Science, 236:1628, 1987.

Berger, P. A., et al.: Behavioral pharmacology of the endorphins. Annu. Rev. Med., 33:397, 1982.

Borbely, A. A., and Tobler, I.: Endogenous sleep-promoting substances and sleep regulation. Physiol. Rev., 69:605, 1989.

Burchfield, S. R. (ed.): Stress, Physiological and Psychological Interactions. Washington, D.C., Hemisphere Publishing Corp., 1985.

Buzsaki, G.: Feed-forward inhibition in the hippocampal formation. Prog. Neurobiol., 22:131, 1984.

Byrne, J. H.: Cellular analysis of associative learning. Physiol. Rev., 67:329, 1987.

Chiba, A., et al.: Synaptic rearrangement during postembryonic development in the cricket. Science, 240:901, 1988.

Chrousos, G. P., et al. (eds.): Mechanisms of Physical and Emotional Stress. New York, Plenum Publishing Corp., 1988.

Clynes, M., and Panksepp, J. (eds.): Emotions and Psychopathology. New York, Plenum Publishing Corp., 1988.

Cohen, D. H., and Randall, D. C.: Classical conditioning of cardiovascular responses. Annu. Rev. Physiol., 46:187, 1984.

De Wied, D., and Jolles, J.: Neuropeptides derived from pro-opiocortin: Behavioral, physiological, and neurochemical effects. Physiol. Rev., 62:976, 1982.

Doane, B. K., and Livingston, K. E. (eds.): The Limbic System. New York, Raven Press, 1986.

Doris, P. A.: Vasopressin and central integrative processes. Neuroendocrinology, 38:75, 1984.

Engel, B. T., and Schneiderman, N.: Operant conditioning and the modulation of cardiovascular function. Annu. Rev. Physiol., 46:199, 1984.

Engel, J., et al. (eds.): Brain Reward Systems and Abuse. New York, Raven Press, 1987.

Fink, M.: Convulsive and drug therapies of depression. Annu. Rev. Med., 32:405, 1981.

Foote, S. L., et al.: Nucleus locus ceruleus: New evidence of anatomical and physiological specificity. Physiol. Rev., 63:844, 1983.

Ganong, W. F.: The brain renin-angiotensin system. Annu. Rev. Physiol., 46:17, 1984.

Givens, J. R.: The Hypothalamus in Health and Disease. Chicago, Year Book Medical Publishers, 1984.

Goldstein, G. (ed.): Neuropsychology Review. New York, Plenum Publishing Corp., 1989.

Iversen, L. L.: Nonopioid neuropeptides in mammalian CNS. Annu. Rev. Pharmacol. Toxicol., 23:1, 1983.

Jones, E. G., and Peters, A. (eds.): Further Aspects of Cortical Function, Including Hippocampus. New York, Plenum Publishing Corp., 1987.

Kandel, E. R.: Molecular Neurobiology in Neurology and Psychiatry. New York, Raven Press, 1987.

Klimov, P. K.: Behavior of the organs of the digestive system. Neurosci. Behav. Physiol., 14:333, 1984.

Malick, J. B. (ed.): Anxiolytics, Neurochemical, Behavioral, and Clinical Perspectives. New York, Raven Press, 1983.

Mariani, J., and Delhaye-Bouchaud, N.: Elimination of functional synapses during development of the nervous system. News Physiol. Sci., 2:93, 1987.

Miller, J. A.: En route to thought: Recognition and recall. Sci. News, 128:373, 1985.

Mishkin, M., and Appenzeller, T.: The Anatomy of Memory. Sci. Am., Special Report, 1987, p. 2.

Nerozzi, D., et al. (eds.): Hypothalamic Dysfunction in Neuropsychiatric Disorders. New York, Raven Press, 1987.

Rescorla, R. A.: Behavioral studies of Pavlovian conditioning. Annu. Rev. Neurosci., 11:329, 1988.

Russell, R. W.: Cholinergic system in behavior: The search for mechanisms of action. Annu. Rev. Pharmacol. Toxicol., 22:435, 1982.

Schatzberg, A. F., and Nemeroff, C. B. (eds.): The Hypothalamic-Pituitary-Adrenal Axis. New York, Raven Press, 1988.

Shepherd, G. M.: Neurobiology, 2nd Ed. New York, Oxford University Press, 1987.

Siddle, D.: Orienting and Habituation. Perspectives in Human Research. New York, John Wiley & Sons, 1982.

Skodol, A. E., and Spitzer, R. L.: The development of reliable diagnostic criteria in psychiatry. Annu. Rev. Med., 33:317, 1982.

Smith, O. A., and DeVito, J. L.: Central neural integration for the control of autonomic responses associated with emotion. Annu. Rev. Neurosci., 7:43, 1984.

Stephenson, R. B.: Modification of reflex regulation of blood pressure by behavior. Annu. Rev. Physiol., 46:133, 1984.

Steriade, M., and Llinas, R. R.: The functional states of the thalamus and the associated neuronal interplay. Physiol. Rev., 68:649, 1988.

Swanson, L. W., and Sawchenko, P. E.: Hypothalamic integration: Organization of the paraventricular and supraoptic nuclei. Annu. Rev. Neurosci., 6:269, 1983.

Tucek, S.: Regulation of acetylcholine synthesis in the brain. J. Neurochem., 44:11, 1985.

Tyrer, P. (ed.): Psychopharmacology of Anxiety. New York, Oxford University Press, 1989.

Usdin, E.: Stress. The Role of Catecholamines and Other Neurotransmitters. New York, Gordon Press Publishers, 1984.

Verrier, R. L., and Lown, B.: Behavioral stress and cardiac arrhythmias. Annu. Rev. Physiol., 46:155, 1984.

Wise, S. P., and Desimone, R.: Behavioral neurophysiology: Insights into seeing and grasping. Science, 242:736, 1988.

Wolman, B. B.: Psychosomatic Disorders. New York, Plenum Publishing Corp., 1988.

Woody, C. D., et al. (eds.): Cellular Mechanisms of Conditioning and Behavioral Plasticity. New York, Plenum Publishing Corp., 1988.

21

States of Brain Activity— Sleep; Brain Waves; Epilepsy; Psychoses

All of us are aware of the many different states of brain activity, including sleep, wakefulness, extreme excitement, and even different levels of mood such as exhilaration, depression, and fear. All these states result from different activating or inhibiting forces generated usually within the brain itself. In the last chapter, we began a partial discussion of this subject when we described different systems that are capable of activating either large or isolated portions of the brain. In this chapter, we present brief surveys of what is known about other states of brain activity, beginning with sleep.

■ SLEEP

Sleep is defined as unconsciousness from which the person can be aroused by sensory or other stimuli. It is to be distinguished from *coma*, which is unconsciousness from which the person cannot be aroused. However, there are multiple stages of sleep, from very light sleep to very deep sleep, and most sleep researchers even divide sleep into two different types of sleep that have different qualities, as follows:

Two Different Types of Sleep—(1) Slow Wave Sleep and (2) REM Sleep. During each night a person goes through stages of two different types of sleep that alternate with each other. These are called (1) *slow wave sleep*, because in this type of sleep the brain waves are very slow, as we discuss later; and (2) *REM sleep*, which stands for *rapid eye movement* sleep, because in this type of sleep the eyes undergo rapid movements despite the fact that the person is still asleep.

Most sleep during each night is of the slow wave variety; this is the deep, restful type of sleep that the person experiences during the first hour of sleep after having been kept awake for many hours. Episodes of REM sleep occur periodically during sleep and occupy about 25 per cent of the sleep time of the young adult; they normally recur about every 90 min. This type of sleep is not so restful, and it is usually associated with dreaming, as we discuss later.

SLOW WAVE SLEEP

Most of us can understand the characteristics of deep slow wave sleep by remembering the last time that we were kept awake for more than 24 hours and then remembering the deep sleep that occurred during the first hour after going to sleep. This sleep is exceedingly restful and is associated with a decrease in both peripheral vascular tone and many other vegetative functions of the body as well. In addition, there is a 10 to 30 per cent decrease in blood pressure, respiratory rate, and basal metabolic rate.

Though slow wave sleep is frequently called "dreamless sleep," dreams do occur often during slow wave sleep, and nightmares even occur during this type of sleep. However, the difference between the dreams occurring in slow wave sleep and those in REM sleep is that those of REM sleep are remembered, whereas those of slow wave sleep usually are not. That is, during slow wave sleep the process of consolidation of the dreams in memory does not occur.

REM SLEEP (PARADOXICAL SLEEP, DESYNCHRONIZED SLEEP)

In a normal night of sleep, bouts of REM sleep lasting 5 to 30 min usually appear on the average every 90 min, the first such period occurring 80 to 100 min after the person falls asleep. When the person is extremely sleepy, the duration of each bout of REM sleep is very short, and it may even be absent. On the other hand, as the person becomes more rested through the night, the duration of the REM bouts greatly increases.

There are several very important characteristics of REM sleep:

1. It is usually associated with active dreaming.
2. The person is even more difficult to arouse by sensory stimuli than during deep slow wave sleep, and yet persons usually awaken in the morning during an episode of REM sleep, not from slow wave sleep.
3. The muscle tone throughout the body is exceedingly depressed, indicating strong inhibition of the spinal projections from the excitatory areas of the brain stem.
4. The heart rate and respiration usually become irregular, which is characteristic of the dream state.
5. Despite the extreme inhibition of the peripheral muscles, a few irregular muscle movements occur. These include, in particular, rapid movements of the eyes; this is the origin of the acronym REM, for "rapid eye movements."
6. The brain is highly active in REM sleep, and the overall brain metabolism may be increased as much as 20 per cent. Also, the electroencephalogram shows a pattern of brain waves similar to those that occur during wakefulness. Therefore, this type of sleep is also frequently called *paradoxical sleep* because it is a paradox that a person can still be asleep despite marked activity in the brain.

In summary, REM sleep is a type of sleep in which the brain is quite active. However, the brain activity is not channeled in the proper direction for persons to be fully aware of their surroundings and therefore to be awake.

BASIC THEORIES OF SLEEP

The Active Theory of Sleep. An earlier theory of sleep was that the excitatory areas of the upper brain stem, which was called the *reticular activating system*, and other parts of the brain simply fatigued over the period of a waking day and therefore became inactive as a result. This was called the *passive theory of sleep*. However, an important experiment changed this view to the current belief that *sleep is probably caused by an active inhibitory process*. This was the experiment in which it was discovered that transecting the brain stem in the midpontile region leads to a brain that never goes to sleep. In other words, there seems to be some center or centers located below the midpontile level of the brain stem that actively cause sleep by inhibiting other parts of the brain. This is called the *active theory* of sleep.

Neuronal Centers, Neurohumoral Substances, and Mechanisms That Can Cause Sleep — A Possible Specific Role for Serotonin

Stimulation of several specific areas of the brain can produce sleep with characteristics very near those of natural sleep. Some of these are the following:

1. The most conspicuous stimulation area for causing almost natural sleep is the raphe nuclei in the lower half of the pons and in the medulla. These are a thin sheet of nuclei located in the midline. Nerve fibers from these nuclei spread widely in the reticular formation and also upward into the thalamus, neocortex, hypothalamus, and most areas of the limbic system. In addition, they extend downward into the spinal cord, terminating in the posterior horns where they can inhibit incoming pain signals, as was discussed in Chapter 10. It is also known that many of the endings of fibers from these raphe neurons secrete *serotonin*. Also, when a drug that blocks the formation of serotonin is administered to an animal, the animal often cannot sleep for the next several days. Therefore, it is assumed that serotonin is the major transmitter substance associated with production of sleep.

2. Stimulation of some areas in the *nucleus of the tractus solitarius*, which is the sensory region of the medulla and pons for the visceral sensory signals entering the brain via the vagus and glossopharyngeal nerves, will also promote sleep. However, this will not occur if the raphe nuclei have been destroyed. Therefore, these regions probably act by exciting the raphe nuclei and the serotonin system.

3. Stimulation of several regions in the diencephalon can also help promote sleep, including (a) the rostral part of the hypothalamus, mainly in the suprachiasmal area, and (b) an occasional area in the diffuse nuclei of the thalamus.

Effect of Lesions in the Sleep-Promoting Centers. Discrete lesions in the raphe nuclei lead to a high state of wakefulness. This is also true of bilateral lesions in the mediorostral suprachiasmal portion of the anterior hypothalamus. In both instances, the excitatory reticular nuclei of the mesencephalon and upper pons seem to become released from inhibition. Indeed, the lesions of the anterior hypothalamus can sometimes cause such intense wakefulness that the animal actually dies of exhaustion.

Other Possible Transmitter Substances Related to Sleep. Experiments have shown that the cerebrospinal fluid and also the blood and urine of animals that have been kept awake for several days contain a substance or substances that will cause sleep when injected into the ventricular system of an animal. One of these substances has been identified as *muramyl peptide*, a low molecular weight substance that accumulates in the cerebrospinal fluid and in the urine in animals kept awake for several days. When only micrograms of this sleep-producing substance are injected into the third ventricle, almost natural sleep occurs within a few minutes, and the animal may then stay asleep for several hours. Another substance that has similar effects in causing sleep is a nonapeptide isolated from the blood of sleeping animals. And still a third and different sleep factor has been isolated from the neuronal tissues of the brain stem of animals kept awake for days. Therefore, it is possible that prolonged wakefulness causes progressive accumulation of a sleep factor in the brain stem or in the cerebrospinal fluid that leads to sleep.

Possible Causes of REM Sleep

Why slow wave sleep is broken periodically by REM sleep is not understood. However, a lesion in the *locus ceruleus* on each side of the brain stem can reduce REM sleep, and if the lesion includes other contiguous areas of the brain stem, REM sleep can be prevented altogether. Therefore, it has been postulated that when stimulated, the norepinephrine-secreting nerve fibers that originate in the locus ceruleus can activate many portions of the brain. This theoretically causes the excess activity that occurs in certain regions of the brain in REM sleep, but the signals are not channeled appropriately in the brain to cause normal conscious awareness that is characteristic of wakefulness.

The Cycle Between Sleep and Wakefulness

The preceding discussions have merely identified neuronal areas, transmitters, and mechanisms that are related to sleep. However, they have not explained the cyclic, reciprocal operation of the sleep-wakefulness cycle. There is, as yet, no explanation. Therefore, we can let our imaginations run wild and suggest the following possible mechanism for causing the rhythmicity of the sleep-wakefulness cycle:

When the sleep centers are not activated, the release from inhibition of the mesencephalic and upper pontile reticular nuclei allows this region to become spontaneously active. This in turn excites both the cerebral cortex and the peripheral nervous system, both of which then send numerous positive feedback signals back to the same reticular nuclei to activate them still further. Thus, once wakefulness begins, it has a natural tendency to sustain itself because of all this positive feedback activity.

However, after the brain remains activated for many hours, even the neurons within the activating system presumably will become fatigued to some extent, and other factors presumably activate the sleep centers. Consequently, the positive feedback cycle between the mesencephalic reticular nuclei and the cortex, and also that between these and the periphery, will fade, and the inhibitory effects of the sleep centers (as well as inhibition by possible sleep-producing chemical transmitter substances) will take over, leading to rapid transition from the wakefulness state to the sleep state.

Then, one could postulate that during sleep the excitatory neurons of the reticular activating system gradually become more and more excitable because of the prolonged rest, while the inhibitory neurons of the sleep centers become less excitable because of their overactivity, thus leading to a new cycle of wakefulness.

This theory can explain the rapid transitions from sleep to wakefulness and from wakefulness to sleep. It can also explain arousal, the insomnia that occurs when a person's mind becomes preoccupied with a thought, the wakefulness that is produced by bodily activity, and many other conditions that affect the person's state of sleep or wakefulness.

PHYSIOLOGICAL EFFECTS OF SLEEP

Sleep causes two major types of physiological effects: first, effects on the nervous system itself, and second, effects on other structures of the body. The first of these seems to be by far the more important, for any person who has a transected spinal cord in the neck shows no harmful effects in the body beneath the level of transection that can be attributed to a sleep and wakefulness cycle; that is, lack of this sleep-wakefulness cycle in the nervous system at any point below the brain causes neither harm to the bodily organs nor even any deranged function. On the other hand, lack of sleep certainly does affect the functions of the central nervous system.

Prolonged wakefulness is often associated with progressive malfunction of the mind and sometimes even causes abnormal behavioral activities of the nervous system. We are all familiar with the increased sluggishness of thought that occurs toward the end of a prolonged wakeful period, but in addition, a person can become irritable or even psychotic following forced wakefulness for prolonged periods of time. Therefore, we can assume that sleep in some way not currently understood restores both normal levels of activity and normal "balance" among the different parts of the central nervous system. This might be likened to the "rezeroing" of electronic analog computers after prolonged use, for all computers of this type gradually lose their "baseline" of operation; it is reasonable to assume that the same effect occurs in the central nervous system, because overuse of some brain areas during wakefulness could easily throw these out of balance with the remainder of the nervous system. Therefore, in the absence of any definitely demonstrated functional value of sleep, we might postulate that the principal value of sleep is to restore the natural balance among the neuronal centers.

Even though, as we pointed out earlier, wakefulness and sleep have not been shown to be harmful to the somatic functions of the body, the cycle of enhanced and depressed nervous excitability that follows the cycle of wakefulness and sleep does have moderate physiological effects on the peripheral body. For instance, there is enhanced sympathetic activity during wakefulness and also enhanced numbers of impulses to the skeletal musculature to increase muscle tone. Conversely, during sleep, sympathetic activity decreases while parasympathetic activity increases. Therefore, arterial blood pressure falls, pulse rate decreases, skin vessels dilate, activity of the gastrointestinal tract sometimes increases, muscles fall into a mainly relaxed state, and the overall basal metabolic rate of the body falls by 10 to 30 per cent.

■ BRAIN WAVES

Electrical recordings from the surface of the brain or from the outer surface of the head demonstrate continuous electrical activity in the brain. Both the intensity and patterns of this electrical activity are determined to a great extent by the overall level of excitation of the brain resulting from *sleep, wakefulness*, and brain diseases such as *epilepsy* and even some *psychoses*. The undulations in the recorded electrical potentials, shown in Figure 21–1, are called *brain waves*, and the entire record is called an *electroencephalogram* (EEG).

The intensities of the brain waves on the surface of the scalp range from 0 to 200 μV, and their frequencies range from once every few seconds to 50 or more per second. The character of the waves is highly dependent on the degree of activity of the cerebral cortex, and the waves change markedly between the states of wakefulness and sleep and coma.

Much of the time, the brain waves are irregular, and no general pattern can be discerned in the EEG. However, at other times, distinct patterns do appear. Some of these are characteristic of specific abnormalities of the brain, such as epilepsy, which is discussed later. Others occur even in normal persons and can be classified as *alpha, beta, theta,* and *delta waves*, which are all illustrated in Figure 21–1.

Alpha waves are rhythmic waves occurring at a frequency of between 8 and 13/sec and are found in the EEGs of almost all normal adult persons when they are awake in a quiet, resting state of cerebration. These waves occur most intensely in the occipital region but can also be recorded from the parietal and frontal regions of the scalp. Their voltage usually is about 50 μV. During deep sleep the alpha waves disappear entirely; and when the awake person's attention is directed to some specific type of mental activity, the alpha waves are replaced by asynchronous, higher-frequency but lower-voltage beta waves. Figure 21–2 illustrates the effect on the alpha waves of simply opening the eyes in bright light and then closing the eyes again. Note that the visual sensations cause immediate cessation of the alpha waves and that these are replaced by low-voltage, asynchronous beta waves.

Beta waves occur at frequencies of more than 14 cycles/sec and as high as 25 and rarely 50 cycles/sec. These are most frequently recorded from the parietal and frontal regions of the scalp during activation of the central nervous system or during tension.

Theta waves have frequencies of between 4 and 7 cycles/sec. These occur mainly in the parietal and temporal regions in children, but they also occur during emotional stress in some adults, particularly during disappointment and frustration. Theta waves also occur in many brain disorders.

Delta waves include all the waves of the EEG below 3.5 cycles/sec and sometimes as low as 1 cycle every 2 to 3 sec. These occur in very deep sleep, in infancy, and in serious organic brain disease. They also occur in the cortex of ani-

Eyes open   Eyes closed

Figure 21–2. Replacement of the alpha rhythm by an asynchronous discharge on opening the eyes.

mals that have had subcortical transections separating the cerebral cortex from the thalamus. Therefore, delta waves can occur strictly in the cortex independent of activities in lower regions of the brain.

ORIGIN IN THE BRAIN OF THE BRAIN WAVES

The discharge of a single neuron or single nerve fiber in the brain can never be recorded from the surface of the head. Instead, many thousands or even millions of neurons or fibers *must fire synchronously*; only then will the potentials from the individual neurons or fibers summate enough to be recorded all the way through the skull. Thus, the intensity of the brain waves from the scalp is determined mainly by the number of neurons and fibers that fire in synchrony with each other, not by the total level of electrical activity in the brain, for even very strong *nonsynchronous* nerve signals will actually nullify each other and cause only weak waves. This was illustrated in Figure 21–2, which showed, when the eyes were closed, synchronous discharge of many neurons in the cerebral cortex at a frequency of about 12/sec, which produced the *alpha waves*. Then, when the eyes were opened, the activity of the brain increased greatly, but the synchronization of the signals decreased to so little that the brain waves decreased to weak waves of generally higher but irregular frequency, called *beta waves*.

The brain wave electrical potentials are generated mainly from cortical layers I and II, especially from the large matt of dendrites that extend into these surface areas from neuronal cells located deeper in the cortex. The potentials generated in the tissue fluids surrounding these dendrites can be either positive or negative, for the following reasons: When the neuronal cell bodies in the deeper layers discharge, negative charges leak out of the cell bodies and cause negativity in the deeper cortical fluids; at the same time this loss of negative charges leaves the interior of the neuronal cell membranes positive. This positivity is conducted electrotonically upward through the dendrites to the surface layers of the brain and then transmitted by a capacitive effect across the dendrite membranes to the fluids surrounding these dendrites. Therefore, stimulation of neurons deep in the cerebral cortex usually causes initial positivity on the surface of the brain. On the other hand, other cortical synapses lie not on the deep cell bodies but instead on the surface dendrites themselves, especially in cortical layers II and III. When these synapses are excited, local depolarization occurs in the dendrites themselves, allowing negative charges to leak outward; then the electrical waves recorded from the surface of the scalp are negative. This difference between positivity and negativity is important because it sometimes allows one to distinguish the depth in the cortex of the neuronal discharges that produce specific types of waves.

Origin of Alpha Waves. Alpha waves will *not* occur in the cortex without connections with the thalamus. Also, stimulation in the nonspecific thalamic nuclei often sets up waves in the generalized thalamocortical system at a frequency of between 8 and 13/sec, the natural frequency of the alpha

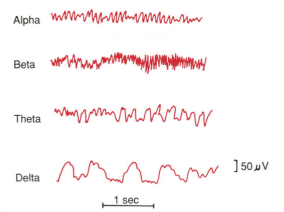

Alpha

Beta

Theta

Delta

] 50 μV

1 sec

Figure 21–1. Different types of normal electroencephalographic waves.

Figure 21–3. Effect of varying degrees of cerebral activity on the basic rhythm of the electroencephalogram (EEG). (From Gibbs and Gibbs: Atlas of Electroencephalography, 2nd. Ed. Vol. I. Reading, Mass., Addison-Wesley, 1974. Reprinted by permission.)

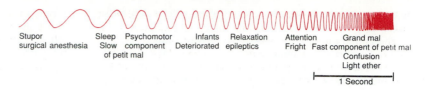

Figure 21–3. Effect of varying degrees of cerebral activity on the basic rhythm of the electroencephalogram (EEG). (From Gibbs and Gibbs: Atlas of Electroencephalography, 2nd. Ed. Vol. I. Reading, Mass., Addison-Wesley, 1974. Reprinted by permission.)

waves. Therefore, it is likely that the alpha waves result from spontaneous activity in the nonspecific thalamocortical system, which causes both the periodicity of the alpha waves and the synchronous activation of literally millions of cortical neurons during each wave.

Origin of Delta Waves. Transection of the fiber tracts from the thalamus to the cortex, which blocks the thalamic activation of the cortex and eliminates the alpha waves, nevertheless causes delta waves in the cortex. This indicates that some synchronizing mechanism can occur in the cortical neurons themselves—entirely independently of lower structures in the brain—to cause the delta waves.

Delta waves also occur in very deep "slow wave" sleep; and this suggests that the cortex then might be released from the activating influences of the lower centers.

EFFECT OF VARYING DEGREES OF CEREBRAL ACTIVITY ON THE BASIC FREQUENCY OF THE ELECTROENCEPHALOGRAM

There is a general relationship between the degree of cerebral activity and the average frequency of the electroencephalographic rhythm, the average frequency increasing progressively with higher degrees of activity. This is illustrated in Figure 21–3, which shows the existence of delta waves in stupor, surgical anesthesia, and sleep; theta waves in psychomotor states and in infants; alpha waves during relaxed states; and beta waves during periods of intense mental activity. *However, during periods of mental activity the waves usually become asynchronous rather than synchronous, so that the voltage falls considerably, despite increased cortical activity,* as illustrated in Figure 21–2.

ELECTROENCEPHALOGRAPHIC CHANGES IN THE DIFFERENT STAGES OF WAKEFULNESS AND SLEEP

Figure 21–4 illustrates the electroencephalogram from a typical person in different stages of wakefulness and sleep. Alert wakefulness is characterized by high-frequency *beta waves,* whereas quiet wakefulness is usually associated with *alpha waves,* as illustrated by the first two electroencephalograms of the figure.

Slow wave sleep is generally divided into four stages. In the first stage, a stage of very light sleep, the voltage of the electroencephalographic waves becomes very low; but this is broken by *"sleep spindles,"* that is, short spindle-shaped bursts of alpha waves that occur periodically. In stages 2, 3, and 4 of slow wave sleep the frequency of the electroencephalogram becomes progressively slower until it reaches a frequency of only 2 to 3 waves/sec; these are typical *delta waves.*

Finally, the bottom record in Figure 21–4 illustrates the electroencephalogram during REM sleep. It is often difficult to tell a difference between this brain wave pattern and that of an alert awake person. The voltage of these waves is con-

siderably lower than the voltage in deep stage 4 slow wave sleep, and the waves are themselves irregular high-frequency beta waves, which is normally suggestive of excess but desynchronized nervous activity as found in the awake state. Therefore, REM sleep is frequently called *desynchronized sleep* because there is a lack of synchrony in the firing of the neurons despite very significant brain activity.

■ EPILEPSY

Epilepsy is characterized by uncontrolled excessive activity of either a part or all of the central nervous system. A person who is predisposed to epilepsy has attacks when the basal level of excitability of the nervous system (or of the part that is susceptible to the epileptic state) rises above a certain critical threshold. But as long as the degree of excitability is held below this threshold, no attack occurs.

Basically, epilepsy can be classified into three major types: *grand mal epilepsy, petit mal epilepsy,* and *focal epilepsy.*

GRAND MAL EPILEPSY

Grand mal epilepsy is characterized by extreme neuronal discharges in all areas of the brain—in the cortex, in the deeper parts of the cerebrum, and even in the brain stem and thalamus. Also, discharges into the spinal cord cause generalized *tonic convulsions* of the entire body, followed toward the end of the attack by alternating tonic and then spasmotic muscular contractions called *tonic-clonic convulsions.* Often the person bites or "swallows" the tongue and usually has difficulty in breathing, sometimes to the extent of develop-

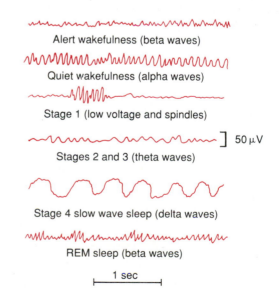

Figure 21–4. Progressive change in the characteristics of the brain waves during different stages of wakefulness and sleep.

ing cyanosis. Also, signals to the viscera frequently cause involutary urination and defecation.

The grand mal seizure lasts from a few seconds to as long as 3 to 4 min and is characterized by postseizure depression of the entire nervous system; the person remains in stupor for one to many minutes after the attack is over and then often remains severely fatigued or even asleep for many hours thereafter.

The top recording of Figure 21–5 illustrates a typical electroencephalogram from almost any region of the cortex during the tonic phase of a grand mal attack. This illustrates that high-voltage, synchronous discharges occur over the entire cortex. Furthermore, the same type of discharge occurs on both sides of the brain at the same time, illustrating that the abnormal neuronal circuitry responsible for the attack strongly involves the basal regions of the brain that drive the cortex.

In experimental animals or even in human beings, grand mal attacks can be initiated by administering neuronal stimulants, such as the drug Metrazol, or they can be caused by insulin hypoglycemia or by the passage of alternating electrical current directly through the brain. Electrical recordings from the thalamus and also from the reticular formation of the brain stem during the grand mal attack show typical high-voltage activity in both of these areas similar to that recorded from the cerebral cortex.

Presumably, therefore, a grand mal attack is caused by abnormal activation in the lower portions of the brain activating system itself.

What Initiates a Grand Mal Attack? Most persons who have grand mal attacks have a hereditary predisposition to epilepsy, a predisposition that occurs in about 1 of every 50 to 100 people. In such persons, some of the factors that can increase the excitability of the abnormal "epileptogenic" circuitry enough to precipitate attacks are (1) strong emotional stimuli, (2) alkalosis caused by overbreathing, (3) drugs, (4) fever, and (5) loud noises or flashing lights. Also, even in persons not genetically predisposed, traumatic lesions in almost any part of the brain can cause excess excitability of local brain areas, as we discuss shortly; and these, too, can transmit signals into the basal activating systems of the brain to elicit grand mal seizures.

What Stops the Grand Mal Attack? The cause of the extreme neuronal overactivity during a grand mal attack is presumed to be massive activation of many reverberating pathways throughout the brain. Presumably, also, the major

factor, or at least one of the major factors, that stops the attack after a few minutes is the phenomenon of neuronal *fatigue*. However, a second factor is probably *active inhibition* by inhibitory neurons that have also been activated by the attack. The stupor and total body fatigue that occur after a grand mal seizure is over are believed to result from the intense fatigue of the neuronal synapses following their intensive activity during the grand mal attack.

PETIT MAL EPILEPSY

Petit mal epilepsy is closely allied to grand mal epilepsy in that it too almost certainly involves the basic brain activating system. It is usually characterized by 3 to 30 sec of unconsciousness during which the person has several twitchlike contractions of the muscles, usually in the head region—especially blinking of the eyes; this is followed by return of consciousness and resumption of previous activities. The patient may have one such attack in many months or in rare instances may have a rapid series of attacks, one following the other. However, the usual course is for the petit mal attacks to appear in late childhood and then to disappear entirely by the age of 30. On occasion, a petit mal epileptic attack will initiate a grand mal attack.

The brain wave pattern in petit mal epilepsy is illustrated by the middle record of Figure 21–5, which is typified by a *spike and dome pattern*. The spike portion of this recording is almost identical to the spikes that occur in grand mal epilepsy, but the dome portion is distinctly different. The spike and dome can be recorded over most or all of the cerebral cortex, illustrating that the seizure involves the entire activating system of the brain.

FOCAL EPILEPSY

Focal epilepsy can involve almost any part of the brain, either localized regions of the cerebral cortex or deeper structures of both the cerebrum and brain stem. And almost always, focal epilepsy results from some localized organic lesion or functional abnormality, such as a scar that pulls on the neuronal tissue, a tumor that compresses an area of the brain, a destroyed area of brain tissue, or congenitally deranged local circuitry. Lesions such as these can promote extremely rapid discharges in the local neurons; and when the discharge rate rises above approximately 1000/sec, synchronous waves begin to spread over the adjacent cortical regions. These waves presumably result from *localized reverberating circuits* that gradually recruit adjacent areas of the cortex into the discharge zone. The process spreads to adjacent areas at a rate as slow as a few millimeters a minute to as fast as several centimeters per second. When such a wave of excitation spreads over the motor cortex, it causes a progressive "march" of muscular contractions throughout the opposite side of the body, beginning most characteristically in the mouth region and marching progressively downward to the legs, but at other times marching in the opposite direction. This is called *jacksonian epilepsy*.

A focal epileptic attack may remain confined to a single area of the brain, but in many instances the strong signals from the convulsing cortex or other part of the brain excite the mesencephalic portion of the brain activating system so greatly that a grand mal epileptic attack ensues as well.

Another type of focal epilepsy is the so-called *psychomotor seizure*, which may cause (1) a short period of amnesia; (2) an

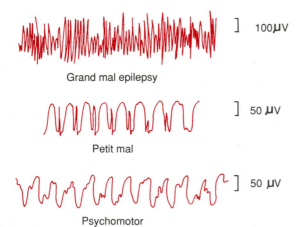

Figure 21–5. Electroencephalograms in different types of epilepsy.

attack of abnormal rage; (3) sudden anxiety, discomfort, or fear; (4) a moment of incoherent speech or mumbling of some trite phrase; or (5) a motor act to attack someone, to rub the face with the hand, and so forth. Sometimes the person cannot remember his activities during the attack, but at other times he will have been conscious of everything that he had been doing but unable to control it. Attacks of this type characteristically involve part of the limbic portion of the brain, such as the hippocampus, the amygdala, the septum, and the temporal cortex.

The lower tracing of Figure 21–5 illustrates a typical electroencephalogram during a psychomotor attack, showing a low-frequency rectangular wave with a frequency of between 2 and 4/sec and with superimposed 14/sec waves.

The electroencephalogram can often be used to localize abnormal spiking waves originating in areas of organic brain disease that might predispose to focal epileptic attacks. Once such a focal point is found, surgical excision of the focus frequently prevents future attacks.

■ PSYCHOTIC BEHAVIOR AND DEMENTIA — ROLES OF SPECIFIC NEUROTRANSMITTER SYSTEMS

Clinical studies of patients with different psychoses and also some types of dementia have suggested that many if not most of these conditions result from diminished function of classes of neurons that secrete specific neurotransmitters. Use of appropriate drugs to counteract the loss of the respective transmitters has been quite successful in treating some patients.

In Chapter 18 we discussed the cause of Parkinson's disease, the loss of the neurons in the substantia nigra whose axons secrete dopamine in the caudate nucleus and putamen. The loss of acetylcholine-secreting neurons in the basal ganglia is associated with the abnormal motor patterns of Huntington's chorea, as well as with the dementia that develops later in the same patients. In the present section, we extend this concept to other abnormalities and other classes of neurons that lead to additional types of psychotic behavior or dementia.

DEPRESSION AND MANIC-DEPRESSIVE PSYCHOSES — DECREASED ACTIVITY OF THE NOREPINEPHRINE AND SEROTONIN NEUROTRANSMITTER SYSTEMS

In the past few years much evidence has accumulated suggesting that the *mental depression psychosis*, which afflicts about 8 million people in the United States at any one time, might be caused by diminished formation of either norepinephrine or serotonin or both. These patients experience symptoms of grief, unhappiness, despair, and misery. In addition, they lose their appetite and sex drive and also have severe insomnia. And associated with all these is a state of psychomotor agitation despite the depression.

In the previous chapter it was pointed out that large numbers of *norepinephrine-secreting neurons* are located in the brain stem, especially in the *locus ceruleus*, and that these send fibers upward to most parts of the limbic system, the thalamus, and the cerebral cortex. Also, many *serotonin-producing neurons* are located in the *midline raphe nuclei* of the lower pons and medulla and also project fibers to many areas of the limbic system and to some other areas of the brain as well.

A principal reason for believing that depression is caused by diminished activity of the norepinephrine and serotonin systems is that drugs that block the secretion of norepinephrine and serotonin, such as the drug reserpine, frequently cause depression. Conversely, about 70 per cent of depressive patients can be treated very effectively with one of two types of drugs that increase especially the excitatory effects of norepinephrine at the nerve endings, and perhaps of serotonin as well: (1) *monoamine oxidase inhibitors,* which block destruction of norepinephrine and serotonin once they are formed; and (2) *tricyclic antidepressants,* which block reuptake of norepinephrine and serotonin by the nerve endings, so that these transmitters remain active for longer periods of time after secretion.

Mental depression can also be treated effectively by electroconvulsive therapy — commonly called "shock therapy." In this therapy an electric shock is used to cause a generalized seizure similar to that of an epileptic attack. This has also been shown to enhance norepinephrine transmission efficiency.

Some patients with mental depression alternate between depression and mania, which is called *manic-depressive psychosis,* and a few persons exhibit only mania without the depressive episodes. Drugs that diminish the formation or action of norepinephrine and serotonin, such as lithium compounds, can be effective in treating the manic condition.

Therefore, it is presumed that the norepinephrine system especially and perhaps the serotonin system as well normally function to provide motor drive to the limbic system to increase a person's sense of well-being, to create happiness, contentment, good appetite, appropriate sex drive, and psychomotor balance, although too much of a good thing can cause mania. In support of this concept is the fact that the pleasure and reward centers of the hypothalamus and surrounding areas receive large numbers of nerve endings from the norepinephrine system.

SCHIZOPHRENIA — EXAGGERATED FUNCTION OF PART OF THE DOPAMINE SYSTEM

Schizophrenia comes in many different varieties. One of the most common is the person who hears voices and has delusions of grandeur, or intense fear, or other types of feelings that are unreal. Schizophrenics are often highly paranoid, with a sense of persecution from outside sources; they may develop incoherent speech, dissociation of ideas, and abnormal sequences of thought; and they are often withdrawn, sometimes with abnormal posture and even rigidity.

There is reason to believe that schizophrenia results from excessive functioning of a group of neurons that secrete dopamine. These neurons are located in the ventral tegmentum of the mesencephalon, medial and superior to the substantia nigra. They give rise to the so-called *mesolimbic dopaminergic system,* which projects nerve fibers mainly into the medial and anterior portions of the limbic system, especially into the nucleus accumbens, the amygdala, the anterior caudate nucleus, and the anterior cingulate gyrus of the cortex, all of which are powerful behavioral control centers.

Some of the reasons for believing the mesolimbic dopaminergic system to be related to schizophrenia are the following: When Parkinson's disease patients are treated with L-dopa, which releases dopamine in the brain, the parkinsonian patient sometimes develops schizophrenic-like symptoms, indicating that excess dopaminergic activity can cause dissociation of a person's drives and thought patterns.

However, an even more compelling reason for believing that schizophrenia might be caused by excess production of dopamine is that those drugs that are effective in treating schizophrenia, such as chlorpromazine, haloperidol, and thiothixene, all decrease the secretion of dopamine by the dopaminergic nerve endings or decrease the effect of dopamine on the subsequent neurons.

Almost certainly there are other factors in schizophrenia besides excess secretion of dopamine; nevertheless, the symptoms of schizophrenia are similar to the behavioral effects of excessive dopamine.

ALZHEIMER'S DISEASE—LOSS OF ACETYLCHOLINE-SECRETING NEURONS

Alzheimer's disease is defined as premature aging of the brain, usually beginning in midadult life and progressing rapidly to extreme loss of mental powers as usually seen in very, very old age. These patients usually require continuous care within a few years after the disease begins.

A consistent finding in Alzheimer's disease is about 75 per cent loss of the neurons in the *nucleus basalis of Meynert*, located beneath the globus pallidus in the substantia innominata. The neurons of this nucleus send *acetylcholine-secreting* fibers to a large share of the neocortex. It is presumed that the acetylcholine in some way activates the neuronal mechanisms for storing and recalling memories. The nucleus basalis in turn receives input signals from multiple portions of the limbic system, which provides the motivational drive for the memory process that was discussed in the previous chapter.

Other neurotransmitters that have also been found deficient in Alzheimer's disease are *somatostatin* and *substance P*. Therefore, the basic cause of the disease might be more global, rather than simply a loss of one specific set of acetylcholine-secreting neurons.

REFERENCES

Ashton, H.: Brain Systems, Disorders and Psychotropic Drugs. New York, Oxford University Press, 1987.
Barnes, D. M.: Biological issues in schizophrenia. Science, 235:430, 1987.
Barnes, D. M.: Debate about epilepsy: What initiates seizures? Science, 234:938, 1986.
Bindman, L.: The Neurophysiology of the Cerebral Cortex. Austin, University of Texas Press, 1981.
Chrousos, G. P., et al. (eds.): Mechanisms of Physical and Emotional Stress. New York, Plenum Publishing Corp., 1988.
Conn, D. K., et al. (eds.): Psychiatric Consequences of Brain Disease in the Elderly. New York, Plenum Publishing Corp., 1989.
Cooper, A. J. L., and Plum, F.: Biochemistry and physiology of brain ammonia. Physiol. Rev., 67:440, 1987.
Dichter, M. A. (ed.): Mechanisms of Epileptogenesis. New York, Plenum Publishing Corp., 1988.
Dichter, M. A., and Ayala, G. F.: Cellular mechanisms of epilepsy: A status report. Science, 237:157, 1987.
DiDonato, S., et al. (eds.): Molecular Genetics of Neurological and Neuromuscular Disease. New York, Raven Press, 1988.
Georgotas, A., and Cancro, R. (eds.): Depression and Mania. New York, Elsevier Science Publishing Co., 1988.
Glaser, G. H., et al. (eds.): Antiepileptic Drugs. New York, Raven Press, 1980.
Hansen, A. J.: Effect of anoxia on ion distribution in the brain. Physiol. Rev., 65:101, 1985.
Hobson, J. A., and Brazier, M. A. B. (eds.): The Reticular Formation Revisited: Specifying Function for a Nonspecific System. New York, Raven Press, 1980.
Hyvarinen, J.: Posterior parietal lobe of the primate brain. Physiol. Rev., 62:1060, 1982.
Jacobs, B. L.: How hallucinogenic drugs work. Am. Sci., 75:386, 1987.
Jones, E. G.: Organization of the thalamocortical complex and its relation to sensory processes. In Darian-Smith, I. (ed.): Handbook of Physiology, Sec. 1, Vol. III. Bethesda, Md., American Physiological Society, 1984, p. 149.
Kaplan, H. I., and Sadock, B. J.: Synopsis of Psychiatry: Behavioral Sciences Clinical Psychiatry, 5th Ed. Baltimore, Williams & Wilkins, 1988.
Klee, M. (ed.): Physiology and Pharmacology of Epileptogenic Phenomena. New York, Raven Press, 1982.
Kryger, M. H., et al.: Principles and Practice of Sleep Medicine. Philadelphia, W. B. Saunders Co., 1986.
Livingston, R. B.: Sensory Processing, Perception, and Behavior. New York, Raven Press, 1978.
Mann, J. J. (ed.): The Phenomenology of Depressive Illness. New York, Plenum Publishing Corp., 1988.
McKinney, W. T.: Models of Mental Disorders. New York, Plenum Publishing Corp., 1988.
Meijer, J. H., and Rietveld, W. J.: Neurophysiology of the suprachiasmatic circadian pacemaker in rodents. Physiol. Rev., 69:671, 1989.
Mesulam, M. M.: The cholinergic connection in Alzheimer's disease. News Physiol. Sci., 1:107, 1986.
Mitzdorf, U.: Current source-density method and application in cat cerebral cortex: Investigation of evoked potentials and EEG phenomena. Physiol. Rev., 65:37, 1985.
Nappi, G., et al. (eds.): Neurodegenerative Disorders. New York, Raven Press, 1988.
Nerozzi, D., et al. (eds.): Hypothalamic Dysfunction in Neuropsychiatric Disorders. New York, Raven Press, 1987.
Newmark, M. E., and Penry, J. K.: Genetics of Epilepsy: A Review. New York, Churchill Livingstone, 1980.
Plum, F., and Posner, J. B.: The Diagnosis of Stupor and Coma, 3rd Ed. Philadelphia, F. A. Davis Co., 1980.
Pollack, M. H., et al.: Propranolol and depression revisited: Three cases and a review. J. Nerv. Ment. Dis., 173:118, 1985.
Stern, R. M., et al.: Psychophysiological Recording. New York, Oxford University Press, 1980.
Strong, R., et al. (eds.): Central Nervous System Disorders of Aging. New York, Raven Press, 1988.
Terry, R. D. (ed.): Aging and the Brain. New York, Raven Press, 1988.
Tucek, S.: Regulation of acetylcholine synthesis in the brain. J. Neurochem., 44:11, 1985.
Wauquier, A., et al.: Slow Wave Sleep: Physiological, Pathophysiological, and Functional Aspects. New York, Raven Press, 1989.
Wolman, B. B.: Psychosomatic Disorders. New York, Plenum Publishing Corp., 1988.

22

The Autonomic Nervous System; The Adrenal Medulla

The portion of the nervous system that controls the visceral functions of the body is called the *autonomic nervous system*. This system helps control arterial pressure, gastrointestinal motility and secretion, urinary bladder emptying, sweating, body temperature, and many other activities, some of which are controlled almost entirely and some only partially by the autonomic nervous system.

One of the most striking characteristics of the autonomic nervous system is the rapidity and intensity with which it can change visceral functions. For instance, within 3 to 5 sec it can increase the heart rate to two times normal, and the arterial pressure can be doubled within as little as 10 to 15 sec; or, at the other extreme, the arterial pressure can be decreased low enough within 4 to 5 sec to cause fainting. Sweating can begin within seconds, and the bladder may empty involuntarily, also within seconds. It is these extremely rapid changes that are measured by the lie detector polygraph, reflecting the innermost feelings of a person.

■ GENERAL ORGANIZATION OF THE AUTONOMIC NERVOUS SYSTEM

The autonomic nervous system is activated mainly by centers located in the *spinal cord, brain stem,* and *hypothalamus*. Also, portions of the cerebral cortex, especially of the limbic cortex, can transmit impulses to the lower centers and in this way influence autonomic control. Often the autonomic nervous system also operates by means of *visceral reflexes*. That is, sensory signals entering the autonomic ganglia, cord, brain stem, or hypothalamus can elicit appropriate reflex responses back to the visceral organs to control their activities.

The efferent autonomic signals are transmitted to the body through two major subdivisions called the *sympathetic nervous system* and the *parasympathetic nervous system*, the characteristics and functions of which follow.

PHYSIOLOGIC ANATOMY OF THE SYMPATHETIC NERVOUS SYSTEM

Figure 22–1 illustrates the general organization of the sympathetic nervous system, showing one of the two *paravertebral sympathetic chains of ganglia* that lie to the two sides of the spinal column, two *prevertebral ganglia* (the *celiac* and *hypogastric*), and nerves extending from the ganglia to the different internal organs. The sympathetic nerves originate in the spinal cord between the segments T-1 and L-2 and pass from here first into the sympathetic chain and thence to the tissues and organs that are stimulated by the sympathetic nerves.

Preganglionic and Postganglionic Sympathetic Neurons

The sympathetic nerves are different from skeletal motor nerves in the following way: Each sympathetic pathway from the cord to the stimulated tissue is composed of two neurons, a *preganglionic neuron* and a *postganglionic neuron*, in contrast to only a single neuron in the skeletal motor pathway. The cell body of each preganglionic neuron lies in the *intermediolateral horn* of the spinal cord; and its fiber passes, as illustrated in Figure 22–2, through an *anterior root* of the cord into the corresponding *spinal nerve*.

Immediately after the spinal nerve leaves the spinal column, the preganglionic sympathetic fibers leave the nerve and pass through the *white ramus* into one of the *ganglia* of the *sympathetic chain.* Then the course of the fibers can be one of the following three: (1) It can synapse with postganglionic neurons in the ganglion that it enters. (2) It can pass upward or downward in the chain and synapse in one of the other ganglia of the chain. Or (3) it can pass for variable distances through the chain and then through one of the *sympathetic nerves* radiating outward from the chain, finally terminating in one of the *prevertebral ganglia*.

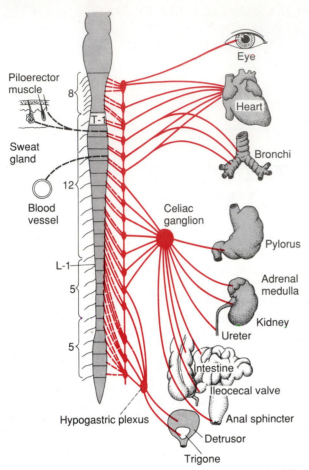

Figure 22–1. The sympathetic nervous system. Dashed lines represent postganglionic fibers in the gray rami leading into the spinal nerves for distribution to blood vessels, sweat glands, and piloerector muscles.

determined partly by the position in the embryo where the organ originates. For instance, the heart receives many sympathetic nerve fibers from the neck portion of the sympathetic chain because the heart originates in the neck of the embryo. Likewise, the abdominal organs receive their sympathetic innervation from the lower thoracic segments because most of the primitive gut originates in this area.

Special Nature of the Sympathetic Nerve Endings in the Adrenal Medullae. Preganglionic sympathetic nerve fibers pass, without synapsing, all the way from the intermediolateral horn cells of the spinal cord, through the sympathetic chains, through the splanchnic nerves, and finally into the adrenal medullae. There they end directly on modified neuronal cells that secrete *epinephrine* and *norepinephrine* into the blood stream. These secretory cells embryologically are derived from nervous tissue and are analogous to postganglionic neurons; indeed, they even have rudimentary nerve fibers, and it is these fibers that secrete the hormones.

PHYSIOLOGIC ANATOMY OF THE PARASYMPATHETIC NERVOUS SYSTEM

The parasympathetic nervous system is illustrated in Figure 22–3, showing that parasympathetic fibers leave the central nervous system through cranial nerves III, VII, IX, and X; the second and third sacral spinal nerves; and occasionally the first and fourth sacral nerves. About 75 per cent of all parasympathetic nerve fibers are in the vagus nerves, passing to the entire thoracic and abdominal regions of the body. Therefore, a physiologist speaking of the parasympathetic nervous system often thinks mainly of the two vagus nerves. The vagus nerves supply parasympathetic nerves to the heart, the lungs, the esophagus, the stomach, the entire small

The postganglionic neuron then originates either in one of the sympathetic chain ganglia or in one of the prevertebral ganglia. From either of these two sources, the postganglionic fibers travel to their destinations in the various organs.

Sympathetic Nerve Fibers in the Skeletal Nerves. Some of the postganglionic fibers pass back from the sympathetic chain into the spinal nerves through *gray rami* at all levels of the cord, illustrated by the dashed fiber in Figure 22–2. These pathways are made up of type C fibers that extend to all parts of the body in the skeletal nerves. They control the blood vessels, sweat glands, and piloerector muscles of the hairs. Approximately 8 per cent of the fibers in the average skeletal nerve are sympathetic fibers, a fact that indicates their importance.

Segmental Distribution of the Sympathetic Nerves. The sympathetic pathways originating in the different segments of the spinal cord are not necessarily distributed to the same part of the body as the spinal nerve fibers from the same segments. Instead, the *sympathetic fibers from cord segment T-1 generally pass up the sympathetic chain to the head; from T-2 into the neck; from T-3, T-4, T-5, and T-6 into the thorax; from T-7, T-8, T-9, T-10, and T-11 into the abdomen; and from T-12, L-1, and L-2 into the legs.* This distribution is only approximate and overlaps greatly.

The distribution of sympathetic nerves to each organ is

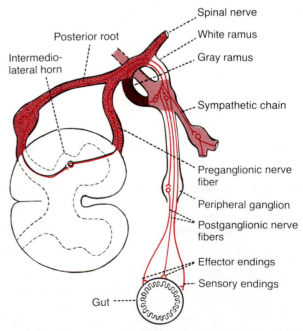

Figure 22–2. Nerve connections between the spinal cord, sympathetic chain, spinal nerves, and peripheral sympathetic nerves.

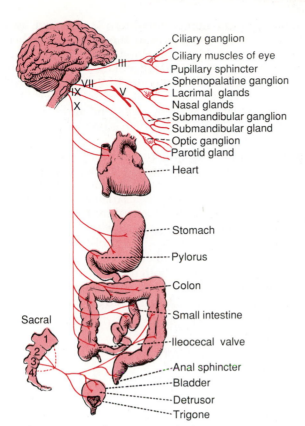

Ciliary ganglion
Ciliary muscles of eye
Pupillary sphincter
Sphenopalatine ganglion
Lacrimal glands
Nasal glands
Submandibular ganglion
Submandibular gland
Optic ganglion
Parotid gland
Heart
Stomach
Pylorus
Colon
Small intestine
Ileocecal valve
Anal sphincter
Bladder
Detrusor
Trigone

Sacral

Figure 22–3. The parasympathetic nervous system.

■ BASIC CHARACTERISTICS OF SYMPATHETIC AND PARASYMPATHETIC FUNCTION

CHOLINERGIC AND ADRENERGIC FIBERS — SECRETION OF ACETYLCHOLINE OR NOREPINEPHRINE

The sympathetic and parasympathetic nerve fibers all secrete one of the two synaptic transmitter substances, *acetylcholine* or *norepinephrine.* Those that secrete acetylcholine are said to be *cholinergic.* Those that secrete norepinephrine are said to be *adrenergic,* a term derived from *adrenaline,* which is the British name for epinephrine.

All *preganglionic neurons* are *cholinergic* in both the sympathetic and parasympathetic nervous systems. Therefore, acetylcholine or acetylcholine-like substances, when applied to the ganglia, will excite both sympathetic and parasympathetic postganglionic neurons.

The *postganglionic neurons of the parasympathetic system are* also *all cholinergic.*

On the other hand, *most of the postganglionic sympathetic neurons are adrenergic,* though this is not entirely true, because the postganglionic sympathetic nerve fibers to the sweat glands, to the piloerector muscles, and to a few blood vessels are cholinergic.

Thus, the terminal nerve endings of the parasympathetic system *all* secrete *acetylcholine,* and *most* of the sympathetic nerve endings secrete *norepinephrine.* These hormones, in turn, act on the different organs to cause the respective parasympathetic and sympathetic effects. Therefore, they are often called, respectively, *parasympathetic* and *sympathetic transmitters.*

These are the molecular structures of acetylcholine and norepinephrine:

Mechanisms of Transmitter Secretion and Removal at the Postganglionic Endings

Secretion of Acetylcholine and Norepinephrine by Postganglionic Nerve Endings. Some of the postganglionic autonomic nerve endings, especially those of the parasympathetic nerves, are similar to but much smaller in size than those of the skeletal neuromuscular junction. However, most of the sympathetic nerve fibers merely touch the effector cells of the organs that they innervate as they pass by; and in some instances they terminate in connective tissue located adjacent to the cells that are to be stimulated. Where these filaments pass over or near the effector cells, they usually have bulbous enlargements called *varicosities;* it is in these varicosities that the transmitter vesicles of acetylcholine or norepinephrine are found. Also in the varicosities are large numbers of mitochondria to sup-

intestine, the proximal half of the colon, the liver, the gallbladder, the pancreas, and the upper portions of the ureters.

Parasympathetic fibers in the *third nerve* flow to the pupillary sphincters and ciliary muscles of the eye. Fibers from the *seventh nerve* pass to the lacrimal, nasal, and submandibular glands, and fibers from the *ninth nerve* pass to the parotid gland.

The sacral parasympathetic fibers congregate in the form of the *nervi erigentes,* also called the *pelvic nerves,* which leave the sacral plexus on each side of the cord and distribute their peripheral fibers to the descending colon, rectum, bladder, and lower portions of the ureters. Also, this sacral group of parasympathetics supplies fibers to the external genitalia to cause sexual stimulation.

Preganglionic and Postganglionic Parasympathetic Neurons. The parasympathetic system, like the sympathetic, has both preganglionic and postganglionic neurons. However, except in the case of a few cranial parasympathetic nerves, the *preganglionic fibers* pass uninterrupted all the way to the organ that is to be controlled. Then, in the wall of the organ are located the *postganglionic neurons.* The preganglionic fibers synapse with these, and short postganglionic fibers, 1 mm to several centimeters in length, leave the neurons to spread through the substance of the organ. This location of the parasympathetic postganglionic neurons in the visceral organ itself is quite different from the arrangement of the sympathetic ganglia, for the cell bodies of the sympathetic postganglionic neurons are almost always located in ganglia of the sympathetic chain or in various other discrete ganglia in the abdomen, rather than in the excited organ itself.

ply the adenosine triphosphate (ATP) required to energize acetylcholine or norepinephrine synthesis.

When an action potential spreads over the terminal fibers, the depolarization process increases the permeability of the fiber membrane to calcium ions, allowing these to diffuse into the nerve terminals. There they interact with the vesicles that are adjacent to the membrane, causing them to fuse with the membrane and to empty their contents to the exterior. Thus, the transmitter substance is secreted.

Synthesis of Acetylcholine, Its Destruction After Secretion, and Duration of Action. Acetylcholine is synthesized in the terminal endings of cholinergic nerve fibers. Most of this synthesis occurs in the axoplasm outside the vesicles, and then the acetylcholine is transported to the interior of the vesicles, where it is stored in a highly concentrated form until it is released. The basic chemical reaction of this synthesis is the following:

$$\text{Acetyl-CoA} + \text{Choline} \xrightarrow{\substack{\text{choline acetyl-} \\ \text{transferase}}} \text{Acetylcholine}$$

Once the acetylcholine has been secreted by the cholinergic nerve ending, it persists in the tissue for a few seconds; then most of it is split into an acetate ion and choline by the enzyme *acetylcholinesterase* bound with collagen and glycosaminoglycans in the local connective tissue. Thus, this is the same mechanism of acetylcholine destruction that occurs at the neuromuscular junctions of skeletal nerve fibers. The choline that is formed is in turn transported back into the terminal nerve ending, where it is used again for synthesis of new acetylcholine.

Synthesis of Norepinephrine, Its Removal, and Duration of Action. Synthesis of norepinephrine begins in the axoplasm of the terminal nerve endings of adrenergic nerve fibers but is completed inside the vesicles. The basic steps are the following:

1. Tyrosine $\xrightarrow{\text{hydroxylation}}$ DOPA

2. DOPA $\xrightarrow{\text{decarboxylation}}$ Dopamine

3. Transport of dopamine into the vesicles

4. Dopamine $\xrightarrow{\text{hydroxylation}}$ Norepinephrine

In the adrenal medulla this reaction goes still one step further to transform about 80 per cent of the norepinephrine into epinephrine, as follows:

5. Norepinephrine $\xrightarrow{\text{methylation}}$ Epinephrine

After secretion of norepinephrine by the terminal nerve endings, it is removed from the secretory site in three different ways: (1) reuptake into the adrenergic nerve endings themselves by an active transport process—accounting for removal of 50 to 80 per cent

of the secreted norepinephrine; (2) diffusion away from the nerve endings into the surrounding body fluids and thence into the blood—accounting for removal of most of the remainder of the norepinephrine; and (3) destruction by enzymes to a slight extent (one of these enzymes is *monoamine oxidase*, which is found in the nerve endings themselves, and another is *catechol-O-methyl transferase*, which is present diffusely in all tissues).

Ordinarily, the norepinephrine secreted directly into a tissue remains active for only a few seconds, illustrating that its reuptake and diffusion away from the tissue are rapid. However, the norepinephrine and epinephrine secreted into the blood by the adrenal medullae remain active until they diffuse into some tissue where they are destroyed by catechol-O-methyl transferase; this occurs mainly in the liver. Therefore, when secreted into the blood, both norepinephrine and epinephrine remain very active for 10 to 30 sec; additional decreasing activity follows for 1 to several minutes.

RECEPTORS OF THE EFFECTOR ORGANS

Before the acetylcholine, norepinephrine, or epinephrine transmitter secreted at the autonomic nerve endings can stimulate the effector organ, it must first bind with highly specific *receptors* of the effector cells. The receptor is usually on the outside of the cell membrane, bound as a prosthetic group to a protein molecule that penetrates all the way through the cell membrane. When the transmitter binds with the receptor, this generally causes a conformational change in the structure of the protein molecule. In turn, the altered protein molecule excites or inhibits the cell, most often by (1) causing a change in the cell membrane permeability to one or more ions or (2) activating or inactivating an enzyme attached to the other end of the receptor protein where it protrudes into the interior of the cell.

Excitation or Inhibition of the Effector Cell by Changing Its Membrane Permeability. Because the receptor protein is an integral part of the cell membrane, a conformational change in the structures of many of these proteins opens or closes *ion channels*, thus altering the permeability of the cell membrane to various ions. For instance, sodium and/or calcium ion channels frequently become opened and allow rapid influx of the respective ions into the cell, usually depolarizing the cell membrane and exciting the cell. At other times, potassium channels are opened, allowing potassium ions to diffuse out of the cell, and this usually inhibits it. Also, in some cells, the ions will cause an internal cell action, such as the direct effect of calcium ions in promoting smooth muscle contraction.

Receptor Action by Altering Intracellular Enzymes. Another way the receptor functions is in activating or inactivating an enzyme (or other intracellular chemical) inside the cell. The enzyme usually is attached to the receptor protein where it protrudes into the interior of the cell. For instance, binding of epinephrine with its receptor on the outside of many cells increases the activity of the enzyme *adenylcyclase* on the inside of the cell, and this then causes the formation of *cyclic adenosine monophosphate (cAMP)*. The cAMP in turn can initiate any one of many different intracellular actions, the exact effect depending on the chemical machinery of the effector cell.

Therefore, it is easy to understand how an autonomic transmitter substance can cause inhibition in some organs or excitation in others. This is usually determined by the nature of the receptor protein in the cell membrane and the effect of receptor binding on its conformational state. In each organ, the resulting effects are likely to be entirely different from those in other organs.

The Acetylcholine Receptors — Muscarinic and Nicotinic Receptors

Acetylcholine activates two different types of receptors. These are called *muscarinic* and *nicotinic* receptors. The reason for these names is that muscarine, a poison from toadstools, activates only the muscarinic receptors but will not activate the nicotinic receptors, whereas nicotine will activate only nicotinic receptors; acetylcholine activates both of them.

The muscarinic receptors are found in all effector cells stimulated by the postganglionic neurons of the parasympathetic nervous system, as well as those stimulated by the postganglionic cholinergic neurons of the sympathetic system.

The nicotinic receptors are found in the synapses between the pre- and postganglionic neurons of both the sympathetic and parasympathetic systems and also in the membranes of skeletal muscle fibers at the neuromuscular junction (discussed in Chapter 25).

An understanding of the two different types of receptors is especially important because specific drugs are frequently used in the practice of medicine to stimulate or to block one or the other of the two types of receptors.

The Adrenergic Receptors — Alpha and Beta Receptors

Research experiments using different drugs that mimic the action of norepinephrine on sympathetic effector organs (called *sympathomimetic drugs*) have shown that there are two major types of adrenergic receptors, *alpha receptors* and *beta receptors*. (The beta receptors in turn are divided into *beta₁* and *beta₂* receptors because certain drugs affect some beta receptors but not all. Also, there is a less distinct division of alpha receptors into *alpha₁* and *alpha₂* receptors.)

Norepinephrine and epinephrine, both of which are secreted by the adrenal medulla, have somewhat different effects in exciting the alpha and beta receptors. Norepinephrine excites mainly alpha receptors but excites the beta receptors to a slight extent as well. On the other hand, epinephrine excites both types of receptors approximately equally. Therefore, the relative effects of norepinephrine and epinephrine on different effector organs is determined by the types of receptors in the organs. Obviously, if they are all beta receptors, epinephrine will be the more effective excitant.

Table 22–1 gives the distribution of alpha and beta receptors in some of the organs and systems controlled by the sympathomimetics. Note that certain alpha functions are excitatory while others are inhibitory. Likewise, certain beta functions are excitatory and others are inhibitory. Therefore, alpha and beta receptors are not necessarily associated with excitation or inhibition but simply with the affinity of the hormone for the receptors in the given effector organ.

A synthetic hormone chemically similar to epinephrine and norepinephrine, *isopropyl norepinephrine*, has an extremely strong action on beta receptors but essentially no action on alpha receptors.

EXCITATORY AND INHIBITORY ACTIONS OF SYMPATHETIC AND PARASYMPATHETIC STIMULATION

Table 22–2 lists the effects on different visceral functions of the body caused by stimulating the parasympathetic and sympathetic nerves. From this table it can be seen again that *sympathetic stimulation causes excitatory effects in some organs but inhibitory effects in others. Likewise, parasympathetic stimulation causes excitation in some but inhibition in others.* Also, when sympathetic stimulation excites a particular organ, parasympathetic stimulation sometimes inhibits it, illustrating that the two systems occasionally act reciprocally to each other. However, most organs are dominantly controlled by one or the other of the two systems.

There is no generalization one can use to explain whether sympathetic or parasympathetic stimulation will cause excitation or inhibition of a particular organ. Therefore, to understand sympathetic and parasympathetic function, one must learn the functions of these two nervous systems as listed in Table 22–2. Some of these functions need to be clarified in still greater detail as follows:

TABLE 22–1 Adrenergic Receptors and Function

Alpha Receptor	Beta Receptor
Vasoconstriction	Vasodilatation (β_2)
Iris dilatation	Cardioacceleration (β_1)
Intestinal relaxation	Increased myocardial strength (β_1)
Intestinal sphincter	Increased myocardial strength (β_1)
Intestinal sphincter contraction	Intestinal relaxation (β_2)
Pilomotor contraction	Uterus relaxation (β_2)
Bladder sphincter contraction	Bronchodilatation (β_2)
	Calorigenesis (β_2)
	Glycogenolysis (β_2)
	Lipolysis (β_1)
	Bladder wall relaxation (β_2)

TABLE 22 – 2 Autonomic Effects on Various Organs of the Body

Organ	Effect of Sympathetic Stimulation	Effect of Parasympathetic Stimulation
Eye		
Pupil	Dilated	Constricted
Ciliary muscle	Slight relaxation (far vision)	Constricted (near vision)
Glands	Vasoconstriction and slight secretion	Stimulation of copious secretion (containing many enzymes for enzyme-secreting glands)
Nasal		
Lacrimal		
Parotid		
Submandibular		
Gastric		
Pancreatic		
Sweat glands	Copious sweating (cholinergic)	Sweating on palms of hands
Apocrine glands	Thick, odoriferous secretion	None
Heart		
Muscle	Increased rate	Slowed rate
	Increased force of contraction	Decreased force of contraction (especially of atria)
Coronaries	Dilated (β_2); constricted (α)	Dilated
Lungs		
Bronchi	Dilated	Constricted
Blood vessels	Mildly constricted	? Dilated
Gut		
Lumen	Decreased peristalsis and tone	Increased peristalsis and tone
Sphincter	Increased tone (most times)	Relaxed (most times)
Liver	Glucose released	Slight glycogen synthesis
Gallbladder and bile ducts	Relaxed	Contracted
Kidney	Decreased output and renin secretion	None
Bladder		
Detrusor	Relaxed (slight)	Contracted
Trigone	Contracted	Relaxed
Penis	Ejaculation	Erection
Systemic arterioles		
Abdominal viscera	Constricted	None
Muscle	Constricted (adrenergic α)	None
	Dilated (adrenergic β_2)	
	Dilated (cholinergic)	
Skin	Constricted	None
Blood		
Coagulation	Increased	None
Glucose	Increased	None
Lipids	Increased	None
Basal metabolism	Increased up to 100%	None
Adrenal medullary secretion	Increased	None
Mental activity	Increased	None
Piloerector muscles	Contracted	None
Skeletal muscle	Increased glycogenolysis	None
	Increased strength	
Fat cells	Lipolysis	None

EFFECTS OF SYMPATHETIC AND PARASYMPATHETIC STIMULATION ON SPECIFIC ORGANS

The Eye. Two functions of the eye are controlled by the autonomic nervous system. These are the pupillary opening and the focus of the lens. Sympathetic stimulation contracts the meridional *fibers of the iris* that dilate the pupil, whereas parasympathetic stimulation contracts the *circular muscle of the iris* to constrict the pupil. The parasympathetics that control the pupil are reflexly stimulated when excess light enters the eyes, which is explained in Chapter 13; this reflex reduces the pupillary opening and decreases the amount of light that strikes the retina. On the other hand, the sympathetics become stimulated during periods of excitement and, therefore, increase the pupillary opening at these times.

Focusing of the lens is controlled almost entirely by the parasympathetic nervous system. The lens is normally held in a flattened state by intrinsic elastic tension of its radial ligaments. Parasympathetic excitation contracts the *ciliary muscle*, which releases this tension and allows the lens to become more convex. This causes the eye to focus on objects near at hand. The focusing mechanism is discussed in Chapters 11 and 13 in relation to the function of the eyes.

The Glands of the Body. The *nasal, lacrimal, salivary,* and many *gastrointestinal glands* are all strongly stimulated by the parasympathetic nervous system, usually resulting in copious quantities of secretion. The glands of the alimentary tract most strongly stimulated by the parasympathetics are those of the upper tract, especially those of the mouth and stomach. The glands of the small and large intestines are controlled principally by local factors in the intestinal tract itself and not by the autonomic nerves.

Sympathetic stimulation has a slight direct effect on glandular cells in causing formation of a concentrated secretion. However, it also causes vasoconstriction of the blood vessels supplying the glands and in this way often reduces their rates of secretion.

The *sweat glands* secrete large quantities of sweat when the sympathetic nerves are stimulated, but no effect is caused by stimulating the parasympathetic nerves. However, the sympathetic fibers to most sweat glands are *cholinergic* (except for a few adrenergic fibers to palms of the hand and the soles of the feet), in contrast to almost all other sympathetic fibers, which are adrenergic. Furthermore, the sweat glands are stimulated primarily by centers in the hypothalamus that are usually considered to be parasympathetic centers. Therefore, sweating could be called a parasympathetic function.

The *apocrine glands* in the axillae secrete a thick, odoriferous secretion as a result of sympathetic stimulation, but they do not react to parasympathetic stimulation. Instead, the apocrine glands, despite their close embryological relationship to sweat glands, are controlled by adrenergic fibers rather than by cholinergic fibers and are controlled by the sympathetic centers of the central nervous system rather than by the parasympathetic centers.

The Gastrointestinal System. The gastrointestinal system has its own intrinsic set of nerves known as the *intramural plexus.* However, both parasympathetic and sympathetic stimulation can affect gastrointestinal activity. Parasympathetic stimulation, in general, increases the overall degree of activity of the gastrointestinal tract by promoting peristalsis and relaxing the sphincters, thus allowing rapid propulsion of contents along the tract. This propulsive effect is associated with simultaneous increases in rates of secretion by many of the gastrointestinal glands, which was described earlier.

Normal function of the gastrointestinal tract is not very dependent on sympathetic stimulation. However, strong sympathetic stimulation inhibits peristalsis and increases the tone of the sphincters. The net result is greatly slowed propulsion of food through the tract and sometimes decreased secretion as well.

The Heart. In general, sympathetic stimulation increases the overall activity of the heart. This is accomplished by increasing both the rate and force of heart contraction. Parasympathetic stimulation causes mainly the opposite effects. To express these effects in another way, sympathetic stimulation increases the effectiveness of the heart as a pump, whereas parasympathetic stimulation decreases its pumping capability.

Systemic Blood Vessels. Most systemic blood vessels, especially those of the abdominal viscera and the skin of the limbs, are constricted by sympathetic stimulation. Parasympathetic stimulation generally has almost no effect on blood vessels but does dilate vessels in certain restricted areas such as in the blush area of the face. Under some conditions, the beta function of the sympathetics causes vascular dilatation, especially when drugs have paralyzed the sympathetic alpha vasoconstrictor effects, which are usually by far dominant over the beta effects.

Effect of Sympathetic and Parasympathetic Stimulation on Arterial Pressure. The arterial pressure is determined by two factors, the propulsion of blood by the heart and the resistance to flow of this blood through the blood vessels. Sympathetic stimulation increases both propulsion by the heart and resistance to flow, which usually causes the pressure to increase greatly.

On the other hand, parasympathetic stimulation decreases the pumping by the heart but has virtually no effect on total peripheral resistance. The usual effect is a slight fall in pressure. Yet very strong vagal parasympathetic stimulation can occasionally stop the heart entirely and cause loss of all arterial pressure.

Effects of Sympathetic and Parasympathetic Stimulation on Other Functions of the Body. Because of the great importance of the sympathetic and parasympathetic control systems, these are discussed many times in this text in relation to a myriad of body functions that are not considered in detail here. In general, most of the entodermal structures, such as the ducts of the liver, the gallbladder, the ureter, the bladder, and the bronchi, are inhibited by sympathetic stimulation but excited by parasympathetic stimulation. Also, sympathetic stimulation has metabolic effects, causing release of glucose from the liver, increase in blood glucose concentration, increase in glycogenolysis in both liver and muscle, increase in muscle strength, increase in basal metabolic rate, and increase in mental activity. Finally, the sympathetics and parasympathetics are involved in the execution of the male and female sexual acts, as are explained in Chapter 29.

FUNCTION OF THE ADRENAL MEDULLAE

Stimulation of the sympathetic nerves to the adrenal medullae causes large quantities of epinephrine and norepinephrine to be released into the circulating blood, and these two hormones in turn are carried in the blood to all tissues of the body. On the average, approximately 80 per cent of the secretion is epinephrine and 20 per cent is norepinephrine, though the relative proportions can change considerably under different physiological conditions.

The circulating epinephrine and norepinephrine have almost the same effects on the different organs as those caused by direct sympathetic stimulation, except that *the effects last 5 to 10 times as long* because these hormones are removed from the blood slowly.

The circulating norepinephrine causes constriction of essentially all the blood vessels of the body; it causes increased activity of the heart, inhibition of the gastrointestinal tract, dilation of the pupils of the eyes, and so forth.

Epinephrine causes almost the same effects as those caused by norepinephrine, but the effects differ in the following respects: First, epinephrine, because of its greater effect in stimulating the beta receptors, has a greater effect on cardiac stimulation than norepinephrine. Second, epinephrine causes only weak constriction of the blood vessels of the muscles, in comparison with the much stronger constriction caused by norepinephrine. Because the muscle vessels represent a major segment of the vessels of the body, this difference is of special importance because norepinephrine greatly increases the total peripheral resistance and thereby greatly elevates arterial pressure, whereas epinephrine raises the arterial pressure to a lesser extent but increases the cardiac output considerably more because of its excitatory effect on the heart.

A third difference between the actions of epinephrine and norepinephrine relates to their effects on tissue metabolism. Epinephrine has up to 5 to 10 times as great a metabolic effect as norepinephrine. Indeed, the epinephrine secreted by the adrenal medullae can increase the metabolic rate of the body often to as much as 100 per cent above normal, in this way increasing the activity and excitability of the whole body. It also increases the rate of other metabolic activities, such as glycogenolysis in the liver and muscle and glucose release into the blood.

In summary, stimulation of the adrenal medullae causes the release of hormones that have almost the same effects throughout the body as direct sympathetic stimulation, except that the effects are greatly prolonged, up to a minute or two after the stimulation is over. The only significant differences are caused by the beta effects of the epinephrine in the secretion, which increase the rate of metabolism and cardiac output to a greater extent than is caused by direct sympathetic stimulation.

Value of the Adrenal Medullae to the Function of the Sympathetic Nervous System. Epinephrine and norepinephrine are almost always released by the adrenal medullae at the same time that the different organs are stimulated directly by generalized sympathetic activation. Therefore, the organs are actually stimulated in two different ways simultaneously, directly by the sympathetic nerves and indirectly by the medullary hormones. The two means of stimulation support each other, and either can usually substitute for the other. For instance, destruction of the direct sympathetic pathways to the organs does not abrogate excitation of the organs, because norepinephrine and epinephrine are still released into the circulating blood and indirectly cause stimulation. Likewise, total loss of the two adrenal medullae usually has little effect on the operation of the sympathetic nervous system because the direct pathways can still perform almost all the necessary duties. Thus, the dual mechanism of sympathetic stimulation provides a safety factor, one mechanism substituting for the other when it is missing.

Another important value of the adrenal medullae is the capability of epinephrine and norepinephrine to stimulate structures of the body that are not innervated by direct sympathetic fibers. For instance, the metabolic rate of every cell of the body is increased by these hormones, especially by epinephrine, even though only a small proportion of all the cells in the body are innervated directly by sympathetic fibers.

RELATIONSHIP OF STIMULUS RATE TO DEGREE OF SYMPATHETIC AND PARASYMPATHETIC EFFECT

A special difference between the autonomic nervous system and the skeletal nervous system is that only a very low frequency of stimulation is required for full activation of autonomic effectors. In general, only one nerve impulse every second or so suffices to maintain normal sympathetic or parasympathetic effect, and full activation occurs when the nerve fibers discharge 10 to 20 times/sec. This compares with full activation in the skeletal nervous system at 50 to 500 or more impulses/sec.

SYMPATHETIC AND PARASYMPATHETIC "TONE"

The sympathetic and parasympathetic systems are continually active, and the basal rates of activity are known, respectively, as *sympathetic tone* or *parasympathetic tone.*

The value of tone is that *it allows a single nervous system to increase or to decrease the activity of a stimulated organ.* For instance, sympathetic tone normally keeps almost all of the systemic arterioles constricted to approximately half their maximum diameter. By increasing the degree of sympathetic stimulation, these vessels can be constricted even more; but, on the other hand, by inhibiting the normal tone, they can be dilated. If it were not for the continual sympathetic tone, the sympathetic system could cause only vasoconstriction, never vasodilatation.

Another interesting example of tone is that of the parasympathetics in the gastrointestinal tract. Surgical removal of the parasympathetic supply to most of the gut by cutting the vagus nerves can cause serious and prolonged gastric and intestinal "atony" with resulting blockage of gastrointestinal propulsion and consequent serious constipation, thus illustrating that parasympathetic tone to the gut is normally very strong. This tone can be decreased by the brain, thereby inhibiting gastrointestinal motility, or it can be increased, thereby promoting increased gastrointestinal activity.

Tone Caused by Basal Secretion of Epinephrine and Norepinephrine by the Adrenal Medullae. The normal resting rate of secretion by the adrenal medullae is about 0.2 μg/kg/min of epinephrine and about 0.05 μg/kg/min of norepinephrine. These quantities are considerable—indeed, enough to maintain the blood pressure almost up to the normal value even if all direct sympathetic pathways to the cardiovascular system are removed. Therefore, it is obvious that much of the overall tone of the sympathetic nervous system results from basal secretion of epinephrine and norepinephrine in addition to the tone resulting from direct sympathetic stimulation.

Effect of Loss of Sympathetic or Parasympathetic Tone Following Denervation. Immediately after a sympathetic or parasympathetic nerve is cut, the innervated organ loses its sympathetic or parasympathetic tone. In the case of the blood vessels, for instance, cutting the sympathetic nerves results immediately in almost maximal vasodilatation. However, over minutes, hours, days, or weeks, *intrinsic tone* in the smooth muscle of the vessels increases, usually eventually restoring almost normal vasoconstriction.

Essentially the same events occur in most effector organs whenever sympathetic or parasympathetic tone is lost. That is, intrinsic compensation soon develops to return the function of the organ almost to its normal basal level. However, in the parasympathetic system, the compensation sometimes requires many months. For instance, loss of parasympathetic tone to the heart increases the heart rate to 160 beats/min in a

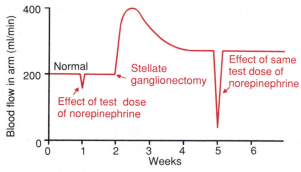

Figure 22–4. Effect of sympathectomy on blood flow in the arm and the effect of a test dose of norepinephrine before and after sympathectomy, showing sensitization of the vasculature to norepinephrine.

dog, and this will still be partially elevated 6 months later.

DENERVATION SUPERSENSITIVITY OF SYMPATHETIC AND PARASYMPATHETIC ORGANS FOLLOWING DENERVATION

During the first week or so after a sympathetic or parasympathetic nerve is destroyed, the innervated organ becomes more and more sensitive to injected norepinephrine or acetylcholine, respectively. This effect is illustrated in Figure 22–4, showing the blood flow in the forearm before removal of the sympathetics to be about 200 ml/min; a test dose of norepinephrine causes only a slight depression in flow. Then the stellate ganglion is removed, and normal sympathetic tone is lost. At first, the blood flow rises markedly because of the lost vascular tone, but over a period of days to weeks the blood flow returns almost to normal because of progressive increase in intrinsic tone of the vascular musculature itself, thus compensating for the loss of sympathetic tone. Another test dose of norepinephrine is then administered; and the blood flow decreases much more than before, illustrating that the blood vessels become about two to four times as responsive to norepinephrine as previously. This phenomenon is called *denervation supersensitivity*. It occurs in both sympathetic and parasympathetic organs and to a far greater extent in some organs than in others, often increasing the response more than tenfold.

Mechanism of Denervation Supersensitivity. The cause of denervation supersensitivity is only partially known. Part of the answer is that the number of receptors in the postsynaptic membranes of the effector cells increases—sometimes manyfold—when norepinephrine or acetylcholine is no longer released at the synapses, a process called "up-regulation" of the receptors. Therefore, when these hormones are injected temporarily into the circulating blood, the effector reaction is vastly enhanced.

■ THE AUTONOMIC REFLEXES

Many of the visceral functions of the body are regulated by *autonomic reflexes*. Throughout this text the functions of these reflexes are discussed in relation to individual organ systems; but to illustrate their importance, a few are presented here briefly.

Cardiovascular Autonomic Reflexes. Several reflexes in the cardiovascular system help to control especially the arterial blood pressure and the heart rate. One of these is the *baroreceptor reflex*, which is described in Chapter 27 along with other cardiovascular reflexes. Briefly, stretch receptors called *baroreceptors* are located in the walls of the major arteries, including the carotid arteries and the aorta. When these become stretched by high pressure, signals are transmitted to the brain stem, where they inhibit the sympathetic impulses to the heart and blood vessels, which allows the arterial pressure to fall back toward normal.

Gastrointestinal Autonomic Reflexes. The uppermost part of the gastrointestinal tract and also the rectum are controlled principally by autonomic reflexes. For instance, the smell of appetizing food or the presence of food in the mouth initiates signals from the nose and mouth to the vagal, glossopharyngeal, and salivary nuclei of the brain stem. These in turn transmit signals through the parasympathetic nerves to the secretory glands of the mouth and stomach, causing secretion of digestive juices even before food enters the mouth. And when fecal matter fills the rectum at the other end of the alimentary canal, sensory impulses initiated by stretching the rectum are sent to the sacral portion of the spinal cord, and a reflex signal is retransmitted through the parasympathetics to the distal parts of the colon; these result in strong peristaltic contractions that empty the bowel.

Other Autonomic Reflexes. Emptying of the bladder is controlled in the same way as emptying the rectum; stretching of the bladder sends impulses to the sacral cord, and this in turn causes contraction of the bladder as well as relaxation of the urinary sphincters, thereby promoting micturition.

Also important are the sexual reflexes, which are initiated both by psychic stimuli from the brain and stimuli from the sexual organs. Impulses from these sources converge on the sacral cord and, in the male, result, first, in erection, mainly a parasympathetic function, and then in ejaculation, a sympathetic function.

Other autonomic reflexes include reflex contributions to the regulation of pancreatic secretion, gallbladder emptying, kidney excretion of urine, sweating, blood glucose concentration, and many other visceral functions, all of which are discussed in detail at other points in this text.

■ STIMULATION OF DISCRETE ORGANS IN SOME INSTANCES AND MASS STIMULATION IN OTHER INSTANCES BY THE SYMPATHETIC AND PARASYMPATHETIC SYSTEMS

The Sympathetic System. In many instances, the sympathetic nervous system discharges almost as a complete unit, a phenomenon called *mass discharge*. This frequently occurs when the hypothalamus is activated by fright or fear or severe pain. The result is a widespread reaction throughout the body called the *alarm* or *stress response*, which we shall discuss shortly.

However, at other times sympathetic activation occurs in isolated portions of the system. The most important of these are the following: (1) In the process of heat regulation, the sympathetics control sweating and blood flow in the skin without affecting other organs innervated by the sympathetics. (2) During muscular activity in some animals, specific cholinergic vasodilator fibers of the skeletal muscles are stimulated independently, apart from the remainder of the sympathetic system. (3) Many "local reflexes" involving sensory afferent fibers that travel centrally in the sympathetic nerves to the sympathetic ganglia and spinal cord cause highly localized reflex responses. For instance, heating a local skin area causes local vasodilatation and enhanced local sweating, whereas cooling causes the opposite effects. (4) Many of the sympathetic reflexes that control gastrointestinal functions are very discrete, operating sometimes by way of nerve pathways that do not even enter the spinal cord, merely passing from the gut to the sympathetic ganglia, mainly the prevertebral ganglia, and then back to the gut through the sympathetic nerves to control motor or secretory activity.

The Parasympathetic System. In contrast to the sympathetic system, control functions of the parasympathetic system are much more likely to be highly specific. For instance, parasympathetic cardiovascular reflexes usually act only on the heart to increase or decrease its rate of beating. Likewise, other parasympathetic reflexes cause secretion mainly in the mouth glands, whereas in other instances secretion mainly in the stomach glands. Finally, the rectal emptying reflex does not affect other parts of the bowel to a major extent.

Yet there is often association between closely allied parasympathetic functions. For instance, though salivary secretion can occur independently of gastric secretion, these two often also occur together, and pancreatic secretion frequently occurs at the same time. Also, the rectal emptying reflex often initiates a bladder emptying reflex, resulting in simultaneous emptying of both the bladder and rectum. Conversely, the bladder emptying reflex can help initiate rectal emptying.

"ALARM" OR "STRESS" RESPONSE OF THE SYMPATHETIC NERVOUS SYSTEM

When large portions of the sympathetic nervous system discharge at the same time—that is, a *mass discharge*—this increases in many different ways the ability of the body to perform vigorous muscle activity. Let us quickly summarize these ways:

1. Increased arterial pressure.
2. Increased blood flow to active muscles concurrent with decreased blood flow to organs such as the gastrointestinal tract and the kidneys that are not needed for rapid motor activity.
3. Increased rates of cellular metabolism throughout the body.
4. Increased blood glucose concentration.
5. Increased glycolysis in the liver and in muscle.
6. Increased muscle strength.
7. Increased mental activity.
8. Increased rate of blood coagulation.

The sum of these effects permits the person to perform far more strenuous physical activity than would otherwise be possible. Because it is mental or physical *stress* that usually excites the sympathetic system, it is frequently said that the purpose of the sympathetic system is to provide extra activation of the body in states of stress: this is often called the sympathetic *stress response.*

The sympathetic system is especially strongly activated in many emotional states. For instance, in the state of *rage*, which is elicited mainly by stimulating the hypothalamus, signals are transmitted downward through the reticular formation and spinal cord to cause massive sympathetic discharge, and all of the sympathetic events listed above ensue immediately. This is called the sympathetic *alarm reaction.* It is also frequently called the *fight or flight reaction* because an animal in this state decides almost instantly whether to stand and fight or to run. In either event, the sympathetic alarm reaction makes the animal's subsequent activities vigorous.

MEDULLARY, PONTINE, AND MESENCEPHALIC CONTROL OF THE AUTONOMIC NERVOUS SYSTEM

Many areas in the reticular substance of the medulla, pons, and mesencephalon, as well as many special nuclei (Fig. 22–5), control different autonomic functions such as arterial pressure, heart rate, glandular secretion in the upper part of the gastrointestinal tract, gastrointestinal peristalsis, the degree of contraction of the urinary bladder, and many others. The control of each of these is discussed at appropriate points in this text. Suffice it to point out here that the most important factors controlled in the lower brain stem are arterial pressure, heart rate, and respiration. Indeed, transection of the brain stem at the midpontine level allows normal basal control of arterial pressure to continue as before but prevents its modulation by higher nervous centers, particularly the hypothalamus. On the other hand, transection immediately below the medulla causes the arterial pressure to fall to about one-half normal for several hours or several days after the transection.

Closely associated with the cardiovascular regulatory centers in the medulla are the medullary and pontine centers for regulation of respiration, which are discussed in Chapter 27. Although this is not considered to be an autonomic function, it is one of the *involuntary* functions of the body.

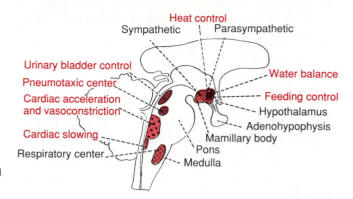

Figure 22–5. Autonomic control areas of the brain stem and hypothalamus.

Control of Lower Brain Stem Autonomic Centers by Higher Areas. Signals from the hypothalamus and even from the cerebrum can affect the activities of almost all the lower brain stem autonomic control centers. For instance, stimulation in appropriate areas of the hypothalamus can activate the medullary cardiovascular control centers strongly enough to increase the arterial pressure to more than double the normal. Likewise, other hypothalamic centers can control body temperature, increase or decrease salivation and gastrointestinal activity, or cause bladder emptying. To some extent, therefore, the autonomic centers in the lower brain stem act as relay stations for control activities initiated at higher levels of the brain.

In the previous two chapters it is pointed out that many of our behavioral responses are mediated through the hypothalamus, the reticular areas of the brain stem, and the autonomic nervous system. Indeed, the higher areas of the brain can alter the function of the whole autonomic nervous system or of portions of it strongly enough to cause severe autonomic-induced disease, such as peptic ulcer, constipation, heart palpitation, and even heart attacks.

■ **PHARMACOLOGY OF THE AUTONOMIC NERVOUS SYSTEM**

DRUGS THAT ACT ON ADRENERGIC EFFECTOR ORGANS — THE SYMPATHOMIMETIC DRUGS

From the foregoing discussion, it is obvious that intravenous injection of norepinephrine causes essentially the same effects throughout the body as sympathetic stimulation. Therefore, norepinephrine is called a *sympathomimetic*, or *adrenergic, drug*. *Epinephrine* and *methoxamine* are also sympathomimetic drugs, and there are many others. These differ from each other in the degree to which they stimulate different sympathetic effector organs and in their durations of action. Norepinephrine and epinephrine have actions as short as 1 to 2 min, whereas the actions of most other commonly used sympathomimetic drugs last 30 min to 2 hours.

Important drugs that stimulate specific adrenergic receptors but not the others are *phenylephrine* — alpha receptors; *isoproterenol* — beta receptors; and *albuterol* — only beta$_2$ receptors.

Drugs That Cause Release of Norepinephrine From Nerve Endings. Certain drugs have an indirect sympathomimetic action rather than directly exciting adrenergic effector organs. These drugs include *ephedrine, tyramine,* and *amphetamine.* Their effect is to cause release of norepinephrine from its storage vesicles in the sympathetic nerve endings. The released norepinephrine in turn causes the sympathetic effects.

Drugs That Block Adrenergic Activity. Adrenergic activity can be blocked at several different points in the stimulatory process as follows:

1. The synthesis and storage of norepinephrine in the sympathetic nerve endings can be prevented. The best known drug that causes this effect is *reserpine.*
2. Release of norepinephrine from the sympathetic endings can be blocked. This is caused by *guanethidine.*
3. The *alpha* receptors can be blocked. Two drugs that cause this effect are *phenoxybenzamine* and *phentolamine.*
4. The beta receptors can be blocked. A drug that blocks all beta receptors is *propranolol.* One that blocks only beta$_1$ receptors is *metoprolol.*
5. Sympathetic activity can be blocked by drugs that block transmission of nerve impulses through the autonomic ganglia. These are discussed in a later section, but the most important drug for blockade of both sympathetic and parasympathetic transmission through the ganglia is *hexamethonium.*

DRUGS THAT ACT ON CHOLINERGIC EFFECTOR ORGANS

Parasympathomimetic Drugs (Muscarinic Drugs). Acetylcholine injected intravenously usually does not cause exactly the same effects throughout the body as parasympathetic stimulation because the acetylcholine is destroyed by cholinesterase in the blood and body fluids before it can reach all the effector organs. Yet a number of other drugs that are not so rapidly destroyed can produce typical parasympathetic effects, and these are called *parasympathomimetic drugs.*

Two commonly used parasympathomimetic drugs are *pilocarpine* and *methacholine.* These act directly on the muscarinic type of cholinergic receptors.

Parasympathomimetic drugs act on the effector organs of cholinergic *sympathetic* fibers also. For instance, these drugs cause profuse sweating. Also, they cause vascular dilatation in some organs, this effect occurring even in vessels not innervated by cholinergic fibers.

Drugs That Have a Parasympathetic Potentiating Effect — Anticholinesterase Drugs. Some drugs do not have a direct effect on parasympathetic effector organs but do potentiate the effects of the naturally secreted acetylcholine at the parasympathetic endings. These are the same drugs as those discussed in Chapter 25 that potentiate the effect of acetylcholine at the neuromuscular junction—that is, *neostigmine, pyridostigmine,* and *ambenonium.* These inhibit acetylcholinesterase, thus preventing rapid destruction of the acetylcholine liberated by the parasympathetic nerve endings. As a consequence, the quantity of acetylcholine acting on the effector organs progressively increases with successive stimuli, and the degree of action also increases.

Drugs That Block Cholinergic Activity at Effector Organs—Antimuscarinic Drugs. *Atropine* and similar drugs, such as *homatropine* and *scopolamine,* block the action of acetylcholine on the muscarinic type of cholinergic effector organs. However, these drugs do not affect the nicotinic action of acetylcholine on the postganglionic neurons or on skeletal muscle.

DRUGS THAT STIMULATE OR BLOCK SYMPATHETIC AND PARASYMPATHETIC POSTGANGLIONIC NEURONS

Drugs That Stimulate Autonomic Ganglia. The preganglionic neurons of both the parasympathetic and sympathetic nervous systems secrete acetylcholine at their endings, and this acetylcholine in turn stimulates the postganglionic neurons. Furthermore, injected acetylcholine can also stimulate the postganglionic neurons of both systems, thereby causing at the same time both sympathetic and parasympathetic effects throughout the body. *Nicotine* is a drug that can also stimulate postganglionic neurons in the same manner as acetylcholine because the membranes of these neurons all contain the *nicotinic type of acetylcholine receptor.* Therefore, drugs that cause autonomic effects by stimulating the postganglionic neurons are frequently called *nicotinic drugs.* Some drugs, such as *acetylcholine* itself and *methacholine,* have both nicotinic and muscarinic actions, whereas pilocarpine has only muscarinic actions.

Nicotine excites both the sympathetic and parasympathetic postganglionic neurons at the same time, resulting in strong sympathetic vasoconstriction in the abdominal organs and limbs, but at the same time resulting in parasympathetic effects, such as increased gastrointestinal activity and, sometimes, slowing of the heart.

Ganglionic Blocking Drugs. Many important drugs block impulse transmission from the preganglionic neurons to the postganglionic neurons, including *tetraethyl ammonium ion, hexamethonium ion,* and *pentolinium.* These inhibit impulse transmission in both the sympathetic and parasympathetic systems simultaneously. They are often used for blocking sympathetic activity but rarely for blocking parasympathetic activity, because the sympathetic blockade usually far overshadows the effects of parasympathetic blockade. The ganglionic blocking drugs can especially reduce the arterial pressure in patients with hypertension, but these drugs are not very useful for this purpose because their effects are difficult to control.

REFERENCES

Abboud, F. M. (ed.): Disturbances in Neurogenic Control of the Circulation. Baltimore, Williams & Wilkins, 1981.
Bannister, Sir R. (ed.): Autonomic Failure. New York, Oxford University Press, 1988.
Buckley, J. P., et al. (eds.): Brain Peptides and Catecholamines in Cardiovascular Regulation. New York, Raven Press, 1987
Burattini, R., and Borgdorff, P.: Closed-loop baroreflex control of total peripheral resistance in the cat: Identification of gains by aid of a model. Cardiovasc. Res., 18:715, 1984.
Burchfield, S. R. (ed.): Stress. Physiological and Psychological Interactions. Washington, D.C., Hemisphere Publishing Corp., 1985.
Christensen, N. J., and Galbo, H.: Sympathetic nervous activity during exercise. Annu. Rev. Physiol., 45:139, 1983.
Cotman, C. W., et al. (eds.): The Neuro-Immune-Endocrine Connection. New York, Raven Press, 1987.
Davies, A. O., and Lefkowitz, R. J.: Regulation of B-adrenergic receptors by steroid hormones. Annu. Rev. Physiol., 46:119, 1984.
Donald, D. E., and Shepherd, J. T.: Autonomic regulation of the peripheral circulation. Annu. Rev. Physiol., 42:429, 1980.
Francis, G. S., and Cohn, J. N.: Catecholamines in Cardiovascular Disease. Current Concepts, February, 1988.
Gillis, C. N., and Pitt, B. R.: The fate of circulating amines within the pulmonary circulation. Annu. Rev. Physiol., 44:269, 1982.
Givens, J. R.: The Hypothalamus in Health and Disease. Chicago, Year Book Medical Publishers, 1984.
Goldstein, D. S., and Eisenhofer, G.: Plasma catechols—What do they mean? News Physiol. Sci., 3:138, 1988.
Guyton, A. C., and Gillespie, W. M., Jr.: Constant infusion of epinephrine: Rate of epinephrine secretion and destruction in the body. Am. J. Physiol., 164:319, 1951.
Guyton, A. C., and Reeder, R. C.: Quantitative studies on the autonomic actions of curare. J. Pharmacol. Exp. Ther., 98:188, 1950.
Hirst, G. D. S., and Edwards, F. R.: Sympathetic neuroeffector transmission in arteries and arterioles. Physiol. Rev., 69:546, 1989.
Hoffman, B. B., and Lefkowitz, R. J.: Radioligand binding studies of adrenergic receptors: New insights into molecular and physiological regulation. Annu. Rev. Pharmacol. Toxicol., 20:581, 1980.
Janig, W.: Pre- and postganglionic vasoconstrictor neurons: Differentiation, types, and discharge properties. Annu. Rev. Physiol., 50:525, 1988.
Kobilka, B. K., et al.: Chimeric α_2, β_2 adrenergic receptors: Delineation of domains involved in effector coupling and ligand binding specificity. Science, 240:1310, 1988.
Kreulen, D. L.: Integration in autonomic ganglia. Physiologist, 27:49, 1984.
Levitzki, A.: β-Adrenergic receptors and their mode of coupling to adenylate cyclase. Physiol. Rev., 66:819, 1986.
Livett, B. G.: Adrenal medullary chromaffin cells in vitro. Physiol. Rev., 64:1103, 1984.
Ludbrook, J., and Evans, R.: Posthemorrhagic syncope. News Physiol. Sci., 4:120, 1989.
Lundberg, J. M., et al.: Neuropeptide Y: Sympathetic cotransmitter and modulator? News Physiol. Sci., 4:13, 1989.
Robinson, R.: Tumours That Secrete Catecholamines: A Study of Their Natural History and Their Diagnosis. New York, John Wiley & Sons, 1980.
Rowell, L. B.: Reflex control of regional circulations in humans. J. Auton. Nerv. Syst., 11:101, 1984.
Simon, P. (ed.): Neurotransmitters. New York, Pergamon Press, 1979.
Stiles, G. L., et al.: β-Adrenergic receptors: Biochemical mechanisms of physiological regulation. Physiol. Rev., 64:661, 1984.
Stjarne, L.: New paradigm: Sympathetic neurotransmission by lateral interaction between secretory units? News Physiol. Sci., 1:103, 1986.
Tauc, L.: Nonvesicular release of neurotransmitter. Physiol. Rev., 62:857, 1982.
Torretti, J.: Sympathetic control of renin release. Annu. Rev. Pharmacol. Toxicol., 22:167, 1982.
Ungar, A., and Phillips, J. H.: Regulation of the adrenal medulla. Physiol. Rev., 63:787, 1983.
Usdin, E.: Stress. The Role of Catecholamines and Other Neurotransmitters. New York, Gordon Press Publishers, 1984.
Vanhoutte, P. M.: Vasodilatation: Vascular Smooth Muscle, Peptides, Autonomic Nerves, and Endothelium. New York, Raven Press, 1988.
Von Euler, U. S.: Noradrenaline. Springfield, Ill., Charles C Thomas, 1956.
Youmans, J. R. (ed.): Neurological Surgery. Philadelphia, W. B. Saunders Co., 1989.

23

Cerebral Blood Flow, the Cerebrospinal Fluid, and Brain Metabolism

Thus far, we have discussed the function of the brain as if it were independent of its blood flow, its metabolism, and its fluids. However, this is far from the truth, for abnormalities of any of these can profoundly affect brain function. For instance, total cessation of blood flow to the brain causes unconsciousness within 5 to 10 sec. This is true because lack of oxygen delivery to the brain cells shuts down most of their metabolism. Also, on a longer time scale, abnormalities of the cerebrospinal fluid, either in its composition or in its fluid pressure, can have equally severe effects on brain function.

■ CEREBRAL BLOOD FLOW

NORMAL RATE OF CEREBRAL BLOOD FLOW

The normal blood flow through the brain tissue of the adult averages 50 to 55 mL/100 g of brain/min. For the entire brain, this is approximately 750 mL/min, or 15 per cent of the total resting cardiac output.

REGULATION OF CEREBRAL BLOOD FLOW

Metabolic Control of Flow

As in most other vascular areas of the body, cerebral blood flow is highly related to the metabolism of the cerebral tissue. At least three different metabolic factors have potent effects in controlling cerebral blood flow. These are carbon dioxide concentration, hydrogen ion concentration, and oxygen concentration. An *increase* in either the carbon dioxide or the hydrogen ion concentration increases cerebral blood flow, whereas a *decrease* in oxygen concentration increases the flow.

Regulation of Cerebral Blood Flow in Response to Excess Carbon Dioxide or Hydrogen Ion Concentration. An increase in carbon dioxide concentration in the arterial blood perfusing the brain greatly increases cerebral blood flow. This is illustrated in Figure 23–1, which shows that a 70 per cent increase in arterial P_{CO_2} approximately doubles the blood flow.

Carbon dioxide is believed to increase cerebral blood flow almost entirely by combining first with water in the body fluids to form carbonic acid, with subsequent dissociation to form hydrogen ions. The hydrogen ions then cause vasodilatation of the cerebral vessels — the dilatation being almost directly proportional to the increase in hydrogen ion concentration.

Any other substance that increases the acidity of the brain tissue, and therefore also increases the hydrogen ion concentration, increases blood flow as well. Such substances include lactic acid, pyruvic acid, and any other acidic material formed during the course of metabolism.

Importance of the Carbon Dioxide and Hydrogen Control of Cerebral Blood Flow. Increased hydrogen ion concentration greatly depresses neuronal activity. Therefore, it is fortunate that an increase in hydrogen ion concentration causes an increase in blood flow, which in turn carries both carbon dioxide and other acidic substances away from the brain tissues. Loss of the carbon dioxide removes carbonic acid from the tissues; and this, along with the removal of other acids, reduces the hydrogen ion concentration back toward normal. Thus, this mechanism helps maintain a constant hydrogen ion concentration in the cerebral fluids and thereby helps to maintain the normal level of neuronal activity.

Oxygen Deficiency as a Regulator of Cerebral Blood Flow. Except during periods of intense brain

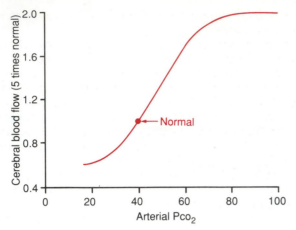

Figure 23–1. Relationship between arterial PCO_2 and cerebral blood flow.

activity, the utilization of oxygen by the brain tissue remains within very narrow limits—within a few percentage points of 3.5 mL of oxygen per 100 g of brain tissue per minute. If the blood flow to the brain ever becomes insufficient and cannot supply this needed amount of oxygen, the oxygen deficiency mechanism for causing vasodilatation, which functions in essentially all tissues of the body, immediately causes vasodilatation, returning the blood flow and transport of oxygen to the cerebral tissues to near normal. Thus, this local blood flow regulatory mechanism is much the same in the brain as in the coronary and skeletal muscle circulation and in many other circulatory areas of the body.

Experiments have shown that a decrease in cerebral *tissue* PO_2 below approximately 30 mm Hg (normal value is 35 to 40 mm Hg) will begin to increase cerebral blood flow. This is very fortuitous because brain function becomes deranged at not much lower values of PO_2, especially so at PO_2 levels below 20 mm Hg. Even coma can result at these low levels. Thus, the oxygen mechanism for local regulation of cerebral blood flow is also a very important protective response against diminished cerebral neuronal activity and, therefore, against derangement of mental capability.

Measurement of Cerebral Blood Flow and Effect of Cerebral Activity on the Flow. A method has recently been developed to record blood flow in as many as two hundred fifty-six isolated segments of the human cerebral cortex simultaneously. A radioactive substance, usually radioactive xenon, is injected into the carotid artery; then the radioactivity of each segment of the cortex is recorded as the radioactive substance passes through the brain tissue. For this to be done, 256 small radioactive scintillation detectors are focused on the same number of separate parts of the cortex; the rapidity of decay of the radioactivity after it once peaks in each tissue segment is a direct measure of the rate of blood flow through the segment.

With this technique, it has become clear that the blood flow in each individual segment of the brain changes within seconds in response to changes in local neuronal activity. For instance, simply making a fist of the hand causes an immediate increase in blood flow in the motor cortex of the opposite side of the brain. Or, reading a book increases the blood flow in multiple areas of the brain, especially in the occipital cortex and in the language areas of the temporal cortex. This measuring procedure can also be used for localizing the origin of epileptic attacks, for the blood flow increases acutely and markedly at the focal point of the attack at its onset.

Illustrating the effect of local neuronal activity on cerebral blood flow, Figure 23–2 shows an increase in occipital blood flow recorded in a cat when intense light was shone into its eyes for a period of 0.5 min.

Autoregulation of Cerebral Blood Flow When the Arterial Pressure Is Changed. Cerebral blood flow is autoregulated extremely well between the pressure limits of 60 and 140 mm Hg. That is, the arterial pressure can be decreased acutely to as low as 60 mm Hg or increased to as high as 140 mm Hg without a significant change in cerebral blood flow. In persons who have hypertension, this autoregulatory range shifts to even higher pressure levels, up to as high as 180 to 200 mm Hg. This effect is illustrated in Figure 23–3, which shows cerebral blood flows measured both in normal human beings and in hypertensive patients. Note the extreme constancy of cerebral blood flow between the limits of 60 and 180 mm Hg mean arterial pressure. On the other hand, if the arterial pressure does fall below 60 mm Hg, cerebral blood flow then becomes severely compromised, and if the pressure rises above the upper limit of autoregulation, the blood flow rises rapidly and can cause severe overstretching of the cerebral blood vessels, sometimes resulting in serious brain edema.

Role of the Sympathetic Nervous System in Regulating Cerebral Blood Flow

The cerebral circulatory system has a strong sympathetic innervation that passes upward from the superior cervical sympathetic ganglia along with the cerebral arteries. This innervation supplies both the large

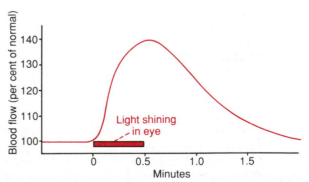

Figure 23–2. Increase in blood flow to the occipital regions of the brain when a light is flashed in the eyes of an animal.

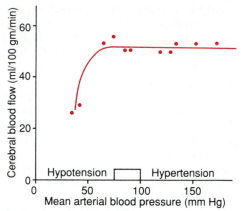

Figure 23–3. Relationship of mean arterial pressure to cerebral blood flow in normotensive, hypotensive, and hypertensive persons. (Modified from Lassen: Physiol. Rev., 39:183, 1959.)

superficial arteries and the small arteries that penetrate into the substance of the brain. However, neither transection of these sympathetic nerves nor mild to moderate stimulation of them normally causes significant change in the cerebral blood flow. Therefore, it has long been stated that the sympathetic nerves play essentially no role in regulating cerebral blood flow.

However, recent experiments have shown that cerebral sympathetic stimulation can, under some conditions, become activated strongly enough to constrict the cerebral arteries markedly. The reason that this usually does not occur is that the local blood flow autoregulatory mechanism is so powerful that it normally compensates almost entirely for the effects of the sympathetic stimulation. Yet in those conditions in which the autoregulatory mechanism fails to compensate enough, sympathetic control of cerebral blood flow becomes quite important. For instance, when the arterial pressure rises to a very high level during strenuous exercise and during other states of excessive circulatory activity, the sympathetic nervous system constricts the large and intermediate-sized arteries and prevents the very high pressures from ever reaching the smaller blood vessels. Experiments have shown that this is important in preventing the occurrence of a vascular hemorrhage into the brain — that is, for preventing the occurrence of cerebral stroke.

Also, sympathetic reflexes are believed to cause vasospasm in the intermediate and large arteries in some instances of brain damage, such as after a cerebral stroke has occurred or in patients with a subdural hematoma or brain tumor.

THE CEREBRAL MICROCIRCULATION

As in almost all other tissues of the body, the density of the blood capillaries in the brain is greatest where the metabolic needs are greatest. The overall metabolic rate of the brain gray matter, where the neuronal cell bodies lie, is about four times as great as that of white matter; correspondingly, the number of capillaries and

rate of blood flow are also about four times as great in the gray matter.

Another important structural characteristic of the brain capillaries is that they are much less "leaky" than the capillaries in almost any other tissue of the body. Most importantly, the capillaries are supported on all sides by "glial feet," which are small projections from the surrounding glia that abut against all surfaces of the capillaries and provide physical support to prevent overstretching of the capillaries in case of high pressure. In addition, the walls of the small arterioles leading to the brain capillaries become greatly thickened in persons who develop high blood pressure, and these arterioles remain significantly constricted all of the time to prevent transmission of the high pressure to the capillaries. We shall see later in the chapter that whenever these systems for protecting against transudation of fluid into the brain break down, serious brain edema ensues, which can lead rapidly to coma and death.

■ THE CEREBROSPINAL FLUID SYSTEM

The entire cavity enclosing the brain and spinal cord has a volume of approximately 1600 mL, and about 150 mL of this volume is occupied by cerebrospinal fluid. This fluid, as shown in Figure 23–4, is found in the *ventricles of the brain*, in the *cisterns around the brain*, and in the *subarachnoid space around both the brain and the spinal cord*. All these chambers are connected with each other, and the pressure of the fluid is regulated at a constant level.

CUSHIONING FUNCTION OF THE CEREBROSPINAL FLUID

A major function of the cerebrospinal fluid is to cushion the brain within its solid vault. Fortunately, the brain and the cerebrospinal fluid have approximately

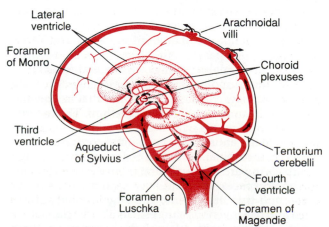

Figure 23–4. Pathway of cerebrospinal fluid flow from the choroid plexuses in the lateral ventricles to the arachnoidal villi protruding into the dural sinuses.

the same specific gravity (only about 4 per cent different), so that the brain simply floats in the fluid. Therefore, a blow to the head moves the entire brain simultaneously, causing no one portion of the brain to be momentarily contorted by the blow.

Contrecoup. When a blow to the head is extremely severe, it usually does not damage the brain on the side of the head where the blow is struck, but, instead, on the opposite side. This phenomenon is known as "contrecoup," and the reason for this effect is the following: When the blow is struck, the fluid on the struck side is so incompressible that, as the skull moves, the fluid pushes the brain at the same time. However, on the opposite side, the sudden movement of the skull causes it to pull away from the brain momentarily because of the brain's inertia, creating for a split second a vacuum space in the cranial vault at this point. Then, when the skull is no longer being accelerated by the blow, the vacuum suddenly collapses and the brain strikes the inner surface of the skull. Because of this effect, the damage to the brain of a boxer usually does not occur in the frontal regions but, instead, in the occipital regions.

FORMATION, FLOW, AND ABSORPTION OF CEREBROSPINAL FLUID

Cerebrospinal fluid is formed at a rate of approximately 500 mL each day, which is about three times as much as the total volume of fluid in the entire cerebrospinal fluid system. Probably two thirds or more of this fluid originates as a secretion from the choroid plexuses in the four ventricles, mainly in the two lateral ventricles. Additional amounts of fluid are secreted by all the ependymal surfaces of the ventricles and the arachnoidal membranes, and a small amount comes from the brain itself through the perivascular spaces that surround the blood vessels entering the brain.

The arrows in Figure 23–4 show the main channel of fluid flow from the choroid plexuses and then through the cerebrospinal fluid system. The fluid secreted in the lateral ventricles and the third ventricle passes along the *aqueduct of Sylvius* into the fourth ventricle, where a small amount of additional fluid is added. It then passes out of the fourth ventricle through three small openings, two lateral *foramina of Luschka* and a midline *foramen of Magendie*, entering the *cisterna magna*, a large fluid space that lies behind the medulla and beneath the cerebellum. The cisterna magna is continuous with the *subarachnoid space* that surrounds the entire brain and spinal cord. Almost all of the cerebrospinal fluid then flows upward through this space toward the cerebrum. From the cerebral subarachnoid spaces, the fluid flows into multiple *arachnoidal villi* that project into the large sagittal venous sinus and other venous sinuses. Finally, the fluid empties into the venous blood through the surfaces of these villi.

Secretion by the Choroid Plexus. The choroid plexus, which is illustrated in Figure 23–5, is a cauliflower-like growth of blood vessels covered by a thin layer of epithelial cells. This plexus projects into (1) the temporal horn of each lateral ventricle, (2) the posterior portion of the third ventricle, and (3) the roof of the fourth ventricle.

The secretion of fluid by the choroid plexus depends mainly on active transport of sodium ions through the epithelial cells that line the outer surfaces of the plexus. The sodium ions in turn pull along large amounts of chloride ions as well because the positive charge of the sodium ion attracts the chloride ion's negative charge. The two of these together increase the quantity of osmotically active substances in the cerebrospinal fluid, which then causes almost immediate osmosis of water through the membrane, thus providing the fluid of the secretion. Less important transport processes move small amounts of glucose into the cerebrospinal fluid and both potassium and bicarbonate ions out of the cerebrospinal fluid into the capillaries. Therefore, the resulting characteristics of the cerebrospinal fluid become approximately the following: osmotic pressure, approximately equal to that of plasma; sodium ion concentration, also approximately equal to that of plasma; chloride, about 15 per cent greater than in plasma; potassium, approximately 40 per cent less; and glucose, about 30 per cent less.

Absorption of Cerebrospinal Fluid Through the Arachnoidal Villi. The *arachnoidal villi* are microscopic fingerlike projections of the arachnoidal membrane through the walls of the venous sinuses. Large conglomerates of these villi are usually found together and form macroscopic structures called *arachnoidal granulations* that can be seen protruding into the sinuses. The endothelial cells covering the villi have been shown by electron microscopy to have large ve-

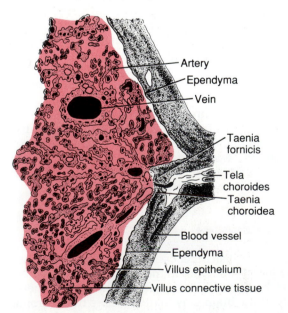

Figure 23–5. The choroid plexus. (Modified from Clara: Das Nervensystem des Menschen. Barth.)

Artery
Ependyma
Vein
Taenia fornicis
Tela choroides
Taenia choroidea
Blood vessel
Ependyma
Villus epithelium
Villus connective tissue

sicular holes directly through the bodies of the cells. It has been proposed that these are large enough to allow relatively free flow of cerebrospinal fluid, protein molecules, and even particles as large as red blood cells into the venous blood.

The Perivascular Spaces and Cerebrospinal Fluid. The blood vessels entering the substance of the brain pass first along the surface of the brain and then penetrate inward, carrying a layer of *pia mater*, the membrane that covers the brain, with them, as shown in Figure 23–6. The pia is only loosely adherent to the vessels, so that a space, the *perivascular space*, exists between it and each vessel. Perivascular spaces follow both the arteries and the veins into the brain as far as the arterioles and venules but not to the capillaries.

The Lymphatic Function of the Perivascular Spaces. As is true elsewhere in the body, a small amount of protein leaks out of the parenchymal capillaries into the interstitial spaces of the brain; and because no true lymphatics are present in brain tissue, this protein leaves the tissue mainly through the perivascular spaces but partly also by direct diffusion through the pia mater into the subarachnoid spaces. On reaching the subarachnoid spaces, the protein flows along with the cerebrospinal fluid to be absorbed through the *arachnoidal villi* into the cerebral veins. Therefore, the perivascular spaces, in effect, are a modified lymphatic system for the brain.

In addition to transporting fluid and proteins, the perivascular spaces also transport extraneous particulate matter from the brain into the subarachnoid space. For instance, whenever infection occurs in the brain, dead white blood cells are carried away through the perivascular spaces.

CEREBROSPINAL FLUID PRESSURE

The normal pressure in the cerebrospinal fluid system when one is lying in a horizontal position averages 130 mm H_2O (10 mm Hg), though this may be as low as 70 mm H_2O or as high as 180 mm H_2O even in the normal person. These values are considerably more positive than the -3 to -5 mm Hg pressure in the interstitial spaces of the subcutaneous tissue.

Regulation of Cerebrospinal Fluid Pressure by the Arachnoidal Villi. The cerebrospinal fluid pressure is regulated almost entirely by absorption of the fluid through the arachnoidal villi. The reason for this is that the rate of cerebrospinal fluid formation is very constant, so that this is rarely a factor in pressure control. On the other hand, the villi function like "valves" that allow the fluid and its contents to flow readily into the venous blood of the sinuses while not allowing the blood to flow backward in the opposite direction. Normally, this valve action of the villi allows cerebrospinal fluid to begin to flow into the blood when its pressure is about 1.5 mm Hg greater than the pressure of the blood in the sinuses. Then, as the cerebrospinal fluid pressure rises still higher, the valves open very widely so that, under normal conditions, the pressure almost never rises more than a few millimeters of mercury more than the pressure in the venous sinuses.

On the other hand, in disease states the villi sometimes become blocked by large particulate matter, by fibrosis, or even by excesses of plasma protein molecules that have leaked into the cerebrospinal fluid in brain diseases. Such blockage can cause very high cerebrospinal fluid pressure, as we discuss later.

Cerebrospinal Fluid Pressure in Pathological Conditions of the Brain. Often a large *brain tumor* elevates the cerebrospinal fluid pressure by decreasing the rate of absorption of fluid. For instance, if the tumor is above the tentorium and becomes so large that it compresses the brain downward, the upward flow of fluid through the subarachnoid space around the brain stem where it passes through the tentorial opening may become blocked and the absorption of fluid by the cerebral arachnoidal villi may become greatly curtailed. As a result, the cerebrospinal fluid pressure can rise to as high as 500 mm H_2O (37 mm Hg) or more.

The pressure also rises considerably when *hemorrhage* or *infection* occurs in the cranial vault. In both of these conditions, large numbers of cells suddenly appear in the cerebrospinal fluid, and these can cause serious blockage of the small channels for absorption through the arachnoidal villi. This sometimes elevates the cerebrospinal fluid pressure to as high as 400 to 600 mm H_2O (about four times normal).

Occasionally babies are born with high cerebrospinal fluid pressure. This is usually caused by abnormally high resistance to fluid reabsorption through the arachnoidal villi, resulting either from too few arachnoidal villi or villi with abnormal absorptive properties. This is discussed later in connection with *hydrocephalus*.

THE BLOOD-CEREBROSPINAL FLUID AND BLOOD-BRAIN BARRIERS

It has already been pointed out that the constituents of the cerebrospinal fluid are not exactly the same as those of the extracellular fluid elsewhere in the body.

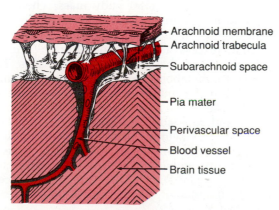

Figure 23–6. Drainage of the perivascular spaces into the subarachnoid space. (From Ranson and Clark: Anatomy of the Nervous System. Philadelphia, W. B. Saunders Company, 1959.)

Labels for figure:
- Arachnoid membrane
- Arachnoid trabecula
- Subarachnoid space
- Pia mater
- Perivascular space
- Blood vessel
- Brain tissue

Furthermore, many large molecular substances hardly pass at all from the blood into the cerebrospinal fluid or into the interstitial fluids of the brain even though these same substances pass readily into the usual interstitial fluids of the body. Therefore, it is said that barriers, called the *blood-cerebrospinal fluid barrier* and the *blood-brain barrier*, exist between the blood and the cerebrospinal fluid and brain fluid, respectively. These barriers exist both in the choroid plexus and at the tissue capillary membranes in essentially all areas of the brain parenchyma *except in some areas of the hypothalamus*, the *pineal gland*, and the *area postrema*, where substances diffuse with ease into the tissue spaces. This ease of diffusion is very important because these areas of the brain have sensory organs that respond to different changes in the body fluids, such as changes in osmolality, glucose concentration, and so forth; these responses provide the signals for feedback regulation of each of the factors.

In general, the blood-cerebrospinal fluid and blood-brain barriers are highly permeable to water, carbon dioxide, oxygen, and most lipid-soluble substances such as alcohol and most anesthetics; slightly permeable to the electrolytes, such as sodium, chloride, and potassium; and almost totally impermeable to plasma proteins and many large organic molecules. Therefore, the blood-cerebrospinal fluid and blood-brain barriers often make it impossible to achieve effective concentrations of either protein antibodies or some non–lipid-soluble drugs in the cerebrospinal fluid or parenchyma of the brain.

The cause of the low permeability of the blood-cerebrospinal fluid and blood-brain barriers is the manner in which the endothelial cells of the capillaries are joined to each other. They are joined by so-called *tight junctions*. That is, the membranes of the adjacent endothelial cells are almost fused with each other, rather than having slit-pores between them, as is the case in most other capillaries of the body.

Diffusion Between the Cerebrospinal Fluid and the Brain Interstitial Fluid. The surfaces of the ventricles are lined with a thin cuboidal epithelium called the *ependyma* and the cerebrospinal fluid on the outer surfaces of the brain is separated from the brain tissue by a thin membrane called the *pia mater*. Both the ependyma and the pia mater are extremely permeable so that almost all substances that enter the cerebrospinal fluid can also diffuse readily into the surface areas of the brain interstitial fluid. Or, likewise, substances in the interstitial fluid can diffuse in the other direction as well. Therefore, some drugs that have no effect at all on the brain when introduced into the blood stream nevertheless can have important effects on the brain when injected into the cerebrospinal fluid.

BRAIN EDEMA

One of the most serious complications of abnormal cerebral hemodynamics and fluid dynamics is the development of brain edema. Because the brain is encased in a solid vault, the accumulation of edema fluid compresses the blood vessels, with eventual depression of blood flow and destruction of brain tissue.

The usual cause of brain edema is either greatly increased capillary pressure or damage to the capillary wall. One cause of excessively high capillary pressure is a sudden increase in the cerebral blood pressure to levels too high for the autoregulatory mechanism to cope with. However, the most common cause is brain concussion, in which the brain tissues and capillaries are traumatized and capillary fluid leaks into the traumatized tissues. Once brain edema begins, it often initiates two vicious circles because of the following positive feedback: (1) Edema compresses the vasculature. This in turn decreases the blood flow and causes brain ischemia. The ischemia causes arteriolar dilatation with increased capillary pressure. The increased capillary pressure then causes more edema fluid, so that the edema becomes progressively worse. (2) The decreased blood flow decreases oxygen delivery. This increases the permeability of the capillaries, allowing more fluid leakage. It also turns off the sodium pumps of the tissue cells, thus allowing these to swell.

Once these two vicious circles have begun, heroic measures must be used to prevent total destruction of the brain. One such measure is to infuse intravenously a concentrated osmotic substance, such as a very concentrated mannitol solution. This pulls fluid by osmosis from the brain tissue and breaks up the vicious circle. Another procedure is to remove fluid quickly from the lateral ventricles of the brain via ventricular puncture, thereby relieving the intracerebral pressure.

■ BRAIN METABOLISM

Like other tissues, the brain requires oxygen and solid nutrients to supply its metabolic needs. However, there are special peculiarities of brain metabolism that need to be mentioned.

Total Metabolic Rate and Metabolic Rate of Neurons. Under resting conditions, the metabolism of the brain accounts for about 15 per cent of the total metabolism in the body, even though the mass of the brain is only 2 per cent of the total body mass. Therefore, under resting conditions brain metabolism is about 7½ times the average metabolism in the rest of the body.

Most of this excess metabolism of the brain occurs in the neurons, not in the glial supportive tissues. The major need for metabolism in the neurons is to pump ions through their membranes, mainly to transport sodium and calcium ions to the outside of the neuronal membrane and potassium and chloride ions to the interior. Each time a neuron conducts an action potential, these ions move through the membranes, increasing the need for membrane transport to restore the proper ionic concentrations. Therefore, during excessive brain activity, neuronal metabolism can increase several times.

Special Requirement of the Brain for Oxygen— Lack of Significant Anaerobic Metabolism. Most tissues of the body can go without oxygen for several minutes and some as long as a half hour. During this time the tissue cells obtain their energy through the processes of anaerobic metabolism, which means the release of energy by partial breakdown of glucose and glycogen but without combining with oxygen. This delivers energy only at the expense of consuming tremendous amounts of glucose and glycogen. However, it does keep the tissues functioning.

Unfortunately, the brain is not capable of much anaerobic metabolism. One of the reasons for this is the very high metabolic rate of the neurons, so that far more energy is required by each brain cell than is required in most tissues. An additional reason is that the amount of glycogen stored in the neurons is very slight. The stores of oxygen in brain tissues are also slight. Therefore, neuronal activity depends on second-by-second delivery of oxygen from the blood.

Putting these various factors together, one can understand why a sudden cessation of blood flow to the brain or a sudden lack of oxygen in the blood can cause unconsciousness within 5 to 10 sec.

Under Normal Conditions, Most Brain Energy Is Supplied by Glucose. Under normal conditions, almost all of the energy used by the brain cells is supplied by glucose derived from the blood. As is true for oxygen, most of this is derived minute by minute and second by second from the capillary blood, with a total of only about a 2-min supply of glucose normally stored as glycogen in the neurons at any given time.

A special feature of glucose delivery to the neurons is that its transport into the neurons through the cell membrane is not dependent on insulin, as is true for almost all other tissue cells. Therefore, even in patients who have serious diabetes with essentially zero secretion of insulin, glucose still diffuses readily into the neurons — which is most fortunate in preventing loss of mental function in diabetic patients. However, when a diabetic patient is overtreated with insulin, the blood glucose concentration can sometimes fall extremely low, because the excess insulin causes almost all of the glucose in the blood to be transported rapidly into the insulin-sensitive nonneural cells throughout the body. When this happens, not enough glucose is left in the blood to supply the neurons, and mental function does then become seriously deranged, leading sometimes to coma but more often to mental imbalances and psychotic disturbances.

REFERENCES

Angerson, W. J., et al. (eds.): Blood Flow in the Brain. New York, Oxford University Press, 1989.

Bevan, J. A., et al.: Sympathetic control of cerebral arteries: specialization in receptor type, reserve, affinity, and distribution. FASEB J., 1:193, 1987.

Cserr, H. F.: Physiology of the choroid plexus. Physiol. Rev., 51:273, 1971.

Daveson, H.: The Physiology of the Cerebrospinal Fluid. Boston, Little, Brown, 1967.

Fenstermacher, J. D., and Rapoport, S. I.: Blood-brain barrier. In Renkin, E. M., and Michel, C. C. (eds.): Handbook of Physiology. Sec. 2, Vol. IV. Bethesda, Md., American Physiological Society, 1984, p. 969.

Finger, S., et al. (eds.): Brain Injury and Recovery. New York, Plenum Publishing Corp., 1988.

Guyton, A. C., et al.: Circulatory Physiology. II. Dynamics and Control of the Body Fluids. Philadelphia, W. B. Saunders Co., 1975.

Hibbard, L. S., et al.: Three-dimensional representation and analysis of brain energy metabolism. Science, 236:1641, 1987.

Hochwald, G. M.: Animal models of hydrocephalus: Recent developments. Proc. Soc. Exp. Biol. Med., 178:1, 1985.

Kazemi, H., and Johnson, D. C.: Regulation of cerebrospinal fluid acid-base balance. Physiol. Rev., 66:953, 1986.

Levin, H. S., and Eisenberg, H. S. (eds.): Mild Head Injury. New York, Oxford University Press, 1989.

Mayhan, W. G., et al.: Cerebral microcirculation. News Physiol. Sci., 3:164, 1988.

McLaurin, R. L., et al. (eds.): Pediatric Neurosurgery, 2nd Ed. Philadelphia, W. B. Saunders Co., 1989.

Neuwelt, E. A. (ed.). Implications of the Blood-Brain Barrier and Its Manipulation. New York, Plenum Publishing Corp., 1989.

Oldendorf, W. H.: Blood-brain barrier permeability to drugs. Annu. Rev. Pharmacol., 14:239, 1974.

Rescigno, A., and Boicelli, A.: Cerebral Blood Flow. New York, Plenum Publishing Corp., 1988.

Saunders, N. R., and Milgard, K.: Development of the blood-brain barrier. J. Dev. Physiol., 6:45, 1984.

Shulman, K. (ed.): Intracranial Pressure IV. New York, Springer-Verlag, 1980.

Siesjo, B. K.: Cerebral circulation and metabolism. J. Neurosurg., 60:883, 1984.

Somjen, G. (ed.): Mechanisms of Cerebral Hypoxia and Stroke. New York, Plenum Publishing Corp., 1988.

Wood, J. H. (ed.): Cerebral Blood Flow: Physiologic and Clinical Aspects. New York, McGraw-Hill Book Co., 1987.

VII

NERVOUS CONTROL OF BODY FUNCTIONS

24

Contraction of Skeletal Muscle

Because the nervous system is the major overall controller of our bodily activities, it is equally as important to understand the ways in which the nervous system interfaces with the peripheral parts of the body as it is to understand the nervous system itself. Therefore, the remaining chapters of this text will help to explain the many ways in which the nervous system controls all our muscular activities as well as what we call the *vegetative functions* of the body, meaning the life processes of the body, such as the control of arterial pressure, respiration, gastrointestinal function, body temperature, and even sexual functions.

The present chapter will address skeletal muscle. Approximately 40 per cent of the body is skeletal muscle, and almost another 10 per cent is smooth and cardiac muscle. Many of the same principles of contraction and control by the nervous system apply to all these different types of muscle, but the specialized functions of smooth muscle are discussed in Chapter 25 and cardiac muscle in Chapter 26.

■ PHYSIOLOGIC ANATOMY OF SKELETAL MUSCLE

THE SKELETAL MUSCLE FIBER

Figure 24–1 illustrates the organization of skeletal muscle, showing that all skeletal muscles are made of numerous fibers ranging between 10 and 80 μm in diameter. Each of these fibers in turn is made up of successively smaller subunits, also illustrated in Figure 24–1, that are described in subsequent paragraphs.

In most muscles the fibers extend the entire length of the muscle; except for about 2 per cent of the fibers, each is innervated by only one nerve ending, located near the middle of the fiber.

The Sarcolemma. The sarcolemma is the cell membrane of the muscle fiber. However, the sarcolemma consists of a true cell membrane, called the *plasma membrane,* and an outer coat made up of a thin layer of polysaccharide material containing numerous thin collagen fibrils. At the end of the muscle fiber, this surface layer of the sarcolemma fuses with a tendon fiber, and the tendon fibers in turn collect into bundles to form the muscle tendons and thence insert into the bones.

Myofibrils; Actin and Myosin Filaments. Each muscle fiber contains several hundred to several thousand *myofibrils,* which are illustrated by the many small open dots in the cross-sectional view of Figure 24–1C. Each myofibril (Figure 24–1D and E) in turn has, lying side-by-side, about 1500 *myosin filaments* and 3000 *actin filaments,* which are large polymerized protein molecules that are responsible for muscle contraction. These can be seen in longitudinal view in the electron micrograph of Figure 24–2 and are represented diagrammatically in Figure 24–1, parts E through L. The thick filaments in the diagrams are *myosin,* and the thin filaments are *actin.*

Note that the myosin and actin filaments partially interdigitate and thus cause the myofibrils to have alternate light and dark bands. The light bands contain only actin filaments and are called *I bands* because they are *isotropic* to polarized light. The dark bands contain the myosin filaments as well as the ends of the actin filaments where they overlap the myosin and are called *A bands* because they are *anisotropic* to polarized light. Note also the small projections from the sides of the myosin filaments. These are called *cross-bridges.* They protrude from the surfaces of the myosin filaments along the entire extent of the filament except in the very center. It is interaction between these cross-bridges and the actin filaments that causes contraction.

Figure 24–1E also shows that the ends of the actin filaments are attached to a so-called Z *disc.* From this disc, these filaments extend in both directions to interdigitate with the myosin filaments. The Z disc, which itself is composed of filamentous proteins different from the actin and myosin filaments, passes from myofibril to myofibril, attaching the myofibrils to each other all the way across the muscle fiber. Therefore,

Skeletal Muscle

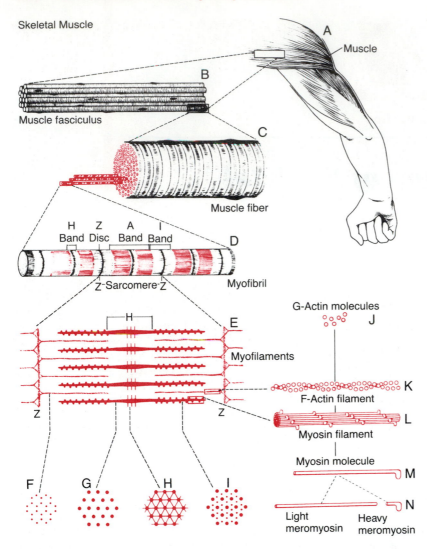

Figure 24—1. Organization of skeletal muscle, from the gross to the molecular level. *F, G, H,* and *I* are cross-sections at the levels indicated. (Drawing by Sylvia Colard Keene. Modified from Fawcett: Bloom and Fawcett: A Textbook of Histology. Philadelphia, W. B. Saunders Company, 1986.)

the entire muscle fiber has light and dark bands, as do the individual myofibrils. These bands give skeletal and cardiac muscle their "striated" appearance.

The portion of a myofibril (or of the whole muscle fiber) that lies between two successive Z discs is called a *sarcomere*. When the muscle fiber is at its normal, fully stretched resting length, the length of the sarcomere is about 2 μm. At this length, the actin filaments completely overlap the myosin filaments and are just beginning to overlap each other. We see later that at this length the sarcomere also is capable of generating its greatest force of contraction.

The Sarcoplasm. The myofibrils are suspended inside the muscle fiber in a matrix called *sarcoplasm,* which is composed of the usual intracellular constituents. The fluid of the sarcoplasm contains large quantities of potassium, magnesium, phosphate, and protein enzymes. Also present are tremendous numbers of *mitochondria* that lie between and parallel to the myofibrils, a condition that is indicative of the great need of the contracting myofibrils for large amounts of adenosine triphosphate (ATP) formed by the mitochondria.

The Sarcoplasmic Reticulum. Also in the sarcoplasm is an extensive endoplasmic reticulum, which in the muscle fiber is called the *sarcoplasmic reticulum.* This reticulum has a special organization that is extremely important in the control of muscle contraction, which is discussed later in the chapter. The electron micrograph of Figure 24–3 illustrates the arrangement of this sarcoplasmic reticulum and shows how extensive it can be. The more rapidly contracting types of muscle have especially extensive sarcoplasmic reticula, indicating that this structure is important in causing rapid muscle contraction, as is also discussed later.

■ THE GENERAL MECHANISM OF MUSCLE CONTRACTION

The initiation and execution of muscle contraction occurs in the following sequential steps:

1. An action potential travels along a motor nerve to its endings on muscle fibers.

2. At each ending, the nerve secretes a small amount of the neurotransmitter substance called *acetylcholine.*

3. The acetylcholine acts on a local area of the muscle fiber membrane to open multiple acetylcholine-gated protein channels in the muscle fiber membrane.

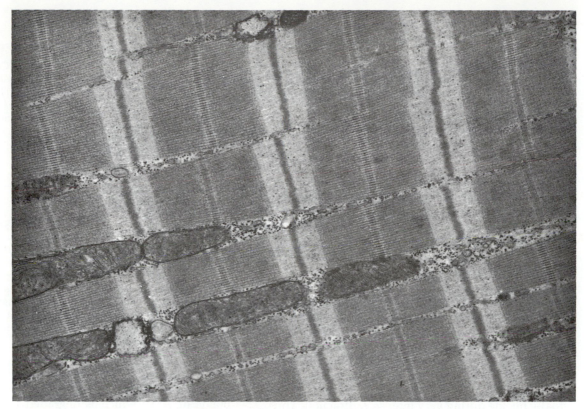

Figure 24 – 2. Electron micrograph of muscle myofibrils, showing the detailed organization of actin and myosin filaments. Note the mitochondria lying between the myofibrils. (From Fawcett: The Cell. Philadelphia, W. B. Saunders Company, 1981.)

4. Opening of the acetylcholine channels allows large quantities of sodium ions to flow to the interior of the muscle fiber membrane at the point of the nerve terminal. This initiates an action potential in the muscle fiber.

5. The action potential travels along the muscle fiber membrane in the same way that action potentials travel along nerve membranes.

6. The action potential depolarizes the muscle fiber membrane and also travels deeply within the muscle fiber. Here it causes the sarcoplasmic reticulum to release into the myofibrils large quantities of calcium ions that have been stored within the reticulum.

7. The calcium ions initiate attractive forces between the actin and myosin filaments, causing them to slide together, which is the contractile process.

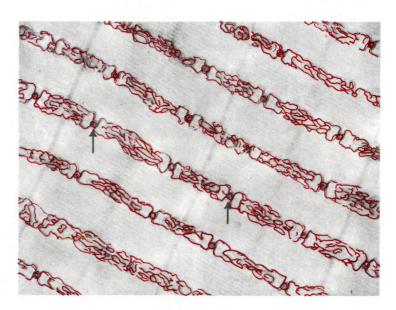

Figure 24 – 3. Sarcoplasmic reticulum surrounding the myofibrils, showing the longitudinal system paralleling the myofibrils. Also shown in cross-section are the T tubules (arrows) that lead to the exterior of the fiber membrane and that contain extracellular fluid. (From Fawcett: The Cell. Philadelphia, W. B. Saunders Company, 1981.)

8. After a fraction of a second, the calcium ions are pumped back into the sarcoplasmic reticulum, where they remain stored until a new muscle action potential comes along; muscle contraction ceases.

We now describe the machinery of the contractile process but return to the details of muscle excitation in the later chapter.

■ MOLECULAR MECHANISM OF MUSCLE CONTRACTION

Sliding Mechanism of Contraction. Figure 24–4 illustrates the basic mechanism of muscle contraction. It shows the relaxed state of a sarcomere (above) and the contracted state (below). In the relaxed state, the ends of the actin filaments derived from two successive Z discs barely begin to overlap each other, while at the same time completely overlapping the myosin filaments. On the other hand, in the contracted state these actin filaments have been pulled inward among the myosin filaments, so that they now overlap each other to a major extent. Also, the Z discs have been pulled by the actin filaments up to the ends of the myosin filaments. Indeed, the actin filaments can be pulled together so tightly that the ends of the myosin filaments actually buckle during very intense contraction. Thus, muscle contraction occurs by a *sliding filament mechanism.*

But what causes the actin filaments to slide inward among the myosin filaments? This is caused by mechanical forces generated by the interaction of the cross-bridges of the myosin filaments with the actin filaments, as we discuss in the following sections. Under resting conditions, these forces are inhibited, but when an action potential travels over the muscle fiber membrane, this causes the release of large quantities of calcium ions into the sarcoplasm surrounding the myofibrils. These calcium ions in turn activate the forces between the filaments, and contraction begins, but energy is also needed for the contractile process to proceed. This energy is derived from the high-energy

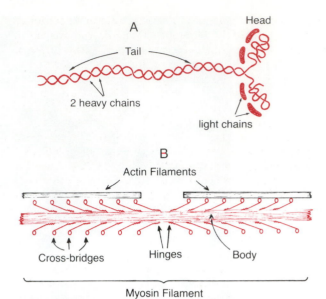

Figure 24–5. *A,* The myosin molecule. *B,* Combination of many myosin molecules to form a myosin filament. Also shown are the cross-bridges and the interaction between the heads of the cross-bridges and adjacent actin filaments.

bonds of ATP, which is degraded to adenosine diphosphate (ADP) to liberate the energy required.

In the next few sections, we describe what is known about the details of the molecular processes of contraction. To begin this discussion, however, we must first characterize in detail the myosin and actin filaments.

MOLECULAR CHARACTERISTICS OF THE CONTRACTILE FILAMENTS

The Myosin Filament. The myosin filament is composed of multiple myosin molecules, each having a molecular weight of about 480,000. Figure 24–5A illustrates an individual molecule; Figure 24–5B illustrates the organization of the molecules to form a myosin filament as well as its interaction on one side with the ends of two actin filaments.

The *myosin molecule* is composed of 6 polypeptide chains, 2 *heavy chains* each with a molecular weight of about 200,000 and 4 *light chains* with molecular weights of about 20,000 each. The 2 heavy chains wrap spirally around each other to form a double helix. However, one end of each of these chains is folded into a globular protein mass called the myosin *head.* Thus, there are 2 free heads lying side by side at one end of the double helix myosin molecule; the elongated portion of the coiled helix is called the *tail.* The 4 light chains are also parts of the myosin heads, 2 to each head. These light chains help control the function of the head during the process of muscle contraction.

The *myosin filament* is made up of 200 or more individual myosin molecules. The central portion of one of these filaments is illustrated in Figure 24–5B, showing

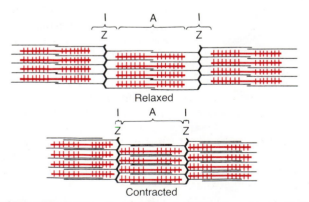

Figure 24–4. The relaxed and contracted states of a myofibril, showing sliding of the actin filaments (black) into the spaces between the myosin filaments (red).

the tails of the myosin molecules bundled together to form the *body* of the filament, while many heads of the molecules hang outward to the sides of the body. Also, part of the helix portion of each myosin molecule extends to the side along with the head, thus providing an *arm* that extends the head outward from the body, as shown in the figure. The protruding arms and heads together are called *cross-bridges*. Each cross-bridge is believed to be flexible at 2 points called *hinges*, one where the arm leaves the body of the myosin filament and the other where the 2 heads attach to the arm. The hinged arms allow the heads either to be extended far outward from the body of the myosin filament or to be brought close to the body. The hinged heads are believed to participate in the actual contraction process, as we discuss in the following sections.

The total length of each myosin filament is very uniform, almost exactly 1.6 μm. However, note that there are no cross-bridge heads in the very center of the myosin filament for a distance of about 0.2 μm because the hinged arms extend toward both ends of the myosin filament away from the center; therefore, in the center there are only tails of the myosin molecules and no heads.

Now, to complete the picture, the myosin filament itself is twisted so that each successive set of cross-bridges is axially displaced from the previous set by 120 degrees. This insures that the cross-bridges extend in all directions around the filament.

ATPase Activity of the Myosin Head. Another feature of the myosin head that is essential for muscle contraction is that it functions as an ATPase enzyme. As we see later, this property allows the head to cleave ATP and to use the energy derived from the ATP's high-energy phosphate bond to energize the contraction process.

The Actin Filament. The actin filament is also complex. It is composed of three different protein components: *actin, tropomyosin,* and *troponin.*

The backbone of the actin filament is a double-stranded F-actin protein molecule, illustrated by the two lighter-colored strands in Figure 24–6. The two strands are wound in a helix in the same manner as the myosin molecule but with a complete revolution every 70 nm.

Each strand of the double F-actin helix is composed of polymerized G-actin molecules, each having a molecular weight of about 42,000. There are approxi-

mately 13 of these molecules in each revolution of each strand of helix. Attached to each one of the G-actin molecules is one molecule of ADP. It is believed that these ADP molecules are the active sites on the actin filaments with which the cross-bridges of the myosin filaments interact to cause muscle contraction. The active sites on the two F-actin strands of the double helix are staggered, giving one active site on the overall actin filament approximately every 2.7 nm.

Each actin filament is approximately 1 μm long. The bases of the actin filaments are inserted strongly into the Z discs, while the other ends protrude in both directions into the adjacent sarcomeres to lie in the spaces between the myosin molecules, as illustrated in Figure 24–4.

Tropomyosin Molecules. The actin filament also contains another protein, *tropomyosin.* Each molecule of tropomyosin has a molecular weight of 70,000 and a length of 40 nm. These molecules are connected loosely with the F-actin strands, wrapped spirally around the sides of the F-actin helix. In the resting state, the tropomyosin molecules are believed to lie on top of the active sites of the actin strands, so that attraction cannot occur between the actin and myosin filaments to cause contraction. Each tropomyosin molecule covers about seven of these active sites.

Troponin and Its Role in Muscle Contraction. Attached near one end of each tropomyosin molecule is still another protein molecule called *troponin.* This is actually a complex of three loosely bound protein subunits, each of which plays a specific role in the control of muscular contraction. One of the subunits (troponin I) has a strong affinity for actin, another (troponin T) for tropomyosin, and a third (troponin C) for calcium ions. This complex is believed to attach the tropomyosin to the actin. The strong affinity of the troponin for calcium ions is believed to initiate the contraction process, as is explained in the following section.

Interaction of Myosin, Actin Filaments, and Calcium Ions to Cause Contraction

Inhibition of the Actin Filament by the Troponin-Tropomyosin Complex; Activation by Calcium Ions. A pure actin filament without the presence of the troponin-tropomyosin complex binds strongly with myosin molecules in the presence of magnesium ions and ATP, both of which are normally abundant in the myofibril. However, if the troponin-tropomyosin complex is added to the actin filament, this binding does not take place. Therefore, it is believed that the active sites on the normal actin filament of the relaxed muscle are inhibited or actually physically covered by the troponin-tropomyosin complex. Consequently, the sites cannot attach to the myosin filaments to cause contraction. Before contraction can take place, the inhibitory effect of the troponin-tropomyosin complex must itself be inhibited.

Now, let us discuss the role of the calcium ions. In the presence of large amounts of calcium ions, the inhibitory effect of the troponin-tropomyosin on the

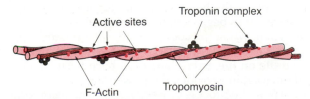

Figure 24–6. The actin filament, composed of two helical strands of F-actin and tropomyosin molecules that fit loosely in the grooves between the actin strands. Attached to one end of each tropomyosin molecule is a troponin complex that initiates contraction.

actin filaments is itself inhibited. The mechanism of this is not known, but one suggestion is the following: When calcium ions combine with troponin C, each molecule of which can bind strongly with up to four calcium ions even when these are present in minute quantities, the troponin complex supposedly undergoes a conformational change that in some way tugs on the tropomyosin molecule and supposedly moves it deeper into the groove between the two actin strands. This "uncovers" the active sites of the actin, thus allowing contraction to proceed. Although this is a hypothetical mechanism, nevertheless it does emphasize that the normal relationship between the tropomyosin-troponin complex and actin is altered by calcium ions, producing a new condition that leads to contraction.

Interaction Between the "Activated" Actin Filament and the Myosin Cross-Bridges — The "Walk-Along" Theory of Contraction. As soon as the actin filament becomes activated by the calcium ions, the heads of the cross-bridges from the myosin filaments immediately become attracted to the active sites of the actin filament, and this in some way causes contraction to occur. Although the precise manner by which this interaction between the cross-bridges and the actin causes contraction is still unknown, a suggested hypothesis for which considerable evidence exists is the "walk-along" theory (or *"ratchet" theory*) of *contraction*.

Figure 24–7 illustrates the postulated walk-along mechanism for contraction. This figure shows the heads of two cross-bridges attaching to and disengaging from the active sites of an actin filament. It is postulated that when the head attaches to an active site, this attachment simultaneously causes profound changes in the intramolecular forces between the head and arm of the cross-bridge. The new alignment of forces causes the head to tilt toward the arm and to drag the actin filament along with it. This tilt of the head is called the *power stroke*. Then, immediately after tilting, the head automatically breaks away from the active site. Next, the head returns to its normal perpendicular direction. In this position it combines with a new active site farther down along the actin filament; then, the head tilts again to cause a new power stroke, and the actin filament moves another step. Thus, the heads of the cross-bridges bend back and forth and

step by step walk along the actin filament, pulling the ends of the actin filaments toward the center of the myosin filament.

Each one of the cross-bridges is believed to operate independently of all others, each attaching and pulling in a continuous but random cycle. Therefore, the greater the number of cross-bridges in contact with the actin filament at any given time, the greater, theoretically, is the force of contraction.

ATP as the Source of Energy for Contraction — Chemical Events in the Motion of the Myosin Heads. When a muscle contracts against a load, work is performed and energy is required. Large amounts of ATP are cleaved to form ADP during the contraction process. Furthermore, the greater the amount of work performed by the muscle, the greater the amount of ATP that is cleaved, which is called the *Fenn effect*. Although it is still not known exactly how ATP is used to provide the energy for contraction, the following is a sequence of events that has been suggested as the means by which this occurs:

1. Before contraction begins, the heads of the cross-bridges bind with ATP. The ATPase activity of the myosin head immediately cleaves the ATP but leaves the cleavage products, ADP plus Pi, bound to the head. In this state, the conformation of the head is such that it extends perpendicularly toward the actin filament but is not yet attached to the actin.

2. Next, when the inhibitory effect of the troponin-tropomyosin complex is itself inhibited by calcium ions, active sites on the actin filament are uncovered, and the myosin heads do then bind with these, as illustrated in Figure 24–7.

3. The bond between the head of the cross-bridge and the active site of the actin filament causes a conformational change in the head, prompting the head to tilt toward the arm of the cross-bridge. This provides the *power stroke* for pulling the actin filament. The energy that activates the power stroke is the energy already stored, like a "cocked" spring, by the conformational change in the head when the ATP molecule had been cleaved.

4. Once the head of the cross-bridge is tilted, this allows release of the ADP and Pi that were previously attached to the head; at the site of release of the ADP, a new molecule of ATP binds. This binding in turn causes detachment of the head from the actin.

5. After the head has detached from the actin, the new molecule of ATP is also cleaved, and the energy again "cocks" the head back to its perpendicular condition ready to begin a new power stroke cycle.

6. Then, when the cocked head, with its stored energy derived from the cleaved ATP, binds with a new active site on the actin filament, it becomes uncocked and once again provides the power stroke.

7. Thus, the process proceeds again and again until the actin filament pulls the Z membrane up against the ends of the myosin filaments or until the load on the muscle becomes too great for further pulling to occur.

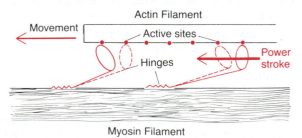

Figure 24 – 7. The "walk-along" mechanism for contraction of the muscle.

DEGREE OF ACTIN AND MYOSIN FILAMENT OVERLAP — EFFECT ON TENSION DEVELOPED BY THE CONTRACTING MUSCLE

Figure 24–8 illustrates the effect of sarcomere length and of myosin-actin filament overlap on the active tension developed by a contracting muscle fiber. To the right are illustrated different degrees of overlap of the myosin and actin filaments at different sarcomere lengths. At point D on the diagram, the actin filament has pulled all the way out to the end of the myosin filament with no overlap at all. At this point, the tension developed by the activated muscle is zero. Then, as the sarcomere shortens and the actin filament begins to overlap the myosin filament, the tension increases progressively until the sarcomere length decreases to about 2.2 μm. At this point, the actin filament has already overlapped all the cross-bridges of the myosin filament but has not yet reached the center of the myosin filament. Upon further shortening, the sarcomere maintains full tension until point B at a sarcomere length of approximately 2.0 μm. At this point, the ends of the two actin filaments begin to overlap each other, in addition to overlapping the myosin filaments. As the sarcomere length falls from 2 μm down to about 1.65 μm, at point A, the strength of contraction decreases. It is at this point that the two Z discs of the sarcomere abut the ends of the myosin filaments. Then, as contraction proceeds to still shorter sarcomere lengths, the ends of the myosin filaments are actually crumpled, and as illustrated in the figure, the strength of contraction also decreases precipitously.

This diagram illustrates that maximum contraction occurs when there is maximum overlap between the actin filaments and the cross-bridges of the myosin

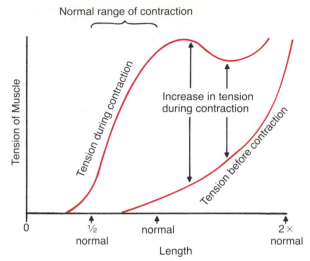

Figure 24–9. Relation of muscle length to force of contraction.

filaments, and it supports the idea that the greater the number of cross-bridges pulling the actin filaments, the greater the strength of contraction.

Effect of Muscle Length on Force of Contraction in the Intact Muscle. The upper curve of Figure 24–9 is similar to that in Figure 24–8, but this illustrates the intact, whole muscle rather than a single muscle fiber. The whole muscle has a large amount of connective tissue in it; also, the sarcomeres in different parts of the muscle do not necessarily contract exactly in unison. Therefore, the curve has somewhat different dimensions from those illustrated for the individual muscle fiber, but it nevertheless exhibits the same form.

Note in Figure 24–9 that when the muscle is at its normal resting length, which is at a sarcomere length of about 2 μm, it contracts with maximum force of contraction. If the muscle is stretched to much greater than normal length prior to contraction, a large amount of *resting tension* develops in the muscle even before contraction takes place; this tension results from the elastic forces of the connective tissue, the sarcolemma, the blood vessels, the nerves, and so forth. However, the *increase* in tension during contraction, called *active tension*, decreases as the muscle is stretched much beyond its normal length — that is, to a sarcomere length greater than about 2.2 μm. This is demonstrated by the decrease in the arrow length in the figure.

RELATION OF VELOCITY OF CONTRACTION TO LOAD

A muscle contracts extremely rapidly when it contracts against no load — to a state of full contraction in approximately 0.1 sec for the average muscle. However, when loads are applied, the velocity of contraction becomes progressively less as the load increases, as illustrated in Figure 24–10. When the load increases to equal the maximum force

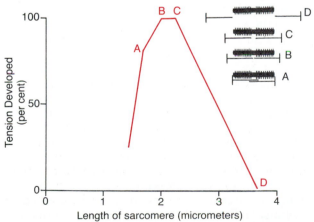

Figure 24–8. Length-tension diagram for a single sarcomere, illustrating maximum strength of contraction when the sarcomere is 2.0 to 2.2 μm in length. At the upper right are shown the relative positions of the actin and myosin filaments at different sarcomere lengths from point A to point D. (Modified from Gordon, Huxley, and Julian: The length-tension diagram of single vertebrate striated muscle fibers. J. Physiol., **171:**28P, 1964.)

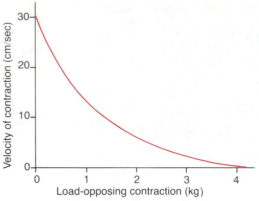

Figure 24–10. Relation of load to velocity of contraction in a skeletal muscle 8 cm long.

that the muscle can exert, the velocity of contraction becomes zero and no contraction at all results, despite activation of the muscle fiber.

This decreasing velocity with load is caused by the fact that a load on a contracting muscle is a reverse force that opposes the contractile force caused by muscle contraction. Therefore, the net force that is available to cause velocity of shortening is correspondingly reduced.

■ INITIATION OF MUSCLE CONTRACTION: EXCITATION-CONTRACTION COUPLING

Initiation of contraction in skeletal muscle begins with action potentials in the muscle fibers. These elicit electrical currents that spread to the interior of the fiber where they cause release of calcium ions from the sarcoplasmic reticulum. It is the calcium ions that in turn initiate the chemical events of the contractile process. This overall process for controlling muscle contraction is called *excitation-contraction coupling.*

THE MUSCLE ACTION POTENTIAL

Almost everything discussed in Chapter 6 regarding initiation and conduction of action potentials in nerve fibers applies equally well to skeletal muscle fibers except for quantitative differences. Some of the quantitative aspects of muscle potentials are the following:

1. Resting membrane potential: approximately −80 to −90 mV in skeletal fibers—the same as in large myelinated nerve fibers.

2. Duration of action potential: 1 to 5 msec in skeletal muscle—about five times as long as large myelinated nerves.

3. Velocity of conduction: 3 to 5 m/sec—about 1/18 the velocity of conduction in the large myelinated nerve fibers that excite skeletal muscle.

Spread of the Action Potential to the Interior of the Muscle Fiber by Way of the Transverse Tubule System

The skeletal muscle fiber is so large that action potentials spreading along its surface membrane cause almost no current flow deep within the fiber. Yet, to cause contraction, these electrical currents must penetrate to the vicinity of all the separate myofibrils. This is achieved by transmission of the action potentials along *transverse tubules* (T tubules) that penetrate all the way through the muscle fiber from one side to the other. The T tubule action potentials in turn cause the sarcoplasmic reticulum to release calcium ions in the immediate vicinity of all the myofibrils, and these calcium ions then cause contraction. This overall process is called *excitation-contraction* coupling. Now, let us describe this in much greater detail.

EXCITATION-CONTRACTION COUPLING

The Transverse Tubule–Sarcoplasmic Reticulum System

Figure 24–11 illustrates several myofibrils surrounded by the transverse tubule–sarcoplasmic reticulum system. The transverse tubules are very small and run transverse to the myofibrils. They begin at the cell membrane and penetrate all the way from one side of the muscle fiber to the opposite side. Not shown in the figure is the fact that these tubules branch among themselves so that they form entire *planes* of T tubules interlacing among all the separate myofibrils. Also, it should be noted that *where the T tubules originate from the cell membrane they are open to the exterior.* Therefore, they communicate with the fluid surrounding the muscle fiber and contain extracellular fluid in their lumens. In other words, the T tubules are internal extensions of the cell membrane. Therefore, when an action potential spreads over a muscle fiber membrane, it spreads along the T tubules to the deep interior of the muscle fiber as well. The action potential currents surrounding these transverse tubules then elicit the muscle contraction.

Figure 24–11 shows the *sarcoplasmic reticulum* as well, shown in red. This is composed of two major parts: (1) long *longitudinal tubules* that run parallel to the myofibrils and terminate in (2) large chambers called *terminal cisternae* that abut the transverse tubules. When the muscle fiber is sectioned longitudinally and electron micrographs are made, one sees this abutting of the cisternae against the transverse tubule, which gives the appearance of a *triad* with a small central tubule and a large cisterna on either side. This is illustrated in Figure 24–11 and is also seen in the electron micrograph of Figure 24–3.

In the muscle of lower animals, such as the frog, there is a single T tubule network for each sarcomere, located at the level of the Z disc, as illustrated in Figure 24–11. Cardiac muscle also has this type of T tubule system. However, in mammalian skeletal muscle there are two T tubule networks for each sarcomere located

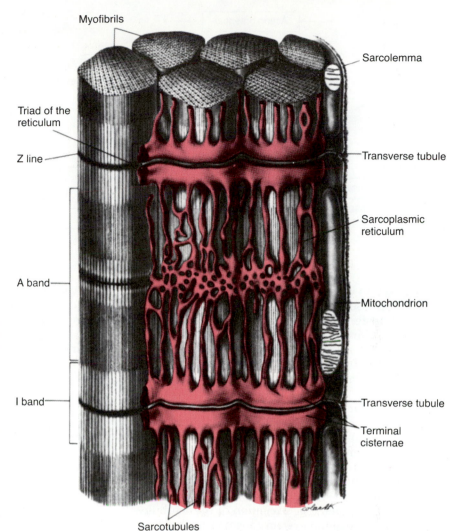

Myofibrils

Sarcolemma

Triad of the reticulum

Z line

Transverse tubule

A band

Sarcoplasmic reticulum

Mitochondrion

I band

Transverse tubule

Terminal cisternae

Sarcotubules

Figure 24–11. The transverse tubule–sarcoplasmic reticulum system. Note the *longitudinal tubules* that terminate in large *cisternae*. The cisternae in turn abut the transverse tubules. Note also that the transverse tubules communicate with the outside of the cell membrane. This illustration was drawn from frog muscle, which has one transverse tubule per sarcomere, located at the Z line. A similar arrangement is found in mammalian heart muscle, but mammalian skeletal muscle has two transverse tubules per sarcomere, located at the A-I junctions. (From Fawcett: Bloom and Fawcett: A Textbook of Histology. Philadelphia, W. B. Saunders Company, 1986. Modified after Peachey: J. Cell Biol. 25:209, 1965. Drawn by Sylvia Colard Keene.)

near the two ends of the myosin filaments, which are the points where the actual mechanical forces of muscle contraction are created. Thus, mammalian skeletal muscle is optimally organized for rapid excitation of muscle contraction.

RELEASE OF CALCIUM IONS BY THE SARCOPLASMIC RETICULUM

One of the special features of the sarcoplasmic reticulum is that it contains calcium ions in very high concentration, and many of these ions are released when the adjacent T tubule is excited.

Figure 24–12 shows that the action potential of the T tubule causes current flow through the tips of the cisternae that abut the T tubule. At these points, each cisterna projects *junctional feet* that attach to the membrane of the T tubule, presumably facilitating passage of some signal from the T tubule to the cisterna. Possibly this signal is electrical current of the action potential itself. However, there are also reasons to believe that it could be some chemical or mechanical signal. Whatever the signal, it causes rapid opening of large numbers of calcium channels through the membranes of the cisternae and their attached longitudinal tubules of the sarcoplasmic reticulum. These channels remain open for a few milliseconds; during this time the calcium ions responsible for muscle contraction are released into the sarcoplasm surrounding the myofibrils.

The calcium ions that are thus released from the sarcoplasmic reticulum diffuse to the adjacent myofibrils, where they bind strongly with troponin C, as discussed in the previous chapter, and this in turn elicits the muscle contraction.

The Calcium Pump for Removing Calcium Ions from the Sarcoplasmic Fluid. Once the calcium ions have been released from the sarcoplasmic tubules and have diffused to the myofibrils, muscle contraction will continue as long as the calcium ions remain in high concentration in the sarcoplasmic fluid. However, a continually active calcium pump located in the walls of the sarcoplasmic reticulum pumps calcium ions out of the sarcoplasmic fluid back into the sarcoplasmic tubules. This pump can concentrate the calcium ions about 10,000-fold inside the sarcoplasmic reticulum. In addition, inside the reticulum is a protein called *calsequestrin* that can bind over 40 times as much cal-

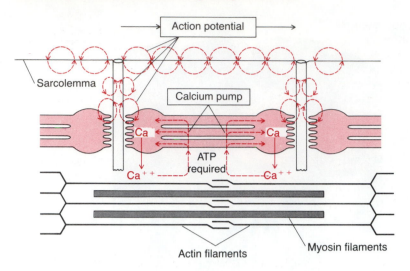

Figure 24–12. Excitation-contraction coupling in the muscle, showing an action potential that causes release of calcium ions from the sarcoplasmic reticulum and then reuptake of the calcium ions by a calcium pump.

cium as that in the ionic state, thus providing another 40-fold increase in the storage of calcium. Thus, this massive transfer of calcium into the sarcoplasmic reticulum causes virtual total depletion of calcium ions in the fluid of the myofibrils. Therefore, except immediately after an action potential, the calcium ion concentration in the myofibrils is kept at an extremely low level.

The Excitatory "Pulse" of Calcium Ions. The normal concentration (less than 10^{-7} M) of calcium ions in the cytosol that bathes the myofibrils is too little to elicit contraction. Therefore, in the resting state, the troponin-tropomyosin complex keeps the actin filaments inhibited and maintains a relaxed state of the muscle.

On the other hand, full excitation of the T tubule–sarcoplasmic reticulum system causes enough release of calcium ions to increase the concentration in the myofibrillar fluid to as high as 2×10^{-4} M concentration, which is about 10 times the level required to cause maximum muscle contraction (about 2×10^{-5} M). Immediately thereafter, the calcium pump depletes the calcium ions again. The total duration of this calcium "pulse" in the usual skeletal muscle fiber lasts about $\frac{1}{20}$ sec, though it may last several times as long as this in some skeletal muscle fibers and be several times shorter in others (in heart muscle, the calcium pulse lasts for as long as $\frac{1}{3}$ sec because of the long duration of the cardiac action potential). It is during this calcium pulse that muscle contraction occurs. If the contraction is to continue without interruption for longer intervals, a series of such pulses must be initiated by a continuous series of repetitive action potentials, as discussed in the previous chapter.

■ ENERGETICS OF MUSCLE CONTRACTION

WORK OUTPUT DURING MUSCLE CONTRACTION

When a muscle contracts against a load, it performs *work*. This means that *energy* is transferred from the muscle to the external load, for example, to lift an object to a greater height or to overcome resistance to movement.

In mathematical terms, work is defined by the following equation:

$$W = L \times D$$

in which W is the work output, L is the load, and D is the distance of movement against the load. The energy required to perform the work is derived from the chemical reactions in the muscle cells during contraction, as we describe in the following sections.

SOURCES OF ENERGY FOR MUSCLE CONTRACTION

We have already seen that muscle contraction depends on energy supplied by ATP. Most of this energy is required to actuate the walk-along mechanism by which the cross-bridges pull the actin filaments, but small amounts are required for (1) pumping calcium from the sarcoplasm into the sarcoplasmic reticulum after the contraction is over and (2) pumping sodium and potassium ions through the muscle fiber membrane to maintain an appropriate ionic environment for the propagation of action potentials.

However, the concentration of ATP present in the muscle fiber, about 4 mM, is sufficient to maintain full contraction for only 1 to 2 sec at most. Fortunately, after the ATP is split into ADP, the ADP is rephosphorylated to form new ATP within a fraction of a second. There are several sources of the energy for this rephosphorylation.

The first source of energy that is used to reconstitute the ATP is the substance *phosphocreatine*, which carries a high-energy phosphate bond similar to those of ATP. The high-energy phosphate bond of the phosphocreatine has a slightly higher amount of free energy than that of the ATP bond. Therefore, phosphocreatine is instantly cleaved, and the released energy causes bonding of a new phosphate ion to ADP to reconstitute the ATP. However, the total amount of phosphocreatine is also very little—only about five times as great as the ATP. Therefore, the combined energy of both the stored ATP and the phosphocreatine in the muscle is still capable of causing maximal muscle contraction for no longer than 7 to 8 sec.

The next important source of energy, which is used to reconstitute both ATP and phosphocreatine, is *glycogen* previously stored in the muscle cells. Rapid enzymatic breakdown of the glycogen to pyruvic acid and lactic acid liberates energy that is used to convert ADP to ATP, and the ATP can then be used directly to energize muscular contraction or to re-form the stores of phosphocreatine. The importance of this "glycolysis" mechanism is twofold. First, the glycolytic reactions occur even in the absence of oxygen, so that muscle contraction can be sustained for a short time when oxygen is not available. Second, the rate of formation of ATP by the glycolytic process is about two and one half times as rapid as ATP formation when the cellular foodstuffs react with oxygen. Unfortunately, though, so many end-products of glycolysis accumulate in the muscle cells that glycolysis alone can sustain maximum muscle contraction for only about 1 minute.

The final source of energy is the process of *oxidative metabolism.* This means the combining of oxygen with the various cellular foodstuffs to liberate ATP. Over 95 per cent of all energy used by the muscles for sustained, long-term contraction is derived from this source. The foodstuffs that are consumed are carbohydrates, fats, and protein. For extremely long-term muscle activity—over a period of hours—by far the greatest proportion of energy comes from fats.

Efficiency of Muscle Contraction. The "efficiency" of an engine or a motor is calculated as the percentage of energy input that is converted into work instead of heat. The percentage of the input energy to the muscle (the chemical energy in the nutrients) that can be converted into work is less than 20 to 25 per cent, the remainder becoming heat. The reason for this low efficiency is that about half of the energy in the foodstuffs is lost during the formation of ATP, and even then only 40 to 45 per cent of the energy in the ATP itself can later be converted into work.

■ CHARACTERISTICS OF WHOLE MUSCLE CONTRACTION

Many features of muscle contraction can be especially well demonstrated by eliciting single *muscle twitches.* This can be accomplished by instantaneously exciting the nerve to a muscle or by passing a short electrical stimulus through the muscle itself, giving rise to a single, sudden contraction lasting for a fraction of a second.

Isometric Versus Isotonic Contraction. Muscle contraction is said to be *isometric* when the muscle does not shorten during contraction and *isotonic* when it shortens with the tension on the muscle remaining constant. Systems for recording the two types of muscle contraction are illustrated in Figure 24–13.

In the isometric system, the muscle contracts against a force transducer without decreasing the muscle length, as illustrated to the right in Figure 24–13. In the isotonic system, the muscle shortens against a fixed load; this is illustrated to the left in the figure, showing a muscle lifting a pan

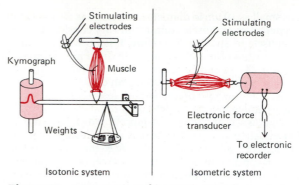

Figure 24–13. Isotonic and isometric recording systems.

of weights. Obviously, the characteristics of isotonic contraction depend on the load against which the muscle contracts as well as on the inertia of the load. On the other hand, the isometric system records strictly changes in force of muscle contraction itself. Therefore, the isometric system is most often used when comparing the functional characteristics of different muscle types.

The Series Elastic Component of Muscle Contraction. When muscle fibers contract against a load, those portions of the muscle that do not contract—the tendons, the sarcolemmal ends of the muscle fibers where they attach to the tendons, and perhaps even the hinged arms of the cross-bridges—stretch slightly as the tension increases. Consequently, the muscle must shorten an extra 3 to 5 per cent to make up for the stretch of these elements. The elements of the muscle that stretch during contraction are called the *series elastic component* of the muscle.

CHARACTERISTICS OF ISOMETRIC TWITCHES RECORDED FROM DIFFERENT MUSCLES

The body has many different sizes of skeletal muscles—from the very small stapedius muscle in the middle ear of only a few millimeters length and a millimeter or so in diameter up to the very large quadriceps muscle, a half million times as large as the stapedius. Furthermore, the fibers may

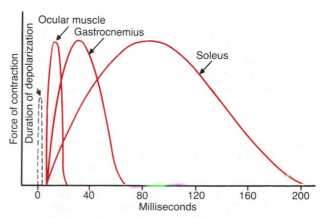

Figure 24–14. Duration of isometric contractions of different types of mammalian muscles, showing also a latent period between the action potential and muscle contraction.

be as small as 10 μm in diameter or as large as 80 μm. Finally, the energetics of muscle contraction vary considerably from one muscle to another. Therefore, it is no wonder that the characteristics of muscle contraction differ among muscles.

Figure 24–14 illustrates isometric contractions of three different types of skeletal muscles: an ocular muscle, which has a duration of contraction of less than $\frac{1}{40}$ sec; the gastrocnemius muscle, which has a duration of contraction of about $\frac{1}{15}$ sec; and the soleus muscle, which has a duration of contraction of about $\frac{1}{5}$ sec. It is interesting that these durations of contractions are adapted to the function of each of the respective muscles. Ocular movements must be extremely rapid to maintain fixation of the eyes on specific objects, and the gastrocnemius muscle must contract moderately rapidly to provide sufficient velocity of limb movement for running and jumping, whereas the soleus muscle is concerned principally with slow contraction for continual support of the body against gravity.

MECHANICS OF SKELETAL MUSCLE CONTRACTION

THE MOTOR UNIT

Each motoneuron that leaves the spinal cord innervates many different muscle fibers, the number depending on the type of muscle. All the muscle fibers innervated by a single motor nerve fiber are called a *motor unit*. In general, small muscles that react rapidly and whose control must be exact have few muscle fibers (as few as two to three in some of the laryngeal muscles) in each motor unit. On the other hand, the large muscles that do not require very fine control, such as the gastrocnemius muscle, may have several hundred muscle fibers in a motor unit. An average figure for all the muscles of the body can be considered to be about 100 muscle fibers to the motor unit.

The muscle fibers in each motor unit are not all bunched together in a muscle but instead are spread out in the muscle in microbundles of 3 to 15 fibers. Therefore, these lie among similar microbundles of other motor units. This interdigitation allows the separate motor units to contract in support of each other rather than entirely as individual segments.

Muscle Contractions of Different Force — Force Summation

Summation means the adding together of individual twitch contractions to increase the intensity of overall muscle contraction. Summation occurs in two different ways: (1) by increasing the number of motor units contracting simultaneously, which is called *multiple fiber summation*; and (2) by increasing the frequency of contraction, which is called *frequency summation* or *tetanization*.

Multiple Fiber Summation. When the central nervous system sends a weak signal to contract a muscle, the motor units in the muscle that contain the smallest and fewest muscle fibers are stimulated in preference to the larger motor units. Then as the strength of the signal increases, larger and larger motor units begin to be excited as well, with the largest motor units often having 50 times as much contractile force as the smallest units. This is called the *size principle*. It is important because it allows the gradations of muscle force during weak contraction to occur in very small steps, while the steps become progressively greater when large amounts of force are required. The cause of this size principle is that the smaller motor units are driven by small motor nerve fibers, and the small motoneurons in the spinal cord are far more excitable than the larger ones, so that they naturally are excited first.

Another important feature of multiple fiber summation is that the different motor units are driven asynchronously by the spinal cord, so that contraction alternates among motor units one after the other, thus providing smooth contraction even at low frequencies of nerve signals.

Frequency Summation and Tetanization. Figure 24–15 illustrates the principles of frequency summation and tetanization. To the left are illustrated individual twitch contractions occurring one after another at low frequency of stimulation. Then as the frequency increases, there comes a point when each new contraction occurs before the preceding one is over. As a result, the second contraction is added partially to the first, so that the total strength of contraction rises progressively with increasing frequency. When the frequency reaches a critical level, the successive contractions are so rapid that they literally fuse together, and the contraction appears to be completely smooth and continuous, as illustrated in the figure. This is called *tetanization*. At a still higher frequency, the strength of contraction reaches its maximum, so that additional increase in frequency beyond that point will have no further effect in increasing contractile force. This occurs because enough calcium ions are then maintained in the muscle sarcoplasm even between action potentials, so that the full contractile state is sustained without allowing any relaxation between the action potentials.

Maximum Strength of Contraction. The maximum strength of tetanic contraction of a muscle operating at a normal muscle length averages between 3 and 4 kg/cm² of muscle, or 50 lb/in.². Because a quadriceps muscle can at times have as much as 16 in.² of muscle belly, as much as 800 lb of tension may at times be applied to the patellar tendon. One can readily understand, therefore, how it is possible for muscles sometimes to pull their tendons out of the insertions in bones.

Muscle Fatigue

Prolonged and strong contraction of a muscle leads to the well-known state of muscle fatigue. Studies in athletes have shown that muscle fatigue increases in almost direct proportion to the rate of depletion of muscle glycogen. Therefore, most fatigue probably results simply from inability of the contractile and metabolic processes of the muscle fibers to continue supplying the same work output. However, experiments have also shown that transmission of the nerve signal

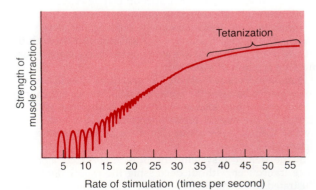

Figure 24–15. Frequency summation and tetanization.

through the neuromuscular junction, which is discussed in the following chapter, can occasionally diminish following prolonged muscle activity, thus further diminishing muscle contraction.

Interruption of blood flow through a contracting muscle leads to almost complete muscle fatigue in a minute or more because of the obvious loss of nutrient supply — especially loss of oxygen.

REMODELING OF MUSCLE TO MATCH FUNCTION

All the muscles of the body are continually being remodeled to match the functions that are required of them. Their diameters are altered, their lengths are altered, their strengths are altered, their vascular supplies are altered, and even the types of muscle fibers are altered at least to a slight extent. This remodeling process is often quite rapid, within a few weeks. Indeed, experiments have shown that even normally the muscle contractile proteins can be totally replaced once every 2 weeks.

Muscle Hypertrophy and Muscle Atrophy

When the total mass of a muscle enlarges, this is called *muscle hypertrophy.* When it decreases, the process is called *muscle atrophy.*

Virtually all muscle hypertrophy results from hypertrophy of the individual muscle fibers, which is called simply *fiber hypertrophy.* This usually occurs in response to contraction of a muscle at maximal or almost maximal force. Hypertrophy occurs to a much greater extent when the muscle is simultaneously stretched during the contractile process. Only a few such strong contractions each day are required to cause almost maximum hypertrophy within 6 to 10 weeks.

Unfortunately, the manner in which forceful contraction leads to hypertrophy is not known. Yet it is known that the rate of synthesis of muscle contractile proteins is far greater during developing hypertrophy than their rate of decay, leading to progressively greater numbers of both actin and myosin filaments in the myofibrils. In turn, the myofibrils themselves split within each muscle fiber to form new myofibrils. Thus, it is mainly this great increase in numbers of additional myofibrils that causes muscle fibers to hypertrophy.

Along with the increasing numbers of myofibrils, the enzyme systems that provide energy also increase. This is especially true of the enzymes for glycolysis, allowing a rapid supply of energy during short-term forceful muscle contraction.

When a muscle remains unused for long periods of time, the rate of decay of the contractile proteins as well as the numbers of myofibrils occurs more rapidly than the rate of replacement. Therefore, muscle atrophy occurs.

Effects of Muscle Denervation

When a muscle loses its nerve supply, it no longer receives the contractile signals that are required to maintain normal muscle size. Therefore, atrophy begins almost immediately. After about 2 months, degenerative changes also begin to appear in the muscle fibers themselves. If the nerve supply grows back to the muscle, full return of function will usually occur up to about 3 months, but from that time onward the capability of functional return becomes less and less, with no return of function after 1 to 2 years.

In the final stage of denervation atrophy, most of the muscle fibers are completely destroyed and replaced by fibrous and fatty tissue. Those fibers that do remain are composed of a long cell membrane with a line-up of muscle cell nuclei but with no contractile properties and with no capability of regenerating myofibrils if a nerve regrows.

Unfortunately, the fibrous tissue that replaces the muscle fibers during denervation atrophy has a tendency to continue shortening for many months, which is called *contracture.* Therefore, one of the most important problems in the practice of physical therapy is to keep atrophying muscles from developing debilitating and disfiguring contracture. This is achieved by daily stretching of the muscles or use of appliances that keep the muscles stretched during the atrophying process.

Recovery of Muscle Contraction in Poliomyelitis: Development of Macromotor Units. When some nerve fibers to a muscle are destroyed but not all of them, as often occurs in poliomyelitis, the remaining nerve fibers sprout forth new axons to form many new branches that then innervate many of the paralyzed muscle fibers. This causes very large motor units called *macromotor units,* containing as many as five times the normal number of muscle fibers for each motoneuron in the spinal cord. This obviously decreases the fineness of control that one has over the muscles but, nevertheless, allows the muscles to regain strength.

REFERENCES

See References, Chapter 25.

25

Neuromuscular Transmission; Function of Smooth Muscle

■ TRANSMISSION OF IMPULSES FROM NERVES TO SKELETAL MUSCLE FIBERS: THE NEUROMUSCULAR JUNCTION

The skeletal muscle fibers are innervated by large, myelinated nerve fibers that originate in the large motoneurons of the anterior horns of the spinal cord. As pointed out in the previous chapter, each nerve fiber normally branches many times and stimulates from three to several hundred skeletal muscle fibers. The nerve ending makes a junction, called the *neuromuscular junction,* with the muscle fiber near the fiber's midpoint, and the action potential in the fiber travels in both directions toward the muscle fiber ends. With the exception of about 2 per cent of the muscle fibers, there is only one such junction per muscle fiber.

Physiologic Anatomy of the Neuromuscular Junction—The "Motor End-Plate." Figure 25–1, parts A and B, illustrates the neuromuscular junction between a large, myelinated nerve fiber and a skeletal muscle fiber. The nerve fiber branches at its end to form a complex of branching nerve *terminals,* which invaginate into the muscle fiber but lie entirely outside the muscle fiber plasma membrane. The entire structure is called the *motor end-plate.* It is covered by one or more Schwann cells that insulate it from the surrounding fluids.

Figure 25–1C shows an electron micrographic sketch of the junction between a single-branch axon terminal and the muscle fiber membrane. The invagination of the membrane is called the *synaptic gutter* or *synaptic trough,* and the space between the terminal and the fiber membrane is called the *synaptic cleft.* The synaptic cleft is 20 to 30 nm wide and is occupied by a basal lamina, which is a thin layer of

spongy reticular fibers through which diffuses extracellular fluid. At the bottom of the gutter are numerous smaller *folds* of the muscle membrane called *subneural clefts,* which greatly increase the surface area at which the synaptic transmitter can act.

In the axon terminal are many mitochondria that supply energy mainly for synthesis of the excitatory transmitter *acetylcholine* that, in turn, excites the muscle fiber. The acetylcholine is synthesized in the cytoplasm of the terminal but is rapidly absorbed into many small synaptic vesicles, approximately 300,000 of which are normally in the terminals of a single endplate. Attached to the matrix of the basal lamina are large quantities of the enzyme *acetylcholinesterase,* which is capable of destroying acetylcholine, to be explained in further detail.

SECRETION OF ACETYLCHOLINE BY THE NERVE TERMINALS

When a nerve impulse reaches the neuromuscular junction, about 300 vesicles of acetylcholine are released from the terminals into the synaptic trough. Figure 25–2 illustrates some of the details of this mechanism, showing an expanded view of a synaptic trough with the neural membrane above and the muscle membrane and its subneural clefts below.

On the inside surface of the neural membrane are linear *dense bars,* shown in cross-section in Figure 25–2. To each side of each dense bar are protein particles that penetrate the membrane, believed to be voltage-gated calcium channels. When the action potential spreads over the terminal, these channels open and allow large quantities of calcium to diffuse to the interior of the terminal. The calcium ions in turn exert an attractive influence on the acetylcholine vesicles, drawing them to the neural membrane adjacent to the

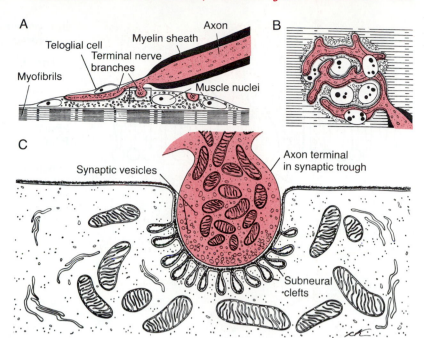

Figure 25–1. Different views of the motor endplate. *A*, Longitudinal section through the endplate. *B*, Surface view of the end-plate. *C*, Electron micrographic appearance of the contact point between one of the axon terminals and the muscle fiber membrane, representing the rectangular area shown in *A*. (From Fawcett, as modified from R. Couteaux: Bloom and Fawcett: A Textbook of Histology. Philadelphia, W. B. Saunders Company, 1986.)

dense bars. Some of the vesicles fuse with the neural membrane and empty their acetylcholine into the synaptic trough by the process of exocytosis.

Although some of the aforementioned details are still speculative, it is known that the effective stimulus for causing acetylcholine release from the vesicles is entry of calcium ions. Furthermore, the vesicles are emptied through the membrane adjacent to the dense bars.

Effect of Acetylcholine to Open Acetylcholine-Gated Ion Channels. Figure 25–2 shows many acetylcholine receptors in the muscle membrane; these are actually acetylcholine-gated ion channels, located almost entirely near the mouths of the subneural clefts

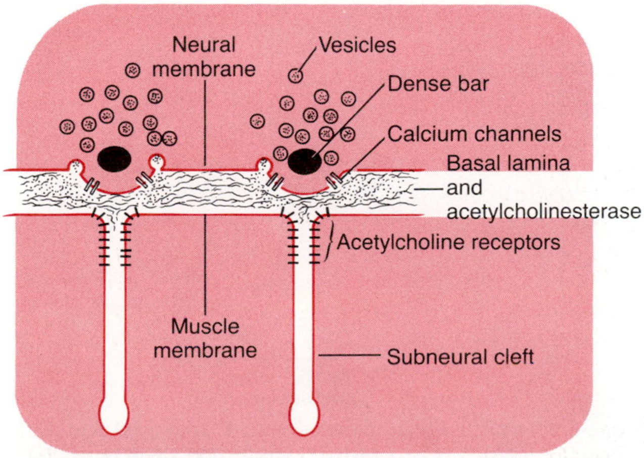

Figure 25–2. Release of acetylcholine from synaptic vesicles at the neural membrane of the neuromuscular junction. Note the close proximity of the release sites to the acetylcholine receptors at the mouths of the subneural clefts.

lying immediately below the dense bar areas, where the acetylcholine vesicles empty into the synaptic trough.

Each receptor is a large protein complex having a total molecular weight of 275,000. The complex is composed of five subunit proteins, which penetrate all the way through the membrane lying side by side in a circle to form a tubular channel. The channel remains constricted until acetylcholine attaches to one of the subunits. This causes a conformational change that opens the channel, as illustrated in Figure 25–3; the upper channel is closed, while the bottom one has been opened by attachment of an acetylcholine molecule.

The acetylcholine channel has a diameter when open of about 0.65 nm, which is large enough to allow all the important positive ions — sodium (Na^+), potassium (K^+), and calcium (Ca^{++}) — to move easily through the opening. On the other hand, negative ions, such as chloride ions, do not pass through because of strong negative charges in the mouth of the channel.

However, in practice, far more sodium ions flow through the acetylcholine channels than any other ions for two reasons. First, there are only two positive ions in great enough concentration to matter greatly, sodium ions in the extracellular fluid and potassium ions in the intracellular fluid. Second, the very negative potential on the inside of the muscle membrane, −80 to −90 mV, pulls the positively charged sodium ions to the inside of the fiber, while at the same time preventing the efflux of the potassium ions when they attempt to pass outward.

Therefore, as illustrated in the lower panel of Figure 25–3, the net effect of opening the acetylcholine-gated channels is to allow large numbers of sodium ions to pour to the inside of the fiber, carrying with them large numbers of positive charges. This creates a

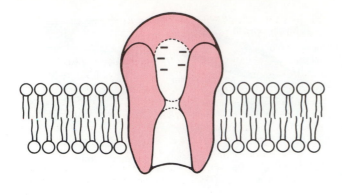

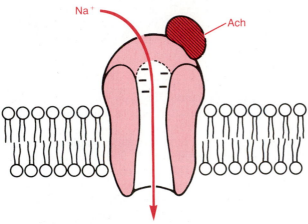

Figure 25–3. The acetylcholine channel: Above, while in the closed state. Below, after acetylcholine has become attached and a conformational change has opened the channel, allowing excess sodium to enter the muscle fiber and excite contraction. Note the negative charges at the channel mouth that prevent passage of negative ions.

line prevents muscle re-excitation after the fiber has recovered from the first action potential.

The "End-Plate Potential" and Excitation of the Skeletal Muscle Fiber. The sudden insurgence of sodium ions into the muscle fiber when the acetylcholine channels open causes the membrane potential in the *local area of the endplate* to increase in the positive direction as much as 50 to 75 mV, creating a *local potential* called the *end-plate potential.* If one recalls from Chapter 6 that a sudden increase in membrane potential of more than 15 to 30 mV is sufficient to initiate the positive feedback effect of sodium channel activation, one can understand that the end-plate potential created by the acetylcholine stimulation is normally far greater than enough to initiate an action potential in the muscle fiber.

Figure 25–4 illustrates the principle of an end-plate potential initiating the action potential. In this figure are shown three separate end-plate potentials. End-plate potentials A and C are too weak to elicit an action potential, but they do nevertheless give the weak local potentials recorded in the figure. In contrast, end-plate potential B is much stronger and causes sodium channels to activate so that the self-regenerative effect of more and more sodium ions flowing to the interior of the fiber initiated an action potential. The weak end-plate potential at point A was caused by poisoning the muscle fiber with *curare,* a drug that blocks the gating action of acetylcholine on the acetylcholine channels by competing with the acetylcholine for the acetylcholine receptor site. The weak end-plate potential at point C resulted from the effect of botulinum toxin, a bacterial toxin that decreases the release of acetylcholine by the nerve terminals.

"Safety Factor" for Transmission at the Neuromuscular Junction; Fatigue of the Junction. Ordinarily, each impulse that arrives at the neuromuscular junction causes about 3 to 4 times as much end-plate potential as that required to stimulate the muscle fiber. Therefore, the normal neuromuscular junction is said to have a very high *safety factor.* However, artificial stimulation of the nerve fiber at rates greater than 100 times/sec for several minutes often diminishes the number of vesicles of acetylcholine released with each impulse so much that impulses then fail to pass into the muscle fiber. This is called *fatigue* of the neuromuscular junction, and it is

local potential inside the fiber called the *end-plate potential* that initiates an action potential at the muscle membrane and thus causes muscle contraction.

Destruction of the Released Acetylcholine by Acetylcholinesterase. The acetylcholine, once released into the synaptic trough, continues to activate the acetylcholine receptors as long as it persists in the trough. However, it is rapidly removed by two means: (1) Most of the acetylcholine is destroyed by the enzyme *acetylcholinesterase* that is attached mainly to the *basal lamina,* a spongy layer of fine connective tissue that fills the synaptic trough between the presynaptic terminal and the postsynaptic muscle membrane. (2) A small amount diffuses out of the synaptic trough and is then no longer available to act on the muscle fiber membrane.

Yet, the very short period of time that the acetylcholine remains in the synaptic trough—a few milliseconds at most—is almost always sufficient to excite the muscle fiber. Then the rapid removal of the acetylcho-

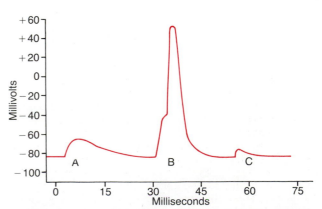

Figure 25–4. End-plate potentials. *A,* A weakened end-plate potential recorded in a curarized muscle, too weak to elicit an action potential; *B,* normal end-plate potential eliciting a muscle action potential; and *C,* weakened end-plate potential caused by botulinum toxin that decreases end-plate release of acetylcholine, again too weak to elicit a muscle action potential.

analogous to fatigue of the synapse in the central nervous system. Under normal functioning conditions, fatigue of the neuromuscular junction probably occurs very rarely and even then only at the most exhausting levels of muscular activity.

MOLECULAR BIOLOGY OF ACETYLCHOLINE FORMATION AND RELEASE

Because the neuromuscular junction is large enough to be easily studied, it is one of the few synapses of the nervous system at which most of the details of chemical transmission have been worked out. The formation and release of acetylcholine at this junction occurs in the following stages:

1. Very small vesicles, about 40 nm in size, are formed by the Golgi apparatus in the cell body of the motoneuron in the spinal cord. These vesicles are then transported by "streaming" of the axoplasm through the core of the axon from the central cell body to the neuromuscular junction at the tips of the nerve fibers. About 300,000 of these small vesicles collect in the nerve terminals of a single end-plate.

2. Acetylcholine is synthesized in the cytosol of the terminal nerve fibers but is then transported through the membranes of the vesicles to their interior, where it is stored in highly concentrated form, with about 10,000 molecules of acetylcholine in each vesicle.

3. Under resting conditions, an occasional vesicle fuses with the surface membrane of the nerve terminal and releases its acetylcholine into the synaptic gutter. When this occurs, a so-called *miniature end-plate* potential, about 1 mV in intensity and lasting for a few milliseconds, occurs in the local area of the muscle fiber because of the action of this "packet" of acetylcholine.

4. When an action potential arrives at the nerve terminal, this opens many calcium channels in the membrane of the terminal because this terminal has an abundance of voltage-gated calcium channels. As a result, the calcium ion concentration in the terminal increases about 100-fold, which in turn increases the rate of fusion of the acetylcholine vesicles with the terminal membrane about 10,000-fold. As each vesicle fuses, its outer surface ruptures through the cell membrane, thus causing *exocytosis* of acetylcholine into the synaptic cleft. Usually about 200 to 300 vesicles rupture with each action potential. The acetylcholine is then split by acetylcholinesterase into acetate ion and choline, and the choline is actively reabsorbed back into the neural terminal to be reused in forming new acetylcholine. This entire sequence of events occurs in 5 to 10 msec.

5. After each vesicle has released its acetylcholine, the membrane of the vesicle becomes part of the cell membrane. However, the number of vesicles available in the nerve ending is sufficient to allow transmission of only a few thousand nerve impulses. Therefore, for continued function of the neuromuscular junction, the vesicles need to be retrieved from the nerve membrane. Retrieval is achieved by the process of *endocytosis*. That is, a few seconds after the action potential is over, "coated pits" appear on the surface of the terminal nerve membrane, caused by contractile proteins of the cytosol, especially the protein *cathrin*, attaching underneath the membrane in the areas of the original vesicles. Within about 20 sec, the proteins contract and cause the pits to break away to the interior of the membrane, thus forming new vesicles. Within another few seconds, acetylcholine is transported to the interior of these vesicles, and they are then ready for a new cycle of acetylcholine release.

Drugs That Affect Transmission at the Neuromuscular Junction

Drugs That Stimulate the Muscle Fiber by Acetylcholine-like Action. Many different compounds, including *methacholine, carbachol,* and *nicotine,* have the same effect on the muscle fiber as does acetylcholine. The difference between these drugs and acetylcholine is that they are not destroyed by cholinesterase or are destroyed very slowly, so that when once applied to the muscle fiber the action persists for many minutes to several hours. These drugs work by causing localized areas of depolarization at the motor end-plate, where the acetylcholine receptors are located. Then, every time the muscle fiber becomes repolarized elsewhere, these depolarized areas, by virtue of their leaking ions, cause new action potentials, thereby causing a state of spasm.

Drugs That Block Transmission at the Neuromuscular Junction. A group of drugs, known as the *curariform drugs,* can prevent passage of impulses from the end-plate into the muscle. Thus, D-tubocurarine affects the membrane by competing with acetylcholine for the receptor sites of the membrane, so that the acetylcholine cannot increase the permeability of the acetylcholine channels sufficiently to initiate a depolarization wave.

Drugs That Stimulate the Neuromuscular Junction by Inactivating Acetylcholinesterase. Three particularly well-known drugs, *neostigmine, physostigmine,* and *diisopropyl fluorophosphate,* inactivate acetylcholinesterase so that the cholinesterase normally in the synapses will not hydrolyze the acetylcholine released at the end-plate. As a result, acetylcholine increases in quantity with successive nerve impulses so that extreme amounts of acetylcholine can accumulate and then repetitively stimulate the muscle fiber. This causes *muscular spasm* when even a few nerve impulses reach the muscle; this can cause death due to laryngeal spasm, which smothers the person.

Neostigmine and physostigmine combine with acetylcholinesterase to inactivate it for up to several hours, after which they are displaced from the acetylcholinesterase so that it once again becomes active. On the other hand, diisopropyl fluorophosphate, which has military potential as a very powerful "nerve" gas, actually inactivates acetylcholinesterase for weeks, which makes this a particularly lethal drug.

MYASTHENIA GRAVIS

The disease *myasthenia gravis,* which occurs in about 1 of every 20,000 persons, causes the person to become paralyzed because of inability of the neuromuscular junctions to transmit signals from the nerve fibers to the muscle fibers. Pathologically, antibodies that attack the acetylcholine-gated transport proteins have been demonstrated in the blood of most of these patients. Therefore, it is believed that myasthenia gravis in most instances is an autoimmune disease in which patients have developed antibodies against their own acetylcholine-activated ion channels.

Regardless of the cause, the end-plate potentials developed in the muscle fibers are too weak to stimulate the muscle fibers adequately. If the disease is intense enough, the patient dies of paralysis—in particular, of paralysis of the respiratory muscles. However, the disease can usually be ameliorated by administering *neostigmine* or some other anticholinesterase drug. This allows for more acetylcholine to accumulate in the synaptic cleft. Within minutes, some of these paralyzed persons can begin to function almost normally.

▪ CONTRACTION OF SMOOTH MUSCLE

In the previous chapter and first part of this chapter, the discussion has been concerned with skeletal muscle. We now turn to smooth muscle, which is composed of far smaller fibers — usually 2 to 5 μm in diameter and only 20 to 500 μm in length — in contrast to the skeletal muscle fibers that are as much as 20 times as large (in diameter) and thousands of times as long. Nevertheless, many of the principles of contraction apply to smooth muscle the same as to skeletal muscle. Most important, essentially the same attractive forces between myosin and actin filaments cause contraction in smooth muscle as in skeletal muscle, but the internal physical arrangement of smooth muscle fibers is entirely different, as we see subsequently.

TYPES OF SMOOTH MUSCLE

The smooth muscle of each organ is distinctive from that of most other organs in several different ways: physical dimensions, organization into bundles or sheets, response to different types of stimuli, characteristics of innervation, and function. Yet, for the sake of simplicity, smooth muscle can generally be divided into two major types, which are illustrated in Figure 25–5: *multiunit smooth muscle* and *single-unit smooth muscle*.

Multiunit Smooth Muscle. This type of smooth muscle is composed of discrete smooth muscle fibers. Each fiber operates entirely independently of the others and is often innervated by a single nerve ending, as occurs for skeletal muscle fibers. Furthermore, the outer surfaces of these fibers, like those of skeletal muscle fibers, are covered by a thin layer of "basement membrane–like" substance, a mixture of fine collagen and glycoprotein fibrillae that helps insulate the separate fibers from each other.

The most important characteristic of multiunit smooth muscle fibers is that each fiber can contract independently of the others, and their control is exerted mainly by nerve signals. This is in contrast to a major share of the control of visceral smooth muscle by non-nervous stimuli. An additional characteristic is that they rarely exhibit spontaneous contractions.

Some examples of multiunit smooth muscle found in the body are the smooth muscle fibers of the ciliary muscle of the eye, the iris of the eye, the nictitating membrane that covers the eyes in some lower animals, and the piloerector muscles that cause erection of the hairs when stimulated by the sympathetic nervous system.

Single-Unit Smooth Muscle. The term "single-unit" is confusing because it does not mean single muscle fibers. Instead, it means a whole mass of hundreds to millions of muscle fibers that contract together as a single unit. The fibers are usually aggregated into sheets or bundles, and their cell membranes are adherent to each other at multiple points so that force generated in one muscle fiber can be transmitted to the next. In addition, the cell membranes are joined by many *gap junctions* through which ions can flow freely from one cell to the next so that action potentials travel from one fiber to the next and cause the muscle fibers all to contract together. This type of smooth muscle is also known as *syncytial smooth muscle* because of its interconnections among fibers. Because such muscle is found in the walls of most viscera of the body — including the gut, the bile ducts, the ureters, the uterus, and many blood vessels — it is also often called *visceral smooth muscle*.

THE CONTRACTILE PROCESS IN SMOOTH MUSCLE

The Chemical Basis for Smooth Muscle Contraction

Smooth muscle contains both *actin* and *myosin filaments*, having chemical characteristics similar to but not exactly the same as those of the actin and myosin filaments in skeletal muscle. However, it does not contain troponin, so that the mechanism for control of contraction is entirely different. This is discussed in detail in a subsequent section of this chapter.

Chemical studies have shown that actin and myosin derived from smooth muscle interact with each other in much the same way that this occurs for actin and myosin derived from skeletal muscle. Furthermore, the contractile process is activated by calcium ions, and adenosine triphosphate (ATP) is degraded to adenosine diphosphate (ADP) to provide the energy for contraction.

On the other hand, there are major differences between the physical organization of smooth muscle and that of skeletal muscle as well as differences in excitation-contraction coupling, control of the contractile process by calcium ions, duration of contraction, and amount of energy required for the contractile process.

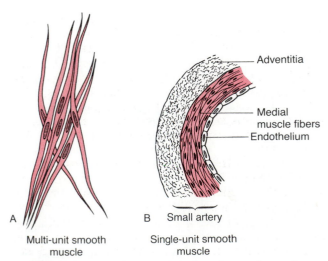

A   Multi-unit smooth muscle

B   Single-unit smooth muscle

Adventitia

Medial muscle fibers

Endothelium

Small artery

Figure 25–5. Multiunit and single-unit smooth muscle.

The Physical Basis for Smooth Muscle Contraction

Smooth muscle does not have the same striated arrangement of the actin and myosin filaments as that found in skeletal muscle. For a long time, it was impossible to discern even in electron micrographs any specific organization in the smooth muscle cell that could account for contraction. However, recent special electron micrographic techniques suggest the physical organization illustrated in Figure 25–6. This shows large numbers of actin filaments attached to so-called *dense bodies.* Some of these bodies are attached to the cell membrane. Others are dispersed inside the cell and are held in place by a scaffold of structural proteins linking one dense body to another. Note in Figure 25–6 that

some of the membrane dense bodies of adjacent cells are also bonded together by intracellular protein bridges. It is mainly through these bonds that the force of contraction is transmitted from one cell to the next.

Interspersed among the many actin filaments in the muscle fiber are a few myosin filaments. These have a diameter over two times as great as that of the actin filaments. When seen in electron micrographic cross-section, one usually finds about 15 times as many actin filaments as myosin filaments. Part of this difference is caused by the fact that the ratio of actin filament length to myosin filament length in smooth muscle is much greater than in skeletal muscle. Therefore, the probability of seeing excess actin filaments is increased. Nevertheless, one is impressed by the relative sparsity of myosin filaments with respect to actin filaments.

To the right in Figure 25–6 is the postulated structure of individual contractile units within smooth muscle cells, showing large numbers of actin filaments radiating from two dense bodies; these filaments overlap a single myosin filament located midway between the dense bodies. Obviously, this contractile unit is similar to the contractile unit of skeletal muscle but without the regularity of the skeletal muscle structure; in fact, the dense bodies of smooth muscle serve the same role as the Z discs in skeletal muscle.

Comparison of Smooth Muscle Contraction with Skeletal Muscle Contraction

Although most skeletal muscle contracts rapidly, most smooth muscle contraction provides prolonged tonic contraction, often lasting hours or even days. Therefore, it is to be expected that both the physical and the chemical characteristics of smooth muscle versus skeletal muscle contraction would differ. The following are some of the differences:

Slow Cycling of the Cross-Bridges. The rapidity of cycling of the cross-bridges in smooth muscle—that is, their attachment to actin, then release from the actin, and attachment again for the next cycle—is much, much slower in smooth muscle than in skeletal muscle, in fact as little as $1/10$ to $1/300$ the frequency in skeletal muscle. Yet, the *fraction of time* that the cross-bridges remain attached to the actin filaments, which is the major factor that determines the force of contraction, is believed to be very greatly increased in smooth muscle. A possible reason for the slow cycling is that the cross-bridge heads have far less ATPase activity than that in skeletal muscle, so that degradation of the ATP that energizes the movements of the heads is greatly reduced, with corresponding slowing of the rate of cycling.

Energy Required to Sustain Smooth Muscle Contraction. Only $1/10$ to $1/300$ as much energy is required to sustain the same tension of contraction in smooth muscle as in skeletal muscle. This, too, is believed to result because of the very slow attachment cycling of the cross-bridges and because only one molecule of ATP is required for each cycle regardless of its duration.

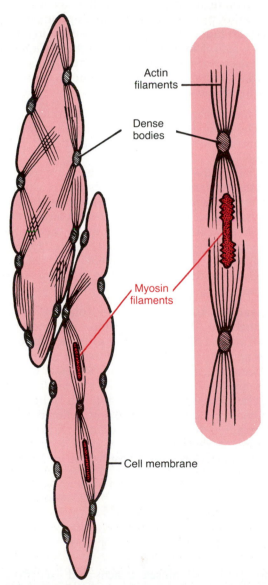

Actin filaments

Dense bodies

Myosin filaments

Cell membrane

Figure 25–6. Physical structure of smooth muscle. The upper left-hand fiber shows actin filaments radiating from "dense bodies." The lower fiber in the right-hand insert demonstrates the relationship of myosin filaments to the actin filaments.

This economy of energy utilization by smooth muscle is exceedingly important to the overall energy economy of the body because organs, such as the intestines, the urinary bladder, the gallbladder, and other viscera must maintain tonic muscle contraction on a daily basis.

Slowness of Onset of Contraction and Relaxation of Smooth Muscle. A typical smooth muscle tissue begins to contract 50 to 100 msec after it is excited, reaches full contraction about ½ sec later, and then declines in contractile force in another 1 to 2 sec, giving a total contraction time of 1 to 3 sec. This is about 30 times as long as a single contraction of an average skeletal muscle. However, because of the many different types of smooth muscle, contraction of some types can be as short as 0.2 sec or as long as 30 sec.

The slow onset of contraction in smooth muscle as well as the prolonged contraction is probably caused by the slowness of attachment and detachment of the cross-bridges. In addition, the initiation of contraction in response to calcium ions, called the excitation-contraction coupling mechanism, is much slower than in skeletal muscle, as we discuss later.

Force of Muscle Contraction. Despite the relatively few myosin filaments in smooth muscle and despite the slow cycling time of the cross-bridges, the maximum force of contraction of smooth muscle is often even greater than that of skeletal muscle — as great as 4 to 6 kg/cm^2 cross-sectional area for smooth muscle in comparison with 3 to 4 kg for skeletal muscle. This great force of attraction is postulated to result from the prolonged period of attachment of the myosin cross-bridges to the actin filaments.

Percentage Shortening of Smooth Muscle During Contraction. A characteristic of smooth muscle that is different from skeletal muscle is its ability to shorten a far greater percentage of its length than can skeletal muscle while still maintaining almost full force of contraction. Skeletal muscle has a useful distance of contraction of only about one third its stretched length, whereas smooth muscle can often contract quite effectively more than two thirds its stretched length. This allows smooth muscle to perform especially important functions in the hollow viscera, allowing the gut, the bladder, the blood vessels, and other internal bodily structures to change their lumen diameters from very large down to almost zero.

Why this difference between smooth muscle and skeletal muscle? The answer to this is not completely known, but there appear to be two probable reasons. First, it is likely that some contractile units of smooth muscle have optimal overlapping of their actin and myosin filaments at one length of the muscle and others at other lengths, rather than all of these being synchronized together, as usually occurs in skeletal muscle. Therefore, a greater distance of contraction can be achieved. Second, the actin filaments in smooth muscle are much longer than those in skeletal muscle. Therefore, these filaments can be pulled along the myosin filaments for a much greater distance during smooth muscle contraction than occurs in skeletal muscle contraction.

REGULATION OF CONTRACTION BY CALCIUM IONS

As is true for skeletal muscle, the initiating event in most smooth muscle contraction is an increase in intracellular calcium ions. This increase can be caused by nerve stimulation of the smooth muscle fiber, hormonal stimulation, stretch of the fiber, or even changes in the chemical environment of the fiber.

Yet, smooth muscle does not contain troponin, the regulatory protein that is activated by calcium ions to cause skeletal muscle contraction. Instead, smooth muscle contraction is activated by an entirely different mechanism, as follows:

Combination of Calcium Ions with "Calmodulin" — Activation of Myosin Kinase and Phosphorylation of the Myosin Head. In place of troponin, smooth muscle cells contain large quantities of another regulatory protein called *calmodulin*. Although this protein is similar to troponin in that it reacts with four calcium ions, it is different in the manner in which it initiates the contraction. Calmodulin does this by activating the myosin cross-bridges. This activation and subsequent contraction occurs in the following sequence:

1. The calcium ions bind with calmodulin.
2. The calmodulin-calcium combination then joins with and activates myosin kinase, a phosphorylating enzyme.
3. One of the light chains of each myosin head, called the *regulatory chain*, becomes phosphorylated in response to the myosin kinase. When this chain is not phosphorylated, the attachment-detachment cycling of the head will not occur. But, when the regulatory chain is phosphorylated, the head has the capability of binding with the actin filament and proceeding through the entire cycling process, thus causing muscle contraction.

Cessation of Contraction — Role of "Myosin Phosphatase." When the calcium ion concentration falls below a critical level, the aforementioned processes all automatically reverse except for the phosphorylation of the myosin head. Reversal of this requires another enzyme, *myosin phosphatase*, which splits the phosphate from the regulatory light chain. Then, the cycling stops and the contraction ceases. The time required for relaxation of muscle contraction, therefore, is determined to a great extent by the amount of active myosin phosphatase in the cell.

■ NEURAL AND HORMONAL CONTROL OF SMOOTH MUSCLE CONTRACTION

Although skeletal muscle is activated exclusively by the nervous system, smooth muscle can be stimulated to contract by multiple types of signals: by nervous signals, by hormonal stimulation, and in several other ways. The principal reason for the difference is that the smooth muscle membrane contains many different types of receptor proteins that can initiate the contractile process. Still other receptor proteins inhibit smooth muscle contraction, which is another difference from skeletal muscle. Therefore, in this section, we discuss, first, neural control of smooth muscle contraction, followed by hormonal control and other means of control.

NEUROMUSCULAR JUNCTIONS OF SMOOTH MUSCLE

Physiologic Anatomy of Smooth Muscle Neuromuscular Junctions. Neuromuscular junctions of the type found on skeletal muscle fibers do not occur in smooth muscle. Instead, the *autonomic nerve fibers* that innervate smooth muscle generally branch diffusely on top of a sheet of muscle fibers, as illustrated in Figure 25–7. In most instances, these fibers do not make direct contact with the smooth muscle fibers at all but instead form so-called *diffuse junctions* that secrete their transmitter substance into the interstitial fluid from a few nanometers to a few microns away from the muscle cells; the transmitter substance then diffuses to the cells. Furthermore, where there are many layers of muscle cells, the nerve fibers often innervate only the outer layer, and the muscle excitation then travels from this outer layer to the inner layers by action potential conduction in the muscle mass or by subsequent diffusion of the transmitter substance.

The axons innervating smooth muscle fibers also do not have typical branching end-feet of the type in the motor end-plate on skeletal muscle fibers. Instead, most of the fine terminal axons have multiple *varicosities* distributed along their axes. At these points the Schwann cells are interrupted so that transmitter substance can be secreted through the walls of the varicosities. In the varicosities are vesicles similar to those in the skeletal muscle end-plate containing transmitter substance. However, in contrast to the vesicles of skeletal muscle junctions that contain only acetylcholine, the vesicles of the autonomic nerve fiber endings contain *acetylcholine* in some fibers and *norepinephrine* in others.

In a few instances, particularly in the multiunit type of smooth muscle, the varicosities lie directly on the muscle fiber membrane with a separation from this membrane of as little as 20 to 30 nm — the same width as the synaptic cleft that occurs in the skeletal muscle junction. These *contact junctions* function in much the same way as the skeletal muscle neuromuscular junction, and the latent period of contraction of these smooth muscle fibers is considerably shorter than of fibers stimulated by the diffuse junctions.

Excitatory and Inhibitory Transmitter Substances at the Smooth Muscle Neuromuscular Junction. Two different transmitter substances known to be secreted by the autonomic nerves innervating smooth muscle are *acetylcholine* and *norepinephrine.* Acetylcholine is an excitatory transmitter substance for smooth muscle fibers in some organs but an inhibitory substance for smooth muscle in other organs. When acetylcholine excites a muscle fiber, norepinephrine ordinarily inhibits it. Conversely, when acetylcholine inhibits a fiber, norepinephrine usually excites it.

But why these different responses? The answer is that both acetylcholine and norepinephrine excite or inhibit smooth muscle by first binding with a *receptor protein* on the surface of the muscle cell membrane. This receptor in turn controls the opening or closing of ion channels or controls some other means for activating or inhibiting the smooth muscle fiber. Furthermore, some of the receptor proteins are *excitatory receptors,* whereas others are *inhibitory receptors.* Thus, it is the type of receptor that determines whether the smooth muscle will be inhibited or excited and also determines which of the two transmitters, acetylcholine or norepinephrine, will be effective in causing the excitation or inhibition. These receptors are discussed in more detail in Chapter 22 in relation to the function of the autonomic nervous system.

MEMBRANE POTENTIALS AND ACTION POTENTIALS IN SMOOTH MUSCLE

Membrane Potentials in Smooth Muscle. The quantitative value of the membrane potential of smooth muscle is variable from one type of smooth muscle to another, and it also depends on the momentary condition of the muscle. However, in the normal resting state, the membrane potential is usually about −50 to −60 mV, or about 30 mV less negative than in skeletal muscle.

Action Potentials in Single-Unit Smooth Muscle. Action potentials occur in single-unit smooth muscle in the same way that they occur in skeletal muscle. However, action potentials do not normally occur in many if not most multiunit types of smooth muscle, as is discussed in a subsequent section.

The action potentials of visceral smooth muscle occur in two different forms: (1) spike potentials and (2) action potentials with plateaus.

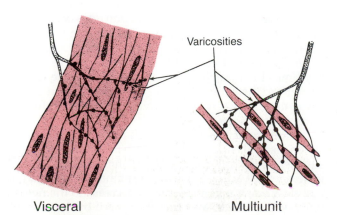

Visceral   Varicosities   Multiunit

Figure 25–7. Innervation of smooth muscle.

Spike Potentials. Typical spike action potentials, such as those seen in skeletal muscle, occur in most types of single-unit smooth muscle. The duration of this type of action potential is 10 to 50 msec, as illustrated in Figure 25–8A and B. Such action potentials can be elicited in many ways, such as by electrical stimulation, by the action of hormones on the smooth muscle, by the action of transmitter substances from nerve fibers, or as a result of spontaneous generation in the muscle fiber itself, as discussed subsequently.

Action Potentials With Plateaus. Figure 25–8C illustrates an action potential with a plateau. The onset of this action potential is similar to that of the typical spike potential. However, instead of rapid repolarization of the muscle fiber membrane, the repolarization is delayed for several hundred to several thousand milliseconds. The importance of the plateau is that it can account for the prolonged periods of contraction that occur in some types of smooth muscle, such as the ureter, the uterus under some conditions, and some types of vascular smooth muscle. (Also, this is the type of action potential seen in cardiac muscle fibers that have a prolonged period of contraction, as we discuss in the next two chapters.)

Importance of Calcium Channels in Generating the Smooth Muscle Action Potential. The smooth muscle cell membrane has far more voltage-gated calcium channels than does skeletal muscle but very, very few voltage-gated sodium channels. Therefore, sodium participates very little if any in the generation of the action potential in most smooth muscle. Instead, the flow of calcium ions to the interior of the fiber is mainly responsible for the action potential. This occurs in the same self-regenerative way as occurs for the sodium channels in nerve fibers and in skeletal muscle fibers. However, calcium channels open many times more slowly than do sodium channels. This accounts in large measure for the slow action potentials of smooth muscle fibers.

Another important feature of calcium entry into the cells during the action potential is that this same calcium acts directly on the smooth muscle contractile mechanism to cause contraction, as discussed earlier. Thus, the calcium performs two tasks at once.

Slow Wave Potentials in Single-Unit Smooth Muscle and Spontaneous Generation of Action Potentials. Some smooth muscle is self-excitatory. That is, action potentials arise within the smooth muscle itself without an extrinsic stimulus. This is usually associated with a basic *slow wave rhythm* of the membrane potential. A typical slow wave of this type in the visceral smooth muscle of the gut is illustrated in Figure 25–8B. The slow wave itself is not an action potential. It is not a self-regenerative process that spreads progressively over the membranes of the muscle fibers. Instead, it is a local property of the smooth muscle fibers that make up the muscle mass.

The cause of the slow wave rhythm is unknown; one suggestion is that the slow waves are caused by waxing and waning of the pumping of sodium ions outward through the muscle fiber membrane; the membrane potential becomes more negative when sodium is pumped rapidly and less negative when the sodium pump becomes less active. Another suggestion is that the conductances of the ion channels increase and decrease rhythmically.

The importance of the slow waves lies in the fact that they can initiate action potentials. The slow waves themselves cannot cause muscle contraction, but when the potential of the slow wave rises above the level of approximately −35 mV (the approximate threshold for eliciting action potentials in most visceral smooth muscle), an action potential develops and spreads over the muscle mass, and then contraction does occur. Figure 25–8B illustrates this effect, showing that at each peak of the slow wave, one or more action potentials occur. This effect can obviously promote a series of rhythmical contractions of the smooth muscle mass. Therefore, the slow waves are frequently called *pacemaker waves*. This type of activity controls the rhythmical contractions of the gut.

Excitation of Visceral Smooth Muscle by Stretch. When visceral (single-unit) smooth muscle is stretched sufficiently, spontaneous action potentials are usually generated. These result from a combination of the normal slow wave potentials plus a decrease in the negativity of the membrane potential caused by the stretch itself. This response to stretch allows a hollow organ

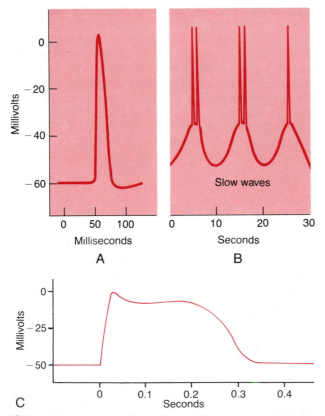

Figure 25–8. *A,* Typical smooth muscle action potential (spike potential) elicited by an external stimulus. *B,* Repetitive spike potentials elicited by slow rhythmical electrical waves that occur spontaneously in the smooth muscle of the intestinal wall. *C,* An action potential with a plateau recorded from a smooth muscle fiber of the uterus.

that is excessively stretched to contract automatically and therefore to resist the stretch. For instance, when the gut is overstretched by intestinal contents, a local automatic contraction often sets up a peristaltic wave that moves the contents away from the excessively stretched intestine.

Depolarization of Multiunit Smooth Muscle Without Action Potentials. The smooth muscle fibers of multiunit smooth muscle normally contract mainly in response to nerve stimuli. The nerve endings secrete acetylcholine in the case of some multiunit smooth muscles and norepinephrine in the case of others. In both instances, these transmitter substances cause depolarization of the smooth muscle membrane, and this response in turn elicits the contraction. However, action potentials most often do not develop. The reason for this is that the fibers are too small to generate an action potential. (When action potentials are elicited in visceral single-unit smooth muscle, as many as 30 to 40 smooth muscle fibers must depolarize simultaneously before a self-propagating action potential ensues.) Yet, even without an action potential in the multiunit smooth muscle fibers, the local depolarization, called the "junctional potential," caused by the nerve transmitter substance itself spreads "electrotonically" over the entire fiber and is all that is needed to cause the muscle contraction.

SMOOTH MUSCLE CONTRACTION WITHOUT ACTION POTENTIALS— EFFECT OF LOCAL TISSUE FACTORS AND HORMONES

Probably half or more of all smooth muscle contraction is initiated not by action potentials but by stimulatory factors acting directly on the smooth muscle contractile machinery. The two types of non-nervous and nonaction potential stimulating factors most often involved are (1) local tissue factors and (2) various hormones.

Smooth Muscle Contraction in Response to Local Tissue Factors. For instance, let us discuss the control of contraction of the arterioles, meta-arterioles, and precapillary sphincters. The smaller of these vessels have little or no nervous supply. Yet, the smooth muscle is highly contractile, responding rapidly to changes in local conditions in the surrounding interstitial fluid. In this way, a powerful local feedback control system controls the blood flow to the local tissue area. Some of the specific control factors are as follows:

1. Lack of oxygen in the local tissues causes smooth muscle relaxation and therefore vasodilatation.
2. Excess carbon dioxide causes vasodilatation.
3. Increased hydrogen ion concentration also causes increased vasodilatation.

Such factors as adenosine, lactic acid, increased potassium ions, diminished calcium ion concentration, and decreased body temperature also cause local vasodilatation.

Effects of Hormones on Smooth Muscle Contraction. Most of the circulating hormones in the body affect smooth muscle contraction at least to some degree, and some have very profound effects. Some of the more important blood-borne hormones that affect contraction are *norepinephrine, epinephrine, acetylcholine, angiotensin, vasopressin, oxytocin, serotonin,* and *histamine.*

A hormone causes contraction of smooth muscle when the muscle cell membrane contains *hormone-gated excitatory receptors* for the respective hormone. However, the hormone causes inhibition instead of contraction if the membrane contains *inhibitory receptors* rather than excitatory receptors.

SOURCE OF CALCIUM IONS THAT CAUSE CONTRACTION BOTH THROUGH THE CELL MEMBRANE AND FROM THE SARCOPLASMIC RETICULUM

Although the contractile process in smooth muscle, as in skeletal muscle, is activated by calcium ions, the source of the calcium ions differs at least partly in smooth muscle; the difference is that the sarcoplasmic reticulum, from which virtually all the calcium ions are derived in skeletal muscle contraction, is often only rudimentary in most smooth muscle. Instead, in most types of smooth muscle, almost all the calcium ions that cause contraction enter the muscle cell from the extracellular fluid at the time of the action potential. There is a reasonably high concentration of calcium ions in the extracellular fluid, greater than 10^{-3} M in comparison with less than 10^{-7} M in the cell sarcoplasm, and as was pointed out earlier, the smooth muscle action potential is caused mainly by influx of calcium ions into the muscle fiber. Because the smooth muscle fibers are extremely small (in contrast to the sizes of the skeletal muscle fibers), these calcium ions can diffuse to all parts of the smooth muscle and elicit the contractile process. The time required for this diffusion to occur is usually 200 to 300 msec and is called the *latent period* before the contraction begins; this latent period is some 50 times as great as that for skeletal muscle contraction.

Still additional calcium can enter the smooth muscle fiber via *hormone-activated calcium channels*; these, too, cause contraction. Usually, the opening of these channels does not cause an action potential, and sometimes not much change in the resting membrane potential either, because the sodium pump in the cell membrane pumps enough sodium ions to the exterior to maintain an almost normal membrane potential. Even so, contraction continues as long as these calcium channels remain open because it is calcium ions, not a change in membrane potential, that cause contraction. This is one means by which contraction is achieved in smooth muscle without significant change in cell membrane potential.

Role of the Sarcoplasmic Reticulum. Some smooth muscle contains a moderately developed sarcoplasmic reticulum. Figure 25–9 shows an example, illustrating separate sarcoplasmic tubules that lie near the cell

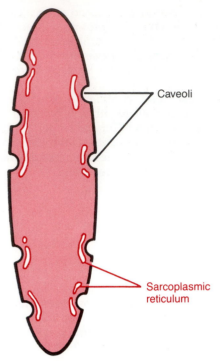

Figure 25–9. Sarcoplasmic tubules in a smooth muscle fiber, showing their relationship to invaginations in the cell membranes called *caveoli.*

membrane. Small invaginations of the membrane, called *caveoli,* abut against the surfaces of these tubules. The caveoli are believed to represent a rudimentary analog of the T tubule system of skeletal muscle. When an action potential is transmitted into the caveoli invaginations, this seems to excite calcium ion release from the sarcoplasmic tubules in the same way that action potentials in skeletal muscle T tubules also cause release of calcium ions.

In general, the more extensive the sarcoplasmic reticulum in the smooth muscle fiber, the more rapidly it contracts, presumably because calcium entry through the cell membrane is much slower than internal release of calcium ions from the sarcoplasmic reticulum.

Effect of Extracellular Calcium Ion Concentration on Smooth Muscle Contraction. Although the extracellular fluid calcium ion concentration has almost no effect on the force of contraction of skeletal muscle, this is not true for most smooth muscle. When the extracellular fluid calcium ion concentration falls to a low level, smooth muscle contraction usually almost ceases. In fact, after several minutes of being immersed in a low calcium medium, even the sarcoplasmic reticulum of smooth muscle fibers loses its calcium supply. Therefore, the force of contraction of smooth muscle is highly dependent on the extracellular fluid calcium ion concentration. We see in the following chapter that this is also true for cardiac muscle.

The Calcium Pump. To cause relaxation of the smooth muscle contractile elements, it is necessary to remove the calcium ions. This removal is achieved by calcium pumps that pump the calcium ions out of the smooth muscle fiber back into the extracellular fluid or pump the calcium ions into the sarcoplasmic reticulum. However, these pumps are very slow acting in comparison with the fast-acting sarcoplasmic reticulum pump in skeletal muscle. Therefore, the duration of smooth muscle contraction is often in the order of seconds rather than hundredths to tenths of a second, as occurs for skeletal muscle.

REFERENCES

SKELETAL MUSCLE AND NEUROMUSCULAR TRANSMISSION

Clausen, T.: Regulation of active Na^+-K^+ transport in skeletal muscle. Physiol. Rev., 66:542, 1986.
DiDonato, S., et al. (eds.): Molecular Genetics of Neurological and Neuromuscular Disease. New York, Raven Press, 1988.
Goldman, Y. E., and Brenner, B. (eds.): General introduction. Annu. Rev. Physiol., 49:629, 1987.
Gowitzke, B. A., et al.: Scientific Bases of Human Movement. Baltimore, Williams & Wilkins, 1988.
Haynes, D. H., and Mandveno, A.: Computer modeling of Ca^{2+}-Mg^{2+}-ATPase of sarcoplasmic reticulum. Physiol. Rev., 67:244, 1987.
Homsher, E.: Muscle enthalpy production and its relationship to actomyosin ATPase. Annu. Rev. Physiol., 49:673, 1987.
Huang, C. L. H.: Intramembrane charge movements in skeletal muscle. Physiol. Rev., 68:1197, 1988.
Huxley, A. F.: Muscular Contraction. Annu. Rev. Physiol., 50:1, 1988.
Huxley, A. F., and Gordon, A. M.: Striation patterns in active and passive shortening of muscle. Nature (Lond.), 193:280, 1962.
Huxley, H. E., and Faruqi, A. R.: Time-resolved x-ray diffraction studies on vertebrate striated muscle. Annu. Rev. Biophys. Bioeng., 12:381, 1983.
Johnson, E. W. (ed.): Practical Electromyography. Baltimore, Williams & Wilkins, 1988.
Kolata, G.: Metabolic catch-22 of exercise regimens. Science, 236:146, 1987.
Korn, E. D., et al.: Actin polymerization and ATP hydrolysis. Science, 238:638, 1987.
Laufer, R., et al.: Regulation of acetylcholine receptor biosynthesis during motor endplate morphogenesis. News Physiol. Sci., 4:5, 1989.
Martonosi, A. N.: Mechanisms of Ca^{2+} release from sarcoplasmic reticulum of skeletal muscle. Physiol. Rev., 64:1240, 1984.
Morkin, E.: Chronic adaptations in contractile proteins: Genetic regulation. Annu. Rev. Physiol., 49:545, 1987.
Oho, S. J.: Electromyography: Neuromuscular Transmission Studies. Baltimore, Williams & Wilkins, 1988.
Pollack, G. H.: The cross-bridge theory. Physiol. Rev., 63:1049, 1983.
Ringel, S. P.: Neuromuscular Disorders: A Guide for Patient and Family. New York, Raven Press, 1987.
Rios, E., and Pizarro, G.: Voltage sensors and calcium channels of excitation-contraction coupling. News Physiol. Sci., 3:223, 1988.
Rowland, L. P., et al. (eds.): Molecular Genetics in Diseases of Brain, Nerve, and Muscle. New York, Oxford University Press, 1989.
Soderberg, G. L.: Kinesiology. Baltimore, Williams & Wilkins, 1986.
Steinbach, J. H.: Structural and functional diversity in vertebrate skeletal muscle nicotinic acetylcholine receptors. Annu. Rev. Physiol., 51:353, 1989.
Sugi, H., and Pollack, G. H. (eds.): Molecular Mechanism of Muscle Contraction. New York, Plenum Publishing Corp., 1988.
Swynghedauw, B.: Developmental and functional adaptation of contractile proteins in cardiac and skeletal muscles. Physiol. Rev., 66:710, 1986.
Thomas, D. D.: Spectroscopic probes of muscle cross-bridge rotation. Annu. Rev. Physiol., 49:691, 1987.
Vergara, J., and Asotra, K.: The chemical transmission mechanisms of excitation-contraction coupling in the skeletal muscle. News Physiol. Sci., 2:182, 1987.

SMOOTH MUSCLE

Butler, T. M., and Siegman, M. J.: High-energy phosphate metabolism in vascular smooth muscle. Annu. Rev. Physiol., 47:629, 1985.
Campbell, J. H., and Campbell, G. R.: Endothelial cell influences on vascular smooth muscle phenotype. Annu. Rev. Physiol., 48:295, 1986.
Furchgott, R. F.: The role of endothelium in the responses of vascular smooth muscle to drugs. Annu. Rev. Pharmacol. Toxicol., 24:175, 1984.
Gabella, G.: Structural apparatus for force transmission in smooth muscle. Physiol. Rev., 64:455, 1984.
Hertzberg, E. L., et al.: Gap junctional communication. Annu. Rev. Physiol., 43:479, 1981.
Hirst, G. D. S., and Edwards, F. R.: Sympathetic neuroeffector transmission in arteries and arterioles. Physiol. Rev., 69:546, 1989.

Homsher, E.: Muscle enthalpy production and its relationship to actomyosin ATPase. Annu. Rev. Physiol., 49:673, 1987.

Kamm, K. E., and Stull, J. T.: Regulation of smooth muscle contractile elements by second messengers. Annu. Rev. Physiol., 51:299, 1989.

Lowenstein, W. R.: Junctional intercellular communication: The cell-to-cell membrane channel. Physiol. Rev., 61:829, 1981.

McKinney, M., and Richelson, E.: The coupling of neuronal muscarinic receptor to responses. Annu. Rev. Pharmacol. Toxicol., 24:121, 1984.

Murphy, R. A.: Muscle cells of hollow organs. News Physiol. Sci., 3:124, 1988.

Paul, R. J.: Smooth muscle energetics. Annu. Rev. Physiol., 51:331, 1989.

Putney, J. W., Jr., et al.: How do inositol phosphates regulate calcium signaling? FASEB J., 3:1899, 1989.

Rasmussen, H., et al.: Protein kinase C in the regulation of smooth muscle contraction. FASEB J., 1:177, 1987.

Rosenthal, W., et al.: Control of voltage-dependent Ca^{2+} channels by G protein-coupled receptors. FASEB J., 2:2784, 1988.

Seidel, C. L., and Schildmeyer, L. A.: Vascular smooth muscle adaptation to increased load. Annu. Rev. Physiol., 49:489, 1987.

Somlyo, A. P.: Ultrastructure of vascular smooth muscle. In Bohr, D. F., et al. (eds.): Handbook of Physiology. Sec. 2, Vol. II. Baltimore, Williams & Wilkins, 1980, p. 33.

Spray, D. C., and Bennett, M. V. L.: Physiology and pharmacology of gap junctions. Annu. Rev. Physiol., 47:281, 1985.

van Breemen, C., and Saida, K.: Cellular mechanisms regulating $[Ca^{2+}]_i$ smooth muscle. Annu. Rev. Physiol., 51:315, 1989.

Vanhoutte, P. M.: Calcium-entry blockers, vascular smooth muscle and systemic hypertension. Am. J. Cardiol., 55:17B, 1985.

26

The Heart: Its Rhythmical Excitation and Nervous Control

The heart is a muscular organ. Like other muscles of the body, it is, in a sense, an extension of the nervous system, for its pumping function is at least partly controlled by nerves.

To perform its pumping function, the heart is divided into four separate chambers, as illustrated in Figure 26-1. The right and left atria pump blood into the right and left ventricles, respectively. The right ventricle then pumps blood through the lungs, and the left ventricle pumps blood through the remainder of the body.

The atria actually function as "primer" pumps for the ventricles. Normally, they contract about one sixth of a second ahead of the ventricles, thus allowing time for extra blood to enter the ventricles prior to their contraction; this greatly increases the effectiveness of ventricular pumping.

The complexity of the heart's pumping cycle requires both rhythmical control of the heart beat and special timing mechanisms for sequential control of the atria and the ventricles. Furthermore, the rate of rhythmicity, as well as the strength of heart beat, can be increased or decreased by signals from the central nervous system. These control mechanisms will be explained later in the chapter. However, first, let us discuss the basic physiology of cardiac muscle itself, especially how it differs from skeletal muscle, which was discussed in Chapter 24.

■ PHYSIOLOGY OF CARDIAC MUSCLE

The heart is composed of three major types of cardiac muscle: atrial muscle, ventricular muscle, and specialized excitatory and conductive muscle fibers. The atrial and ventricular types of muscle contract in much the same way as skeletal muscle except that the duration of contraction is much longer. On the other hand, the specialized excitatory and conductive fibers contract only feebly because they contain few contractile fibrils; instead, they exhibit rhythmicity and varying rates of conduction, providing an excitatory system for the heart and a transmission system for controlled conduction of the cardiac excitatory signal throughout the heart.

PHYSIOLOGIC ANATOMY OF CARDIAC MUSCLE

Figure 26-2 illustrates a typical histological picture of cardiac muscle, showing the cardiac muscle fibers arranged in a latticework, the fibers dividing, then recombining, and then spreading again. One notes immediately from this figure that cardiac muscle is *striated* in the same manner as typical skeletal muscle. Furthermore, cardiac muscle has typical myofibrils that contain *actin* and *myosin filaments* almost identical to those found in skeletal muscle, and these filaments interdigitate and slide along each other during the process of contraction in the same manner as occurs in skeletal muscle. (See Chapter 24.)

Cardiac Muscle as a Syncytium. The angulated dark areas crossing the cardiac muscle fibers in Figure 26-2 are called *intercalated discs;* they are actually cell membranes that separate individual cardiac muscle cells from each other. That is, cardiac muscle fibers are made up of many individual cells connected in series with each other. Yet electrical resistance through the intercalated disc is only $1/400$ the resistance through the outside membrane of the cardiac muscle fiber because the cell membranes fuse with each other and form very permeable "communicating" junctions (gap junctions) that allow relatively free diffusion of ions. Therefore, from a functional point of view, ions move with ease along the axes of the cardiac muscle fibers, so that action potentials travel from one cardiac muscle cell to another, past the intercalated discs, with only slight hindrance. Therefore, cardiac muscle is a *syncytium* of many heart muscle cells, in which

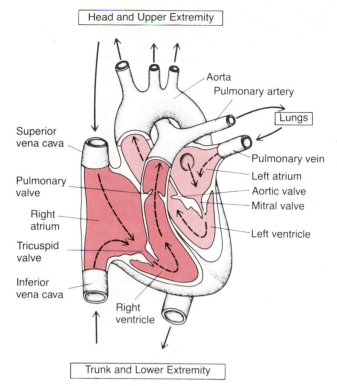

Figure 26–1. Structure of the heart and course of blood flow through the heart chambers.

the cardiac cells are so interconnected that when one of these cells becomes excited, the action potential spreads to all of them, spreading from cell to cell and spreading throughout the latticework interconnections.

The heart is composed of two separate syncytiums: the *atrial syncytium* that constitutes the walls of the two atria and the *ventricular syncytium* that constitutes the walls of the two ventricles. The atria are separated from the ventricles by fibrous tissue that surrounds the valvular openings between the atria and ventricles. Normally, action potentials can be conducted from the atrial syncytium into the ventricular syncytium only by way of a specialized conductive system, the *A-V bundle*, which is discussed in detail in the following chapter. This division of the muscle mass of the heart into two separate functional syncytiums allows the atria to con-

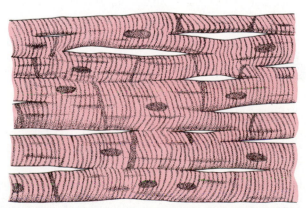

Figure 26–2. The "syncytial," interconnecting nature of cardiac muscle.

tract a short time ahead of ventricular contraction, which is important for the effectiveness of heart pumping.

ACTION POTENTIALS IN CARDIAC MUSCLE

The *resting membrane potential* of normal cardiac muscle is approximately −85 to −95 mV and approximately −90 to −100 mV in the specialized conductive fibers, the Purkinje fibers, which are discussed in the following chapter.

The *action potential* recorded in ventricular muscle, shown by the bottom record of Figure 26–3, is 105 mV, which means that the membrane potential rises from its normally very negative value to a slightly positive value of about +20 mV. The positive portion is called the *overshoot potential*. Then, after the initial *spike*, the membrane remains depolarized for about 0.2 sec in atrial muscle and about 0.3 sec in ventricular muscle, exhibiting a *plateau* as illustrated in Figure 26–3, followed at the end of the plateau by abrupt repolarization. The presence of this plateau in the action potential causes muscle contraction to last 3 to 15 times as long in cardiac muscle as in skeletal muscle.

At this point, we must ask the question: Why is the action potential of cardiac muscle so long, and why does it have a plateau, whereas that of skeletal muscle does not? The basic biophysical answers to these questions are presented in Chapters 5 and 6, but they merit summarizing again.

At least two major differences between the membrane properties of cardiac and skeletal muscle account for the prolonged action potential and the plateau in cardiac muscle.

First, the action potential of skeletal muscle is caused almost entirely by sudden opening of large numbers of so-called *fast sodium channels* that allow tremendous numbers of sodium ions to enter the skeletal muscle fiber. These channels are called "fast" channels be-

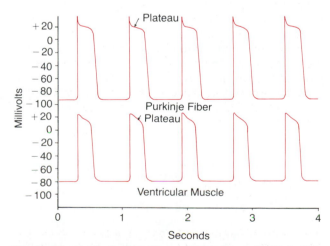

Figure 26–3. Rhythmic action potentials from a Purkinje fiber and from a ventricular muscle fiber, recorded by means of microelectrodes.

cause they remain open for only a few 10,000ths of a second and then abruptly close. At the end of this closure, the process of repolarization occurs, and the action potential is over within another 10,000th of a second or so. In cardiac muscle, on the other hand, the action potential is caused by the opening of two types of channels: (1) the same *fast sodium channels* as those in skeletal muscle and (2) another entire population of so-called *slow calcium channels,* also called *calcium-sodium channels.* This second population of channels differs from the fast sodium channels in being slower to open, but, more importantly, they remain open for several tenths of a second. During this time, large amounts of both calcium and sodium ions flow through these channels to the interior of the cardiac muscle fiber, and this maintains a prolonged period of depolarization, causing the plateau in the action potential. Furthermore, the calcium ions that enter the muscle during this action potential play an important role in helping excite the muscle contractile process, which is another difference between cardiac muscle and skeletal muscle, as we discuss later in this chapter.

The second major functional difference between cardiac muscle and skeletal muscle that helps account for both the prolonged action potential and its plateau is this: Immediately after the onset of the action potential, the permeability of the cardiac muscle membrane for potassium *decreases* about fivefold, an effect that does not occur in skeletal muscle. It is possible that this decreased potassium permeability is caused in some way by the excess calcium influx through the calcium channels just noted. However, regardless of the cause, the decreased potassium permeability greatly decreases the outflux of potassium ions during the action potential plateau and thereby prevents early recovery. When the slow calcium-sodium channels close at the end of 0.2 to 0.3 sec and the influx of calcium and sodium ions ceases, the membrane permeability for potassium increases very rapidly, and the rapid loss of potassium from the fiber returns the membrane potential to its resting level, thus ending the action potential.

Velocity of Conduction in Cardiac Muscle. The velocity of conduction of the action potential in both atrial and ventricular muscle fibers is about 0.3 to 0.5 m/sec, or about 1/250 the velocity in very large nerve fibers and about 1/10 the velocity in skeletal muscle fibers. The velocity of conduction in the specialized conductive system varies from 0.02 to 4 m/sec in different parts of the system, as is explained in the following chapter.

Refractory Period of Cardiac Muscle. Cardiac muscle, like all excitable tissue, is refractory to restimulation during the action potential. Therefore, the refractory period of the heart is the interval of time, as shown to the left in Figure 26–4, during which a normal cardiac impulse cannot re-excite an already excited area of cardiac muscle. The normal refractory period of the ventricle is 0.25 to 0.3 sec, which is approximately the duration of the action potential. There is an additional *relative refractory period* of about 0.05 sec during which the muscle is more difficult than

normal to excite but nevertheless can be excited, as illustrated by the early premature contraction in the second example of Figure 26–4.

The refractory period of atrial muscle is much shorter than that for the ventricles (about 0.15 sec), and the relative refractory period is another 0.03 sec. Therefore, the rhythmical rate of contraction of the atria can be much faster than that of the ventricles.

CONTRACTION OF CARDIAC MUSCLE

Excitation-Contraction Coupling—Function of Calcium Ions and of the T Tubules. The term "excitation-contraction coupling" means the mechanism by which the action potential causes the myofibrils of muscle to contract. This is discussed for skeletal muscle in Chapter 24. However, once again there are differences in this mechanism in cardiac muscle that have important effects on the characteristics of cardiac muscle contraction.

As is true for skeletal muscle, when an action potential passes over the cardiac muscle membrane, the action potential also spreads to the interior of the cardiac muscle fiber along the membranes of the T tubules. The T tubule action potentials in turn act on the membranes of the longitudinal sarcoplasmic tubules to cause instantaneous release of very large quantities of calcium ions into the muscle sarcoplasm from the sarcoplasmic reticulum. In another few thousandths of a second, these calcium ions diffuse into the myofibrils and catalyze the chemical reactions that promote sliding of the actin and myosin filaments along each other; this in turn produces the muscle contraction.

Thus far, this mechanism of excitation-contraction coupling is the same as that for skeletal muscle, but there is a second effect that is quite different. In addition to the calcium ions released into the sarcoplasm from the cisternae of the sarcoplasmic reticulum, large quantities of extra calcium ions also diffuse into the sarcoplasm from the T tubules at the time of the action

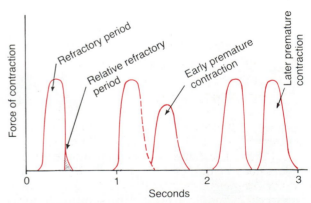

Figure 26–4. Contraction of the heart, showing the durations of the refractory period and the relative refractory period, the effect of an early premature contraction, and the effect of a later premature contraction. Note that the premature contractions do not cause wave summation, as occurs in skeletal muscle.

potential. Indeed, without this extra calcium from the T tubules, the strength of cardiac muscle contraction would be considerably reduced because the sarcoplasmic reticulum of cardiac muscle is less well developed than that of skeletal muscle and does not store enough calcium to provide full contraction. On the other hand, the T tubules of cardiac muscle have a diameter 5 times as great as that of the skeletal muscle tubules and a volume 25 times as great; also, inside the T tubules is a large quantity of mucopolysaccharides that are electronegatively charged and bind an abundant store of calcium ions, keeping this always available for diffusion to the interior of the cardiac muscle fiber when the T tubule action potential occurs.

The strength of contraction of cardiac muscle depends to a great extent on the concentration of calcium ions in the extracellular fluids. The reason for this is that the ends of the T tubules open directly to the outside of the cardiac muscle fibers, allowing the same extracellular fluid that is in the cardiac muscle interstitium to percolate through the T tubules as well. Consequently, the quantity of calcium ions in the T tubule system as well as the availability of calcium ions to cause cardiac muscle contraction depends directly on the extracellular fluid calcium ion concentration.

By way of contrast, the strength of skeletal muscle contraction is hardly affected by the extracellular fluid calcium concentration because its contraction is caused almost entirely by calcium ions released from the sarcoplasmic reticulum inside the skeletal muscle fiber itself.

At the end of the plateau of the action potential, the influx of calcium ions to the interior of the muscle fiber is suddenly cut off, and the calcium ions in the sarcoplasm are rapidly pumped back into both the sarcoplasmic reticulum and the T tubules. As a result, the contraction ceases until a new action potential occurs.

Duration of Contraction. Cardiac muscle begins to contract a few milliseconds after the action potential begins and continues to contract for a few milliseconds after the action potential ends. Therefore, the duration of contraction of cardiac muscle is mainly a function of the duration of the action potential—about 0.2 sec in atrial muscle and 0.3 sec in ventricular muscle.

■ THE SPECIALIZED EXCITATORY AND CONDUCTIVE SYSTEM OF THE HEART

Figure 26–5 illustrates the specialized excitatory and conductive system of the heart that controls cardiac contractions. The figure shows the *sinus node* (also called *sinoatrial* or *S-A node*), in which the normal rhythmic impulse is generated; the *internodal pathways* that conduct the impulse from the sinus node to the A-V node; the *A-V node* (also called *atrioventricular node*), in which the impulse from the atria is delayed before passing into the ventricles; the *A-V bundle*, which conducts the impulse from the atria into the

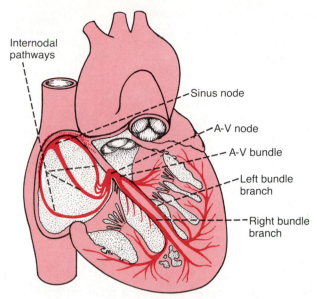

Figure 26–5. The sinus node and the Purkinje system of the heart, showing also the A-V node, the atrial internodal pathways, and the ventricular bundle branches.

ventricles; and the *left* and *right bundles of Purkinje fibers*, which conduct the cardiac impulse to all parts of the ventricles.

THE SINUS NODE

The sinus node is a small, flattened, ellipsoid strip of specialized muscle approximately 3 mm wide, 15 mm long, and 1 mm thick; it is located in the superior lateral wall of the right atrium immediately below and lateral to the opening of the superior vena cava. The fibers of this node have almost no contractile filaments and are each 3 to 5 μm in diameter, in contrast to a diameter of 10 to 15 μm for the surrounding atrial muscle fibers. However, the sinus fibers are continuous with the atrial fibers, so that any action potential that begins in the sinus node spreads immediately into the atria.

Automatic Rhythmicity of the Sinus Fibers

Many cardiac fibers have the capability of *self-excitation*, a process that can cause automatic rhythmical contraction. This is especially true of the fibers of the heart's specialized conducting system; the portion of this system that displays self-excitation to the greatest extent is the fibers of the sinus node. For this reason, the sinus node ordinarily controls the rate of beat of the entire heart, as is discussed in detail later in this chapter. First, however, let us describe this automatic rhythmicity.

Mechanism of Sinus Nodal Rhythmicity. Figure 26–6 illustrates action potentials recorded from a sinus nodal fiber for three heartbeats and, by comparison, a single ventricular muscle fiber action poten-

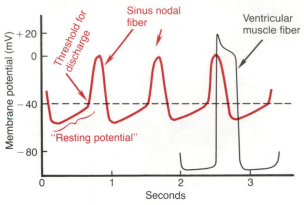

Figure 26–6. Rhythmic discharge of a sinus nodal fiber and comparison of the sinus nodal action potential with that of a ventricular muscle fiber.

tial, shown to the right. Note that the potential of the sinus nodal fiber between discharges has a negativity of only −55 to −60 mV in comparison with −85 to −90 mV for the ventricular fiber. The cause of this reduced negativity is that the cell membranes of the sinus fibers are naturally leaky to sodium ions.

Before attempting to explain the rhythmicity of the sinus nodal fibers, first recall from earlier discussions that in cardiac muscle three different types of membrane ion channels play important roles in causing the voltage changes of the action potential. These are (1) the *fast sodium channels*, (2) the *slow calcium-sodium channels*, and (3) the *potassium channels*. The opening of the fast sodium channels for a few 10,000ths of a second is responsible for the very rapid spikelike onset of the action potential observed in ventricular muscle because of rapid influx of positive sodium ions to the interior of the fiber. Then the plateau of the ventricular action potential is caused primarily by slower opening of the slow calcium-sodium channels, which lasts for a few tenths of a second. Finally, increased opening of the potassium channels and diffusion of large amounts of positive potassium ions out of the fiber return the membrane potential to its resting level.

But there is a difference in the function of these channels in the sinus nodal fiber because of the much lesser negativity of the "resting" potential — only −55 mV. At this level of negativity, the fast sodium channels have mainly become "inactivated," which means that they have become blocked. The cause of this is that any time the membrane potential remains less negative than about −60 mV for more than a few milliseconds, the inactivation gates on the inside of the cell membrane that close these channels become closed and remain so. Therefore, only the slow calcium-sodium channels can open (that is, can become "activated") and thereby cause the action potential. Therefore, the action potential is slower to develop than that of the ventricular muscle and also recovers with a slow decrement of the potential rather than the abrupt recovery that occurs for the ventricular fiber.

Self-Excitation of Sinus Nodal Fibers. Sodium ions naturally tend to leak to the interior of the sinus nodal

fibers through multiple membrane channels, and this influx of positive charges also causes a rising membrane potential. Thus, as illustrated in Figure 26–6, the "resting" potential gradually rises between each two heartbeats. When it reaches a *threshold voltage* of about −40 mV, the calcium-sodium channels become activated, leading to very rapid entry of both calcium and sodium ions, thus causing the action potential. Therefore, basically, the inherent leakiness of the sinus nodal fibers to sodium ions causes their self-excitation.

Why does this leakiness to sodium ions not cause the sinus nodal fibers to remain depolarized all the time? The answer is that two events occur during the course of the action potential. First, the calcium-sodium channels become inactivated (that is, they close) within about 100 to 150 msec after opening, and second, at about the same time greatly increased numbers of potassium channels open. Therefore, now the influx of calcium and sodium ions through the calcium-sodium channels ceases simultaneously, while large quantities of positive potassium ions diffuse *out* of the fiber, thus terminating the action potential. Furthermore, the potassium channels remain open for another few tenths of a second, carrying a great excess of positive potassium charges out of the cell, which temporarily causes considerable excess negativity inside the fiber; this is called *hyperpolarization.* This hyperpolarization initially carries the "resting" membrane potential down to about −55 to −60 mV at the termination of the action potential.

Last, we must explain why the state of hyperpolarization also is not maintained forever. The reason is that during the next few tenths of a second after the action potential is over, progressively more and more of the potassium channels begin to close. Now the inward-leaking sodium ions once again overbalance the outward flux of potassium ions, which causes the "resting" potential to drift upward, finally reaching the threshold level for discharge at a potential of about −40 mV. Then the entire process begins again: self-excitation, recovery from the action potential, hyperpolarization after the action potential is over, upward drift of the "resting" potential, then re-excitation still again to elicit another cycle. This process continues indefinitely throughout the life of the person.

INTERNODAL PATHWAYS AND TRANSMISSION OF THE CARDIAC IMPULSE THROUGH THE ATRIA

The ends of the sinus nodal fibers fuse with the surrounding atrial muscle fibers, and action potentials originating in the sinus node travel outward into these fibers. In this way, the action potential spreads through the entire atrial muscle mass and eventually also to the A-V node. The velocity of conduction in the atrial muscle is approximately 0.3 m/sec. However, conduction is somewhat more rapid in several small bundles of atrial muscle fibers. One of these, called the *anterior interatrial band,* passes through the anterior walls of the atria to the left atrium and conducts the cardiac impulse at a velocity of about 1 m/sec. In addition, three other small bundles curve through the atrial walls and terminate in the A-V node, also conducting the cardiac impulse at this rapid velocity. These three small bundles, illustrated in Figure 26–5, are called, respectively, the *anterior, middle,* and *posterior internodal pathways.* The cause of the more rapid velocity of conduction in these bundles is the presence of a number of specialized conduction fibers mixed with the atrial muscle. These fibers are similar to the very rapidly conducting Purkinje fibers of the ventricles, which are discussed subsequently.

THE A-V NODE AND DELAY IN IMPULSE CONDUCTION

Fortunately, the conductive system is organized so that the cardiac impulse will not travel from the atria into the ventricles too rapidly; this allows time for the atria to empty their contents into the ventricles before ventricular contraction begins. It is primarily the A-V node and its associated conductive fibers that delay this transmission of the cardiac impulse from the atria into the ventricles.

The A-V node is located in the posterior septal wall of the right atrium immediately behind the tricuspid valve and adjacent to the opening of the coronary sinus, as illustrated in Figure 26–5. Figure 26–7 shows diagrammatically the different parts of this node and its connections with the atrial internodal pathway fibers and the A-V bundle. The figure also shows the approximate intervals of time in fractions of a second between the genesis of the cardiac impulse in the sinus node and its appearance at different points in the A-V nodal system. Note that the impulse, after traveling through the internodal pathway, reaches the A-V node approximately 0.03 sec after its origin in the sinus node. Then there is a further delay of 0.09 sec in the A-V node itself before the impulse enters the *penetrating portion of the A-V bundle.* A final delay of another 0.04 sec occurs mainly in this penetrating A-V bundle, which is composed of multiple small fascicles passing through the fibrous tissue separating the atria from the ventricles.

Thus, the total delay in the A-V nodal and A-V bundle system is approximately 0.13 sec. About a quarter of this time lapse occurs in the *transitional fibers,* which are very small fibers that connect the fibers of the atrial internodal pathways with the A-V node (see Fig. 26–7). The velocity of conduction in these fibers is as little as 0.02 to 0.05 m/sec (about 1/12 that in normal cardiac muscle), which greatly delays entrance of the impulse into the A-V node. After entering the node proper, the velocity of conduction in the nodal fibers is still quite low, only 0.05 m/sec, about one eighth the conduction velocity in normal cardiac muscle. This low velocity of conduction is also approximately true for the penetrating portion of the A-V bundle.

Cause of the Slow Conduction. The cause of the

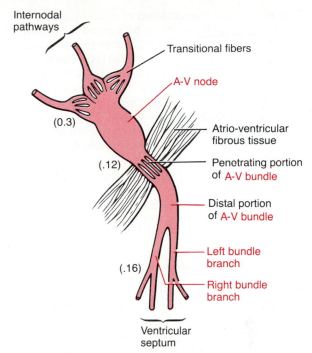

Figure 26–7. Organization of the A-V node. The numbers represent the interval of time from the origin of the impulse in the sinus node. The values have been extrapolated to the human being.

extremely slow conduction in the transitional as well as the nodal and penetrating A-V bundle fibers is partly that their sizes are considerably smaller than the sizes of the normal atrial muscle fibers. However, most of the slow conduction is probably caused by two entirely different factors. First, all these fibers have resting membrane potentials that are much less negative than the normal resting potential of other cardiac muscle. Second, very few gap junctions connect the successive fibers in the pathway, so that there is great resistance to the conduction of excitatory ions from one fiber to the next. Thus, with both low voltage to drive the ions and great resistance to the movement of the ions, it is easy to see why each succeeding fiber is slow to be excited.

TRANSMISSION IN THE PURKINJE SYSTEM

The *Purkinje fibers* lead from the A-V node through the A-V bundle into the ventricles. Except for the initial portion of these fibers where they penetrate the atrioventricular fibrous barrier, they have functional characteristics quite the opposite of those of the A-V nodal fibers; they are very large fibers, even larger than the normal ventricular muscle fibers, and they transmit action potentials at a velocity of 1.5 to 4.0 m/sec, a velocity about 6 times that in the usual cardiac muscle and 150 times that in some transitional fibers. This allows almost immediate transmission of the cardiac impulse throughout the entire ventricular system.

The very rapid transmission of action potentials by Purkinje fibers is believed to be caused by increased permeability of the gap junctions at the intercalated discs between the successive cardiac cells that make up the Purkinje fibers. At these discs, ions are transmitted easily from one cell to the next, thus enhancing the velocity of transmission. The Purkinje fibers also have very few myofibrils, which means that they barely contract during the course of impulse transmission.

One-Way Conduction Through the A-V Bundle. A special characteristic of the A-V bundle is the inability except in abnormal states of action potentials to travel backward in the bundle from the ventricles to the atria. This prevents re-entry of cardiac impulses by this route from the ventricles into the atria, allowing only forward conduction from the atria to the ventricles.

Furthermore, it should be recalled that the atrial muscle is separated from the ventricular muscle by a continuous fibrous barrier, a portion of which is illustrated in Figure 26–7. This barrier normally acts as an insulator to prevent passage of the cardiac impulse between the atria and the ventricles through any other route besides forward conduction through the A-V bundle itself. (However, in rare instances an abnormal muscle bridge does penetrate through the fibrous barrier elsewhere besides at the A-V bundle. Under such conditions, the cardiac impulse can then re-enter the atria from the ventricles and cause serious cardiac arrhythmia.)

Distribution of the Purkinje Fibers in the Ventricles. After penetrating through the atrioventricular fibrous tissue, the distal portion of the A-V bundle passes downward in the ventricular septum for 5 to 15 mm toward the apex of the heart, as shown in Figures 26–5 and 26–7. Then the bundle divides into the *left* and *right bundle branches* that lie beneath the endocardium of the two respective sides of the septum. Each branch spreads downward to the apex of the ventricle, dividing into smaller branches that course around each ventricular chamber and back toward the base of the heart. The terminal Purkinje fibers penetrate about one third of the way into the muscle mass and then become continuous with the cardiac muscle fibers.

From the time that the cardiac impulse first enters the bundle branches until it reaches the terminations of the Purkinje fibers, the total time that elapses averages only 0.03 sec; therefore, once the cardiac impulse enters the Purkinje system, it spreads almost immediately to the entire endocardial surface of the ventricular muscle.

TRANSMISSION OF THE CARDIAC IMPULSE IN THE VENTRICULAR MUSCLE

Once the impulse has reached the ends of the Purkinje fibers, it is then transmitted through the ventricular muscle mass by the ventricular muscle fibers them-selves. The velocity of transmission is now only 0.3 to 0.5 m/sec, one sixth that in the Purkinje fibers.

The cardiac muscle wraps around the heart in a double spiral with fibrous septa between the spiraling layers; therefore, the cardiac impulse does not necessarily travel directly outward toward the surface of the heart but instead angulates toward the surface along the directions of the spirals. Because of this, transmission from the endocardial surface to the epicardial surface of the ventricle requires as much as another 0.03 sec, approximately equal to the time required for transmission through the entire Purkinje system. Thus, the total time for transmission of the cardiac impulse from the initial bundle branches to the last of the ventricular muscle fibers in the normal heart is about 0.06 sec.

SUMMARY OF THE SPREAD OF THE CARDIAC IMPULSE THROUGH THE HEART

Figure 26–8 illustrates in summary form the transmission of the cardiac impulse through the human heart. The numbers on the figure represent the intervals of time in hundredths of a second that lapse between the origin of the cardiac impulse in the sinus node and its appearance at each respective point in the heart. Note that the impulse spreads at moderate velocity through the atria but is delayed more than 0.1 sec in the A-V nodal region before appearing in the ventricular septal A-V bundle. Once it has entered this bundle, it spreads rapidly through the Purkinje fibers to the entire endocardial surfaces of the ventricles. Then the impulse

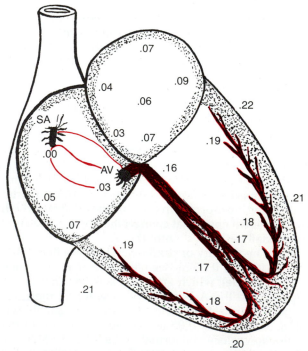

Figure 26–8. Transmission of the cardiac impulse through the heart, showing the time of appearance (in fractions of a second) of the impulse in different parts of the heart.

once again spreads slowly through the ventricular muscle to the epicardial surfaces.

It is extremely important that the reader learn in detail the course of the cardiac impulse through the heart and the times of its appearance in each separate part of the heart, for a quantitative knowledge of this process is essential to the understanding of electrocardiography, which is discussed in the following three chapters.

■ CONTROL OF EXCITATION AND CONDUCTION IN THE HEART

THE SINUS NODE AS THE PACEMAKER OF THE HEART

In the previous discussion of the genesis and transmission of the cardiac impulse through the heart, we have noted that the impulse normally arises in the sinus node. However, this often is not the case under abnormal conditions, for other parts of the heart can exhibit rhythmic contraction in the same way that the sinus nodal fibers can; this is particularly true of the A-V nodal and Purkinje fibers.

The A-V nodal fibers, when not stimulated from some outside source, discharge at an intrinsic rhythmic rate of 40 to 60 times/min, and the Purkinje fibers discharge at a rate of somewhere between 15 and 40 times/min. These rates are in contrast to the normal rate of the sinus node of 70 to 80 times/min.

Therefore, the question that we must ask is: Why does the sinus node control the heart's rhythmicity rather than the A-V node or the Purkinje fibers? The answer to this derives from the fact that the discharge rate of the sinus node is considerably greater than that of either the A-V node or the Purkinje fibers. Each time the sinus node discharges, its impulse is conducted into both the A-V node and the Purkinje fibers, discharging their excitable membranes. Then these tissues as well as the sinus node recover from the action potential and become hyperpolarized. But the sinus node loses its hyperpolarization much more rapidly than does either of the other two and emits a new impulse before either one of them can reach its own threshold for self-excitation. The new impulse again discharges both the A-V node and Purkinje fibers. This process continues on and on, the sinus node always exciting these other potentially self-excitatory tissues before self-excitation can actually occur.

Thus, the sinus node controls the beat of the heart because its rate of rhythmic discharge is greater than that of any other part of the heart. Therefore, the sinus node is the normal *pacemaker* of the heart.

Abnormal Pacemakers — The Ectopic Pacemaker. Occasionally some other part of the heart develops a rhythmic discharge rate that is more rapid than that of the sinus node. For instance, this often occurs in the A-V node or in the Purkinje fibers. In either of these cases, the pacemaker of the heart shifts from the sinus node to the A-V node or to the excitable Purkinje fibers. Under more rare conditions, a point in the atrial or ventricular muscle develops excessive excitability and becomes the pacemaker.

A pacemaker elsewhere than the sinus node is called an *ectopic pacemaker*. Obviously, an ectopic pacemaker causes an abnormal sequence of contraction of the different parts of the heart.

Another cause of shift of the pacemaker is blockage of transmission of the impulses from the sinus node to the other parts of the heart, this occurring most frequently at the A-V node or in the penetrating portion of the A-V bundle on the way to the ventricles. When A-V block occurs, the atria continue to beat at the normal rate of rhythm of the sinus node, while a new pacemaker develops in the Purkinje system of the ventricles and drives the ventricular muscle at a new rate somewhere between 15 and 40 beats/min. However, after a sudden block, the Purkinje system does not begin to emit its rhythmical impulses until 5 to 30 sec later because up to that point it has been "overdriven" by the rapid sinus impulses and consequently is in a suppressed state. During this 5 to 30 sec, the ventricles fail to pump any blood, and the person faints after the first 4 to 5 secs because of lack of blood flow to the brain. This delayed pickup of the heartbeat is called *Stokes-Adams syndrome*. If the period is too long, it can lead to death.

ROLE OF THE PURKINJE SYSTEM IN CAUSING SYNCHRONOUS CONTRACTION OF THE VENTRICULAR MUSCLE

It is clear from the previous description of the Purkinje system that the cardiac impulse arrives at almost all portions of the ventricles within a very narrow span of time, exciting the first ventricular muscle fiber only 0.06 sec ahead of excitation of the last ventricular muscle fiber. This causes all portions of the ventricular muscle in both ventricles to begin contracting at almost exactly the same time. Effective pumping by the two ventricular chambers requires this synchronous type of contraction. If the cardiac impulse traveled through the ventricular muscle very slowly, much of the ventricular mass would contract prior to contraction of the remainder, in which case the overall pumping effect would be greatly depressed. Indeed, in some types of cardiac debilities, slow transmission does occur, and the pumping effectiveness of the ventricles is decreased perhaps as much as 20 to 30 per cent.

CONTROL OF HEART RHYTHMICITY AND CONDUCTION BY THE SYMPATHETIC AND PARASYMPATHETIC NERVES

The heart is supplied with both sympathetic and parasympathetic nerves, as illustrated in Figure 26–9. The

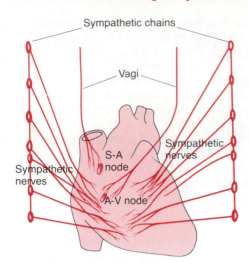

Figure 26–9. The cardiac nerves.

parasympathetic nerves (the vagi) are distributed mainly to the sinus and A-V nodes, to a lesser extent to the muscle of the two atria, and even less to the ventricular muscle. The sympathetic nerves, on the other hand, are distributed to all parts of the heart, with a strong representation to the ventricular muscle as well as to all the other areas.

Effect of Parasympathetic (Vagal) Stimulation on Cardiac Rhythm and Conduction — Ventricular Escape. Stimulation of the parasympathetic nerves to the heart (the vagi) causes the hormone *acetylcholine* to be released at the vagal endings. This hormone has two major effects on the heart. First, it decreases the rate of rhythm of the sinus node, and, second, it decreases the excitability of the A-V junctional fibers between the atrial musculature and the A-V node, thereby slowing transmission of the cardiac impulse into the ventricles. Very strong stimulation of the vagi can completely stop the rhythmic contraction of the sinus node or completely block transmission of the cardiac impulse through the A-V junction. In either case, rhythmic impulses are no longer transmitted into the ventricles. The ventricles stop beating for 4 to 10 sec, but then some point in the Purkinje fibers, usually in the ventricular septal portion of the A-V bundle, develops a rhythm of its own and causes ventricular contraction at a rate of 15 to 40 beats/min. This phenomenon is called *ventricular escape.*

Mechanism of the Vagal Effects. The acetylcholine released at the vagal nerve endings greatly increases the permeability of the fiber membranes to potassium, which allows rapid leakage of potassium to the exterior. This causes increased negativity inside the fibers, an effect called *hyperpolarization*, which makes excitable tissue much less excitable, as was explained in Chapter 6.

In the sinus node, the state of hyperpolarization decreases the "resting" membrane potential of the sinus nodal fibers to a level considerably more negative than the normal value, to a level as low as -65 to -75 mV rather than the normal level of -55 to -60 mV.

Therefore, the upward drift of the resting membrane potential caused by sodium leakage requires much longer to reach the threshold potential for excitation. Obviously, this greatly slows the rate of rhythmicity of these nodal fibers. And, if the vagal stimulation is strong enough, it is possible to stop completely the rhythmical self-excitation of this node.

In the A-V node, the state of hyperpolarization makes it difficult for the minute junctional fibers, which can generate only small quantities of current during the action potential, to excite the nodal fibers. Therefore, the *safety factor* for transmission of the cardiac impulse through the junctional fibers and into the nodal fibers decreases. A moderate decrease in this simply delays conduction of the impulse, but a decrease in safety factor below unity (which means a level so low that the action potential of a fiber cannot cause an action potential in the successive portion of the fiber) completely blocks conduction.

Effect of Sympathetic Stimulation on Cardiac Rhythm and Conduction. Sympathetic stimulation causes essentially the opposite effects on the heart to those caused by vagal stimulation as follows: First, it increases the rate of sinus nodal discharge. Second, it increases the rate of conduction as well as the level of excitability in all portions of the heart. Third, it increases greatly the force of contraction of all the cardiac musculature, both atrial and ventricular.

In short, sympathetic stimulation increases the overall activity of the heart. Maximal stimulation can almost *triple the rate of heartbeat* and can *increase the strength of heart contraction as much as twofold.*

Mechanism of the Sympathetic Effect. Stimulation of the sympathetic nerves releases the hormone *norepinephrine* at the sympathetic nerve endings. The precise mechanism by which this hormone acts on cardiac muscle fibers is still somewhat doubtful, but the present belief is that it increases the permeability of the fiber membrane to sodium and calcium. In the sinus node, an increase of sodium permeability causes a more positive resting potential and an increased rate of upward drift of the membrane potential to the threshold level for self-excitation, both of which obviously accelerate the onset of self-excitation and therefore increase the heart rate.

In the A-V node, increased sodium permeability makes it easier for the action potential to excite the succeeding portion of the conducting fiber, thereby decreasing the conduction time from the atria to the ventricles.

The increase in permeability to calcium ions is at least partially responsible for the increase in contractile strength of the cardiac muscle under the influence of sympathetic stimulation because calcium ions play a powerful role in exciting the contractile process of the myofibrils.

REFERENCES

Akera, T., and Brody, T. M.: Myocardial membranes: Regulation and function of the sodium pump. Annu. Rev. Physiol., 44:375, 1982.
Brown, H. F.: Electrophysiology of the sinoatrial node. Physiol. Rev., 62:505, 1982.

Brutsaert, D. L., and Paulus, W. J.: Contraction and relaxation of the heart as muscle and pump. In Guyton, A. C., and Young, D. B. (eds.): International Review of Physiology: Cardiovascular Physiology III. Vol. 18. Baltimore, University Park Press, 1979, p. 1.

DiFrancesco, D., and Noble, D.: A model of cardiac electrical activity incorporating ionic pumps and concentration changes. Phil. Trans. R. Soc. Lond. (Biol.), 307:353, 1985.

Ellis, D.: Na-Ca exchange in cardiac tissues. Adv. Myocardiol., 5:295, 1985.

Farah, A. E., et al.: Positive inotropic agents. Annu. Rev. Pharmacol. Toxicol., 24:275, 1985.

Fozzard, H. A.: Heart: Excitation-contraction coupling. Annu. Rev. Physiol., 39:201, 1977.

Fozzard, H. A., et al.: The Heart and Cardiovascular System: Scientific Foundations. New York, Raven Press, 1986.

Geddes, L. A.: Cardiovascular Medical Devices. New York, John Wiley & Sons, 1984.

Gilmour, R. F., Jr., and Zipes, D. P.: Slow inward current and cardiac arrhythmias. Am. J. Cardiol., 55:89B, 1985.

Glitsch, H. G.: Electrogenic Na pumping in the heart. Annu. Rev. Physiol., 44:389, 1982.

Gravanis, M. B. (ed.): Cardiovascular Pathophysiology. New York, McGraw-Hill Book Co., 1987.

Guyton, A. C., and Satterfield, J.: Factors concerned in electrical defibrillation of the heart, particularly through the unopened chest. Am. J. Physiol., 167:81, 1951.

Herbette, L., et al.: The interaction of drugs with the sarcoplasmic reticulum. Annu. Rev. Pharmacol. Toxicol., 22:413, 1982.

Hondeghem, L. M., and Katzung, B. G.: Antiarrhythmic agents: The modulated receptor mechanism of action of sodium and calcium channel-blocking drugs. Annu. Rev. Pharmacol. Toxicol., 24:387, 1984.

Irisawa, H.: Comparative physiology of the cardiac pacemaker mechanism. Physiol. Rev., 58:461, 1984.

Josephson, M. E., and Singh, B. N.: Use of calcium antagonists in ventricular dysfunction. Am. J. Cardiol., 55:81B, 1985.

Langer, G. A.: Sodium-calcium exchange in the heart. Annu. Rev. Physiol., 44:435, 1982.

Latorre, R., et al.: K^+ channels gated by voltage and ions. Annu. Rev. Physiol., 46:485, 1984.

Lazdunski, M., and Renaud, J. F.: The action of cardiotoxins on cardiac plasma membranes. Annu. Rev. Physiol., 44:463, 1982.

Levy, M. N., and Martin, P. J.: Neural control of the heart. In Berne, R. M., et al. (eds.): Handbook of Physiology. Sec. 2, Vol. I. Baltimore, Williams & Wilkins, 1979, p. 581.

Levy, M. N., et al.: Neural regulation of the heart beat. Annu. Rev. Physiol., 43:443, 1981.

Loewenstein, W. R.: Junctional intercellular communication: the cell-to-cell membrane channel. Physiol. Rev., 61:829, 1981.

Mazzanti, M., and DeFelice, L. J.: K channel kinetics during the spontaneous heart beat in embryonic chick ventricle cells. Biophys. J., 54:1139, 1988.

Mazzanti, M., and DeFelice, L. J.: Na channel kinetics during the spontaneous heart beat in embryonic chick ventricle cells. Biophys. J., 52:95, 1987.

Mazzanti, M., and DeFelice, L. J.: Regulation of the Na-conducting Ca channel during the cardiac action potential. Biophys. J., 51:115, 1987.

McAnulty, J., and Rahimtoola, S.: Prognosis in bundle branch block. Annu. Rev. Med., 32:499, 1981.

McDonald, T. F.: The slow inward calcium current in the heart. Annu. Rev. Physiol., 44:425, 1982.

Meijler, F. L.: Atrioventricular conduction versus heart size from mouse to whale. J. Am. Coll. Cardiol., 5:363, 1985.

Meijler, F. L., and Janse, M. J.: Morphology and electrophysiology of the mammalian atrioventricular node. Physiol. Rev., 68:608, 1988.

Orrego F.: Calcium and the mechanism of action of digitalis. Gen. Pharmacol., 15:273, 1984.

Reuter, H.: Ion channels in cardiac cell membranes. Annu. Rev. Physiol., 44:473, 1984.

Sheridan, J. D., and Atkinson, M. M.: Physiological roles of permeable junctions: Some possibilities. Annu. Rev. Physiol., 47:337, 1985.

Spear, J. F., and Moore, E. N.: Mechanisms of cardiac arrhythmias. Annu. Rev. Physiol., 44:485, 1982.

Sperelakis, N.: Hormonal and neurotransmitter regulation of Ca^{++} influx through voltage-dependent slow channels in cardiac muscle membrane. Membr. Biochem., 5:131, 1984.

Vasselle, M.: Electrogenesis of the plateau and pacemaker potential. Annu. Rev. Physiol., 41:425, 1979.

Verrier, R. L., and Lown, B.: Behavioral stress and cardiac arrhythmias. Annu. Rev. Physiol., 46:155, 1984.

27

Nervous Regulation of the Circulation and of Respiration

■ NERVOUS REGULATION OF THE CIRCULATION

The circulation is regulated partly by nonnervous mechanisms that are intrinsic to the circulation itself and partly by extrinsic mechanisms, especially by the nervous system. The intrinsic mechanisms consist of such functions as intrinsic control of rhythmicity within the heart muscle itself and local control of blood flow by each tissue of the body when the tissue needs more or less nutrients.

Nervous control of the circulation has two very important features: First, nervous regulation can function extremely rapidly, some of the nervous effects beginning to occur within 1 sec and reaching full development within 5 to 30 sec. Second, the nervous system provides a means for controlling large parts of the circulation simultaneously. For instance, when it is important to raise the arterial pressure temporarily, the nervous system can arbitrarily cut off, or at least greatly decrease, blood flow to major segments of the circulation despite the fact that the local blood flow regulatory mechanisms oppose this.

AUTONOMIC CONTROL OF THE CIRCULATION

The autonomic nervous system was discussed in detail in Chapter 22. However, it is so important to the regulation of the circulation that its specific anatomical and functional characteristics relating to the circulation deserve special attention here.

By far the most important part of the autonomic nervous system for regulation of the circulation is the *sympathetic nervous system.* The *parasympathetic nervous system* is important only for its regulation of heart function, as was discussed in detail in the previous chapter.

The Sympathetic Nervous System. Figure 27–1 illustrates the anatomy of sympathetic nervous control of the circulation. Sympathetic vasomotor nerve fibers leave the spinal cord through all the thoracic and the first one to two lumbar spinal nerves. These pass into the sympathetic chain and thence by two routes to the circulation: (1) through specific *sympathetic nerves* that innervate mainly the vasculature of the internal viscera and the heart and (2) through the *spinal nerves* that innervate mainly the vasculature of the peripheral areas. The precise pathways of these fibers in the spinal cord and in the sympathetic chains are discussed in Chapter 22.

Sympathetic Innervation of the Blood Vessels. Figure 27–2 illustrates the distribution of sympathetic nerve fibers to the blood vessels, showing that all the vessels except the capillaries, precapillary sphincters, and most of the metarterioles are innervated.

The innervation of the *small arteries* and *arterioles* allows sympathetic stimulation to increase the *resistance* and thereby to change the rate of blood flow through the tissues.

The innervation of large vessels, particularly of the *veins,* makes it possible for sympathetic stimulation to change the volume of these vessels and thereby to alter the volume of the peripheral circulatory system. This can translocate blood into the heart and thereby play a major role in the regulation of cardiovascular function, as we see later in this and subsequent chapters.

Sympathetic Nerve Fibers to the Heart. In addition to sympathetic nerve fibers supplying the blood vessels, other sympathetic fibers go to the heart, as was discussed in the last chapter. It should be recalled that sympathetic stimulation markedly increases the activity of the heart, increasing the heart rate and enhancing its strength of pumping.

Parasympathetic Control of Heart Function, Especially Heart Rate. Although the parasympathetic nervous system is exceedingly important for many other

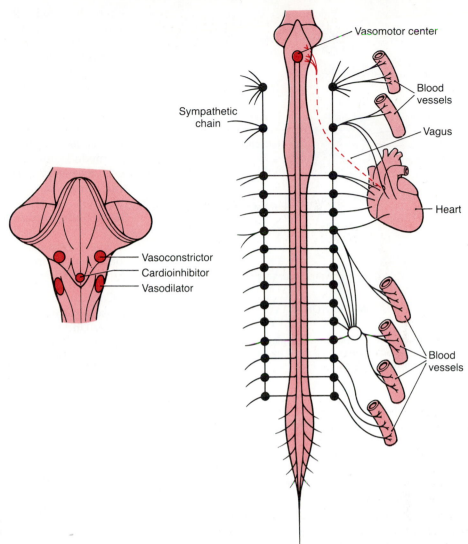

Figure 27–1. Anatomy of sympathetic nervous control of the circulation.

autonomic functions of the body, it plays only a minor role in regulation of the circulation. Its only really important circulatory effect is its control of heart rate by way of parasympathetic fibers carried to the heart in the *vagus nerves,* shown in Figure 27–1 by the dashed nerve from the medulla directly to the heart.

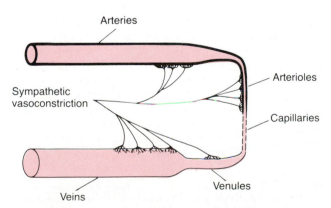

Figure 27–2. Sympathetic innervation of the systemic circulation.

The effects of parasympathetic stimulation on heart function are discussed in detail in Chapter 26. Principally, parasympathetic stimulation causes a marked *decrease* in heart rate and a slight decrease in contractility.

The Sympathetic Vasoconstrictor System and Its Control by the Central Nervous System

The sympathetic nerves carry tremendous numbers of *vasoconstrictor fibers* and only a very few vasodilator fibers. The vasoconstrictor fibers are distributed to essentially all segments of the circulation. However, this distribution is greater in some tissues than in others. It is especially powerful in the kidneys, the gut, the spleen, and the skin but less potent in skeletal muscle and in the brain.

The Vasomotor Center and Its Control of the Vasoconstrictor System. Located bilaterally in the reticular substance of the medulla and lower third of the pons, illustrated in Figure 27–3, is an area called the

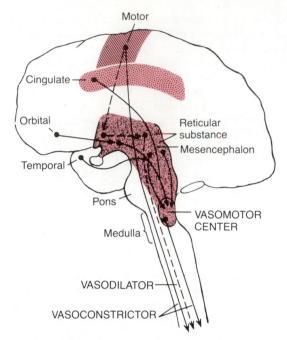

Figure 27–3. Areas of the brain that play important roles in the nervous regulation of the circulation. The dashed lines represent inhibitory pathways.

vasomotor center. The center transmits impulses downward through the cord and thence through the sympathetic vasoconstrictor fibers to all or almost all the blood vessels of the body.

Although the total organization of the vasomotor center is still unclear, experiments have made it possible to identify certain important areas in the center, as follows:

1. A *vasoconstrictor area,* called area "C-1," located bilaterally in the anterolateral portions of the upper medulla. The neurons in this area secrete *norepinephrine;* their fibers are distributed throughout the cord,

where they excite the vasoconstrictor neurons of the sympathetic nervous system.

2. A *vasodilator area,* called area "A-1," located bilaterally in the anterolateral portions of the lower half of the medulla. The fibers from these neurons project upward to the vasoconstrictor area (C-1) and inhibit the vasoconstrictor activity of that area, thus causing vasodilation.

3. A *sensory area,* area "A-2," located bilaterally in the *tractus solitarius* in the posterolateral portions of the medulla and lower pons. The neurons of this area receive sensory nerve signals mainly from the vagus and glossopharyngeal nerves, and the output signals from this sensory area then help to control the activities of both the vasoconstrictor and vasodilator areas, thus providing "reflex" control of many circulatory functions. An example is the baroreceptor reflex for controlling arterial pressure, which we describe later in the chapter.

Continuous Partial Constriction of the Blood Vessels Caused by Sympathetic Vasoconstrictor Tone. Under normal conditions, the vasoconstrictor area of the vasomotor center transmits signals continuously to the sympathetic vasoconstrictor nerve fibers, causing continuous slow firing of these fibers at a rate of about ½ to 2 impulses/sec. This continual firing is called *sympathetic vasoconstrictor tone.* These impulses maintain a partial state of contraction in the blood vessels, a state called *vasomotor tone.*

Figure 27–4 demonstrates the significance of vasoconstrictor tone. In the experiment of this figure, total spinal anesthesia was administered to an animal, which completely blocked all transmission of nerve impulses from the central nervous system to the periphery. As a result, the arterial pressure fell from 100 to 50 mm Hg, illustrating the effect of loss of vasoconstrictor tone throughout the body. A few minutes later a small amount of the hormone norepinephrine was injected intravenously—norepinephrine is the sub-

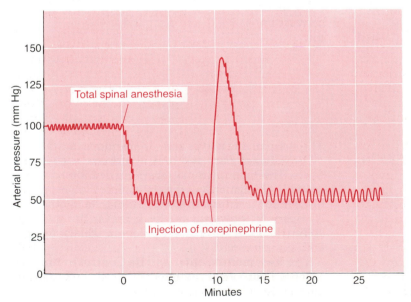

Figure 27–4. Effect of total spinal anesthesia on the arterial pressure, showing a marked fall in pressure resulting from loss of vasomotor tone.

stance secreted at the endings of sympathetic nerve fibers throughout the body. As this hormone was transported in the blood to all the blood vessels, the vessels once again became constricted, and the arterial pressure rose to a level even greater than normal for a minute or two until the norepinephrine was destroyed.

Control of Heart Activity by the Vasomotor Center. At the same time that the vasomotor center is controlling the degree of vascular constriction, it also controls heart activity. The lateral portions of the vasomotor center transmit excitatory impulses through the sympathetic nerve fibers to the heart to increase heart rate and contractility, whereas the medial portion of the vasomotor center, which lies in immediate apposition to the *dorsal motor nucleus of the vagus nerve,* transmits impulses through the vagus nerve to the heart to decrease heart rate. Therefore, the vasomotor center can either increase or decrease heart activity, this ordinarily increasing at the same time that vasoconstriction occurs throughout the body and ordinarily decreasing at the same time that vasoconstriction is inhibited.

Control of the Vasomotor Center by Higher Nervous Centers. Large numbers of areas throughout the *reticular substance* of the *pons, mesencephalon,* and *diencephalon* can either excite or inhibit the vasomotor center. This reticular substance is illustrated in Figure 27–3 by the diffuse shaded area. In general, the more lateral and superior portions of the reticular substance cause excitation, whereas the more medial and inferior portions cause inhibition.

The *hypothalamus* plays a special role in the control of the vasoconstrictor system, for it can exert powerful excitatory or inhibitory effects on the vasomotor center. The *posterolateral portions* of the hypothalamus cause mainly excitation, whereas the *anterior part* can cause mild excitation or inhibition, depending on the precise part of the anterior hypothalamus stimulated.

Many different parts of the *cerebral cortex* can also excite or inhibit the vasomotor center. Stimulation of the *motor cortex,* for instance, excites the vasomotor center because of impulses transmitted downward into the hypothalamus and thence to the vasomotor center. Also, stimulation of the *anterior temporal lobe,* the *orbital areas of the frontal cortex,* the *anterior part of the cingulate gyrus,* the *amygdala,* the *septum,* and the *hippocampus* can all either excite or inhibit the vasomotor center, depending on the precise portion of these areas that is stimulated and on the intensity of the stimulus.

Thus, widespread areas of the brain can have profound effects on the cardiovascular function.

Norepinephrine—The Sympathetic Vasoconstrictor Transmitter Substance. The substance secreted at the endings of the vasoconstrictor nerves is norepinephrine. Norepinephrine acts directly on the so-called alpha receptors of the vascular smooth muscle to cause vasoconstriction, as is discussed in Chapter 22.

The Adrenal Medullae and Their Relationship to the Sympathetic Vasoconstrictor System. Sympathetic impulses are transmitted to the adrenal medullae at the same time that they are transmitted to all the blood vessels. These cause the medullae to secrete both epinephrine and norepinephrine into the circulating blood. These two hormones are carried in the bloodstream to all parts of the body, where they act directly on the blood vessels usually to cause vasoconstriction, but sometimes the epinephrine causes vasodilation because it has a potent "beta" receptor stimulatory effect, which often dilates vessels, as is discussed in Chapter 22.

■ ROLE OF THE NERVOUS SYSTEM FOR RAPID CONTROL OF ARTERIAL PRESSURE

One of the most important functions of nervous control of the circulation is its capability to cause very rapid increases in arterial pressure. For this purpose, the entire vasoconstrictor and cardioaccelerator functions of the sympathetic nervous system are stimulated as a unit. At the same time there is reciprocal inhibition of the normal parasympathetic vagal inhibitory signals to the heart. In consequence, three major changes occur simultaneously, each of which helps to increase the arterial pressure. These are as follows:

1. *Almost all arterioles of the body are constricted.* This greatly increases the total peripheral resistance, impeding the run-off of blood from the arteries and thereby increasing the arterial pressure.

2. *The veins especially but the other large vessels of the circulation as well are strongly constricted.* This displaces blood out of the circulation toward the heart, thus increasing the volume of blood in the heart chambers. This then causes the heart to beat with far greater force and therefore to pump increased quantities of blood. This, too, increases the arterial pressure.

3. Finally, *the heart itself is directly stimulated by the autonomic nervous system, further enhancing cardiac pumping.* Much of this is caused by an increase in the heart rate sometimes to as great as three times normal. In addition, sympathetic nervous signals have a direct effect to increase the contractile force of the heart muscle, this too increasing the capability of the heart to pump larger volumes of blood. Therefore, under strong sympathetic stimulation, the heart can pump for several minutes at least two times as much blood as under normal conditions. Thus, this too contributes still more to the rise in arterial pressure.

Rapidity of Nervous Control of Arterial Pressure. An especially important characteristic of nervous control of arterial pressure is its rapidity of response, beginning within seconds and often increasing the pressure to two times normal within 5 to 15 secs. Conversely, sudden inhibition of nervous stimulation can decrease the arterial pressure to as little as one half normal within 10 to 40 secs. Therefore, nervous control of arterial pressure is by far the most rapid of all our mechanisms for pressure control.

INCREASE IN ARTERIAL PRESSURE DURING MUSCLE EXERCISE AND OTHER TYPES OF STRESS

An important example of the ability of the nervous system to increase the arterial pressure is the increased pressure during muscle exercise. During heavy exercise, the muscles require greatly increased blood flow. Part of this increase results from local vasodilation of the muscle vasculature caused by increased metabolism of the muscle cells, as explained in the previous chapter. However, still additional increase results from simultaneous elevation of arterial pressure during the exercise. In most heavy exercise, the arterial pressure rises about 30 to 40 per cent, which will increase blood flow by approximately an additional twofold.

The increase in arterial pressure during exercise is believed to result mainly from the following effect: At the same time that the motor areas of the nervous system become activated to cause exercise, most of the reticular activating system of the brain stem is also activated, which includes greatly increased stimulation of the vasoconstrictor and cardioacceleratory areas of the vasomotor center. These raise the arterial pressure instantaneously to keep pace with the increase in muscle activity.

In many other types of stress besides muscle exercise, a similar rise in pressure can also take place. For instance, during extreme fright, the arterial pressure often rises to as high as double normal within a few seconds. This is called the *alarm reaction,* and it obviously provides a head of pressure that can immediately supply blood to any or all muscles of the body that might wish to respond instantly to cause flight from danger or to stay and fight.

THE REFLEX MECHANISMS FOR MAINTAINING NORMAL ARTERIAL PRESSURE

Aside from the exercise and stress functions of the autonomic nervous system to raise the arterial pressure, there are also multiple subconscious nervous mechanisms for maintaining the arterial pressure at or near its normal operating level. Almost all of these are *negative feedback reflex mechanisms,* which we explain in the following sections.

The Arterial Baroreceptor Control System — Baroreceptor Reflexes

By far the best known of the mechanisms for arterial pressure control is the *baroreceptor reflex.* Basically, this reflex is initiated by stretch receptors, called either *baroreceptors* or *pressoreceptors,* which are located in the walls of the large systemic arteries. A rise in pressure stretches the baroreceptors and causes them to transmit signals into the central nervous system, and "feedback" signals are then sent back through the autonomic nervous system to the circulation to reduce arterial pressure downward toward the normal level.

Physiologic Anatomy of the Baroreceptors, and Their Innervation. Baroreceptors are spray-type nerve endings lying in the walls of the arteries; they are stimulated when stretched. A few baroreceptors are located in the wall of almost every large artery of the thoracic and neck regions; but, as illustrated in Figure 27–5, baroreceptors are extremely abundant, in (1) the wall of each internal carotid artery slightly above the carotid bifurcation, an area known as the *carotid sinus,* and (2) the wall of the aortic arch.

Figure 27–5 also shows that signals are transmitted from each carotid sinus through the very small *Hering's nerve* to the glossopharyngeal nerve and thence to the *tractus solitarius* in the medullary area of the brain stem. Signals from the arch of the aorta are transmitted through the vagus nerves also into this area of the medulla.

Response of the Baroreceptors to Pressure. Figure 27–6 illustrates the effect of different arterial pressures on the rate of impulse transmission in a Hering's nerve. Note that the carotid sinus baroreceptors are not stimulated at all by pressures between 0 and 60 mm Hg, but above 60 mm Hg they respond progressively more and more rapidly and reach a maximum at about 180 mm Hg. The responses of the aortic baroreceptors are similar to those of the carotid receptors except that they operate, in general, at pressure levels about 30 mm Hg higher.

Note especially that in the normal operating range of arterial pressure, around 100 mm Hg, even a slight change in pressure causes strong autonomic reflexes to readjust the arterial pressure back toward normal. Thus, the baroreceptor feedback mechanism functions most effectively in the very pressure range where it is most needed.

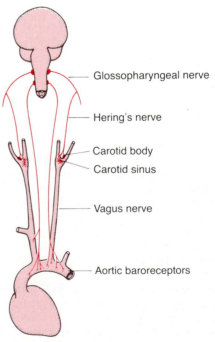

Glossopharyngeal nerve

Hering's nerve

Carotid body

Carotid sinus

Vagus nerve

Aortic baroreceptors

Figure 27–5. The baroreceptor system.

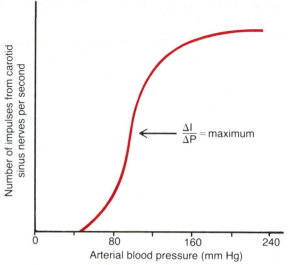

Figure 27–6. Response of the baroreceptors at different levels of arterial pressure.

The baroreceptors respond extremely rapidly to changes in arterial pressure; in fact, the rate of impulse firing increases during systole and decreases again during diastole. Furthermore, the baroreceptors *respond much more to a rapidly changing pressure* than to a stationary pressure. That is, if the mean arterial pressure is 150 mm Hg but at that moment is rising rapidly, the rate of impulse transmission may be as much as twice that when the pressure is stationary at 150 mm Hg. On the other hand, if the pressure is falling, the rate might be as little as one quarter that for the stationary pressure.

The Reflex Initiated by the Baroreceptors. After the baroreceptor signals have entered the tractus solitarius of the medulla, secondary signals *inhibit the vasoconstrictor center* of the medulla and *excite the vagal center*. The net effects are (1) *vasodilation* of the veins and the arterioles throughout the peripheral circulatory system and (2) *decreased heart rate* and *strength of heart contraction*. Therefore, excitation of the baroreceptors by pressure in the arteries reflexly *causes the arterial pressure to decrease* because of both a decrease in peripheral resistance and a decrease in cardiac output. Conversely, low pressure has opposite effects, reflexly causing the pressure to rise back toward normal.

Figure 27–7 illustrates a typical reflex change in arterial pressure caused by occluding the common carotid arteries. This reduces the carotid sinus pressure; as a result, the baroreceptors become inactive and lose their inhibitory effect on the vasomotor center. The vasomotor center then becomes much more active than usual, causing the arterial pressure to rise and to remain elevated during the 10 min that the carotids are occluded. Removal of the occlusion allows the pressure to fall immediately to slightly below normal as a momentary overcompensation and then to return to normal in another minute or so.

Function of the Baroreceptors During Changes in Body Posture. The ability of the baroreceptors to maintain relatively constant arterial pressure is extremely important when a person sits or stands after having been lying down. Immediately upon standing, the arterial pressure in the head and upper part of the body obviously tends to fall, and marked reduction of this pressure can cause loss of consciousness. Fortunately, however, the falling pressure at the baroreceptors elicits an immediate reflex, resulting in strong sympathetic discharge throughout the body, and this minimizes the decrease in pressure in the head and upper body.

The "Buffer" Function of the Baroreceptor Control System. Because the baroreceptor system opposes either increases or decreases in arterial pressure, it is often called a *pressure buffer system*, and the nerves from the baroreceptors are called *buffer nerves*.

Figure 27–8 illustrates the importance of this buffer function of the baroreceptors. The upper record in this figure shows an arterial pressure recording for 2 hours from a normal dog and the lower record from a dog whose baroreceptor nerves from both the carotid sinuses and the aorta had previously been removed. Note the extreme variability of pressure in the denervated dog caused by simple events of the day, such as lying down, standing, excitement, eating, defecation, noises, and so forth.

Figure 27–9 illustrates the frequency distributions of the arterial pressures recorded for a full 24-hour day in both the normal dog and the denervated dog. Note that when the baroreceptors were normal, the arterial pressure remained throughout the day within a narrow range between 85 and 115 mm Hg—indeed, during most of the day at almost exactly 100 mm Hg. On the other hand, after denervation of the baroreceptors, the frequency distribution curve became the broad, low curve of the figure, showing that the pressure range increased 2.5-fold, frequently falling to as low as 50 mm Hg or rising to over 160 mm Hg. Thus, one can see the extreme variability of pressure in the absence of the arterial baroreceptor system.

In summary, a primary purpose of the arterial baro-

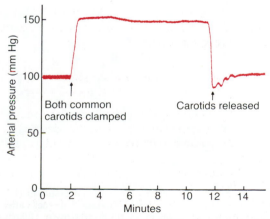

Figure 27–7. Typical carotid sinus reflex effect on arterial pressure caused by clamping both common carotids (after the two vagus nerves have been cut).

receptor system is to reduce the daily variation in arterial pressure to about one half to one third that which would occur were the baroreceptor system not present.

Unimportance of the Baroreceptor System for Long-Term Regulation of Arterial Pressure—"Resetting" of the Baroreceptors. The baroreceptor control system is probably of little or no importance in long-term regulation of arterial pressure for a very simple reason: The baroreceptors themselves reset in 1 to 2 days to whatever pressure level they are exposed. That is, if the pressure rises from the normal value of 100 mm Hg to 200 mm Hg, extreme numbers of baroreceptor impulses are at first transmitted. During the next few seconds, the rate of firing diminishes considerably; then it diminishes much more slowly during the next 1 to 2 days, at the end of which time the rate will have returned essentially to the normal level despite the fact that the arterial pressure now remains 200 mm Hg. Conversely, when the arterial pressure falls to a very low level, the baroreceptors at first transmit no impulses at all, but gradually over a day or two the rate of baroreceptor firing returns again to the original control level.

This "resetting" of the baroreceptors obviously prevents the baroreceptor reflex from functioning as a control system for arterial pressure changes that last longer than a few days at a time. In fact, referring again to Figure 27–8 and 27–9, one can see that the average arterial pressure over any prolonged period of time is

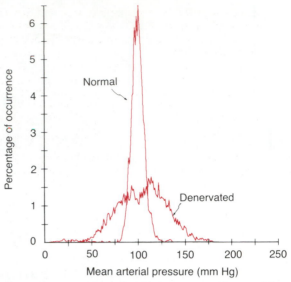

Figure 27–9. Frequency distribution curves of the arterial pressure for a 24-hour period in a normal dog and in the same dog several weeks after the baroreceptors had been denervated. (From Cowley, Liard, and Guyton: Circ. Res., 32:564, 1973. By permission of the American Heart Association, Inc.)

almost exactly the same whether the baroreceptors are present or not. This illustrates the *unimportance of the baroreceptor system for long-term regulation of the arterial pressure* even though it is a potent mechanism for preventing the rapid changes of arterial pressure that occur moment by moment or hour by hour. Prolonged regulation of arterial pressure requires other control systems, principally the renal-body fluid-pressure control system (along with its associated hormonal mechanisms.)

Control of Arterial Pressure by the Carotid and Aortic Chemoreceptors—Effect of Oxygen Lack on Arterial Pressure

Closely associated with the baroreceptor pressure control system is a chemoreceptor reflex that operates in much the same way as the baroreceptor reflex except that instead of stretch receptors initiating the response, chemoreceptors do this.

The chemoreceptors are chemosensitive cells sensitive to oxygen lack, carbon dioxide excess, or hydrogen ion excess. They are located in several small organs 1 to 2 mm in size: two *carotid bodies,* one of which lies in the bifurcation of each common carotid artery, and several *aortic bodies* adjacent to the aorta. The chemoreceptors excite nerve fibers that pass along with the baroreceptor fibers through Hering's nerves and the vagus nerves into the vasomotor center.

Each carotid or aortic body is supplied with an abundant blood flow through a small nutrient artery, so that the chemoreceptors are always in close contact with the arterial blood. Whenever the arterial pressure falls below a critical level, the chemoreceptors become stimulated because of diminished blood flow to the bodies and therefore diminished availability of oxygen and excess buildup of carbon dioxide and hydrogen ions that are not removed by the slow flow of blood.

The signals transmitted from the chemoreceptors into the

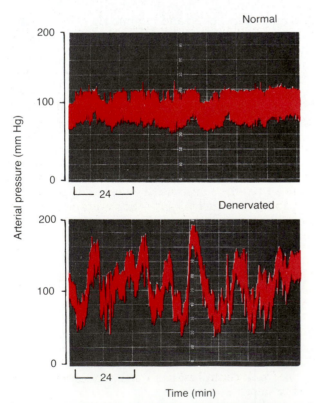

Figure 27–8. Two-hour records of arterial pressure in a normal dog (above) and in the same dog (below) several weeks after the baroreceptors had been denervated. (From Cowley, Liard, and Guyton: Circ. Res., 32:564, 1973. By permission of the American Heart Association, Inc.)

vasomotor center *excite* the vasomotor center, and this elevates the arterial pressure. Obviously, this reflex helps to return the arterial pressure back toward the normal level whenever it falls too low.

However, the chemoreceptor reflex is not a powerful arterial pressure controller in the normal arterial pressure range because the chemoreceptors themselves are not stimulated strongly until the arterial pressure falls below 80 mm Hg. Therefore, it is at the lower pressures that this reflex becomes important to help prevent still further fall in pressure.

The chemoreceptors are discussed in much more detail later in the chapter in relation to respiratory control, in which they play an even more important role than in pressure control.

THE CENTRAL NERVOUS SYSTEM ISCHEMIC RESPONSE—CONTROL OF ARTERIAL PRESSURE BY THE VASOMOTOR CENTER IN RESPONSE TO DIMINISHED BRAIN BLOOD FLOW

Normally, most nervous control of blood pressure is achieved by reflexes originating in the baroreceptors, the chemoreceptors, and the low pressure receptors, all of which are located in the peripheral circulation outside the brain. However, when blood flow to the vasomotor center in the lower brain stem becomes decreased enough to cause nutritional deficiency, that is, to cause *cerebral ischemia,* the neurons in the vasomotor center itself respond directly to the ischemia and become strongly excited. When this occurs, the systemic arterial pressure often rises to a level as high as the heart can possibly pump. This effect is believed to be caused by failure of the slowly flowing blood to carry carbon dioxide away from the vasomotor center; the local concentration of carbon dioxide then increases greatly and has an extremely potent effect in stimulating the sympathetic nervous system. It is possible that other factors, such as the buildup of lactic acid and other acidic substances, also contribute to the marked stimulation of the vasomotor center and to the elevation in pressure. This arterial pressure elevation in response to cerebral ischemia is known as the *central nervous system ischemic response* or simply *CNS ischemic response.*

The magnitude of the ischemic effect on vasomotor activity is tremendous; it can elevate the mean arterial pressure for as long as 10 min sometimes to as high as 250 mm Hg. *The degree of sympathetic vasoconstriction caused by intense cerebral ischemia is often so great that some of the peripheral vessels become totally or almost totally occluded.* The kidneys, for instance, often entirely cease their production of urine because of arteriolar constriction in response to the sympathetic discharge. Therefore, *the CNS ischemic response is one of the most powerful of all the activators of the sympathetic vasoconstrictor system.*

Importance of the CNS Ischemic Response as a Regulator of Arterial Pressure. Despite the extremely powerful nature of the CNS ischemic response, it does not become very active until the arterial pressure falls far below normal, down to 60 mm Hg and below, reaching its greatest degree of stimulation at a pressure of 15 to 20 mm Hg. Therefore, it is not one of the usual mechanisms for regulating normal arterial pressure. Instead, it operates principally as an *emergency arterial pressure control system that acts rapidly and extremely powerfully to prevent further decrease in arterial pressure whenever blood flow to the brain decreases dangerously close to the lethal level.* It is sometimes called the "last ditch stand" pressure control mechanism.

■ NERVOUS CONTROL OF BODY WATER AND BODY FLUID OSMOLALITY

Another important function of the nervous system in controlling the overall environment of the body is its capability to control the amount of water in the body. This is achieved by two separate centers located in the anterior and lateral hypothalamus: (1) a center for controlling the rate of excretion of water by the kidneys and (2) a center for controlling the rate of intake of water by mouth. The first of these centers is called the *antidiuretic center* because it controls the secretion of the hormone *antidiuretic hormone,* which in turn acts on the kidneys to reduce water excretion, thus retaining water in the body. The second of these centers is the *thirst* center, which controls one's drive to drink liquids.

At the same time that the nervous system controls total body water, it also controls the concentration of the dissolved constituents in both the extracellular and intracellular fluid. The normal concentration of these fluids is about 300 mOsm of solute in each liter of fluid. In the extracellular fuid, almost half of these milliosmoles are comprised of sodium ions, and most of the remaining one half are negative ions that balance the sodium ions. Therefore, the total osmolality of the fluids is mainly determined either directly or indirectly by the sodium concentration itself.

The basic stimulus to the receptors in both the antidiuretic center and the thirst center is the osmolality of the extracellular fluid, but since this is determined almost entirely by the sodium ion concentration, these receptors are in effect sodium receptors as well and can be called either *osmoreceptors* or *osmosodium receptors.* When the osmolality of the extracellular fluid becomes too great, this activates the antidiuretic system to retain water and the thirst system to increase water intake, thus diluting the fluids and correcting the hyperosmotic state. Conversely, a hypo-osmotic state inactivates the antidiuretic mechanism as well as thirst, so that excessive amounts of water are now excreted in the urine and at the same time the person drinks little if any water until the hypo-osmotic state is corrected. In the following sections we will explain these two nervous mechanisms for body water control.

THE OSMORECEPTOR–ANTIDIURETIC HORMONE FEEDBACK CONTROL SYSTEM

Figure 27–10 illustrates the osmoreceptor–antidiuretic hormone system for control of extracellular fluid sodium concentration and osmolality. It is a typical feedback control system that operates by the following steps:

1. An increase in osmolality (mainly excess sodium and the negative ions that go with it) excites *osmoreceptors* located in the anterior hypothalamus near to the supraoptic nuclei.

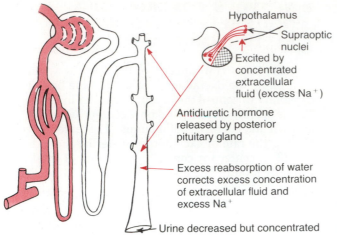

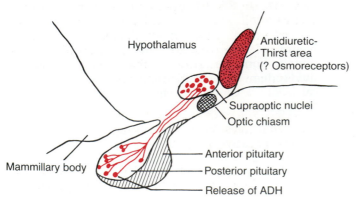

Figure 27–10. Control of extracellular fluid osmolality and sodium ion concentration by the osmosodium receptor-antidiuretic hormone feedback control system.

2. Excitation of the osmoreceptors in turn stimulates the supraoptic nuclei, which then cause the posterior pituitary gland to release ADH.

3. The ADH increases the permeability of the late distal tubules, the cortical collecting ducts, and the collecting ducts to water and therefore *causes increased conservation of water by the kidneys.*

4. The conservation of water but *loss of sodium and other osmolar substances in the urine* causes dilution of the sodium and other substances in the extracellular fluid, thus correcting the initial, excessively concentrated extracellular fluid.

Conversely, when the extracellular fluid becomes too dilute (hypo-osmotic), less ADH is formed, and excess water is lost in comparison with the extracellular fluid solutes, thus concentrating the body fluids back toward normal.

The Osmoreceptors (or Osmosodium Receptors) —The "AV3V Region" of the Brain. Figure 27–11 illustrates the hypothalamus and the pituitary gland. The hypothalamus contains two areas that are important in controlling ADH secretion and also in controlling thirst. One of these is the *supraoptic nuclei.* Here about five sixths of the ADH is formed in the cell bodies of large neuronal cells; the remaining one sixth is formed in the nearby *paraventricular nuclei.* This hormone is transported down the axons of the neurons to their tips, terminating in the posterior pituitary gland. When the supraoptic and paraventricular nuclei are stimulated, nerve impulses pass to these nerve endings and cause ADH to be released into the capillary blood of the posterior pituitary gland.

A second neuronal area important in controlling osmolality is a broad area located along the anteroventral border of the third ventricle, called the *AV3V region,* also illustrated in Figure 27–11. At the superior part of this area is a special structure called the *subfornical organ* and at the inferior part another called the *organum vasculosum of the lamina terminalis.* In between these two "organs" is the *median preoptic nucleus,* which has multiple nervous connections with the two organs and also with both the supraoptic nuclei and the blood pressure control centers in the medulla of the brain. Lesions in this AV3V region cause multiple deficits in the control of ADH secretion, thirst, sodium appetite, and blood pressure. Also, electrical stimulation as well as stimulation by the hormone angiotensin II can alter ADH secretion, thirst, and sodium appetite.

In the vicinity of the AV3V region and the supraoptic nuclei are other neuronal cells that are excited by very minute increases in extracellular fluid osmolality and, conversely, inhibited by decreases in osmolality. These neurons are called *osmoreceptors.* They in turn send nerve signals to the supraoptic nuclei to control the secretion of ADH. It is likely that they also induce thirst as well.

Both the subfornical organ and the organum vasculosum of the lamina terminalis have vascular supplies

Figure 27–11. The supraopticopituitary antidiuretic system and its relationship to the thirst center in the hypothalamus.

that lack the typical blood-brain barrier present elsewhere in the brain that impedes diffusion of most ions from the blood into the brain tissue. Therefore, this makes it possible for ions and other solutes to cross with ease between the blood and the local interstitial fluid. In this way the osmoreceptors respond rapidly to changes in the osmolality of the fluid in the blood, exerting powerful control over the secretion of ADH and probably over thirst as well.

Summary of the Antidiuretic Hormone Mechanism for Controlling Extracellular Fluid Osmolality and Extracellular Fluid Sodium Concentration. From these discussions, we can reiterate once again the importance of the ADH mechanism for controlling at the same time both extracellular fluid osmolality and extracellular fluid sodium concentration. That is, an increase in sodium concentration causes almost an exactly parallel increase in osmolality, which in turn excites the osmoreceptors of the hypothalamus. These receptors then cause the secretion of ADH, which markedly increases the reabsorption of water in the renal tubules. Consequently, very little water is lost into the urine, but the urinary solutes continue to be lost. Therefore, the relative proportion of water in the extracellular fluid increases, whereas the proportion of solutes decreases. In this way, the sodium ion concentration of the extracellular fluid, and the osmolality as well, decrease toward the normal level. This is a very powerful mechanism for controlling both the extracellular fluid osmolality and the extracellular fluid sodium concentration.

THIRST AND ITS ROLE IN CONTROLLING EXTRACELLULAR FLUID OSMOLALITY AND SODIUM CONCENTRATION

The phenomenon of thirst is equally as important for regulating body water, osmolality, and sodium concentration as is the osmoreceptor-renal mechanism previously discussed because the amount of water in the body at any one time is determined by the balance between both *intake* and *output* of water. Thirst, the primary regulator of the intake of water, is defined as the *conscious desire for water.*

Neural Integration of Thirst — The "Thirst" Center

Referring again to Figure 27–11, note that the same area along the anteroventral wall of the third ventricle that promotes antidiuresis can also cause thirst. Also located anterolaterally in the preoptic hypothalamus are other small areas that, when stimulated electrically, will cause immediate onset of drinking that continues as long as the stimulation lasts. All these areas together are called the *thirst center.*

Injection of hypertonic salt solution into portions of the center causes osmosis of water out of the neuronal cells, and this causes drinking. Therefore, these cells function as *osmoreceptors* to activate the thirst mechanism. Probably these are the same osmoreceptors that activate the antidiuretic system as well.

In addition, an increase in osmotic pressure of the cerebrospinal fluid in the third ventricle has essentially the same effect to promote drinking. Some experiments suggest that the site of this effect is the *organum vasculosum of the lamina terminalis,* which lies immediately beneath the ventricular surface at the inferiormost end of the AV3V region. Neuronal cells found here have been shown to be excited by increased osmolality.

Basic Stimulus for Exciting Thirst — Intracellular Dehydration. Any factor that will cause *intracellular dehydration* will in general cause the sensation of thirst. The most common cause of this is increased osmolar concentration of the extracellular fluid, especially increased sodium concentration, which causes osmosis of fluid from the neuronal cells of the thirst center. However, another important cause is excessive loss of potassium from the body, which reduces the intracellular potassium of the thirst cells and therefore decreases their volume.

Temporary Relief of Thirst Caused by the Act of Drinking

A thirsty person receives relief from thirst immediately after drinking water, even before the water has been absorbed from the gastrointestinal tract. In fact, in persons who have an esophageal opening to the exterior so that the water is lost to the outside and never goes into the gastrointestinal tract, partial relief of thirst still occurs, but this relief is only temporary, and the thirst returns after 15 or more min. If the water does enter the stomach, distension of the stomach and other portions of the upper gastrointestinal tract provides still further temporary relief from thirst. Indeed, even simple inflation of a balloon in the stomach can relieve thirst for 5 to 30 min.

One might wonder what the value of this temporary relief from thirst could be, but there is good reason for its occurrence. After a person has drunk water, as long as ½ to 1 hour may be required for all the water to be absorbed and distributed throughout the body. Were the thirst sensation not temporarily relieved after drinking water, the person would continue to drink more and more. When all this water should finally become absorbed, the body fluids would be far more diluted than normal, and an abnormal condition opposite to that which the person was attempting to correct would have been created. It is well known that a thirsty animal almost never drinks more than the amount of water needed to relieve its state of dehydration. Indeed, it is uncanny that the animal usually drinks almost exactly the right amount.

Role of Thirst in Controlling Extracellular Fluid Osmolality and Sodium Concentration

Threshold for Drinking — The Tripping Mechanism. The kidneys are continually excreting fluid; also water is lost by evaporation from the skin and lungs. Therefore, a person is continually being dehydrated, causing the volume of extracellular fluid to decrease and its concentration of sodium and other osmolar elements to rise. When the sodium concentration rises approximately 2 mEq/L above normal (or the osmolality rises approximately 4 mOsm/L above normal), the drinking mechanism becomes "tripped"; that is, the person then reaches a level of thirst that is strong enough to activate the necessary motor effort to cause drinking. This is called the *threshold for drinking.* The person ordinarily drinks precisely the required amount of fluid to bring the extracellular fluids back to normal —that is, to a state of *satiety.* Then the process of dehydration and sodium concentration begins again, and after a period of time the drinking act is tripped again, the process continuing on and on indefinitely.

In this way, both the osmolality and the sodium concentration of the extracellular fluid are very precisely controlled.

COMBINED ROLES OF THE ANTIDIURETIC AND THIRST MECHANISMS FOR CONTROL OF EXTRACELLULAR FLUID OSMOLALITY AND SODIUM CONCENTRATION

When either the ADH mechanism or the thirst mechanism fails, the other ordinarily can still control both extracellular fluid osmolality and sodium concentration with reasonable effectiveness. On the other hand, if both of them fail simultaneously, neither sodium nor osmolality is then adequately controlled.

Figure 27–12 illustrates dramatically the overall capability of the ADH-thirst system to control extracellular fluid sodium concentration (and therefore osmolality as well). This figure demonstrates the ability of the same animals to control their extracellular fluid sodium concentrations in two different conditions: (1) in the normal state and (2) after both the ADH and thirst mechanisms had been blocked. Note that in the normal state (the solid curve), a sixfold increase in sodium intake caused the sodium concentration to change only two thirds of 1 per cent (from 142 mEq/L to 143 mEq/L)—an excellent degree of sodium concentration control. Now note the dashed curve of the figure, which shows the change in sodium concentration when the ADH-thirst system was blocked. In this case, the sodium concentration increased 10 per cent with only a fivefold increase in sodium intake (a change in sodium concentration from 137 mEq/L to 151 mEq/L), which is an extreme change in sodium concentration when one realizes that the normal sodium concentration rarely rises or falls more than 1 per cent from day to day.

Therefore, the major feedback mechanism for control of sodium concentration (and also for extracellular osmolality) is the ADH-thirst mechanism. In the absence of this dual mechanism there is no feedback mechanism that will cause the body to increase water ingestion or water conservation by the kidneys when

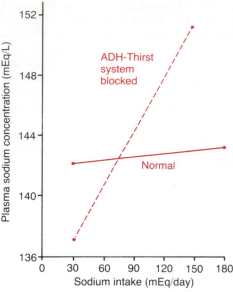

Figure 27–12. Effect on the extracellular fluid sodium concentration in dogs caused by tremendous changes in sodium intake (1) under normal conditions, and (2) after the antidiuretic hormone and thirst feedback systems had been blocked. This figure shows lack of sodium ion control in the absence of these systems. (Courtesy of Dr. David B. Young.)

excess sodium enters the body. Therefore, the sodium concentration simply increases.

Effect of Cardiovascular Reflexes on the ADH-Thirst Control System

Two cardiovascular reflexes also have powerful effects on the ADH-thirst mechanism: (1) the *arterial barore-*

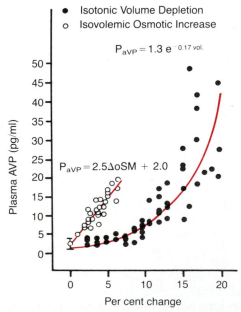

Figure 27–13. Effect of changes in plasma osmolality or blood volume on the level of plasma antidiuretic hormone (ADH) (arginine vasopressin [AVP]). (From Dunn et al.: J. Clin. Invest., 52:3212, 1973. By copyright permission of the American Society for Clinical Investigation.)

ceptor reflex and (2) the *volume receptor reflex,* (a reflex initiated by decreased stretch of the atria of the heart). When the blood volume falls, both of these cause increased secretion of ADH and increased thirst. That is, a decrease in blood volume causes the arterial pressure to fall and activates the arterial baroreceptor reflex. And the volume receptor reflex is activated when the pressures in the two atria, in the pulmonary artery, and in other low pressure areas of the lesser circulation fall below normal, all a common result of too little volume in the circulation. The net result is to activate the ADH-thirst system and thereby to increase the body fluid volume.

To compare the osmolality effects in activating the ADH system with the circulatory reflex effects, Figure 27–13 illustrates by the open circles the effect of increases in body fluid osmolality to cause ADH secretion and by the solid circles the effect of decreased blood volume.

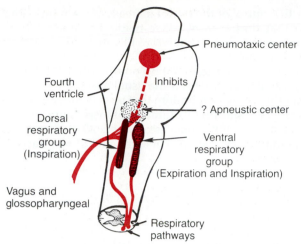

Figure 27–14. Organization of the respiratory center.

■ REGULATION OF RESPIRATION

The nervous system adjusts the rate of alveolar ventilation almost exactly to the demands of the body so that the arterial blood oxygen pressure (Po_2) and carbon dioxide pressure (Pco_2) are hardly altered even during strenuous exercise and most other types of respiratory stress.

The present section describes the operation of this neurogenic system for regulation of respiration.

THE RESPIRATORY CENTER

The "respiratory center" is composed of several widely dispersed groups of neurons located *bilaterally* in the medulla oblongata and pons, as illustrated for one side in Figure 27–14. It is divided into three major collections of neurons: (1) a *dorsal respiratory group,* located in the dorsal portion of the medulla, which mainly causes inspiration; (2) a *ventral respiratory group,* located in the ventrolateral part of the medulla, which can cause either expiration or inspiration, depending upon which neurons in the group are stimulated; and (3) the *pneumotaxic center,* located dorsally in the superior portion of the pons, which helps control both the rate and pattern of breathing. The dorsal respiratory group of neurons plays the fundamental role in the control of respiration. Therefore, let us discuss its function first.

The Dorsal Respiratory Group of Neurons — Its Inspiratory and Rhythmical Functions

The dorsal respiratory group of neurons extends most of the length of the medulla. Either all or most of its neurons are located within the *nucleus of the tractus solitarius,* though additional neurons in the adjacent reticular substance of the medulla probably also play

important roles in respiratory control. The nucleus of the tractus solitarius is also the sensory termination of both the vagal and glossopharyngeal nerves, which transmit sensory signals into the respiratory center from the peripheral chemoreceptors, the baroreceptors, and several different types of receptors in the lung. All the signals from these peripheral areas help in the control of respiration, as we discuss in subsequent sections of this chapter.

Rhythmical Inspiratory Discharges from the Dorsal Respiratory Group. The basic rhythm of respiration is generated mainly in the dorsal respiratory group of neurons. Even when all the peripheral nerves entering the medulla are sectioned and the brain stem is transected both above and below the medulla, this group of neurons still emits repetitive bursts of *inspiratory* action potentials. Unfortunately, though, the basic cause of these repetitive discharges is still unknown. In primitive animals, neural networks have been found in which activity of one set of neurons excites a second set, which in turn inhibits the first. Then after a period of time the mechanism repeats itself, continuing throughout the life of the animal. Therefore, most respiratory physiologists believe that some similar network of neurons located entirely within the medulla, probably involving not only the dorsal respiratory group but adjacent areas of the medulla as well, is responsible for the basic rhythm of respiration.

The Inspiratory "Ramp" Signal. The nervous signal that is transmitted to the inspiratory muscles is not an instantaneous burst of action potentials. Instead, in normal respiration, it begins very weakly at first and increases steadily in a ramp fashion for about 2 sec. It abruptly ceases for approximately the next 3 sec, then begins again for still another cycle, and again and again. Thus, the inspiratory signal is said to be a *ramp signal.* The obvious advantage of this is that it causes a steady increase in the volume of the lungs during inspiration, rather than inspiratory gasps.

There are two ways in which the inspiratory ramp is controlled:

1. Control of the rate of increase of the ramp signal, so that during very active respiration the ramp increases rapidly and therefore fills the lungs rapidly as well.

2. Control of the limiting point at which the ramp suddenly ceases. This is the usual method for controlling the rate of respiration; that is, the earlier the ramp ceases, the shorter the duration of inspiration. For reasons not presently understood, this also shortens the duration of expiration. Thus, the rate of respiration is increased.

The Pneumotaxic Center — Its Function in Limiting the Duration of Inspiration and Increasing Respiratory Rate

The pneumotaxic center, located dorsally in the *nucleus parabrachialis* of the upper pons, transmits impulses continuously to the inspiratory area. The primary effect of these is to control the "switch-off" point of the inspiratory ramp, thus controlling the duration of the filling phase of the lung cycle. When the pneumotaxic signals are strong, inspiration might last for as little as 0.5 sec; but when weak, for as long as 5 or more secs, thus filling the lungs with a great excess of air.

Therefore, the function of the pneumotaxic center is primarily to limit inspiration. However, this has a secondary effect of increasing the rate of breathing because limitation of inspiration also shortens expiration and the entire period of respiration. Thus, a strong pneumotaxic signal can increase the rate of breathing up to 30 to 40 breaths/min, whereas a weak pneumotaxic signal may reduce the rate to only a few breaths per minute.

The Ventral Respiratory Group of Neurons — Its Function in Both Inspiration and Expiration

Located about 5 mm anterior and lateral to the dorsal respiratory group of neurons is the ventral respiratory group of neurons, found in the *nucleus ambiguus* rostrally and the *nucleus retroambiguus* caudally. The function of this neuronal group differs from that of the dorsal respiratory group in several important ways:

1. The neurons of the ventral respiratory group remain almost totally *inactive* during normal quiet respiration. Therefore, normal quiet breathing is caused only by repetitive inspiratory signals from the dorsal respiratory group transmitted mainly to the diaphragm, and expiration results from elastic recoil of the lungs and thoracic cage.

2. There is no evidence that the ventral respiratory neurons participate in the basic rhythmic oscillation that controls respiration.

3. When the respiratory drive for increased pulmonary ventilation becomes greater than normal, respiratory signals then spill over into the ventral respiratory neurons from the basic oscillating mechanism of the dorsal respiratory area. As a consequence, the ventral respiratory area then does contribute its share to the respiratory drive as well.

4. Electrical stimulation of some of the neurons in the ventral group causes inspiration, whereas stimulation of others causes expiration. Therefore, these neurons contribute to both inspiration and expiration. However, they are especially important in providing the powerful expiratory signals to the abdominal muscles during expiration. Thus, this area operates more or less as an overdrive mechanism when high levels of pulmonary ventilation are required.

Reflex Limitation of Inspiration by Lung Inflation Signals — The Hering-Breuer Inflation Reflex

In addition to the neural mechanisms operating entirely within the brain stem, reflex signals from the periphery also help to control respiration. Most important, located in the walls of the bronchi and bronchioles throughout the lungs are *stretch receptors* that transmit signals through the *vagi* into the dorsal respiratory group of neurons when the lungs become overstretched. These signals affect inspiration in much the same way as signals from the pneumotaxic center; that is, when the lungs become overly inflated, the stretch receptors activate an appropriate feedback response that "switches off" the inspiratory ramp and thus stops further inspiration. This is called the *Hering-Breuer inflation reflex*. This reflex also increases the rate of respiration, the same as is true for signals from the pneumotoxic center.

However, in human beings, the Hering-Breuer reflex probably is not activated until the tidal volume increases to greater than approximately 1.5 L. Therefore, this reflex appears to be mainly a protective mechanism for preventing excess lung inflation rather than an important ingredient in the normal control of ventilation.

Control Of Overall Respiratory Center Activity

Up to this point we have discussed the basic mechanisms for causing inspiration and expiration, but it is also important to know how the intensity of the respiratory control signals is increased or decreased to match the ventilatory needs of the body. For example, during very heavy exercise, the rates of oxygen utilization and carbon dioxide formation are often increased to as much as 20 times normal, requiring commensurate increases in pulmonary ventilation.

The major purpose of the remainder of this chapter is to discuss this control of ventilation in response to the needs of the body.

CHEMICAL CONTROL OF RESPIRATION

The ultimate goal of respiration is to maintain proper concentrations of oxygen, carbon dioxide, and hydrogen ions in the tissues. It is fortunate, therefore, that respiratory activity is highly responsive to changes in each of these.

Excess carbon dioxide or hydrogen ions mainly stimulate the respiratory center itself, causing greatly increased strength of both the inspiratory and expiratory signals to the respiratory muscles.

Oxygen, on the other hand, does not have a significant *direct* effect on the respiratory center of the brain in controlling respiration. Instead, it acts almost entirely on peripheral chemoreceptors located in the carotid and aortic bodies, and these in turn transmit appropriate nervous signals to the respiratory center for control of respiration.

Let us discuss first the stimulation of the respiratory center itself by carbon dioxide and hydrogen ions.

Direct Chemical Control Of Respiratory Center Activity By Carbon Dioxide And Hydrogen Ions

The Chemosensitive Area of the Respiratory Center. We have discussed mainly three different areas of the respiratory center: the dorsal respiratory group of neurons, the ventral respiratory group, and the pneumotaxic center. However, it is believed that none of these are affected directly by changes in blood carbon dioxide concentration or hydrogen ion concentration. Instead, an additional neuronal area, a very sensitive *chemosensitive area,* illustrated in Figure 27–15, is located bilaterally lying less than 1 mm beneath the ventral surface of the medulla. This area is highly sensitive to changes in either blood P_{CO_2} or hydrogen ion concentration, and it in turn excites the other portions of the respiratory center.

Response of the Chemosensitive Neurons to Hydrogen Ions—The Primary Stimulus. The sensor neurons in the chemosensitive area are especially excited by hydrogen ions; in fact, it is believed that hydrogen ions are perhaps the only important direct stimulus for these neurons. Unfortunately, though, hydrogen ions do not easily cross either the blood-brain barrier or the blood–cerebrospinal fluid barrier. For this reason, changes in hydrogen ion concentration in the blood actually have considerably less effect in stimulating the chemosensitive neurons than do changes in carbon dioxide, even though carbon dioxide stimulates these neurons indirectly, as is explained below.

Effect of Blood Carbon Dioxide on Stimulating the Chemosensitive Area. Though carbon dioxide has very little direct effect to stimulate the neurons in the chemosensitive area, it does have a very potent indirect effect. It does this by reacting with the water of the tissues to form carbonic acid. This in turn dissociates into hydrogen and bicarbonate ions; the hydrogen ions then have a potent direct stimulatory effect. These reactions are illustrated in Figure 27–15.

Why is it that blood carbon dioxide has a more potent effect in stimulating the chemosensitive neurons than do blood hydrogen ions? The answer is that the blood-brain barrier and the blood–cerebrospinal fluid barrier are both almost completely impermeable to hydrogen ions, whereas carbon dioxide passes through both these barriers almost as if they did not exist. Consequently, whenever the blood P_{CO_2} increases, so also does the P_{CO_2} of both the interstitial fluid of the medulla and of the cerebrospinal fluid. In both of these fluids the carbon dioxide immediately reacts with the water to form hydrogen ions. Thus, paradoxically, more hydrogen ions are released into the respiratory chemosensitive sensory area when the blood carbon dioxide concentration increases than when the blood hydrogen ion concentration increases. For this reason, respiratory center activity is affected considerably more by changes in blood carbon dioxide than by changes in blood hydrogen ions, a fact that we subsequently discuss quantitatively.

Importance of Cerebrospinal Fluid P_{CO_2} in Stimulating the Chemoreceptive Area. Changing the P_{CO_2} in the cerebrospinal fluid that bathes the surface of the brainstem chemoreceptive area will excite respiration in the same way that increased P_{CO_2} in the medullary interstitial fluids also excites respiration. However, the excitation occurs far more rapidly. The reason for this is believed to be that the cerebrospinal fluid has very little protein acid-base buffers. Therefore, the hydrogen ion concentration increases almost instantly when carbon dioxide enters the cerebrospinal fluid from the extensive arachnoid blood vessels. By contrast, the brain tissues have great quantities of protein buffers, so that the change there in hydrogen ion concentration in response to carbon dioxide is greatly delayed. Consequently, the initial rapid excitation of the respiratory system by carbon dioxide entering the cerebrospinal fluid occurs within seconds, in comparison with as long as a minute or more for the stimulation through the brain interstitial fluid.

Decreased Stimulatory Effect of Carbon Dioxide After the First 1 to 2 Days. The excitation of the respiratory center by carbon dioxide is very great the first few hours but then gradually declines over the next 1 to 2 days, decreasing to about one fifth the initial effect. Part of this decline results from renal readjustment of the hydrogen ion concentration back toward normal after the carbon dioxide first increases the hy-

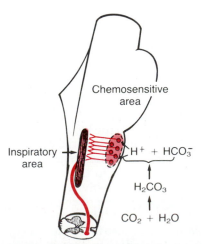

Figure 27–15. Stimulation of the inspiratory area by the *chemosensitive area* located bilaterally in the medulla, lying only a few microns beneath the ventral medullary surface. Note also that hydrogen ions stimulate the chemosensitive area, whereas carbon dioxide in the fluid gives rise to most of the hydrogen ions.

drogen concentration. The kidneys achieve this by increasing the blood bicarbonate. This binds with the hydrogen ions in the cerebrospinal fluid to reduce their concentration. Furthermore, over a period of hours, the bicarbonate ions also slowly diffuse through the blood-brain and blood–cerebrospinal fluid barriers and reduce the hydrogen ions around the respiratory neurons as well.

Therefore, a change in blood carbon dioxide concentration has a very potent *acute* effect on controlling respiratory but only a weak *chronic* effect after a few days' adaptation.

Quantitative Effects of Blood P_{CO_2} and Hydrogen Ion Concentration on Alveolar Ventilation. Figure 27–16 illustrates quantitatively the approximate effects of blood P_{CO_2} and blood pH (which is an inverse logarithmic measure of hydrogen ion concentration) on alveolar ventilation. Note the marked increase in ventilation caused by the increase in P_{CO_2}. But note also the much smaller effect of increased hydrogen ion concentration (that is, decreased pH).

Finally, note the very great change in alveolar ventilation in the normal blood P_{CO_2} range between 35 and 60 mm Hg. This illustrates the tremendous effect that carbon dioxide changes have in controlling respiration. By contrast, the change in respiration in the normal pH range between 7.3 and 7.5 is more than ten times less. The likely reason for this very great differ-

ence is the slight permeability of the blood-brain barrier to hydrogen ions in comparison with its extreme permeability to carbon dioxide. After the carbon dioxide crosses the barrier, however, it reacts with water to form large numbers of hydrogen ions that then stimulate respiration strongly. Yet the hydrogen ions already formed before crossing the barrier cannot cross in large enough numbers to be very effective.

Unimportance of Oxygen for Direct Control of the Respiratory Center. Changes in oxygen concentration have virtually no *direct* effect on the respiratory center itself to alter respiratory drive (although it does have an indirect effect, as explained in the next section). Nevertheless, the respiratory control system is quite poor in controlling the P_{O_2} in the arterial blood going from the lungs to the peripheral tissues. Yet, fortunately, the hemoglobin oxygen buffer system will deliver almost exactly normal amounts of oxygen to the tissues even when the pulmonary P_{O_2} changes from a value as low as 60 mm Hg up to as high as 1000 mm Hg. Therefore, except under special conditions, proper delivery of oxygen can occur despite changes in lung ventilation ranging from slightly below one half normal to as high as 20 or more times normal. On the other hand, this is not true for carbon dioxide, for both the blood and tissue P_{CO_2} change almost exactly inversely with the rate of pulmonary ventilation; thus, evolution has made carbon dioxide the major controller of respiration, not oxygen.

Yet for those special conditions where the tissues do get into trouble for lack of oxygen, the body has a special mechanism for respiratory control located outside the brain respiratory center; this responds when the blood oxygen falls too low, as is explained in the following section.

THE PERIPHERAL CHEMORECEPTOR SYSTEM FOR CONTROL OF RESPIRATORY ACTIVITY — ROLE OF OXYGEN IN RESPIRATORY CONTROL

Aside from the direct control of respiratory activity by the respiratory center itself, still another accessory mechanism is also available for controlling respiration. This is the *peripheral chemoreceptor system*, illustrated in Figure 27–17. Special nervous chemical receptors, called *chemoreceptors,* are located in several areas outside the brain and are especially important for detecting changes in oxygen in the blood, although they also respond to changes in carbon dioxide and hydrogen ion concentrations, too. The chemoreceptors in turn transmit nervous signals to the respiratory center to help regulate respiratory activity.

By far the largest number of chemoreceptors are located in the *carotid bodies.* However, a sizable number are in the *aortic bodies,* also illustrated in Figure 27–17, and a few are located elsewhere in association with other arteries of the thoracic and abdominal regions of the body. The *carotid bodies* are located bilaterally in the bifurcations of the common carotid arteries, and

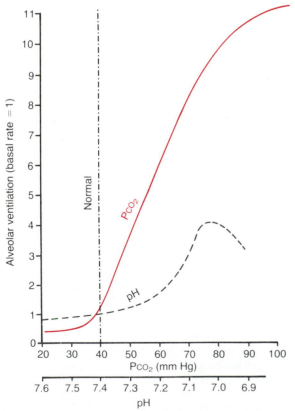

Figure 27–16. Effects of increased arterial P_{CO_2} and decreased arterial pH on the rate of alveolar ventilation.

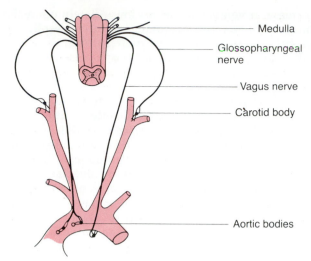

Figure 27–17. Respiratory control by the carotid and aortic bodies.

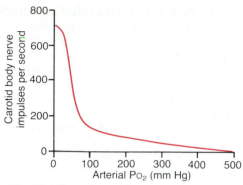

Figure 27–18. Effect of arterial P_{O_2} on impulse rate from the carotid body of a cat. (Curve drawn from data from several sources, but primarily from Von Euler.)

their afferent nerve fibers pass through Hering's nerves to the *glossopharyngeal nerves* and thence to the dorsal respiratory area of the medulla. The *aortic bodies* are located along the arch of the aorta; their afferent nerve fibers pass through the *vagi* also to the dorsal respiratory area. Each of these chemoreceptor bodies receives a special blood supply through a minute artery directly from the adjacent arterial trunk. Furthermore, the blood flow through these bodies is extreme, 20 times the weight of the bodies themselves each minute. Therefore, the percentage removal of oxygen is virtually 0. This means that *the chemoreceptors are exposed at all times to arterial blood,* not venous blood, and their P_{O_2}s are arterial P_{O_2}s.

Stimulation of the Chemoreceptors by Decreased Arterial Oxygen. Changes in arterial oxygen concentration have *no* direct stimulatory effect on the respiratory center itself, but when the oxygen concentration in the arterial blood falls below normal, the chemoreceptors become strongly stimulated. This effect is illustrated in Figure 27–18, which shows the effect of different levels of *arterial* P_{O_2} on the rate of nerve impulse transmission from a carotid body. Note that the impulse rate is particularly sensitive to changes in arterial P_{O_2} in the range between 60 and 30 mm Hg, the range in which the arterial hemoglobin saturation with oxygen decreases rapidly.

Effect of Carbon Dioxide and Hydrogen Ion Concentration on Chemoreceptor Activity. An increase in either carbon dioxide concentration or hydrogen ion concentration also excites the chemoreceptors and in this way indirectly increases respiratory activity. However, the direct effects of both these factors in the respiratory center itself are so much more powerful than their effects mediated through the chemoreceptors (about seven times as powerful) that for most practical purposes the indirect effects through the chemoreceptors do not need to be considered. Yet there is one difference between the peripheral and central effects of carbon dioxide: the peripheral stimulation of the chemoreceptors occurs as much as five times as rapidly as central stimulation,

so that the peripheral chemoreceptors might increase the rapidity of response to carbon dioxide at the onset of exercise.

Basic Mechanism of Stimulation of the Chemoreceptors by Oxygen Deficiency. The exact means by which low P_{O_2} excites the nerve endings in the carotid and aortic bodies is still unknown. However, these bodies have two different, highly characteristic glandular-like cells in them. For this reason, some investigators have suggested that these cells might function as the chemoreceptors and then in turn stimulate the nerve endings. However, other studies suggest that the nerve endings themselves are directly sensitive to the low P_{O_2}.

REGULATION OF RESPIRATION DURING EXERCISE

In strenuous exercise, oxygen consumption and carbon dioxide formation can increase as much as 20-fold. Yet alveolar ventilation ordinarily increases almost exactly in step with the increased level of metabolism, as illustrated by the relationship between oxygen consumption and ventilation in Figure 27–19. Therefore, the arterial P_{O_2}, P_{CO_2}, and pH all remain *almost exactly normal.*

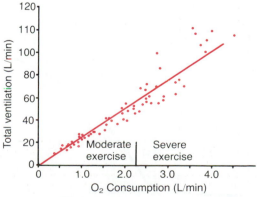

Figure 27–19. Effect of exercise on oxygen consumption and ventilation rate. (From Gray: Pulmonary Ventilation and Its Physiological Regulation. Springfield, Ill., Charles C Thomas.)

In trying to analyze the factors that cause increased ventilation during exercise, one is tempted immediately to ascribe this to the chemical alterations in the body fluids during exercise, including increase of carbon dioxide, increase of hydrogen ions, and decrease of oxygen. However, this is not valid, for measurements of arterial P_{CO_2}, pH, and P_{O_2} show that none of these usually changes significantly.

Therefore, the question must be asked: What is it during exercise that causes the intense ventilation? This question has not been answered, but at least two different effects seem to be predominantly concerned:

1. The brain, on transmitting impulses to the contracting muscles, is believed to transmit collateral impulses into the brain stem to excite the respiratory center. This is analogous to the stimulatory effect of the higher centers of the brain on the vasomotor center of the brain stem during exercise, causing a rise in arterial pressure as well as an increase in ventilation.

2. During exercise, the body movements, especially of the limbs, are believed to increase pulmonary ventilation by exciting joint proprioceptors that then transmit excitatory impulses to the respiratory center. The reason for believing this is that even passive movements of the limbs often increase pulmonary ventilation severalfold.

It is possible that still other factors are also important in increasing pulmonary ventilation during exercise. For instance, some experiments even suggest that hypoxia developing in the muscles during exercise elicits afferent nerve signals to the respiratory center to excite respiration. However, because a large share of the total increase in ventilation begins immediately upon the initiation of exercise, most of the increase in respiration probably results from the two neurogenic factors noted above, namely, *stimulatory impulses from the higher centers of the brain and proprioceptive stimulatory reflexes.*

Interrelationship Between Chemical Factors and Nervous Factors in the Control of Respiration During Exercise. Figure 27–20 summarizes the control of respiration in still another way, this time more quantitatively. The lower curve of this figure shows the effect of different levels of arterial P_{CO_2} on alveolar ventilation when the body is at rest—that is, not exercising. The upper curve shows the approximate shift of this ventilatory curve caused by the neurogenic drive to the respiratory center that occurs during very heavy exercise. The crosses on the two curves show the arterial P_{CO_2}s first in the resting state and then in the exercising state. Note in both instances that the P_{CO_2} is exactly at the normal level of 40 mm Hg. In other words, the neurogenic factor shifts the curve more than 20-fold in the upward direction so that the ventilation almost exactly matches the rate of oxygen consumption and rate of carbon dioxide release, thus keeping the arterial P_{O_2} and P_{CO_2} very near to their normal values.

Yet the upper curve of Figure 27–20 also illustrates that if the arterial P_{CO_2} does change from the normal value of 40 mm Hg, it has its usual stimulatory effect on ventilation at P_{CO_2}s greater than 40, and its usual depressant effect at P_{CO_2}s less than 40 mm Hg.

Possibility That the Neurogenic Factor for Control of Ventilation During Exercise Is a Learned Response. Many experiments suggest that the brain's

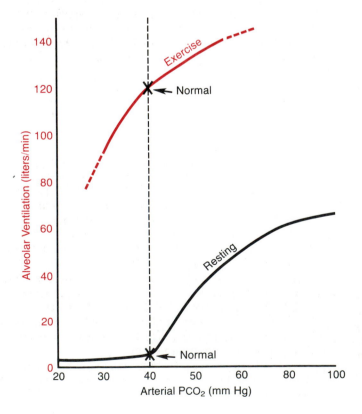

Figure 27–20. Approximate effect of maximum exercise to shift the alveolar P_{CO_2}-ventilation response curve to a much higher than normal level. The shift, believed to be caused by neurogenic factors, is almost exactly the right amount to maintain the arterial P_{CO_2} at the normal level of 40 mm Hg both in the resting state and during very heavy exercise.

ability to shift the ventilatory response curve during exercise, as illustrated in Figure 27–20, is mainly a *learned* response. That is, with repeated exercise, the brain becomes progressively more able to provide the proper amount of brain signal required to keep the blood chemical factors at their normal levels. Also, there is much reason to believe that some of the higher learning centers of the brain are important in this neurogenic respiratory control factor — probably even the cerebral cortex. One important reason for believing this is that when the cerebral cortex is anesthetized, the respiratory control system loses its special ability to maintain the arterial blood gases near normal during exercise.

REFERENCES

THE CIRCULATION

Blessing, W. W.: Central neurotransmitter pathways for baroreceptor initiated secretion of vasopressin. News Physiol. Sci., 1:90, 1986.

Blix, A. S., and Folkow, B.: Cardiovascular adjustments to diving in mammals and birds. In Sheperd, J. T., and Abboud, F. M. (eds.): Handbook of Physiology. Sec. 2, Vol. III. Bethesda, Md., American Physiological Society, 1983, p. 917.

Bohr, D. F., and Webb, R. C.: Vascular smooth muscle function and its changes in hypertension. Am. J. Med., 1984.

Brown, A. J., et al.: Cardiovascular and renal responses to chronic vasopressin infusion. Am. J. Physiol., 250:H584, 1986.

Buckley, J. P., et al. (eds.): Brain Peptides and Catecholamines in Cardiovascular Regulation. New York, Raven Press, 1987.

Buratini, R., and Borgdorff, P.: Closed-loop baroreflex control of total peripheral resistance in the cat: Identification of gains by aid of a model. Cardiovas. Res., 18:715, 1984.

Calaresu, F. R., and Yardley, C. P.: Medullary basal sympathetic tone. Annu. Rev. Physiol., 50:511, 1988.

Coleman, T. G., et al.: Angiotensin and the hemodynamics of chronic salt deprivation. Am. J. Physiol., 229:167, 1975.

Coleridge, H. M., and Coleridge, J. C. G.: Cardiovascular afferents involved in regulation of peripheral vessels. Annu. Rev. Physiol., 42:413, 1980.

Cowley, A. W., Jr., and Guyton, A. C.: Baroreceptor reflex contribution in angiotensin II–induced hypertension. Circulation, 50:61, 1974.

Cowley, A. W., Jr., et al.: Interaction of vasopressin and the baroreceptor reflex system in the regulation of arterial pressure in the dog. Circ. Res., 34:505, 1974.

Cushing, H.: Concerning a definite regulatory mechanism of the vasomotor center which controls blood pressure during cerebral compression. Bull. Johns Hopkins Hosp., 12:290, 1901.

Guyton, A. C.: Acute hypertension in dogs with cerebral ischemia. Am. J. Physiol., 154:45, 1948.

Guyton, A. C.: Arterial Pressure and Hypertension. Philadelphia, W. B. Saunders Co., 1980.

Herd, J. A.: Cardiovascular response to stress in man. Annu. Rev. Physiol., 46:177, 1984.

Lisney, S. J. W., and Bharali, L. A. M.: The axon reflex: An outdated idea or a valid hypothesis? News Physiol. Sci., 4:45, 1989.

Ludbrook, J.: Reflex control of blood pressure during exercise. Annu. Rev. Physiol., 45:155, 1983.

Mancia, G., and Mark, A. L.: Arterial baroreflexes in humans. In Shepherd, J. T., and Abboud, F. M. (eds.): Handbook of Physiology. Sec. 2, Vol. III. Bethesda, Md., American Physiological Society, 1983, p. 755.

Mathias, C. J., and Frankel, H. L.: Cardiovascular control in spinal man. Annu. Rev. Physiol., 50:577, 1988.

Mitchell, J. H., and Schmidt, R. F.: Cardiovascular reflex control by afferent fibers from skeletal muscle receptors. In Sheperd, J. T., and Abboud, F. M. (eds.): Handbook of Physiology. Sec. 2, Vol. III. Bethesda, Md., American Physiological Society, 1983, p. 623.

Opie, L. H. (ed.): Calcium Antagonists and Cardiovascular Disease. New York, Raven Press, 1984.

Persson, P. B., et al.: Cardiopulmonary-arterial baroreceptor interaction in control of blood pressure. News Physiol. Sci., 4:56, 1989.

Randall, W. C. (ed.): Nervous Control of Cardiovascular Function. New York, Oxford University Press, 1984.

Regoli, D.: Neurohumoral regulation of precapillary vessels: The kallikrein-kinin system. J. Cardiovasc. Pharmacol., 6:(Suppl. 2) S401, 1984.

Reid, J. L., and Rubin, P. C.: Peptides and central neural regulation of the circulation. Physiol. Rev., 67:725, 1987.

Sagawa, K.: Baroreflex control of systemic arterial pressure and vascular bed. In Sheperd, J. T., and Abboud, F. M. (eds.): Handbook of Physiology. Sec. 2, Vol. III. Bethesda, Md., American Physiological Society, 1983, p. 453.

Share, L.: Role of vasopressin in cardiovascular regulation. Physiol. Rev., 68:1246, 1988.

Stiles, G. L., et al.: β-Adrenergic receptors: Biochemical mechanisms of physiological regulation. Physiol. Rev., 64:661, 1984.

Vanhoutte, P. M.: Vasodilation: Vascular Smooth Muscle, Peptides, Autonomic Nerves, and Endothelium. New York, Raven Press, 1988.

Vanhoutte, P. M.: Calcium-entry blockers, vascular smooth muscle and systemic hypertension. Am. J. Cardiol., 55:17B, 1985.

THE RESPIRATION

Acker, H.: PO_2 chemoreception in arterial chemoreceptors. Annu. Rev. Physiol., 51:835, 1989.

Cohen, M. I.: Central determinants of respiratory rhythm. Annu. Rev. Physiol., 43:91, 1981.

Eyzaguirre, C., et al.: Arterial chemoreceptors. In Shepherd, J. T., and Abboud, F. M. (eds.): Handbook, of Physiology. Sec. 2, Vol. III. Bethesda, Md., American Physiological Society, 1983, p. 557.

Feldman, J. L., and Ellenberger, H. H.: Central coordination of respiratory and cardiovascular control in mammals. Annu. Rev. Physiol., 50:593, 1988.

Guyton, A. C., et al.: Basic oscillating mechanism of Cheyne-Stokes breathing. Am. J. Physiol., 187:395, 1956.

Honig, A.: Salt and water metabolism in acute high-altitude hypoxia: Role of peripheral arterial chemoreceptors. News Physiol. Sci., 4:109, 1989.

Kalia, M. P.: Anatomical organization of central respiratory neurons. Annu. Rev. Physiol., 43:105, 1981.

Karczewski, W. A., et al.: Control of Breathing During Sleep and Anesthesia. New York, Plenum Publishing Corp., 1988.

Masuyama, H., and Honda, Y.: Differences in overall "gain" of CO_2-feedback system between dead space and CO_2 ventilations in man. Bull. Eur. Physiopathol. Respir., 20:501, 1984.

Milhorn, H. T., Jr., and Guyton, A. C.: An analog computer analysis of Cheyne-Stokes breathing. J. Appl. Physiol., 20:328, 1965.

Milhorn, H. T., Jr., et al.: A mathematical model of the human respiratory control system. Biophys. J., 5:27, 1965.

Mitchell, G. S., et al.: Changes in the V_I-V_{CO} relationship during exercise in goats: Role of carotid bodies. J. Appl. Physiol., 57:1894, 1984.

Rigatto, H.: Control of ventilation in the newborn. Annu. Rev. Physiol., 46:661, 1984.

Rowell, L. B., and Sheriff, D. D.: Are muscle "chemoreflexes" functionally important? News Physiol. Sci., 3:250, 1988.

Sinclair, J. D.: Respiratory drive in hypoxia: Carotid body and other mechanisms compared. News Physiol. Sci., 2:57, 1987.

Von Euler, C., and Lagercrantz, H.: Neurobiology of the Control of Breathing. New York, Raven Press, 1987.

West, J. B.: Pulmonary Pathophysiology — The Essentials. Baltimore, Williams & Wilkins, 1987.

Whipp, B. J.: Ventilatory control during exercise in humans. Annu. Rev. Physiol., 45:393, 1983.

28

Regulation of Gastrointestinal Function, Food Intake, Micturition, and Body Temperature

■ NEURAL CONTROL OF GASTROINTESTINAL FUNCTION

The gastrointestinal tract has a nervous system all its own called the *enteric nervous system.* It lies entirely in the wall of the gut, beginning in the esophagus and extending all the way to the anus. The number of neurons in this enteric system is about 100,000,000, almost exactly equal to the number in the entire spinal cord; this illustrates the importance of the enteric system for controlling gastrointestinal function. It especially controls gastrointestinal movements and secretion.

The enteric system is composed mainly of two plexuses, as illustrated in Figure 28–1: an outer plexus lying between the longitudinal and circular muscular layers, called the *myenteric plexus* or *Auerbach's plexus;* and (2) an inner plexus, called the *submucosal plexus* or *Meissner's plexus,* that lies in the submucosa. The myenteric plexus controls mainly the gastrointestinal movements, and the submucosal plexus controls mainly gastrointestinal secretion and local blood flow.

Note in Figure 28–1 the sympathetic and parasympathetic fibers that connect with both the myenteric and submucosal plexuses. Although the enteric nervous system can function on its own, independently of these extrinsic nerves, stimulation of the parasympathetic and sympathetic systems can further activate or inhibit gastrointestinal functions, as we discuss later.

Also shown in Figure 28–1 are sensory nerve endings that originate in the gastrointestinal epithelium or gut wall and then send afferent fibers to both plexuses of the enteric system and also to (1) the prevertebral ganglia of the sympathetic nervous system, (2) the spinal cord, and (3) some traveling in the parasympathetic nerves (the vagi, for instance) all the way to the brain stem. These sensory nerves elicit local reflexes within the gut itself and also reflexes that are relayed back to the gut from either the prevertebral ganglia or the central nervous system.

AUTONOMIC CONTROL OF THE GASTROINTESTINAL TRACT

Parasympathetic Innervation. The parasympathetic supply to the gut is divided into *cranial* and *sacral divisions,* which are discussed in Chapter 22. Except for a few parasympathetic fibers to the mouth and pharyngeal regions of the alimentary tract, the cranial parasympathetics are transmitted almost entirely in the *vagus nerves.* These fibers provide extensive innervation to the esophagus, stomach, pancreas, and first half of the large intestine (but rather little innervation to the small intestine). The sacral parasympathetics originate in the second, third, and fourth sacral segments of the spinal cord and pass through the *pelvic nerves* to the distal half of the large intestine. The sigmoidal, rectal, and anal regions of the large intestine are considerably better supplied with parasympathetic fibers than are the other portions. These fibers function especially in the defecation reflexes, which are discussed later in the chapter.

The *postganglionic neurons* of the parasympathetic system are located in the myenteric and submucosal plexuses, and stimulation of the parasympathetic nerves causes a general increase in activity of the entire enteric nervous system. This in turn enhances the activity of most gastrointestinal functions, but not all, for some of the enteric neurons are inhibitory and therefore inhibit certain of the functions.

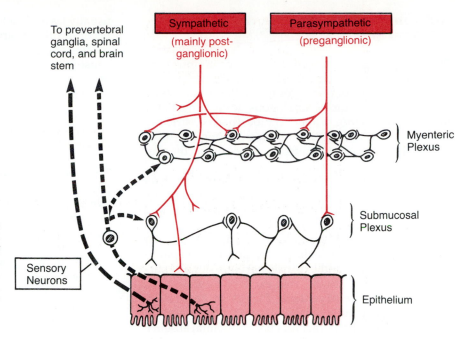

Figure 28–1. Neural control of the gut wall, showing (1) the myenteric and submucosal plexuses; (2) extrinsic control of these plexuses by the sympathetic and parasympathetic nervous systems; and (3) sensory fibers passing from the luminal epithelium and gut wall to the enteric plexuses and from there to the prevertebral ganglia, spinal cord, and brain stem.

Sympathetic Innervation. The sympathetic fibers to the gastrointestinal tract originate in the spinal cord between the segments T-5 and L-2. The preganglionic fibers, after leaving the cord, enter the sympathetic chains and pass through the chains to outlying ganglia, such as the *celiac ganglion* and various *mesenteric ganglia*. Here, most of the *postganglionic neuron bodies* are located, and postganglionic fibers spread from them along with the blood vessels to all parts of the gut, terminating principally on neurons in the enteric nervous system. The sympathetics innervate essentially all portions of the gastrointestinal tract, rather than being more extensively supplied to the most orad and most analward portions, as is true of the parasympathetics. The sympathetic nerve endings secrete *norepinephrine*.

In general, stimulation of the sympathetic nervous system inhibits activity in the gastrointestinal tract, causing effects essentially opposite to those of the parasympathetic system. It exerts its effects in two different ways: (1) to a slight extent by direct effect of the norepinephrine on the smooth muscle to inhibit this (except the muscularis mucosa, which it excites), and (2) to a major extent by an inhibitory effect of the norepinephrine on the neurons of the enteric nervous system. Thus, strong stimulation of the sympathetic system can totally block movement of food through the gastrointestinal tract.

THE GASTROINTESTINAL REFLEXES

The anatomical arrangement of the enteric nervous system and its connections with the sympathetic and parasympathetic systems supports three different types of gastrointestinal reflexes that are essential to gastrointestinal control. These are the following:

1. *Reflexes that occur entirely within the enteric nervous system.* These include reflexes that control gastrointestinal secretion, peristalsis, mixing contractions, local inhibitory effects, and so forth.

2. *Reflexes from the gut to the prevertebral sympathetic ganglia and then back to the gastrointestinal tract.* These reflexes transmit signals for long distances in the gastrointestinal tract, such as signals from the stomach to cause evacuation of the colon (the *gastrocolic reflex*), signals from the colon and small intestine to inhibit stomach motility and stomach secretion (the *enterogastric reflexes*), and reflexes from the colon to inhibit emptying of ileal contents into the colon (the *colonoileal reflex*).

3. *Reflexes from the gut to the spinal cord or brain stem and then back to the gastrointestinal tract.* These include especially (a) reflexes from the stomach and duodenum to the brain stem and back to the stomach to control gastric motor and secretory activity; (b) pain reflexes that cause general inhibition of the entire gastrointestinal tract; and (c) defecation reflexes that travel to the spinal cord and back again to produce the powerful colonic, rectal, and abdominal contractions required for defecation (the defecation reflexes).

■ FUNCTIONAL TYPES OF MOVEMENTS IN THE GASTROINTESTINAL TRACT

Two basic types of movements occur in the gastrointestinal tract: (1) *propulsive movements*, which cause food to move forward along the tract at an appropriate rate for digestion and absorption; and (2) *mixing movements*, which keep the intestinal contents thoroughly mixed at all times.

THE PROPULSIVE MOVEMENTS — PERISTALSIS

The basic propulsive movement of the gastrointestinal tract is *peristalsis,* which is illustrated in Figure 28 – 2. A contractile ring appears around the gut and then moves forward; this is analogous to putting one's fingers around a thin distended tube, then constricting the fingers and sliding them forward along the tube. Obviously, any material in front of the contractile ring is moved forward.

The usual stimulus for peristalsis is *distention.* That is, if a large amount of food collects at any point in the gut, the stretching of the gut wall stimulates the gut 2 to 3 cm above this point, and a contractile ring appears that initiates a peristaltic movement. Other stimuli that can initiate peristalsis include irritation of the epithelium lining the gut and extrinsic nervous signals, particularly parasympathetic, that excite the gut.

Function of the Myenteric Plexus in Peristalsis. Peristalsis occurs only weakly, if at all, in any portion of the gastrointestinal tract that has congenital absence of the myenteric plexus. Therefore, *effectual* peristalsis requires an active myenteric plexus.

Directional Movement of Peristaltic Waves Toward the Anus. Peristalsis, theoretically, can occur in either direction from a stimulated point, but it normally dies out rapidly in the orad direction while continuing for a considerable distance analward. The exact cause of this directional transmission of peristalsis has never been ascertained, though it probably results mainly from the fact that the myenteric plexus itself is "polarized" in the anal direction.

THE MIXING MOVEMENTS

The mixing movements are quite different in different parts of the alimentary tract. In some areas, the peristaltic contractions themselves cause most of the mixing. This is especially true when forward progression of the intestinal contents is blocked by a sphincter, so that a peristaltic wave can then only churn the intestinal contents, rather than propelling them forward. At other times, *local constrictive contractions* occur every few centimeters in the gut wall. These constrictions

usually last only a few seconds; then new constrictions occur at other points in the gut, thus "chopping" the contents first here and then there. These peristaltic and constrictive movements are modified in different parts of the gastrointestinal tract for proper propulsion and mixing.

■ INGESTION OF FOOD

The amount of food that a person ingests is determined principally by the intrinsic desire for food called *hunger.* The type of food that a person preferentially seeks is determined by *appetite.* These mechanisms in themselves are extremely important automatic regulatory systems for maintaining an adequate nutritional supply for the body, and they are discussed later in the chapter in relation to nutrition of the body.

MASTICATION (CHEWING)

The teeth are admirably designed for chewing, the anterior teeth (incisors) providing a strong cutting action and the posterior teeth (molars) a grinding action. All the jaw muscles working together can close the teeth with a force as great as 55 lb on the incisors and 200 lb on the molars.

Most of the muscles of chewing are innervated by the motor branch of the fifth cranial nerve, and the chewing process is controlled by nuclei in the brain stem. Stimulation of the reticular formation near the brain stem centers for taste can cause continual rhythmic chewing movements. Also, stimulation of areas in the hypothalamus, amygdala, and even in the cerebral cortex near the sensory areas for taste and smell can cause chewing.

Much of the chewing process is caused by the *chewing reflex,* which may be explained as follows: The presence of a bolus of food in the mouth causes reflex inhibition of the muscles of mastication, which allows the lower jaw to drop. The drop in turn initiates a stretch reflex of the jaw muscles that leads to *rebound* contraction. This automatically raises the jaw to cause closure of the teeth, but it also compresses the bolus again against the linings of the mouth, which inhibits the jaw muscles once again, allowing the jaw to drop and rebound another time; and this is repeated again and again.

SWALLOWING (DEGLUTITION)

Swallowing is a complicated mechanism, principally because the pharynx most of the time subserves several other functions besides swallowing and is converted for only a few seconds at a time into a tract for propulsion of food. It is especially important that respiration not be compromised because of swallowing.

In general, swallowing can be divided into (1) the *voluntary stage,* which initiates the swallowing pro-

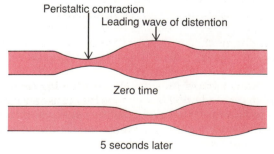

Peristaltic contraction
Leading wave of distention

Zero time

5 seconds later

Figure 28 – 2. Peristalsis.

cess, (2) the *pharyngeal stage,* which is involuntary and constitutes the passage of food through the pharynx into the esophagus, and (3) the *esophageal stage,* another involuntary phase that promotes passage of food from the pharynx to the stomach.

Voluntary Stage of Swallowing. When the food is ready for swallowing, it is "voluntarily" squeezed or rolled posteriorly into the pharynx by pressure of the tongue upward and backward against the palate, as shown in Figure 28–3. From here on, the process of swallowing becomes entirely, or almost entirely, automatic and ordinarily cannot be stopped.

Pharyngeal Stage of Swallowing. As the bolus of food enters the pharynx, it stimulates *swallowing receptor areas* all around the opening of the pharynx, especially on the tonsillar pillars, and impulses from these pass to the brain stem to initiate a series of automatic pharyngeal muscular contractions as follows:

1. The soft palate is pulled upward to close the posterior nares, in this way preventing reflux of food into the nasal cavities.

2. The palatopharyngeal folds on either side of the pharynx are pulled medially to approximate each other. In this way these folds form a sagittal slit through which the food must pass into the posterior pharynx. This slit performs a selective action, allowing food that has been masticated properly to pass with ease while impeding the passage of large objects. Because this stage of swallowing lasts less than 1 sec, any large object is usually impeded too much to pass through the pharynx into the esophagus.

3. The vocal cords of the larynx are strongly approximated, and the larynx is pulled upward and anteriorly by the neck muscles. This action, combined with the presence of ligaments that prevent upward movement of the epiglottis, causes the epiglottis to swing backward over the opening of the larynx. Both effects prevent passage of food into the trachea. Especially important is the approximation of the vocal cords, but the epiglottis helps to prevent food from ever getting as far as the vocal cords. Destruction of the vocal cords or of the muscles that approximate them can cause strangulation. On the other hand, removal of the epiglottis usually does not cause serious debility in swallowing.

4. The upward movement of the larynx also enlarges the opening of the esophagus. At the same time, the upper 3 to 4 cm of the esophageal muscular wall, an area called the *upper esophageal sphincter* or the *pharyngoesophageal sphincter,* relaxes, thus allowing food to move easily and freely from the posterior pharynx into the upper esophagus. This sphincter, between swallows, remains strongly contracted, thereby preventing air from going into the esophagus during respiration. The upward movement of the larynx also lifts the glottis out of the main stream of food flow so that the food usually passes on either side of the epiglottis rather than over its surface; this adds still another protection against entry of food into the trachea.

5. At the same time that the larynx is raised and the pharyngoesophageal sphincter is relaxed, the entire muscular wall of the pharynx contracts, beginning in the superior part of the pharynx and spreading downward as a rapid peristaltic wave over the middle and inferior pharyngeal muscles and thence into the esophagus, which propels the food into the esophagus.

To summarize the mechanics of the pharyngeal stage of swallowing: the trachea is closed, the esophagus is opened, and a fast peristaltic wave originating in the pharynx forces the bolus of food into the upper esophagus, the entire process occurring in 1 to 2 sec.

Nervous Control of the Pharyngeal Stage of Swallowing. The most sensitive tactile areas of the pharynx for initiation of the pharyngeal stage of swallowing lie in a ring around the pharyngeal opening, with greatest sensitivity in the tonsillar pillars. Impulses are transmitted from these areas through the sensory portions of the trigeminal and glossopharyngeal nerves into a region of the medulla oblongata closely associated with the *tractus solitarius,* which receives essentially all sensory impulses from the mouth.

The successive stages of the swallowing process are then automatically controlled in orderly sequence by neuronal areas distributed throughout the reticular substance of the medulla and lower portion of the pons. The sequence of the swallowing reflex is the same from one swallow to the next, and the timing of the entire cycle also remains constant from one swallow to the next. The areas in the medulla and lower pons that control swallowing are collectively called the *deglutition* or *swallowing center.*

The motor impulses from the swallowing center to the pharynx and upper esophagus that cause swallowing are transmitted by the fifth, ninth, tenth, and twelfth cranial nerves and even a few of the superior cervical nerves.

In summary, the pharyngeal stage of swallowing is

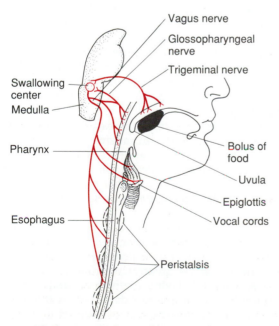

Vagus nerve
Glossopharyngeal nerve
Trigeminal nerve
Swallowing center
Medulla
Pharynx
Bolus of food
Uvula
Epiglottis
Vocal cords
Esophagus
Peristalsis

Figure 28–3. The swallowing mechanism.

principally a reflex act. It is almost never initiated by direct stimuli to the swallowing center from higher regions of the central nervous system. Instead, it is almost always initiated by voluntary movement of food into the back of the mouth, which, in turn, elicits the swallowing reflex.

Esophageal Stage of Swallowing

The esophagus functions primarily to conduct food from the pharynx to the stomach, and its movements are organized specifically for this function.

Normally the esophagus exhibits two types of peristaltic movements—*primary peristalsis* and *secondary peristalsis*. Primary peristalsis is simply a continuation of the peristaltic wave that begins in the pharynx and spreads into the esophagus during the pharyngeal stage of swallowing. This wave passes all the way from the pharynx to the stomach in approximately 8 to 10 sec. However, food swallowed by a person who is in the upright position is usually transmitted to the lower end of the esophagus even more rapidly than the peristaltic wave itself, in about 5 to 8 sec, because of the additional effect of gravity pulling the food downward. If the primary peristaltic wave fails to move all the food that has entered the esophagus into the stomach, *secondary peristaltic waves* result from distention of the esophagus by the retained food, and they continue until all the food has emptied into the stomach. These secondary waves are initiated partly by intrinsic neural circuits in the esophageal enteric nervous system and partly by reflexes that are transmitted through *vagal afferent fibers* from the esophagus to the medulla and then back again to the esophagus through *vagal efferent fibers.*

NERVOUS CONTROL OF FOOD MOVEMENT THROUGH THE STOMACH, SMALL INTESTINE, AND COLON

Movement of food through the stomach, small intestine, and colon is caused by various forms of peristaltic propulsive movements. Most of these movements are controlled by the enteric nervous system of the gastrointestinal wall. That is, when a segment of bowel becomes overfilled, the stretch of the nerve endings elicits a local peristaltic reflex, as explained earlier in the chapter, causing propulsion of the food forward.

In general, parasympathetic stimulation by way of the vagi and sacral nerves increases the rate of peristalsis, and sympathetic stimulation inhibits it.

Intrinsic GI Reflexes That Inhibit Rate of Food Movement. At several points in the gastrointestinal tract special reflex mechanisms prevent too rapid movement of food along the gastrointestinal tract. For instance, when the stomach empties too much food into the upper portions of the small intestine, stretch of the intestinal walls transmits signals backwards along the myenteric plexus to the stomach to inhibit its peri-

staltic movements. This obviously allows the small intestine to receive food at a rate slow enough for it to process the food appropriately. Another reflex occurs from the colon to the lower end of the small intestine; when the colon becoms overfilled, myenteric reflex signals inhibit peristalsis in the small intestine and thereby prevent movement of the intestinal contents into the colon at a rate more rapid than these can be processed.

DEFECATION

Most of the time, the rectum is empty of feces. This results partly from the fact that a weak functional sphincter exists approximately 20 cm from the anus at the juncture between the sigmoid and the rectum. There is also a sharp angulation here that contributes additional resistance to filling of the rectum. However, when a mass movement forces feces into the rectum, the desire for defecation is normally initiated, including reflex contraction of the rectum and relaxation of the anal sphincters.

Continual dribble of fecal matter through the anus is prevented by tonic constriction of (1) the *internal anal sphincter,* a thickening of the intestinal circular smooth muscle that lies immediately inside the anus, and (2) the *external anal sphincter,* composed of striated voluntary muscle that both surrounds the internal sphincter and also extends distal to it; the external sphincter is controlled by nerve fibers in the pudendal nerve, which is part of the somatic nervous system and therefore is under *voluntary, conscious control.*

The Defecation Reflexes. Ordinarily, defecation is initiated by *defecation reflexes.* One of these reflexes is an *intrinsic reflex* mediated by the local enteric nervous system. This can be described as follows: When the feces enter the rectum, distention of the rectal wall initiates afferent signals that spread through the *myenteric plexus* to initiate peristaltic waves in the descending colon, sigmoid, and rectum, forcing feces toward the anus. As the peristaltic wave approaches the anus, the internal anal sphincter is relaxed by inhibitory signals from the myenteric plexus; and if the external anal sphincter is voluntarily relaxed at the same time, defecation will occur.

However, the intrinsic defecation reflex itself is weak; and to be effective in causing defecation it usually must be fortified by another type of defecation reflex, a *parasympathetic defecation reflex* that involves the sacral segments of the spinal cord, as illustrated in Figure 28–4. When the nerve endings in the rectum are stimulated, signals are transmitted into the spinal cord and thence, reflexly, back to the descending colon, sigmoid, rectum, and anus by way of parasympathetic nerve fibers in the *pelvic nerves.* These parasympathetic signals greatly intensify the peristaltic waves as well as relaxing the internal anal sphincter and thus convert the intrinsic defecation reflex from an ineffectual weak movement into a powerful process of defecation that is sometimes effective in emptying the

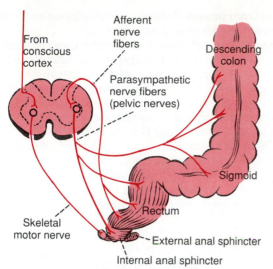

Figure 28–4. The afferent and efferent pathways of the parasympathetic mechanism for enhancing the defecation reflex.

large bowel in one movement all the way from the splenic flexure of the colon to the anus. Also, the afferent signals entering the spinal cord initiate other effects, such as taking a deep breath, closure of the glottis, contraction of the abdominal wall muscles to force the fecal contents of the colon downward, and at the same time cause the pelvic floor to extend downward and pull outward on the anal ring to evaginate the feces.

However, despite the defecation reflexes, other effects are also necessary before actual defecation occurs. In the toilet-trained human being, relaxation of the internal sphincter and forward movement of feces toward the anus normally initiate an instantaneous contraction of the external sphincter, which still temporarily prevents defecation. Except in babies and mentally inept persons, the conscious mind then takes over voluntary control of the external sphincter and either relaxes it to allow defecation to occur or further contracts it if the moment is not socially acceptable for defecation. If the external sphincter is kept contracted, the defecation reflexes die out after a few minutes, and they remain quiescent for several hours or until additional amounts of feces enter the rectum.

When it becomes convenient for the person to defecate, the defecation reflexes can sometimes be excited by taking a deep breath to move the diaphragm downward and then contracting the abdominal muscles to increase the pressure in the abdomen, thus forcing fecal contents into the rectum to elicit new reflexes. Unfortunately, reflexes initiated in this way are almost never as effective as those that arise naturally, for which reason people who too often inhibit their natural reflexes are likely to become severely constipated.

In the newborn baby and in some persons with transected spinal cords, the defecation reflexes cause automatic emptying of the lower bowel without the normal control exercised through contraction of the external anal sphincter.

■ NERVOUS CONTROL OF GASTROINTESTINAL SECRETION

The nervous system of the gastrointestinal tract not only controls peristaltic movement of food but also controls secretion by many of the gastrointestinal glands, especially salivary secretion in the mouth, gastric juice secretion in the stomach, and mucus secretion in the distal colon and in the sigmoid.

Nervous Regulation of Salivary Secretion. Figure 28–5 illustrates the parasympathetic nervous pathways for regulation of salivation, showing that the salivary glands are controlled mainly by *parasympathetic nervous signals* from the *salivatory nuclei.* The salivary nuclei are located approximately at the juncture of the medulla and pons and are excited by both taste and tactile stimuli from the tongue and other areas of the mouth. Many taste stimuli, especially the sour taste, elicit copious secretion of saliva — often as much as 5 to 8 ml/min, or 8 to 20 times the basal rate of secretion. Also, certain tactile stimuli, such as the presence of smooth objects in the mouth (a pebble, for instance), cause marked salivation; whereas rough objects cause less salivation and occasionally even inhibit salivation.

Salivation can also be stimulated or inhibited by impulses arriving in the salivary nuclei from higher centers of the central nervous system. For instance, when a person smells or eats favorite foods, salivation is greater than when disliked food is smelled or eaten. The *appetite area* of the brain, which partially regulates these effects, is located in close proximity to the parasympathetic centers of the anterior hypothalamus, and it functions to a great extent in response to signals from

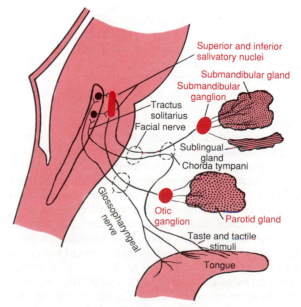

Figure 28–5. Parasympathetic nervous regulation of salivary secretion.

the taste and smell areas of the cerebral cortex or amygdala.

Salivation also occurs in response to reflexes originating in the stomach and upper intestines—particularly when very irritating foods are swallowed or when a person is nauseated because of some gastrointestinal abnormality.

REGULATION OF GASTRIC SECRETION BY NERVOUS AND HORMONAL MECHANISMS

Basic Factors That Stimulate Gastric Secretion: Acetylcholine, Gastrin, and Histamine

The basic neurotransmitters or hormones that directly stimulate secretion by the gastric glands are *acetylcholine, gastrin,* and *histamine.* All of these function by binding first with receptors on the secretory cells. Then the receptors activate the secretory processes. Acetylcholine excites secretion by all of the secretory cell types in the gastric glands, including secretion of *pepsinogen* by the *peptic cells, hydrochloric acid* by the *parietal cells, mucus* by the *mucus cells,* and *gastrin* by the *gastrin cells.* On the other hand, both *gastrin* and *histamine* stimulate very strongly the *secretion of acid* by the parietal cells but have much less effect in stimulating the other cells.

Stimulation of Acid Secretion

Nervous Stimulation. About half of the nerve signals to the stomach that cause gastric secretion originate in the *dorsal motor nuclei of the vagi* and pass via the *vagus nerves* first to the *enteric nervous system* of the stomach wall and then to the gastric glands. The other half of the secretory signals are generated by local reflexes that occur entirely within the enteric nervous system itself. All the secretory nerves release acetylcholine as the neurotransmitter at their endings on the glandular cells, with one exception: for those signals that go to the gastrin-secreting cells in the pyloric glands, an intermediate neuron serves as the final path and secretes *gastrin-releasing peptide (GRP)* as the neurotransmitter.

Nerve stimulation of gastric secretion can be initiated by signals that originate either in the brain, especially in the limbic system, or in the stomach itself. And the stomach-initiated signals can activate two different types of reflexes: (1) *long reflexes* that are transmitted from the stomach mucosa all the way to the brain stem and then back to the stomach through the vagus nerves and (2) *short reflexes* that originate locally and are transmitted entirely through the local enteric nervous system.

The types of stimuli that can initiate the reflexes are (1) distention of the stomach, (2) tactile stimuli on the surface of the stomach mucosa, and (3) chemical stimuli, including especially *amino acids* and *peptides* derived from protein foods or *acid* that has already been secreted by the gastric glands.

Stimulation of Acid Secretion by Gastrin. Both the nerve signals from the vagus nerves and those from the local enteric reflexes, aside from causing direct stimulation of glandular secretion of stomach juices, also cause the mucosa in the stomach antrum to secrete the hormone *gastrin.* This hormone is secreted by the *gastrin cells,* also called *G cells,* in the pyloric glands.

Gastrin is absorbed into the blood and carried to the *oxyntic glands* in the body of the stomach; there it stimulates the *parietal cells* very strongly and perhaps the peptic cells as well, but to a lesser extent. Thus, the really important effect is to increase the rate of hydrochloric acid secretion, often as much as eight times. In turn, the hydrochloric acid excites still additional enteric reflex activity that not only further increases hydrochloric acid secretion but also stimulates secondarily the secretion of enzymes by the peptic cells to increase as much as two to four times.

Role of Histamine in Controlling Gastric Secretion. *Histamine,* an amino acid derivative, also stimulates acid secretion by the *parietal cells.* A small amount of histamine is formed continually in the gastric mucosa, either in response to acid in the stomach or for other reasons. This amount, acting by itself, causes very little acid secretion. However, whenever acetylcholine or gastrin stimulates the parietal cells at the same time, then even the small normal amounts of histamine greatly enhance acid secretion.

The Three Phases of Gastric Secretion

Gastric secretion is said to occur in three separate phases (as illustrated in Fig. 28–6): a *cephalic phase,* a *gastric phase,* and an *intestinal phase.* However, as will be apparent in the following discussion, these three phases in reality fuse together.

The Cephalic Phase. The cephalic phase of gastric secretion occurs even before food enters the stomach or while it is being eaten. It results from the sight, smell, thought, or taste of food; and the greater the appetite, the more intense is the stimulation. Neurogenic signals causing the cephalic phase of secretion can originate in the cerebral cortex or in the appetite centers of the amygdala or hypothalamus. They are transmitted through the dorsal motor nuclei of the vagi to the stomach. This phase of secretion normally accounts for less than one fifth of the gastric secretion associated with eating a meal.

The Gastric Phase. Once the food enters the stomach, it excites the long vagovagal reflexes, the local enteric reflexes, and the gastrin mechanism, which in turn cause secretion of gastric juice that continues throughout the several hours that the food remains in the stomach.

The gastric phase of secretion accounts for at least two thirds of the total gastric secretion associated with eating a meal and, therefore, accounts for most of the total daily gastric secretion of about 1500 mL.

The Intestinal Phase. The presence of food in the upper portion of the small intestine, particularly in the duodenum, can cause the stomach to secrete small amounts of gastric juice, probably partly because of the small amounts of gastrin that are also released by the duodenal mucosa in response to distention or chemical stimuli of the same type as those that stimulate the stomach gastrin mechanism. In ad-

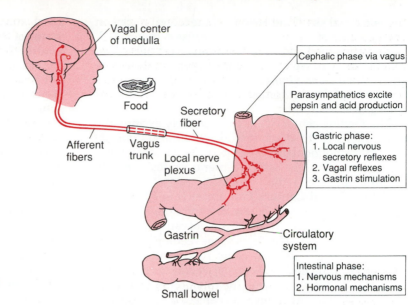

Figure 28–6. The phases of gastric secretion and their regulation.

dition, amino acids absorbed into the blood, as well as several other hormones or reflexes, play minor roles in causing secretion of gastric juice.

However, several intestinal factors can also inhibit gastric secretion, and these often completely override the excitatory factors.

NERVOUS CONTROL OF SECRETION IN THE SMALL AND LARGE INTESTINES

The intestinal mucosa contains several million minute tubular glands each a millimeter or so in length. In addition, mucous cells line the entire inner surface of the intestinal tract from the small intestine to the anus and secrete intestinal mucus. Both the tubular glands and the mucous cells are controlled almost entirely by local control mechanisms within the gastrointestinal tract itself and only to a very slight extent by the parasympathetic and sympathetic nerves.

The mucous cells respond mainly to direct contact with the intestinal contents; the mucus that they secrete serves as a lubricant for movement of the intestinal matter along the intestinal tube. The tubular glands, on the other hand, secrete large amounts of electrolytic solution that provides a vehicle for moving food along the intestinal tract. Also, this fluid serves as a transport medium for absorbing the digestive products from the food. The tubular glands are controlled by both hormonal and nervous stimuli. The hormonal stimuli are similar to the stimulation of the gastric mucosa by gastrin, though the hormones responsible are relatively uncharted. The nervous stimuli control intestinal secretion through local enteric nervous reflexes. As the food passes through the intestines, either contact with the epithelial surfaces or the action of chemical substances from the foods elicits local nerve signals that excite the submucosal and myenteric plexuses, and these in turn stimulate the tubular glands. This enteric reflex mechanism is probably responsible for most of the intestinal secretion.

■ REGULATION OF FOOD INTAKE

Hunger. The term "hunger" means a craving for food, and it is associated with a number of objective sensations. For instance, in a person who has not had food for many hours, the stomach undergoes intense rhythmic contractions called *hunger contractions.* These cause a tight or gnawing feeling in the pit of the stomach and sometimes actually cause pain called *hunger pangs.* However, even after the stomach is completely removed, the psychic sensations of hunger still occur, and craving for food still makes the person search for an adequate food supply.

Appetite. The term "appetite" is often used in the same sense as hunger except that it usually implies desire for specific types of food instead of food in general. Therefore, appetite helps a person choose the quality of food to eat.

Satiety. Satiety is the opposite of hunger. It means a feeling of fulfillment in the quest for food. Satiety usually results from a filling meal, particularly when the person's nutritional storage depots, the adipose tissue and glycogen stores, are already filled.

NEURAL CENTERS FOR REGULATION OF FOOD INTAKE

Hunger and Satiety Centers. Stimulation of the *lateral hypothalamus* causes an animal to eat voraciously, which is called *hyperphagia.* On the other hand, stimulation of the *ventromedial nuclei of the hypothalamus* causes complete satiety; and even in the presence of highly appetizing food, the animal will still refuse to eat, which is *aphagia.* Conversely, destructive lesions of the two areas cause results exactly opposite to those caused by stimulation. That is, ventromedial lesions cause voracious and continued eating until the animal becomes extremely obese, sometimes as large as four times normal size. And lesions of the lateral nuclei on the two sides of the hypothalamus cause complete lack of desire for food and progressive inanition of the animal. Therefore, we can label the lateral nuclei of the hypothalamus as a *hunger center* or *feeding center,* and we can label the ventromedial nuclei of the hypothalamus as a *satiety center.*

The feeding center operates by directly exciting the emotional drive to search for food (while also stimulating other emotional drives as well). On the other hand, it is believed that the satiety center operates primarily by inhibiting the feeding center.

Other Neural Centers That Enter Into Feeding. If the brain is sectioned below the hypothalamus but above the mesencephalon, the animal can still perform the basic mechanical features of the feeding process. It can salivate, lick its lips, chew food, and swallow. Therefore, *the actual mechanics of feeding are controlled by centers in the brain stem.* The function of the hypothalamus in feeding, then, is to control the quantity of food intake and to excite the lower centers to activity.

Higher centers than the hypothalamus also play important roles in the control of feeding, particularly in the control of appetite. These centers include especially the *amygdala* and the *prefrontal cortex,* which are closely coupled with the hypothalamus. It will be recalled from the discussion of the sense of smell that portions of the amygdala are a major part of the olfactory nervous system. Destructive lesions in the amygdala have demonstrated that some of its areas greatly increase feeding, whereas others inhibit feeding. In addition, stimulation of some areas of the amygdala elicits the mechanical act of feeding. However, the most important effect of destruction of the amygdala on both sides of the brain is a "psychic blindness" in the choice of foods. In other words, the animal (and presumably the human being as well) loses or at least partially loses the mechanism of appetite control of the type and quality of food that it eats.

FACTORS THAT REGULATE FOOD INTAKE

We can divide the regulation of food into (1) *nutritional regulation* (or *long-term regulation*), which is concerned primarily with long-term maintenance of normal quantities of nutrient stores in the body, and (2) *alimentary regulation* (or *short-term regulation*), which is concerned primarily with preventing overeating at the time of each meal.

Nutritional Regulation (Long-Term Regulation)

An animal that has been starved for a long time and is then presented with unlimited food eats a far greater quantity than does an animal that has been on a regular diet. Conversely, an animal that has been force-fed for several weeks eats little when allowed to eat according to its own desires. Thus, the feeding control

mechanism of the body is geared to the nutritional status of the body. Some of the nutritional factors that control the degree of activity of the feeding center are the following:

Effect of Blood Concentrations of Glucose, Amino Acids, and Lipids on Hunger and Feeding — The Glucostatic, Aminostatic, and Lipostatic Theories. It has long been known that a decrease in blood glucose concentration causes hunger, which has led to the so-called *glucostatic theory of hunger and feeding regulation.* Similar studies have more recently demonstrated the same effect for blood amino acid concentration and blood concentration of breakdown products of lipids such as the keto acids and some fatty acids, leading to the *aminostatic* and *lipostatic* theories of regulation. That is, when the availability of any of the three major types of food decreases, the animal automatically increases its feeding, which eventually returns the blood metabolite concentrations back toward normal.

Neurophysiological studies of function in the hypothalamus have also substantiated the glucostatic, aminostatic, and lipostatic theories by the following observations: (1) An increase in blood glucose level *increases* the rate of firing of *glucoreceptor neurons* in the *satiety center in the ventromedial nucleus of the hypothalamus.* (2) The same increase in blood glucose level simultaneously *decreases* the firing of neurons called *glucosensitive neurons* in the *hunger center of the lateral hypothalamus.* In addition, some amino acids and lipid substances also affect the rates of firing of these same neurons.

Still other neurons, found in the *dorsomedial nuclei of the hypothalamus,* respond to the rate of utilization of all the foodstuffs that provide energy for the cells. This has led to a more global theory of hunger and feeding regulation based on *power generation* within these cells.

Summary of Long-Term Regulation. Even though our information on the different feedback factors in long-term feeding regulation is imprecise, we can make the following general statement: When the nutrient stores of the body fall below normal, the feeding center of the hypothalamus becomes highly active, and the person exhibits increased hunger; on the other hand, when the nutrient stores are abundant, the person loses the hunger and develops a state of satiety.

Alimentary Regulation of Feeding (Short-Term Regulation)

When a person is driven by hunger to eat, what turns off the eating when he or she has eaten enough? It is not the nutritional feedback mechanisms that we have discussed above, because all of them take an hour to several hours before enough quantities of the nutritional factors are absorbed into the blood to cause the necessary inhibition of eating. Yet it is very important that the person not overeat and even that he eat an amount of food that approximates his nutritional needs. The following are several different types of signals that are important for this purpose:

Gastrointestinal Filling. When the gastrointestinal tract becomes distended, especially the stomach and the duodenum, inhibitory signals are transmitted mainly by way of the vagi to suppress the feeding center, thereby reducing the desire for food.

Humoral and Hormonal Factors That Suppress Feeding — Cholecystokinin, Glucagon, and Insulin. The gastrointestinal hormone *cholecystokinin,* released mainly in response to fat entering the duodenum, has a strong direct effect on the feeding center to reduce further eating. In addition, for reasons that are not entirely understood, the presence of food in the stomach and duodenum causes the pancreas to secrete significant quantities of both *glucagon* and *insulin,* both of which also suppress the hypothalamic feeding center.

Metering of Food by Oral Receptors. When a person with an esophageal fistula is fed large quantities of food, even though this food is immediately lost again to the exterior, the degree of hunger is decreased after a reasonable quantity of food has passed through the mouth. This effect occurs despite the fact that the gastrointestinal tract does not become the least bit filled. Therefore, it is postulated that various "oral factors" relating to feeding, such as chewing, salivation, swallowing, and tasting, "meter" the food as it passes through the mouth, and after a certain amount has passed, the hypothalamic feeding center becomes inhibited. However, the inhibition caused by this metering mechanism is considerably less intense and less lasting, usually lasting only 20 to 40 min, than is the inhibition caused by gastrointestinal filling.

Importance of Having Both Long- and Short-Term Regulatory Systems for Feeding

The long-term regulatory system for feeding, which includes all the metabolite feedback mechanisms, obviously helps to maintain constant stores of nutrients in the tissues, preventing these from becoming too low or becoming too high. The short-term regulatory stimuli, on the other hand, serve two other purposes. First, they make the person or animal eat smaller quantities at a time, thus allowing food to pass through the gastrointestinal tract at a steadier pace so that its digestive and absorptive mechanisms can work at more optimal rates rather than becoming excessively overburdened only when the animal needs food. Second, they prevent the person or animal from eating amounts of food at each meal that would be too much for the metabolic storage systems once the food has all been absorbed.

■ OBESITY

Energy Input Versus Energy Output. When greater quantities of energy (in the form of food) enter the body than are expended, the body weight increases. Therefore, obesity is obviously caused by excess energy input over energy output. For each 9.3 Cal excess energy entering the body, 1 g fat is stored.

Excess energy input occurs *only during the developing phase of obesity.* Once a person has become obese, all that is required to remain obese is that the energy input equal the energy output. For the person to reduce in weight, the input must be *less* than the output. Indeed, studies of obese persons have shown that the intake of food of most of them in the static stage of obesity (after the obesity has already been attained) is approximately the same as that for normal persons.

Effect of Muscular Activity on Energy Output. About one third of the energy used each day by the normal person goes into muscular activity, and in the laborer as much as two thirds or occasionally three fourths is used in this way. Because muscular activity is by far the most important means by which energy is expended in the body, it is frequently said that obesity in the otherwise normal person results from *too high a ratio of food intake to daily exercise.*

ABNORMAL FEEDING REGULATION AS A PATHOLOGICAL CAUSE OF OBESITY

We have already emphasized that the rate of feeding is normally regulated in proportion to the nutrient stores in the body. When these stores begin to approach an optimal level in a normal person, feeding is automatically reduced to prevent overstorage. However, in many obese persons this is not true, for feeding does not slacken until body weight is far above normal. Therefore, in effect, obesity is often caused by an abnormality of the feeding regulatory mechanism. This can result from either psychogenic factors that affect the regulation or actual abnormalities of the hypothalamus itself.

Psychogenic Obesity. Studies of obese patients show that a large proportion of obesity results from psychogenic factors. Perhaps the most common psychogenic factor contributing to obesity is the prevalent idea that healthy eating habits require three meals a day and that each meal must be filling. Many young children are forced into this habit by overly solicitous parents, and the children continue to practice it throughout life. In addition, persons are known often to gain large amounts of weight during or after stressful situations, such as the death of a parent, a severe illness, or even mental depression. It seems that eating is often a means of release from tension.

Hypothalamic Abnormalities as a Cause of Obesity. In the preceding discussion of feeding regulation, it was pointed out that lesions in the ventromedial nuclei of the hypothalamus cause an animal to eat excessively and become obese. It has also been discovered that such lesions cause excess insulin production, which in turn increases fat deposition. In addition, many persons with hypophysial tumors that encroach on the hypothalamus develop progressive obesity, illustrating that obesity in the human being, too, can definitely result from damage to the hypothalamus.

Yet in the normal obese person hypothalamic damage is almost never found. Nevertheless, it is possible that the functional organization of the feeding center is different in the obese person from that of the nonobese person. For instance, a normally obese person who has reduced to normal weight by strict dietary measures usually develops intense hunger that is demonstrably far greater than that of the normal person. This indicates that the "set-point" of the obese person's feeding center is at a much higher level of nutrient storage than that of the normal person.

Genetic Factors in Obesity. Obesity definitely runs in families. Furthermore, identical twins usually maintain weight levels within 2 lb of each other throughout life if they live under similar conditions, or within 5 lb of each other if their conditions of life differ markedly. This might result partly from eating habits engendered during childhood, but it is generally believed that this close similarity between twins is genetically controlled.

The genes can direct the degree of feeding in several different ways, including (1) a genetic abnormality of the feeding center to set the level of nutrient storage high or low and (2) abnormal hereditary psychic factors that either whet the appetite or cause the person to eat as a "release" mechanism.

Genetic abnormalities in the *chemistry of fat storage* are also known to cause obesity in certain strains of rats and mice. In one strain of rats, fat is easily stored in the adipose tissue, but the quantity of hormone-sensitive lipase in the adipose tissue is greatly reduced, so that little of the fat can be removed. This obviously results in a one-way path, the fat continually being deposited but never being released. In a strain of obese mice that has been studied, there is an excess of fatty acid synthetase, which causes excess synthesis of fatty acids. Therefore, similar genetic mechanisms are possible causes of obesity in human beings.

■ MICTURITION

Micturition is the process by which the urinary bladder empties when it becomes filled. Basically the bladder (1) progressively fills until the tension in its walls rises above a threshold value, at which time (2) a nervous reflex called the "micturition reflex" occurs that either causes micturition or, if it fails in this, at least causes a conscious desire to urinate.

PHYSIOLOGIC ANATOMY AND NERVOUS CONNECTIONS OF THE BLADDER

The urinary bladder, illustrated in Figure 28–7, is a smooth muscle chamber composed of two principal

parts: (1) the *body*, which is the major part of the bladder, in which the urine collects, and (2) the *neck*, which is a funnel-shaped extension of the body, passing inferiorly and anteriorly into the urogenital triangle and connecting with the urethra. The lower part of the bladder neck is also called the *posterior urethra* because of its relationship to the urethra.

The smooth muscle of the bladder is known as the *detrusor muscle*. Its muscle fibers extend in all directions and, when contracted, can increase the pressure in the bladder sometimes to as high as 40 to 60 mm Hg. Thus, it is the detrusor muscle that empties the bladder. The smooth muscle cells of the detrusor muscle fuse with each other so that low resistance electrical pathways exist from one to the other. Therefore, an action potential can spread throughout the detrusor muscle to cause contraction of the entire bladder at once.

On the posterior wall of the bladder, lying immediately above the bladder neck, is a small triangular area called the *trigone*. At the lowermost apex of the trigone is the opening of the bladder through the *bladder neck* into the *posterior urethra,* and the two ureters enter the bladder at the uppermost angles of the trigone. The trigone can be identified by the fact that its mucosa is very smooth, in contrast to the remainder of the bladder mucosa, which is folded to form *rugae*. Where each ureter enters the bladder, it courses obliquely through the detrusor muscle and then passes still another 1 to 2 cm underneath the bladder mucosa before emptying into the bladder.

The bladder neck (posterior urethra) is 2 to 3 cm long, and its wall is composed of detrusor muscle interlaced among a large amount of elastic tissue. The muscle in this area is frequently called the *internal sphincter*. Its natural tone normally keeps the bladder neck and posterior urethra empty of urine, and therefore prevents emptying of the bladder until the pres-

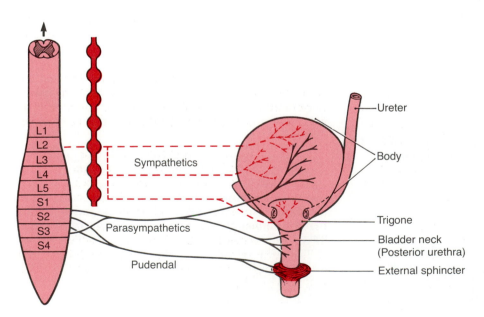

Figure 28–7. The urinary bladder and its innervation.

sure in the body of the bladder rises above a critical threshold.

Beyond the posterior urethra, the urethra passes through the *urogenital diaphragm,* which contains a layer of muscle called the *external sphincter* of the bladder. This muscle is a voluntary skeletal muscle in contrast to the muscle of the bladder body and bladder neck, which is entirely smooth muscle. This external muscle is under voluntary control of the nervous system and can be used to prevent urination even when the involuntary controls are attempting to empty the bladder.

Innervation of the Bladder. The principal nerve supply to the bladder is by way of the *pelvic nerves,* which connect with the spinal cord through the sacral plexus, mainly connecting with cord segments S-2 and S-3. Coursing through the pelvic nerves are both *sensory nerve fibers* and *motor fibers.* The sensory fibers mainly detect the degree of stretch of the bladder wall. Stretch signals from the posterior urethra are especially strong and are mainly responsible for initiating the reflexes that cause bladder emptying.

The motor nerve fibers transmitted in the pelvic nerves are *parasympathetic fibers.* These terminate on ganglion cells located in the wall of the bladder. Short postganglionic nerves then innervate the *detrusor muscle.*

Aside from the pelvic nerves, two other types of innervation are important to bladder function. Most important are the skeletal motor fibers transmitted through the *pudendal nerve* to the *external bladder sphincter.* These are *somatic* nerve fibers that innervate and control the voluntary skeletal muscle of this sphincter. In addition, the bladder receives sympathetic innervation from the sympathetic chain through the hypogastric nerves, connecting mainly with the L-2 segment of the spinal cord. These sympathetic fibers probably stimulate mainly the blood vessels and have very little to do with bladder contraction. Some sensory nerve fibers also pass by way of the sympathetic nerves and may be important for the sensation of fullness and pain in some instances.

THE MICTURITION REFLEX

As the bladder fills, many *micturition contractions* begin to appear. These are the result of a stretch reflex initiated by sensory stretch receptors in the bladder wall, especially by the receptors in the posterior urethra when it begins to fill with urine at the higher bladder pressures. *Sensory signals* are conducted to the sacral segments of the cord through the *pelvic nerves* and then back again to the bladder through the *parasympathetic fibers* in these same nerves.

Once a micturition reflex begins, it is "self-regenerative." That is, initial contraction of the bladder further activates the receptors to cause still further increase in sensory impulses from the bladder and posterior urethra, which causes further increase in reflex contraction of the bladder, the cycle thus repeating itself again and again until the bladder has reached a strong degree of contraction. Then, after a few seconds to more than a minute, the reflex begins to fatigue, and the regenerative cycle of the micturition reflex ceases, allowing rapid reduction in bladder contraction. In other words, the micturition reflex is a single complete cycle of (1) progressive and rapid increase in pressure, (2) a period of sustained pressure, and (3) return of the pressure to the basal tonic pressure of the bladder. Once a micturition reflex has occurred but has not succeeded in emptying the bladder, the nervous elements of this reflex usually remain in an inhibited state for at least a few minutes to sometimes as long as an hour or more before another micturition reflex occurs. However, as the bladder becomes more and more filled, micturition reflexes occur more and more often and more and more powerfully.

Once the micturition reflex becomes powerful enough, this causes still another reflex, which passes through the *pudendal nerves* to the *external sphincter* to inhibit it. If this inhibition is more potent than the voluntary constrictor signals to the external sphincter from the brain, urination will occur. If not, urination will not occur until the bladder fills still more and the micturition reflex becomes more powerful.

Control of Micturition by the Brain. The micturition reflex is a completely automatic cord reflex, but it can be inhibited or facilitated by centers in the brain. These include (1) strong *facilitatory and inhibitory centers in the brain stem,* probably located in the pons, and (2) several *centers located in the cerebral cortex* that are mainly inhibitory but can at times become excitatory.

The micturition reflex is the basic cause of micturition, but the higher centers normally exert final control of micturition by the following means:

1. The higher centers keep the micturition reflex partially inhibited all the time except when micturition is desired.

2. The higher centers prevent micturition, even if a micturition reflex occurs, by continual tonic contraction of the external bladder sphincter until a convenient time presents itself.

3. When the time to urinate arrives, the cortical centers can (a) facilitate the sacral micturition centers to help initiate a micturition reflex and (b) inhibit the external urinary sphincter so that urination can occur.

However, even more important, voluntary urination is usually initiated in the following way: First, the person contracts his abdominal muscles, which increases the pressure of the urine in the bladder and allows extra urine to enter the bladder neck and posterior urethra under pressure, thus stretching their walls. This then excites the stretch receptors, which excites the micturition reflex and simultaneously inhibits the external urethral sphincter. Ordinarily, all the urine will then be emptied, with rarely more than 5 to 10 mL left in the bladder.

ABNORMALITIES OF MICTURITION

The Atonic Bladder. Destruction of the sensory nerve fibers from the bladder to the spinal cord prevents transmission of stretch signals from the bladder and, therefore, also prevents micturition reflex contractions. Therefore, the person loses all bladder control despite intact efferent fibers from the cord to the bladder and despite intact neurogenic connections with the brain. Instead of emptying periodically, the bladder fills to capacity and overflows a few drops at a time through the urethra. This is called *outflow incontinence,* or simply *overflow dribbling.*

The atonic bladder was a common occurrence when syphilis was widespread, because syphilis frequently causes constrictive fibrosis around the dorsal nerve root fibers where they enter the spinal cord and, subsequently, destroys these fibers. This condition is called *tabes dorsalis,* and the resulting bladder condition is called a *tabetic bladder.* Another common cause of this condition is crushing injuries to the sacral region of the cord.

The Automatic Bladder. If the spinal cord is damaged above the sacral region but the sacral segments are still intact, typical micturition reflexes still occur. However, they are no longer controllable by the brain. During the first few days to several weeks after the damage to the cord has occurred, the micturition reflexes are completely suppressed because of the state of "spinal shock" caused by the sudden loss of facilitatory impulses from the brain stem and cerebrum. However, if the bladder is emptied periodically by catheterization to prevent physical bladder injury, the excitability of the micturition reflex gradually increases until typical micturition reflexes return.

It is especially interesting that stimulating the skin in the genital region can sometimes elicit a micturition reflex in this condition, thus providing a means by which some patients can still control urination.

■ REGULATION OF BODY TEMPERATURE—ROLE OF THE HYPOTHALAMUS

Figure 28–8 illustrates approximately what happens to the temperature of the nude body after a few hours' exposure to *dry* air ranging from 30 to 170°F. Obviously, the precise dimensions of this curve depend on the movement of the air, the amount of moisture in the air, and even the nature of the surroundings. However, in general, between approximately 60° and 130°F in dry air the nude body is capable of maintaining a normal body core temperature somewhere between 97° and 100°F.

The temperature of the body is regulated almost entirely by nervous feedback mechanisms, and almost all of these operate through *temperature-regulating centers* located in the *hypothalamus.* However, for these feedback mechanisms to operate, there must also

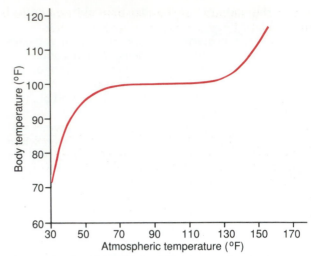

Figure 28–8. Effect of high and low atmospheric temperature for several hours' duration on the internal body temperature, showing that the internal body temperature remains stable despite wide changes in atmospheric temperature.

exist temperature detectors, dicussed next, to determine when the body temperature becomes either too hot or too cold.

THERMOSTATIC DETECTION OF TEMPERATURE IN THE HYPOTHALAMUS—ROLE OF THE ANTERIOR HYPOTHALAMUS-PREOPTIC AREA

In recent years, experiments have been performed in which minute areas in the brain have been either heated or cooled by use of a so-called *thermode.* This small, needle-like device is heated by electrical means or by passing hot water through it, or it is cooled by cold water. The principal area in the brain in which heat from a thermode affects body temperature control consists of the preoptic and anterior hypothalamic nuclei of the hypothalamus.

Using the thermode, the anterior hypothalamic-preoptic area has been found to contain large numbers of heat-sensitive neurons and about a third as many cold-sensitive neurons that seem to function as temperature sensors for controlling body temperature. The heat-sensitive neurons increase their firing rate as the temperature rises, twofold to tenfold with an increase in body temperature of 10°C. The cold-sensitive neurons, by contrast, increase their firing rate when body temperature falls.

When the preoptic area is heated, the skin immediately breaks out into a profuse sweat while at the same time the skin blood vessels over the entire body become greatly vasodilated. Thus, this is an immediate reaction to cause the body to lose heat, thereby helping to return the body temperature toward the normal level. In addition, excess body heat production is inhibited. Therefore, it is clear that the preoptic area of the hypothalamus has the capability of serving as a thermostatic body temperature control center.

DETECTION OF TEMPERATURE BY RECEPTORS IN THE SKIN AND DEEP BODY TISSUES

Though the signals generated by the temperature receptors of the hypothalamus are extremely powerful in controlling body temperature, receptors in other parts of the body also play important roles in temperature regulation. This is especially true of temperature receptors in the skin and in a few specific deep tissues of the body.

The skin is endowed with both *cold* and *warmth* receptors. However, there are far more cold receptors than warmth receptors; in fact, ten times as many in many parts of the skin. Therefore, peripheral detection of temperature mainly concerns detecting cool and cold instead of warm temperatures.

When the skin is chilled over the entire body, immediate reflex effects are invoked to increase the temperature of the body in several ways: (1) by providing a strong stimulus to cause shivering, with resultant increase in the rate of body heat production; (2) by inhibiting the process of sweating if this should be occurring; and (3) by promoting skin vasoconstriction to diminish the transfer of body heat to the skin.

ROLE OF THE POSTERIOR HYPOTHALAMUS IN INTEGRATING PERIPHERAL AND CENTRAL TEMPERATURE SIGNALS

Even though a large share of the signals for temperature detection arise in peripheral receptors, these signals help control body temperature mainly through the hypothalamus. However, the area of the hypothalamus that they stimulate is not the anterior hypothalamus-preoptic area but instead an area located bilaterally in the posterior hypothalamus approximately at the level of the mammary bodies. The thermostatic signals from the anterior hypothalamus–preoptic area are also transmitted into this posterior hypothalamus area. Here the signals from the preoptic area and the signals from the body periphery are combined to provide mainly the heat-producing and heat-conserving reactions of the body.

NEURONAL EFFECTOR MECHANISMS THAT DECREASE OR INCREASE BODY TEMPERATURE

When the hypothalamic temperature centers detect that the body temperature is either too hot or too cold, they institute appropriate temperature-decreasing or temperature-increasing procedures. The student is familiar with most of these from personal experience, but special features are the following:

Temperature-Decreasing Mechanisms When the Body Is Too Hot

The temperature control system employs three important mechanisms to reduce body heat when the body temperature becomes too great:

1. *Vasodilatation.* In almost all areas of the body the skin blood vessels become intensely dilated. This is caused by *inhibition of the sympathetic centers in the posterior hypothalamus that cause vasoconstriction.* Full vasodilatation can increase the rate of heat transfer to the skin as much as eightfold.

2. *Sweating.* The effect of increased temperature on causing sweating is illustrated by the solid curve in Figure 28–9, which shows a sharp increase in the rate of evaporative heat loss resulting from sweating when the body core temperature rises above the critical temperature level of 37°C (98.6° F). An additional 1°C increase in body temperature causes enough sweating to remove 10 times the basal rate of body heat production.

3. *Decrease in heat production.* Those mechanisms that cause excess heat production, such as shivering and chemical thermogenesis, are strongly inhibited.

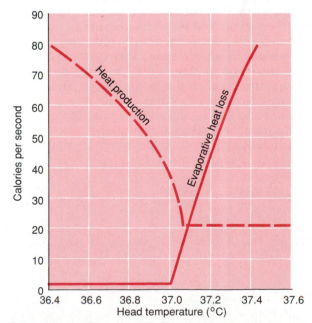

Figure 28–9. Effect of hypothalamic temperature on (1) evaporative heat loss from the body and (2) heat production caused primarily by muscular activity and shivering. This figure demonstrates the extremely critical temperature level at which increased heat loss begins and increased heat production stops. (Drawn from data in Benzinger, Kitzinger, and Pratt, in Hardy [ed.]: Temperature. Part 3. Reinhold Publishing Corp., p. 637.)

Temperature-Increasing Mechanisms When the Body Is Too Cold

When the body is too cold, the temperature control system institutes exactly opposite procedures. These are:

1. *Skin vasoconstriction throughout the body.* This is caused by stimulation of the posterior hypothalamic sympathetic centers.

2. *Piloerection.* Piloerection means hairs "standing on end." Sympathetic stimulation causes the arrector pili muscles attached to the hair follicles to contract, which brings the hairs to an upright stance. This is not important in the human being, but in lower animals upright projection of the hairs allows them to entrap a thick layer of "insulator air" next to the skin so that the transfer of heat to the surroundings is greatly depressed.

3. *Increase in heat production.* Heat production by the metabolic systems is increased by promoting (1) shivering, (2) sympathetic excitation of heat production, and (3) thyroxine secretion. These require additional explanation, as follows:

Hypothalamic Stimulation of Shivering. Located in the dorsomedial portion of the posterior hypothalamus near the wall of the third ventricle is an area called the *primary motor center for shivering.* This area is normally inhibited by signals from the heat center in the anterior hypothalamic-preoptic area but is excited by cold signals from the skin and spinal cord. Therefore, as illustrated by the dashed curve in Figure 28–9, this center becomes activated when the body temperature falls even a fraction of a degree below a critical temperature level. It then transmits signals that cause shivering through bilateral tracts down the brain stem, into the lateral columns of the spinal cord, and, finally, to the anterior motor neurons. These signals are nonrhythmic and do not cause the actual muscle shaking. Instead, they increase the tone of the skeletal muscles throughout the body. When the tone rises above a certain critical level, shivering begins. This probably results from feedback oscillation of the muscle spindle stretch reflex mechanism discussed in chapter 16. During maximum shivering, body heat production can rise to as high as four to five times normal.

Sympathetic "Chemical" Excitation of Heat Production. Either sympathetic stimulation or circulating norepinephrine and epinephrine in the blood can cause an immediate increase in the rate of cellular metabolism; this effect is called *chemical thermogenesis,* and it results at least partially from the ability of norepinephrine and epinephrine to *uncouple* oxidative phosphorylation, which means that excess foodstuffs are oxidized and thereby release energy in the form of heat but do not cause adenosine triphosphate to be formed.

The degree of chemical thermogenesis that occurs in an animal is almost directly proportional to the amount of *brown* fat that exists in the animal's tissues. This is a type of fat that contains large numbers of special mitochondria where the uncoupled oxidation occurs; and these cells are supplied by a strong sympathetic innervation.

The process of acclimatization greatly affects the intensity of chemical thermogenesis; some animals that have been exposed for several weeks to a very cold environment exhibit as much as a 100 to 500 per cent increase in heat production when acutely exposed to cold, in contrast to the unacclimatized animal, which responds with an increase of perhaps one third as much.

In adult human beings, who have almost no brown fat, it is rare that chemical thermogenesis increases the rate of heat production more than 10 to 15 per cent. However, in infants, who *do* have a small amount of brown fat in the interscapular space, chemical thermogenesis can increase the rate of heat production as much as 100 per cent, which is probably a very important factor in maintaining normal body temperature in the neonate.

Increased Thyroxine Output as a Cause of Increased Heat Production. Cooling the anterior hypothalamic-preoptic area of the hypothalamus also increases the production of the neurosecretory hormone *thyrotropin-releasing hormone* by the hypothalamus. This hormone is carried by way of the hypothalamic portal veins to the anterior pituitary gland, where it stimulates the secretion of *thyroid-stimulating hormone.* Thyroid-stimulating hormone, in turn, stimulates increased output of *thyroxine* by the thyroid gland. The increased thyroxine increases the rate of cellular metabolism throughout the body, which is yet another mechanism of *chemical thermogenesis.* However, this increase in metabolism does not occur immediately but requires several weeks for the thyroid gland to hypertrophy before it reaches its new level of thyroxine secretion.

Exposure of animals to extreme cold for several weeks can cause their thyroid glands to increase in size as much as 20 to 40 per cent. However, human beings rarely allow themselves to be exposed to the same degree of cold as that to which animals have been subjected. Therefore, we still do not know, quantitatively, how important the thyroid method of adaptation to cold is in the human being. Yet isolated measurements have shown that military personnel residing for several months in the Arctic develop increased metabolic rates; Eskimos also have abnormally high basal metabolic rates. Also, the continuous stimulatory effect of cold on the thyroid gland can probably explain the much higher incidence of toxic thyroid goiters in persons living in colder climates than in those living in warmer climates.

THE CONCEPT OF A "SET-POINT" FOR TEMPERATURE CONTROL

In the example of Figure 28–9, it is clear that at a very critical body core temperature, at a level of almost exactly 37.1°C, drastic changes occur in the rates of both heat loss and heat production. At temperatures above this level, the rate of heat loss is greater than that of heat production, so that the body temperature falls and reapproaches the 37.1°C level. At temperatures below this level, the rate of heat production is greater than heat loss, so that now the body temperature rises and again approaches the 37.1°C level. Therefore, this critical temperature level is called the "set-point" of the temperature control mechanism. That is, all the temperature control mechanisms continually attempt to bring the body temperature back to this set-point level.

The Feedback Gain for Body Temperature Control. Feedback gain is a measure of the effectiveness of a control system. In the case of body temperature control, it is important for the internal core temperature to change as little as possible despite marked changes in the environmental temperature, and the gain of the temperature control system is approximately equal to the ratio of the change in environmental temperature to the change in body temperature that this causes. Experiments have shown that the body temperature of man changes about 1°C for each 25 to 30°C change in environmental temperature. Therefore, the feedback gain of the total mechanism for control of body temperature averages about 27, which is an extremely high gain for a biological control system (the baroreceptor arterial pressure control system, for instance, has a gain of less than 2).

■ ABNORMALITIES OF BODY TEMPERATURE REGULATION

FEVER

Fever, which means a body temperature above the usual range of normal, may be caused by abnormalities in the brain itself or by toxic substances that affect the temperature-regulating centers. The causes of fever include bacterial diseases, brain tumors, and environmental conditions that may terminate in heat stroke.

RESETTING THE HYPOTHALAMIC TEMPERATURE REGULATING CENTER IN FEBRILE DISEASES — EFFECT OF PYROGENS

Many proteins, breakdown products of proteins, and certain other substances, especially lipopolysaccharide toxins secreted by bacteria, can cause the set-point of the hypothalamic thermostat to rise. Substances that cause this effect are called *pyrogens*. It is pyrogens secreted by toxic bacteria or pyrogens released from degenerating tissues of the body that cause fever during disease conditions. When the set-point of the hypothalamic temperature regulating center becomes increased to a higher level than normal, all the mechanisms for raising the body temperature are brought into play, including heat conservation and increased heat production. Within a few hours after the set-point

has been increased to a higher level, the body temperature also approaches this level.

Mechanism of Action of Pyrogens in Causing Fever — Role of Interleukin-1. Experiments in animals have shown that some pyrogens, when injected into the hypothalamus, can act directly on the hypothalamic temperature regulating center to increase its set-point, though still other pyrogens function indirectly and also may require several hours of latency before causing their effects. This is true of many of the bacterial pyrogens, especially the *endotoxins* from gram-negative bacteria, as follows:

When bacteria or breakdown products of bacteria are present in the tissues or in the blood, these are *phagocytized by the blood leukocytes, the tissue macrophages,* and the *large granular killer lymphocytes.* All of these cells in turn digest the bacterial products and then release into the body fluids the *substance interleukin-1,* which is also called *leukocyte pyrogen* or *endogenous pyrogen.* The interleukin-1, on reaching the hypothalamus, immediately produces fever, increasing the body temperature in as little as 8 to 10 min. As little as one ten millionth of a gram of endotoxin lipopolysaccharide acting in this manner in concert with the blood leukocytes, tissue macrophages, and killer lymphocytes can cause fever. The amount of interleukin-1 that is formed in response to the lipopolysaccharide to cause the fever is only a few nanograms.

Several recent experiments have suggested that interleukin-1 causes fever by first inducing the formation of one of the prostaglandins or a similar substance and this in turn acting in the hypothalamus to elicit the fever reaction. When prostaglandin formation is blocked by drugs, the fever is either completely abrogated or at least reduced. In fact, this may be the explanation for the manner in which aspirin reduces the degree of fever, because aspirin impedes the formation of prostaglandins from arachidonic acid. It also would explain why aspirin does not lower the body temperature in a normal person, because a normal person does not have any interleukin-1. Drugs such as aspirin that reduce the level of fever are called *antipyretics.*

Fever Caused by Brain Lesions. When a brain surgeon operates in the region of the hypothalamus, severe fever almost always occurs; rarely, however, the opposite effect occurs, thus illustrating both the potency of the hypothalamic mechanisms for body temperature control and also the ease with which abnormalities of the hypothalamus can alter the set-point of temperature control. Another condition that frequently causes prolonged high temperature is compression of the hypothalamus by brain tumors.

REFERENCES

GASTROINTESTINAL MOTILITY

Berk, J. E., et al.: Bockus Gastroenterology, 4th Ed. Philadelphia, W. B. Saunders Co., 1985.
Hunt, J. N.: Mechanisms and disorders of gastric emptying. Annu. Rev. Med., 34:219, 1983.
Johnson, L. R., et al.: Physiology of the Gastrointestinal Tract, 2nd Ed. New York, Raven Press, 1987.

Klimov, P. K.: Behavior of the organs of the digestive system. Neurosci. Behav. Physiol., 14:333, 1984.

Luschei, E. S., and Goldberg, L. J.: Neural mechanisms of mandibular control: Mastication and voluntary biting. In Brooks, V. B. (ed.): Handbook of Physiology. Sec. 1, Vol. II. Bethesda, Md., American Physiological Society, 1981, p. 1237.

Magee, D. F.: Interdigestive activity in the gastrointestinal tract. News in Physiol. Sci., 2:101, 1987.

Mei, N.: Intestinal chemosensitivity. Physiol. Rev., 65:211, 1985.

Miller, A. J.: Deglutition. Physiol. Rev., 62:129, 1982.

Murphy, R. A.: Muscle cells of hollow organs. News Physiol. Sci., 3:124, 1988.

Sternini, C.: Structural and chemical organization of the myenteric plexus. Annu. Rev. Physiol., 50:81, 1988.

Thompson, J. C., et al. (eds.): Gastrointestinal Endocrinology. New York, McGraw-Hill Book Co., 1987.

Weems, W. A.: The intestine as a fluid propelling system. Annu. Rev. Physiol., 43:9, 1981.

Weisbrodt, N. W.: Patterns of intestinal motility. Annu. Rev. Physiol., 43:21, 1981.

GASTROINTESTINAL SECRETION

Burnham, D. B., and Williams, J. A.: Stimulus-secretion coupling in pancreatic acinar cells. J. Pediatr. Gastroenterol. Nutr., 3 (Suppl. 1):S1, 1984.

Cheli, R.: Gastric Secretion: A Physiological and Pharmacological Approach. New York, Raven Press, 1986.

Cooke, H. J.: Role of the "little brain" in the gut in water and electrolyte homeostasis. FASEB J., 3:127, 1989.

Fushiki, T., and Iwai, K.: Two hypotheses on the feedback regulation of pancreatic enzyme secretion. FASEB J., 3:121, 1989.

Johnson, L. R., et al.: Physiology of the Gastrointestinal Tract, 2nd Ed. New York, Raven Press, 1987.

Machen, T. E., and Paradiso, A. M.: Regulation of intracellular pH in the stomach. Annu. Rev. Physiol., 49:19, 1987.

Petersen, O. H., and Findlay, I.: Electrophysiology of the pancreas. Physiol. Rev., 67:1054, 1987.

Putney, J. W., Jr.: Identification of cellular activation mechanisms associated with salivary secretion. Annu. Rev. Physiol., 48:75, 1986.

Streebny, L. M.: The Salivary System. Boca Raton, CRC Press Inc., 1987.

Tache, Y.: CNS peptides and regulation gastric acid secretion. Annu. Rev. Physiol., 50:19, 1988.

Thompson, J. C., et al.: Gastrointestinal Endocrinology. New York, McGraw-Hill Book Co., 1987.

Walsh, J. H.: Peptides as regulators of gastric acid secretion. Annu. Rev. Physiol., 50:41, 1988.

Williams, J. A.: Regulatory mechanisms in pancreas and salivary acini. Annu. Rev. Physiol., 46:361, 1984.

FOOD INTAKE

Brownell, K. D.: The psychology and physiology of obesity: Implications for screening and treatment. J. Am. Diet. Assoc., 84:406, 1984.

Flint, D. J., et al.: Can obesity be controlled? News Physiol. Sci., 2:1, 1987.

Guthrie, H. A.: Introductory Nutrition, 7th Ed. St. Louis, C. V. Mosby Co., 1988.

Katch, F. I., and McArdle, W. D.: Nutrition, Weight Control, and Exercise. Philadelphia, Lea & Febiger, 1988.

Magnen, J. L.: Body energy balance and food intake: A neuroendocrine regulatory mechanism. Physiol. Rev., 63:314, 1983.

Nicolaidis, S.: What determines food intake? The ischymetric theory. News Physiol. Sci., 2:104, 1987.

Oomura, Y.: Regulation of feeding by neural responses to endogenous factors. News Physiol. Sci., 2:199, 1987.

Storlien, L. H.: The role of the ventromedial hypothalamic area in periprandial glucoregulation. Life. Sci., 360:505, 1985.

Williams, S. R.: Nutrition and Diet Therapy, 6th Ed. St. Louis, C. V. Mosby Co., 1989.

MICTURITION

Bricker, N. S.: The Kidney: Diagnosis and Management. New York, John Wiley & Sons, 1984.

Charlton, C. A.: The Urological System. New York, Churchill Livingstone, 1983.

Dirks, J. H., and Sutton, R. A.: Diuretics: Physiology, Pharmacology and Clinical Use. Philadelphia, W. B. Saunders Co., 1986.

Earley, L. E., and Gottschalk, C. W. (eds.): Strauss and Welt's Diseases of the Kidney, 3rd Ed. Boston. Little, Brown, 1979.

Tanagho, E. A., and McAninch, J. W.: Smith's General Urology. East Norwalk, Conn., Appleton & Lange, 1988.

TEMPERATURE CONTROL

Boulant, J. A., and Dean, J. B.: Temperature receptors in the central nervous system. Annu. Rev. Physiol., 48:639, 1986.

Boulant, J. A., and Scott, I. M.: Effects of leukocytic pyrogen on hypothalamic neurons in tissue slices. Environment, Drugs and Thermoregulation, 5th Int. Symp. Pharmacol. Thermoregulation, Saint-Paul-de-Vence, 1982, p. 125.

Brengelmann, G. L.: Circulatory adjustments to exercise and heat stress. Annu. Rev. Physiol., 45:191, 1983.

Crawshaw, L. I.: Temperature regulation in vertebrates. Annu. Rev. Physiol., 42:473, 1980.

Felig, P., et al. (eds.): Endocrinology and Metabolism, 2nd Ed. New York, McGraw-Hill Book Co., 1987.

Galanter, E.: Detection and discrimination and environmental change. In Darian-Smith, I. (ed.): Handbook of Physiology, Sec. 1, Vol. III. Bethesda, Md., American Physiological Society, 1984, p. 103.

Gordon, C. J., and Heath, J. E.: Integration and central processing in temperature regulation. Annu. Rev. Physiol., 48:595, 1986.

Hales, J. E. (ed.): Thermal Physiology. New York, Raven Press, 1984.

Hellon, R.: Thermoreceptors. In Shepherd, J. T., and Abboud, F. M. (eds.): Handbook of Physiology. Sec. 1, Vol. III. Bethesda, Md., American Physiological Society, 1983., p. 659.

Hong, S. K., et al.: Humans can acclimatize to cold: A lesson from Korean women divers. News Physiol. Sci., 2:79, 1987.

Kelso, S. R., et al.: Thermosensitive single-unit activity of in vitro hypothalamic slices. Am. J. Physiol., 242:R77, 1982.

Kluger, M. J.: Fever: A hot topic. News Physiol. Sci., 1:25, 1986.

Lipton, J. M., and Clark, W. G.: Neurotransmitters in temperature control. Annu. Rev. Physiol., 48:613, 1986.

Myers, R. D.: Neurochemistry of thermoregulation. Physiologist, 27:41, 1984.

Nicholls, D. G., and Locke, R. M.: Thermogenic mechanisms in brown fat. Physiol. Rev., 64:1, 1984.

Quinton, P. M.: Sweating and its disorders. Annu. Rev. Med., 34:453, 1983.

Rowell, L. B.: Cardiovascular adjustments to thermal stress. In Shepherd, J. T., and Abboud, F. M. (eds.): Handbook of Physiology. Sec. 2, Vol. III. Bethesda, Md. American Physiological Society, 1983, p. 967.

Scott, I. M., and Boulant, J. A.: Dopamine effects on thermosensitive neurons in hypothalamic tissue slices. Brain Res., 306:157, 1984.

Simon, E., et al.: Central and peripheral thermal control of effectors in homeothermic temperature regulation. Physiol. Rev., 66:235, 1986.

Spray, D. C.: Cutaneous temperature receptors. Annu. Rev. Physiol., 48:625, 1986.

29

Hypothalamic and Pituitary Control of Hormones and Reproduction

One of the means by which the nervous system controls bodily activity is by increasing or decreasing the secretion of many of the body's hormones. In previous chapters we have already discussed part of this function, such as control of the secretion of epinephrine and norepinephrine by the autonomic nervous system and control of different local hormones in the gastrointestinal tract.

In this chapter we will discuss much more global means by which the nervous system controls the secretion of hormones by most of the body's endocrine glands. These hormones in turn control many if not most of the body's metabolic functions as well as reproduction.

The nervous system's central control area for this global system is the *hypothalamus,* which controls the secretions of at least eight important hormones by the pituitary gland. These pituitary hormones in turn control the secretion of still other hormones by the thyroid gland, the adrenal glands, and the ovaries or testes. The function of the hypothalamus, and therefore its control effects on this pyramidal hormonal system, is itself controlled by nerve signals from almost all other parts of the brain, especially signals elicited by such subconscious nervous effects as the emotions, sex drives, body temperature, hunger, thirst, and even the crying of a baby.

It will not be possible to describe in the limited pages of this text this entire system and its multiple effects on the body. Yet, in this chapter we will discuss the most important features of the system, (1) the relationship of the hypothalamus to the pituitary gland and (2) the neurohormonal factors that promote the onset of reproduction.

■ THE PITUITARY GLAND AND ITS RELATION TO THE HYPOTHALAMUS

The *pituitary gland* (Fig. 29–1), also called the *hypophysis,* is a small gland — about 1 cm in diameter and 0.5 to 1 g in weight — that lies in the *sella turcica* at the base of the brain and is connected with the hypothalamus by the *pituitary* (or *hypophysial*) stalk. Physiologically, the pituitary gland is divisible into two distinct portions: the *anterior pituitary,* also known as the *adenohypophysis,* and the *posterior pituitary,* also known as the *neurohypophysis.* Between these is a small, relatively avascular zone called the *pars intermedia,* which is almost absent in the human being while much larger and much more functional in some lower animals.

Embryologically, the two portions of the pituitary originate from different sources, the anterior pituitary from *Rathke's pouch,* which is an embryonic invagination of the pharyngeal epithelium, and the posterior pituitary from an outgrowth of the hypothalamus. The origin of the anterior pituitary from the pharyngeal epithelium explains the epithelioid nature of its cells, while the origin of the posterior pituitary from neural tissue explains the presence of large numbers of glial-type cells in this gland.

Six very important hormones plus several less important ones are secreted by the *anterior* pituitary, and two important hormones are secreted by the *posterior* pituitary. The hormones of the anterior pituitary play major roles in the control of metabolic functions throughout the body, as shown in Figure 29–2. (1) *Growth hormone* promotes growth of the animal by

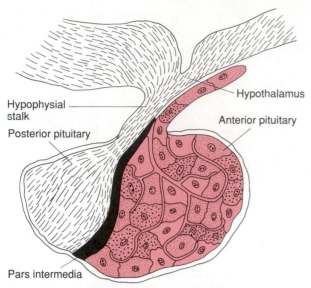

Figure 29–1. The pituitary gland.

affecting protein formation, cell multiplication, and cell differentiation. (2) *Adrenocorticotropin (corticotropin)* controls the secretion of some of the adrenocortical hormones, which in turn affect the metabolism of glucose, proteins, and fats. (3) *Thyroid-stimulating hormone (thyrotropin)* controls the rate of secretion of thyroxine by the thyroid gland, and thyroxine in turn controls the rates of most chemical reactions of the entire body. (4) *Prolactin* promotes the mammary gland development and milk production. And two separate gonadotropic hormones, (5) *follicle-stimulating hormone* and (6) *luteinizing hormone,* control growth of the gonads as well as their reproductive activities.

The two hormones secreted by the posterior pitui-

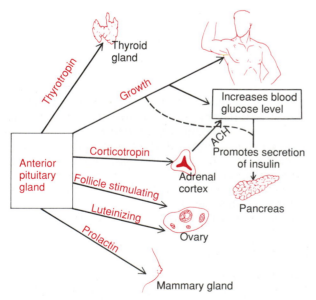

Figure 29–2. Metabolic functions of the anterior pituitary hormones.

tary play other roles. (1) *Antidiuretic hormone* (also called *vasopressin*) controls the rate of water excretion into the urine and in this way helps control the concentration of water in the body fluids. (2) *Oxytocin* (a) helps deliver milk from the glands of the breast to the nipples during suckling and (b) helps in the delivery of the baby at the end of gestation.

CONTROL OF PITUITARY SECRETION BY THE HYPOTHALAMUS

Almost all secretion by the pituitary is controlled by either hormonal or nervous signals from the hypothalamus. Indeed, when the pituitary gland is removed from its normal position beneath the hypothalamus and transplanted to some other part of the body, its rates of secretion of the different hormones (except for prolactin) fall to low levels — in the case of some of the hormones, to zero.

Secretion from the posterior pituitary is controlled by nerve signals originating in the hypothalamus and terminating in the posterior pituitary. In contrast, secretion by the anterior pituitary is controlled by hormones called *hypothalamic releasing* and *inhibitory hormones* (or *factors*) secreted within the hypothalamus itself and then conducted to the anterior pituitary through minute blood vessels called *hypothalamic-hypophysial portal vessels.* In the anterior pituitary these releasing and inhibitory hormones act on the glandular cells to control their secretion. This system of control will be discussed in detail later in the chapter.

The hypothalamus in turn receives signals from almost all possible sources in the nervous system. Thus, when a person is exposed to pain, a portion of the pain signal is transmitted into the hypothalamus. Likewise, when a person experiences some powerful depressing or exciting thought, a portion of the signal is transmitted into the hypothalamus. Olfactory stimuli denoting pleasant or unpleasant smells transmit strong signal components directly and through the amygdaloid nuclei into the hypothalamus. *Even the concentrations of nutrients, electrolytes, water, and various hormones* in the blood excite or inhibit various portions of the hypothalamus. Thus, the hypothalamus is a collecting center for information concerned with the internal well-being of the body, and in turn much of this information is used to control secretions of the many globally important pituitary hormones.

The Hypothalamic-Hypophysial Portal System

The anterior pituitary is a highly vascular gland with extensive capillary sinuses among the glandular cells. Almost all the blood that enters these sinuses passes first through another capillary bed in the lower tip of the hypothalamus and then through small *hypothalamic-hypophysial portal vessels* into the anterior pituitary sinuses. Thus, Figure 29–3 illustrates a small artery supplying the lowermost portion of the hypo-

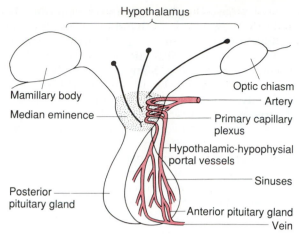

Figure 29–3. The hypothalamic-hypophysial portal system.

thalamus called the *median eminence* that connects inferiorly with the pituitary stalk. Small blood vessels penetrate into the substance of the median eminence and then return to its surface, coalescing to form the hypothalamic-hypophysial portal vessels. These in turn pass downward along the pituitary stalk to supply blood to the anterior pituitary sinuses.

Secretion of Hypothalamic Releasing and Inhibitory Hormones Into the Median Eminence. Special neurons in the hypothalamus synthesize and secrete the *hypothalamic releasing* and *inhibitory hormones* (or *releasing* and *inhibitory factors*) that control the secretion of the anterior pituitary hormones. These neurons originate in various parts of the hypothalamus and send their nerve fibers into the median eminence and the tuber cinereum, an extension of hypothalamic tissue that extends into the pituitary stalk. The endings of these fibers are different from most endings in the central nervous system in that their function is not to transmit signals from one neuron to another but merely to secrete the hypothalamic releasing and inhibitory hormones (factors) into the tissue fluids. These hormones are immediately absorbed into the hypothalamic-hypophysial portal system and carried directly to the sinuses of the anterior pituitary gland.

(To avoid confusion, the student needs to know the difference between a "factor" and a "hormone." A substance that has the actions of a hormone but that has not been purified and identified as a distinct chemical compound is called a *factor*. Once it has been so identified, it is thereafter known as a *hormone* instead of simply a *factor*.)

Function of the Releasing and Inhibitory Hormones. The function of the releasing and inhibitory hormones is to control the secretion of the anterior pituitary hormones. For most of the anterior pituitary hormones it is the releasing hormones that are important; but, for prolactin, an inhibitory hormone probably exerts the most control. The hypothalamic releasing and inhibitory hormones (or factors) that are of major importance are the following:

1. *Thyroid-stimulating hormone releasing hormone* (TRH), which causes release of thyroid-stimulating hormone.

2. *Corticotropin-releasing hormone* (CRH), which causes release of adrenocorticotropin.

3. *Growth hormone releasing hormone* (GHRH), which causes release of growth hormone, and *growth hormone inhibitory hormone* (GHIH), which is the same as the hormone *somatostatin* and which inhibits the release of growth hormone.

4. *Gonadotropin-releasing hormone* (GnRH), which causes release of the two gonadotropic hormones, luteinizing hormone and follicle-stimulating hormone.

5. *Prolactin inhibitory factor* (PIF), which causes inhibition of prolactin secretion.

In addition to these more important hypothalamic hormones, still another probably excites the secretion of prolactin, and several possibly inhibit some of the other anterior pituitary hormones. Each of the more important hypothalamic hormones will be discussed in detail at the time that the specific hormonal system controlled by them is presented.

Specific Areas in the Hypothalamus That Control Secretion of Specific Hypothalamic Releasing and Inhibitory Factors. All or most of the hypothalamic hormones are secreted at nerve endings in the median eminence before being transported to the anterior pituitary gland. Electrical stimulation of this region excites these nerve endings and therefore causes release of essentially all the hypothalamic hormones. However, the neuronal cell bodies that give rise to these median eminence nerve endings are located in other discrete areas of the hypothalamus or in closely related areas of the basal brain. Unfortunately, the specific loci of the neuronal cell bodies that form the different hypothalamic releasing or inhibitory hormones are so incompletely known that it would be misleading to attempt a delineation here.

PHYSIOLOGICAL FUNCTIONS OF THE ANTERIOR PITUITARY HORMONES

All the major anterior pituitary hormones besides growth hormone exert their principal effects by stimulating target glands—such as the thyroid gland, the adrenal cortex, the ovaries, the testicles, and the mammary glands. We will discuss briefly the functions of four of the pituitary hormones, *growth hormone, thyroid-stimulating hormone, adrenocorticotropin,* and *prolactin.* The functions of the two gonadotropic hormones, *luteinizing hormone* and *follicle-stimulating hormone,* will be discussed later in this chapter.

Growth Hormone

Growth hormone is secreted by the anterior pituitary gland throughout the entire life of a person. Its rate of secretion is controlled by *growth hormone–releasing hormone,* which is formed in the hypothalamus and transmitted to the anterior pituitary gland through the hypothalamic-hypophysial portal system.

Growth hormone has two major functions: The first of these is to promote growth of the infant into the child and then of the child into the adult. One of the growth-promoting effects is to cause growth of the bones, thus making the skeleton grow progressively larger. However, the body stops increasing in height at adolescence because the long bones of the body then lose their capability for growing longer because the growing cells of the bone epiphyses (where growth occurs) become used up. In addition to causing the bones to grow, this hormone promotes growth of essentially all other tissues of the body.

The second function of growth hormone is to control several of the metabolic functions of the body. Though these have not been defined in great detail, it is clear that growth hormone is essential for protein formation and for its maintenance in essentially all cells of the body. This, presumably, is the principal cause of growth. Growth hormone also enhances fat utilization but at the same time decreases carbohydrate use by the cells.

When growth hormone is not secreted by the anterior pituitary gland of a child, growth fails to occur, and the result is a so-called *pituitary dwarf* who retains childish features and often attains a height no greater than two to three feet. On the other hand, excess production of growth hormone by the anterior pituitary gland, caused usually by a tumor of the pituitary's growth hormone–producing cells, will cause the person to become a *giant*.

Thyroid-Stimulating Hormone and Its Control of Thyroid Gland Secretion

The thyroid gland secretes two hormones, *thyroxin* and *diiodothyronine,* that are very important in controlling the overall rate of metabolism of almost all tissues of the body—that is in controlling how rapidly the chemical reactions occur in all the tissues. When the rate of metabolism is great, a large amount of heat is also formed in the body, so that these thyroid hormones also play an important role in controlling body temperature.

Control of thyroid hormone secretion is vested in the hypothalamus and anterior pituitary gland. The hypothalamus secretes *thyroid-stimulating hormone–releasing hormone.* This then passes by way of the hypothalamic-hypophysial portal system to the anterior pituitary gland where it causes the release of *thyroid-stimulating hormone;* the thyroid-stimulating hormone, in turn, is carried to the thyroid gland in the blood and is the controller of thyroid hormone secretion. Therefore, damage to either the hypothalamus or the anterior pituitary gland, especially the latter, can greatly reduce the rate of production of the thyroid hormones.

In the absence of the thyroid hormones, the person develops *hypothyroidism,* in which all the bodily activities become very sluggish, and the person often becomes obese. On the other hand, overactivity of the hypothalamus or anterior pituitary gland can cause *hyperthyroidism,* in which the bodily functions become greatly overactive. Many of the body's tissues can actually be seriously damaged because of this overactivity. The person also becomes very excitable and nervous. Hyperthyroidism is often caused by psychic stress that elicits excessive signals from the hypothalamus.

Fortunately, in the normal person there is a feedback mechanism for control of thyroid secretion. That is, when thyroid secretion becomes too great, the hypothalamic-pituitary control system itself becomes inhibited, thus returning the body to normal thyroid secretion.

Adrenocorticotropin and Its Control of the Adrenocortical Hormones

Each person has two adrenal glands, lying respectively on top of the superior poles of the two kidneys. The cortical portions of these glands secrete several different steroid hormones, the two most important of which are *cortisol* and *aldosterone.* The rate of secretion of aldosterone is controlled mainly by such factors as plasma potassium ion concentration, plasma sodium ion concentration, and by the hormone angiotensin. On the other hand, the secretion of cortisol by the adrenal cortex is controlled almost entirely by the rate of secretion of *adrenocorticotropin* by the anterior pituitary. And the rate of secretion of adrenocorticotropin is itself controlled by *corticotropin-releasing hormone,* which is formed in the hypothalamus and transported to the pituitary gland through the hypothalamic-hypophysial portal system.

Cortisol has powerful controlling effects on many metabolic functions of the body, including protein metabolism, carbohydrate metabolism, and fat metabolism. For instance, it has an opposite effect on protein metabolism to that of growth hormone—that is, it reduces the amount of protein in most of the tissues. In doing so, it releases amino acids into the circulating blood that can be used elsewhere in the body when they are required to repair damaged tissues. For this reason as well as others, cortisol is a very valuable hormone to help the body repair tissue destruction during stress. Cortisol also enhances the conversion of amino acids into carbohydrates, which in turn are used for energy. It increases the utilization of fats for energy as well.

The hypothalamic control of cortisol secretion is especially important. Within minutes after a person experiences serious physical stress, such as a broken bone, excessive heat, hemorrhage, or almost any other life-threatening occurrence, the level of cortisol circulating in the body fluids increases many times, sometimes 20-fold or more. And the cortisol in turn plays a very significant role in helping the cells resist the destructive effects of the stress.

Prolactin and Its Control of Milk Secretion

Under normal conditions, the rate of prolactin secretion is relatively slight. However, during pregnancy its secretion increases progressively until it becomes about ten times normal by the time the baby is born. The prolactin helps to promote growth of protein tissues in the fetus. But another important function is its ability to cause development of the breasts during pregnancy and to cause milk secretion after birth of the baby.

Even though prolactin stimulates development of the breasts during pregnancy, the large quantities of estrogens and progesterone that are secreted by the placenta during pregnancy inhibit milk production until the baby is born. But after birth, when the placenta is no longer available to secrete the estrogens and progesterone, the prolactin then causes rapid secretion of milk by the breasts. Furthermore, everytime the baby suckles a breast, nerve signals pass from the breast, up the spinal cord, and into the hypothalamus to cause an increase in prolactin secretion and therefore production of more milk.

The hypothalamus controls prolactin secretion somewhat differently from the way that it controls the secretion of other anterior pituitary hormones. It does so by an inhibitory hormone, *prolactin inhibitory hormone*, rather than by a releasing hormone. That is, the hypothalamus normally secretes an excess of prolactin inhibitory hormone, which keeps the rate of secretion of prolactin at a relatively low level. However, during pregnancy and during periods of milk production, the hypothalamus *decreases* its rate of formation of prolactin inhibitory hormone, in this way allowing increased production of prolactin.

Because of the essential role of the hypothalamus in milk secretion, strong psychic stimuli, particularly emotional states of anguish, can actually cause the mother's milk-secreting apparatus to dry up.

▪ THE POSTERIOR PITUITARY GLAND AND ITS RELATION TO THE HYPOTHALAMUS

The *posterior pituitary gland,* also called the *neurohypophysis,* is composed mainly of glial-like cells called *pituicytes.* However, the pituicytes do not secrete hormones; they act simply as a supporting structure for large numbers of *terminal nerve fibers* and *terminal nerve endings* from nerve tracts that originate in the *supraoptic* and *paraventricular nuclei* of the hypothalamus, as shown in Figure 29–4. These tracts pass to the neurohypophysis through the *pituitary stalk* (hypophysial stalk). The nerve endings are bulbous knobs containing many secretory granules that lie on the surfaces of capillaries onto which they secrete the two

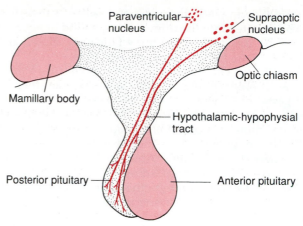

Figure 29–4. Hypothalamic control of the posterior pituitary.

posterior pituitary hormones: (1) *antidiuretic hormone* (ADH), also called *vasopressin*; and (2) *oxytocin.*

If the pituitary stalk is cut above the pituitary gland but the entire hypothalamus is left intact, the posterior pituitary hormones continue, after a transient decrease for a few days, to be secreted almost normally; but they are then secreted by the cut ends of the fibers within the hypothalamus and not by the nerve endings in the posterior pituitary. The reason for this is that the hormones are initially synthesized in the cell bodies of the supraoptic and paraventricular nuclei and are then transported in combination with "carrier" proteins called *neurophysins* down to the nerve endings in the posterior pituitary gland, requiring several days to reach the gland.

ADH is formed primarily in the supraoptic nuclei, whereas oxytocin is formed primarily in the paraventricular nuclei. However, each of these two nuclei can synthesize approximately one sixth as much of the second hormone as of its primary hormone.

When nerve impulses are transmitted downward along the fibers from the supraoptic or paraventricular nuclei, the hormone is immediately released from the secretory granules in the nerve endings by the usual secretory mechanism of *exocytosis* and is absorbed into adjacent capillaries.

FUNCTION OF THE POSTERIOR PITUITARY HORMONES

Antidiuretic Hormone. The function of antidiuretic hormone was discussed in detail in Chapter 27. It was pointed out that special neuronal cells called osmoreceptors located in the supraoptic nuclei are stimulated when the osmotic concentration of the extracellular fluids becomes too great. These cells then secrete antidiuretic hormone at their endings in the posterior pituitary gland. The antidiuretic hormone in turn acts on the tubules of the kidneys to cause increased water reabsorption, which retains water in the body fluids while allowing sodium and other dissolved substances to be excreted into the urine. Over a period of hours to days, by this process the body fluids are diluted and their osmotic concentration returns to normal.

Oxytocin. The hormone oxytocin powerfully stimulates the pregnant uterus, especially so toward the end of gestation. Therefore, this hormone is at least partially responsible for causing birth. In a hypophysectomized animal, the duration of labor is prolonged. Also, the amount of oxytocin in the plasma increases during the latter stages of labor. And, finally, stimulation of the cervix in a pregnant animal elicits nerve signals that pass through the hypothalamus to the posterior pituitary gland to cause increased secretion of oxytocin.

Therefore, there are many reasons to believe that the intense stretching of the cervix of the uterus during labor elicits nervous reflexes which excite the anterior hypothalamus to cause increased secretion of the hormone oxytocin. The oxytocin in turn promotes additional contraction of the uterus, which expels the baby much more rapidly than would otherwise occur. In other words, once labor begins, this oxytocin feedback mechanism helps to create a cycle of progressively increasing uterine contractions that ordinarily will not stop until the baby is delivered.

Effect of Oxytocin on Milk Ejection. Oxytocin also plays an important role in the process of lactation. In lactation, oxytocin causes milk to be expressed from the alveoli of the breasts into the ducts that lead to the nipple. This makes the milk available to the baby by suckling. This mechanism works as follows: The suckling stimuli on the nipple cause signals to be transmitted through sensory nerves to the brain. The signals pass upward through the brain stem and finally reach the oxytocin neurons in the paraventricular and supraoptic nuclei in the hypothalamus, to cause release of oxytocin. The oxytocin then is carried by the blood to the breasts. Here it causes contraction of *myoepithelial cells* that lie outside of and form a lattice-work around the alveoli of the mammary glands. In less than a minute after the beginning of suckling, milk begins to flow. Therefore, this mechanism is frequently called *milk let-down* or *milk ejection.*

■ CONTROL OF MALE SEXUAL FUNCTIONS BY THE GONADOTROPIC HORMONES —FSH AND LH

The anterior pituitary gland secretes two major gonadotropic hormones: (1) *follicle-stimulating hormone* (FSH); and (2) *luteinizing hormone* (LH). These are glycoprotein hormones that play major roles in the control of both male and female sexual function.

Regulation of Testosterone Production by LH. Testosterone is produced by the interstitial cells of Leydig in the testes but only when they are stimulated by LH from the pituitary gland. The quantity of testosterone secreted increases approximately in direct proportion to the amount of LH available.

Injection of purified LH into a child causes fibroblasts in the interstitial areas of the testes to develop into interstitial cells of Leydig, though mature Leydig cells are not normally found in the child's testes until after the age of approximately 10 years.

REGULATION OF PITUITARY SECRETION OF LH AND FSH BY THE HYPOTHALAMUS

The gonadotropins, like corticotropin and thyrotropin, are secreted by the anterior pituitary gland mainly in response to nervous activity in the hypothalamus. For instance, in the female rabbit, coitus with a male rabbit elicits nervous activity in the hypothalamus that in turn stimulates the anterior pituitary to secrete FSH and LH. These hormones then cause rapid ripening of follicles in the rabbit's ovaries, followed a few hours later by ovulation.

Many other types of nervous stimuli are also known to affect gonadotropin secretion. For instance, in sheep, goats, and deer, nervous stimuli in response to changes in weather and amount of light in the day increase the quantities of gonadotropins during one season of the year, the mating season, thus allowing birth of the young during an appropriate period for survival. Also, psychic stimuli can affect fertility of the male animal, as exemplified by the fact that transporting a bull under uncomfortable conditions can often cause almost complete temporary sterility. In the human being, too, it is known that various psychic stimuli feeding into the hypothalamus can cause marked excitatory or inhibitory effects on gonadotropin secretion, in this way sometimes greatly altering the degree of fertility.

Gonadotropin-Releasing Hormone (GnRH), the Hypothalamic Hormone That Stimulates Gonadotropin Secretion. In both the male and the female the hypothalamus controls gonadotropin secretion by way of the hypothalamic-hypophysial portal system, as was discussed earlier in the chapter. Though there are two different gonadotropic hormones, luteinizing hormone and follicle-stimulating hormone, only one hypothalamic-releasing hormone has been discovered; this is *gonadotropin-releasing hormone* (GnRH). This hormone has an especially strong effect on inducing luteinizing hormone secretion by the anterior pituitary gland. However, it also has a potent effect in causing follicle-stimulating hormone secretion as well.

GnRH plays a similar role in controlling gonadotropin secretion in the female, where the interrelationships are far more complex. Therefore, its nature and its functions will be discussed in much more detail later in the chapter.

Reciprocal Inhibition of Hypothalamic-Anterior Pituitary Secretion of Gonadotropic Hormones by Testicular Hormones. *Feedback Control of Testosterone Secretion.* Injection of testosterone into either a male or a female animal strongly inhibits the secretion of luteinizing hormone but only slightly inhibits the secretion of the follicle-stimulating hormone. This inhibition depends on normal function of the hypothalamus; therefore, it is quite clear that the following negative feedback control system operates continuously to control very precisely the rate of testosterone secretion:

1. The hypothalamus secretes *gonadotropin-releasing hormone,* which stimulates the anterior pituitary gland to secrete *luteinizing hormone.*

2. Luteinizing hormone in turn stimulates *hyperplasia of the Leydig cells* of the testes and also stimulates production of *testosterone* by these cells.

3. The testosterone in turn feeds back negatively to the hypothalamus, inhibiting production of gonadotropin-releasing hormone. This obviously limits the rate at which testosterone will be produced. On the other hand, when testosterone production is too low, lack of inhibition of the hypothalamus leads to subsequent return of testosterone secretion to the normal level.

Feedback Control of Spermatogenesis — Role of "Inhibin." It is known, too, that spermatogenesis by the testes inhibits the secretion of FSH. Conversely, failure of spermatogenesis causes markedly increased secretion of FSH; this is especially true when the seminiferous tubules are destroyed, including destruction of the Sertoli cells in addition to the germinal cells. Therefore, it is believed that the Sertoli cells secrete a hormone that has a direct inhibitory effect mainly on the anterior pituitary gland (but perhaps slightly on the hypothalamus as well) to inhibit the secretion of FSH. A glycoprotein hormone having a molecular weight between 10,000 and 30,000 called *inhibin* has been isolated from cultured Sertoli cells and is probably responsible for most of the feedback control of FSH secretion and of spermatogenesis. This feedback cycle is the following:

1. Follicle-stimulating hormone stimulates the Sertoli cells that provide nutrition for the developing spermatozoa.

2. The Sertoli cells release inhibin that in turn feeds back negatively to the anterior pituitary gland to inhibit the production of FSH. Thus, this feedback cycle maintains a constant rate of spermatogenesis, without underproduction or overproduction that is required for male reproductive function.

PUBERTY AND REGULATION OF ITS ONSET

Initiation of the onset of puberty has long been a mystery. In the earliest history of humanity, the belief was simply that the testicles "ripened" at this time. With the discovery of the gonadotropins, ripening of the anterior pituitary gland was considered responsible. Now it is known from experiments in which both testicular and pituitary tissues have been transplanted from infant animals into adult animals that both the testes and the anterior pituitary of the infant are capable of performing adult functions if appropriately stimulated. Therefore, it is now certain that *during childhood the hypothalamus simply does not secrete GnRH.*

For reasons not understood, some maturation process in the brain causes the hypothalamus to begin secreting GnRH at the time of puberty. This secretion will not occur if the neuronal connections between the hypothalamus and other parts of the brain are not intact. Therefore, the present belief is that the maturation process probably occurs elsewhere in the brain instead of in the hypothalamus. One suggested locus is the amygdala.

■ REGULATION OF THE FEMALE MONTHLY RHYTHM — INTERPLAY BETWEEN THE OVARIAN AND HYPOTHALAMIC-PITUITARY HORMONES

FUNCTION OF THE HYPOTHALAMUS IN THE REGULATION OF GONADOTROPIN SECRETION — GONADOTROPIN-RELEASING HORMONE

As pointed out earlier in the chapter, secretion of most of the anterior pituitary hormones is controlled by releasing hormones formed in the hypothalamus and then transported to the anterior pituitary gland by way of the hypothalamic-hypophysial portal system. In the case of the gonadotropins, at least one releasing hormone, *gonadotropin-releasing hormone (GnRH)*, also called *luteinizing hormone – releasing hormone (LHRH)*, is important. This has been purified and has been found to be a decapeptide having the following formula:

$$\text{Glu-His-Trp-Ser-Tyr-Gly-Leu-Arg-Pro-Gly-NH}_2$$

Hypothalamic Centers for Release of GnRH. The neuronal activity that causes release of GnRH occurs primarily in the mediobasal hypothalamus, especially in the arcuate nuclei of this area. Therefore, it is believed that these arcuate nuclei control most female sexual activity, though other neurons located in the preoptic area of the anterior hypothalamus also secrete GnRH in moderate amounts, the function of which is unclear. Multiple neuronal centers in the brain's limbic system transmit signals into the arcuate nuclei to modify both the intensity of GnRH release and the frequency of the pulses, thus providing a possible explanation of why psychic factors very often modify female sexual function.

NEGATIVE FEEDBACK EFFECT OF ESTROGEN, PROGESTERONE, AND INHIBIN ON SECRETION OF FOLLICLE-STIMULATING HORMONE AND LUTEINIZING HORMONE

Estrogen in small amounts and progesterone in large amounts inhibit the production of FSH and LH. These feedback effects seem to operate mainly directly on the anterior pituitary gland but to a lesser extent on the hypothalamus to decrease secretion of GnRH, especially by altering the frequency of the GnRH pulses.

In addition to the feedback effects of estrogen and progesterone, still another hormone also seems to be involved. This is *inhibin,* which is secreted along with the steroid sex hormones by the corpus luteum in the same way that the Sertoli cells secrete this same hormone in the male testes. This inhibin has the same effect in the female as in the male of inhibiting the secretion of FSH by the anterior pituitary gland and LH to a lesser extent as well. Therefore, it is believed that inhibin might be especially important in causing the decrease in secretion of FSH and LH toward the end of the female sexual month.

POSITIVE FEEDBACK EFFECT OF ESTROGEN BEFORE OVULATION— THE PREOVULATORY LUTEINIZING HORMONE SURGE

For reasons not completely understood, the anterior pituitary gland secretes greatly increased amounts of LH for a period of 1 to 2 days beginning 24 to 48 hours before ovulation. This effect is illustrated in Figure 29–5, and the figure shows a much smaller preovulatory surge of FSH as well.

Experiments have shown that infusion of estrogen into a female above a critical rate for a period of 2 to 3 days during the first half of the ovarian cycle will cause

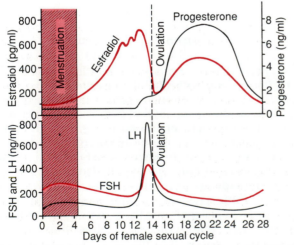

Figure 29–5. Approximate plasma concentrations of the gonadotropins and ovarian hormones during the normal female sexual cycle.

rapidly accelerating growth of the follicles and also rapidly accelerating secretion of ovarian estrogens. During this period the secretion of both FSH and LH by the anterior pituitary gland is at first suppressed slightly. Then abruptly the secretion of LH increases six- to eightfold, and the secretion of FSH increases about twofold. The cause of this abrupt increase in secretion of the gonadotropins is not known. However, several possible causes are as follows: (1) It has been suggested that estrogen at this point in the cycle has a peculiar *positive feedback effect* to stimulate pituitary secretion of the gonadotropins; this is in sharp contrast to its normal negative feedback effect that occurs during the remainder of the female monthly cycle. (2) The granulosa cells of the follicles begin to secrete small but increasing quantities of progesterone a day or so prior to the preovulatory LH surge, and it has been suggested that this might be the factor that stimulates the excess LH secretion.

Without this normal preovulatory surge of LH, ovulation will not occur.

FEEDBACK OSCILLATION OF THE HYPOTHALAMIC-PITUITARY-OVARIAN SYSTEM

Now, after discussing much of the known information about the interrelationships of the different components of the female hormonal system, we can digress from the area of proven fact into the realm of speculation and attempt to explain the feedback oscillation that controls the rhythm of the female sexual cycle. It seems to operate in approximately the following sequence of three successive events:

1. The Postovulatory Secretion of the Ovarian Hormones and Depression of Gonadotropins. The easiest part of the cycle to explain is the events that occur during the postovulatory phase—between ovulation and the beginning of menstruation. During this time the corpus luteum secretes large quantities of both progesterone and estrogen and probably the hormone inhibin as well. All these hormones together have a combined negative feedback effect on the anterior pituitary gland and the hypothalamus to cause suppression of both FSH and LH, decreasing these to their lowest levels at about three to four days before the onset of menstruation. These effects are illustrated in Figure 29–5.

2. The Follicular Growth Phase. Two to 3 days before menstruation the corpus luteum involutes, and the secretion of estrogen, progesterone, and inhibin decreases to a low ebb. This releases the hypothalamus and anterior pituitary from the feedback effect of these hormones; and a day or so later, at about the time that menstruation begins, FSH increases as much as twofold; then several days after menstruation begins, LH secretion increases as much as twofold as well. These hormones initiate new follicular growth and progressive increase in the secretion of estrogen, reaching a peak estrogen secretion at about 12.5 to 13 days after the onset of menstruation. During the first 11 to 12 days of this follicular growth the rates of secretion of the gonadotropins FSH and LH decrease very slightly because of the negative feedback effect mainly of estrogen on the anterior pituitary gland. Then comes a sudden increase in secretion of both of these hormones, leading to the preovulatory surge of LH, followed by ovulation.

3. Preovulatory Surge of LH and FSH; Ovulation. At approximately 11.5 to 12 days after the onset of menstruation, the decline in secretion of FSH and LH comes to an abrupt halt. It is believed that the high level of estrogens at this time (or the beginning secretion of progesterone by the follicles) causes a positive feedback effect principally on the anterior pituitary, as explained earlier, which leads to a terrific surge of secretion of LH and to a lesser extent of FSH. Whatever the cause of this preovulatory LH and FSH surge, the LH leads to both ovulation and subsequent secretion by the corpus luteum. Thus, the hormonal system begins a new round of the cycle until the next ovulation.

PUBERTY AND MENARCHE

Puberty means the onset of adult sexual life, and menarche means the onset of menstruation. The period of puberty is caused by a gradual increase in gonadotropic hormone secretion by the pituitary, beginning approximately in the eighth year of life and usually culminating in the onset of menstruation between the ages of 11 and 16 years (average, 13 years).

In the female, as in the male, the infantile pituitary gland and ovaries are capable of full function if appropriately stimulated. However, as is also true in the male and for reasons not yet understood, the hypothalamus does not secrete significant quantities of GnRH during childhood. Experiments have shown that the hypothalamus itself is perfectly capable of secreting this hormone, but there is lack of the appropriate signal from some other brain area to cause the secretion. Therefore, it is now believed that the onset of puberty is initiated by some maturation process occurring elsewhere in the brain, perhaps somewhere in the limbic system.

Figure 29–6 illustrates (1) the increasing levels of estrogen secretion at puberty, (2) the cyclic variation during the monthly sexual cycles, (3) the further increase in estrogen secretion during the first few years of reproductive life, (4) then progressive decrease in estrogen secretion toward the end of reproductive life, and (5) finally almost no estrogen secretion beyond the menopause.

THE MENOPAUSE

At the age of 40 to 50 years the sexual cycles usually become irregular, and ovulation fails to occur during many of the cycles. After a few months to a few years,

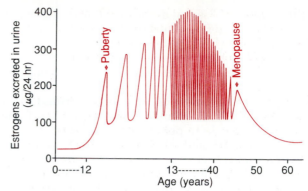

Figure 29–6. Estrogen secretion throughout sexual life.

the cycles cease altogether, as illustrated in Figure 29–6. This period during which the cycles cease and the female sex hormones diminish rapidly to almost none at all is called the *menopause.*

The cause of the menopause is "burning out" of the ovaries. Throughout a woman's reproductive life about 400 of the primordial follicles grow into vesicular follicles and ovulate, while literally hundreds of thousands of the ova degenerate. At the age of about 45 years only a few primordial follicles still remain to be stimulated by FSH and LH, and the production of estrogens by the ovary decreases as the number of primordial follicles approaches zero (also illustrated in Figure 29–6). When estrogen production falls below a critical value, the estrogens can no longer inhibit the production of FSH and LH; nor can they cause an ovulatory surge of LH and FSH to cause ovulatory cycles.

■ THE MALE SEXUAL ACT

NEURONAL STIMULUS FOR PERFORMANCE OF THE MALE SEXUAL ACT

The most important source of impulses for initiating the male sexual act is the glans penis, for the glans contains an especially sensitive sensory end-organ system that transmits into the central nervous system that special modality of sensation called *sexual sensation.* The massaging action of intercourse on the glans stimulates the sensory end-organs, and the sexual sensations in turn pass through the pudendal nerve, thence through the sacral plexus into the sacral portion of the spinal cord, and finally up the cord to undefined areas of the cerebrum. Impulses may also enter the spinal cord from areas adjacent to the penis to aid in stimulating the sexual act. For instance, stimulation of the anal epithelium, the scrotum, and perineal structures in general can all send signals into the cord that add to the sexual sensation. Sexual sensations can even originate in internal structures, such as irritated areas of the urethra, the bladder, the prostate, the seminal vesicles, the testes, and the vas deferens. Indeed, one of the causes of "sexual drive" is filling of the sexual organs with secretions. Infection and inflammation of these sexual organs sometimes cause almost continual sexual desire, and "aphrodisiac" drugs, such as cantharides, increase the sexual desire by irritating the bladder and urethral mucosa.

The Psychic Element of Male Sexual Stimulation. Appropriate psychic stimuli can greatly enhance the ability of a person to perform the sexual act. Simply thinking sexual thoughts or even dreaming that the act of intercourse is being performed can cause the male sexual act to occur and to culminate in ejaculation. Indeed, *nocturnal emissions* during dreams occur in many males during some stages of sexual life, especially during the teens.

Integration of the Male Sexual Act in the Spinal Cord. Though psychic factors usually play an important part in the male sexual act and can actually initiate or inhibit it, the cerebrum is probably not absolutely necessary for its performance, because appropriate genital stimulation can cause ejaculation in some animals and occasionally in a human being after their spinal cords have been cut above the lumbar region. Therefore, the male sexual act results from inherent reflex mechanisms integrated in the sacral and lumbar spinal cord, and these mechanisms can be initiated by either psychic stimulation or actual sexual stimulation but most likely a combination of both.

STAGES OF THE MALE SEXUAL ACT

Erection; Role of the Parasympathetic Nerves. Erection is the first effect of male sexual stimulation, and the degree of erection is proportional to the degree of stimulation, whether psychic or physical.

Erection is caused by parasympathetic impulses that pass from the sacral portion of the spinal cord through the *nervi erigentes* to the penis. These parasympathetic impulses dilate the arteries of the penis, thus allowing arterial blood to build up under high pressure in the *erectile tissue* of the *corpus cavernosum* and *corpus spongiosum* in the shaft of the penis, illustrated in Figure 29–7. This erectile tissue is nothing more than large, cavernous, venous sinusoids, which are normally relatively empty but which become dilated tremendously when arterial blood flows rapidly into them under pressure, since the venous outflow is partially occluded. Also, the erectile bodies, especially the two corpora cavernosa, are surrounded by strong fibrous coats; therefore, high pressure within the sinusoids causes ballooning of the erectile tissue to such an extent that the penis becomes hard and elongated.

Lubrication, a Parasympathetic Function. During sexual stimulation, the parasympathetic impulses, in addition to promoting erection, cause the urethral glands and the bulbourethral glands to secrete mucus. This mucus flows through the urethra during intercourse to aid in the lubrication of coitus. However,

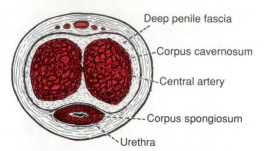

Figure 29–7. Erectile tissue of the penis.

most of the lubrication of coitus is provided by the female sexual organs rather than by the male. Without satisfactory lubrication, the male sexual act is rarely successful because unlubricated intercourse causes grating, painful sensations that inhibit rather than excite sexual sensations.

Emission and Ejaculation; Function of the Sympathetic Nerves. Emission and ejaculation are the culmination of the male sexual act. When the sexual stimulus becomes extremely intense, the reflex centers of the spinal cord begin to emit *sympathetic impulses* that leave the cord at L-1 and L-2 and pass to the genital organs through the hypogastric and pelvic plexuses to initiate emission, the forerunner of ejaculation.

Emission begins with contraction of the vas deferens and the ampulla to cause expulsion of sperm into the internal urethra. Then, contractions of the muscular coat of the prostate gland followed finally by contraction of the seminal vesicles expel prostatic fluid and seminal fluid, forcing the sperm forward. All of these fluids mix in the internal urethra with the mucus already secreted by the bulbourethral glands to form the semen. The process to this point is *emission*.

The filling of the internal urethra then elicits sensory signals that are transmitted through the pudendal nerves to the sacral regions of the cord, giving the feeling of sudden fullness in the internal genital organs. Also, these sensory signals further excite the rhythmic contraction of the internal genital organs and also cause contraction of the ischiocavernosus and bulbocavernosus muscles that compress the bases of the penile erectile tissue. These effects together cause rhythmic, wavelike increases in pressure in the genital ducts and urethra, which "ejaculate" the semen from the urethra to the exterior. The process is called *ejaculation*. At the same time, rhythmic contractions of the pelvic muscles and even of some of the muscles of the body trunk cause thrusting movements of the pelvis and penis, which also help propel the semen into the deepest recesses of the vagina and perhaps even slightly into the cervix of the uterus.

This entire period of emission and ejaculation is called the *male orgasm*. At its termination, the male sexual excitement disappears almost entirely within 1 to 2 min, and erection ceases.

■ THE FEMALE SEXUAL ACT

Stimulation of the Female Sexual Act. As is true in the male sexual act, successful performance of the female sexual act depends on both psychic stimulation and local sexual stimulation.

As is also true in the male, the thinking of erotic thoughts can lead to female sexual desire, and this aids greatly in the performance of the female sexual act. Such desire is probably based as much on one's background training as on physiological drive, though sexual desire does increase in proportion to the level of secretion of the sex hormones. Desire also changes during the sexual month, reaching a peak near the time of ovulation, probably because of the high levels of estrogen secretion during the preovulatory period.

Local sexual stimulation in women occurs in more or less the same manner as in men, for massage, irritation, or other types of stimulation of the vulva, vagina, and other perineal regions, and even of the urinary tract, create sexual sensations. The glans of the *clitoris* is especially sensitive for initiating sexual sensations. As in the male, the sexual sensory signals are mediated to the sacral segments of the spinal cord through the pudendal nerve and sacral plexus. Once these signals have entered the spinal cord, they are transmitted thence to the cerebrum. Also, local reflexes integrated in the sacral and lumbar spinal cord are at least partly responsible for female sexual reactions.

Female Erection and Lubrication. Located around the introitus and extending into the clitoris is erectile tissue almost identical with the erectile tissue of the penis. This erectile tissue, like that of the penis, is controlled by the parasympathetic nerves that pass through the nervi erigentes from the sacral plexus to the external genitalia. In the early phases of sexual stimulation, parasympathetic signals dilate the arteries of the erectile tissues, and this allows rapid accumulation of blood in the erectile tissue so that the introitus tightens around the penis; this aids the male greatly in his attainment of sufficient sexual stimulation for ejaculation to occur.

Parasympathetic signals also pass to the bilateral Bartholin's glands located beneath the labia minora to cause secretion of mucus immediately inside the introitus. This mucus is responsible for much of the lubrication during sexual intercourse, though much is also provided by mucus secreted by the vaginal epithelium as well and a small amount from the male urethral glands. The lubrication in turn is necessary for establishing during intercourse a satisfactory massaging sensation rather than an irritative sensation, which may be provoked by a dry vagina. A massaging sensation constitutes the optimal type of sensation for evoking the appropriate reflexes that culminate in both the male and female climaxes.

The Female Orgasm. When local sexual stimulation reaches maximum intensity, and especially when the local sensations are supported by appropriate psychic conditioning signals from the cerebrum, reflexes are initiated that cause the female orgasm, also called the *female climax.* The female orgasm is analogous to emission and ejaculation in the male, and it perhaps helps promote fertilization of the ovum. Indeed, the human female is known to be somewhat more fertile when inseminated by normal sexual intercourse rather than by artificial methods, thus indicating an important function of the female orgasm. Possible reasons for this are as follows:

First, during the orgasm the perineal muscles of the female contract rhythmically, which results from spinal cord reflexes similar to those that cause ejaculation in the male. It is possible that these same reflexes increase uterine and fallopian tube motility during the orgasm, thus helping transport the sperm toward the ovum, but information on this subject is scanty. Also, the orgasm seems to cause dilation of the cervical canal for up to half an hour, thus allowing easy transport of the sperm.

Second, in many lower animals, copulation causes the posterior pituitary gland to secrete oxytocin; this effect is probably mediated through the amygdaloid nuclei and then through the hypothalamus to the pituitary. The oxytocin in turn causes increased rhythmical contractions of the uterus, which have been postulated to cause rapid transport of the sperm. Sperm have been shown to traverse the entire length of the fallopian tube in the cow in approximately 5 min, a rate at least 10 times as fast as that which the swimming motions of the sperm themselves could possibly achieve. Whether or not this occurs in the human female is unknown.

In addition to the possible effects of the orgasm on fertilization, the intense sexual sensations that develop during the orgasm also pass to the cerebrum and cause intense muscle tension throughout the body. But after culmination of the sexual act, this gives way during the succeeding minutes to a sense of satisfaction characterized by relaxed peacefulness, an effect called *resolution.*

REFERENCES

HYPOTHALAMUS-PITUITARY

Bercu, B. B. (ed.): Basic and Clinical Aspects of Growth Hormone. New York, Plenum Publishing Corp., 1988.

Campion, D. R., et al. (eds.): Animal Growth Regulation. New York, Plenum Publishing Corp., 1989.

DeGroot, L. J. (ed.): Endocrinology, 2nd Ed. Philadelphia, W. B. Saunders Co., 1989.

Gann, D. S., et al.: Neural interaction in control of adrenocorticotropin. Fed. Proc., 44:161, 1985.

Kannan, C. R.: The Pituitary Gland. New York, Plenum Publishing Corp., 1987.

Kudlow, J. E., et al. (eds.): Biology of Growth Factors. New York, Plenum Publishing Corp., 1988.

Muller, E. E.: Neural control of somatotropic function. Physiol. Rev., 67:962, 1987.

MALE FUNCTIONS

Beyer, C., and Feder, H. H.: Sex steroids and afferent input: Their roles in brain sexual differentiation. Annu. Rev. Physiol., 49:349, 1987.

Conn, P. M., et al.: Mechanism of action of gonadotropin releasing hormone. Annu. Rev. Physiol., 48:495, 1986.

Diczfalusy, E., and Bygdeman, M.: Fertility Regulation Today & Tomorrow. New York, Raven Press, 1987.

Knobil, E.: A hypothalamic pulse generator governs mammalian reproduction. News Physiol. Sci., 2:42, 1987.

Knobil, E., et al. (eds.): The Physiology of Reproduction. New York, Raven Press, 1988.

Mahesh, V. B., et al. (eds.): Regulation of Ovarian and Testicular Function. New York, Plenum Publishing Corp., 1987.

Marx, J. L.: Sexual responses are—almost—all in the brain. Science, 241:903, 1988.

FEMALE FUNCTIONS

Beyer C., and Feder, H. H.: Sex steroids and afferent input: Their roles in brain sexual differentiation. Annu. Rev. Physiol., 49:349, 1987.

Diczfalusy, E., and Bygdeman, M.: Fertility Regulation Today & Tomorrow. New York, Raven Press, 1987.

Karsch, F. J.: Central actions of ovarian steroids in the feedback regulation of pulsatile secretion of luteinizing hormone. Annu. Rev. Physiol., 49:365, 1987.

Knobil, E.: A hypothalamic pulse generator governs mammalian reproduction. News Physiol. Sci., 2:42, 1987.

Millar, R. P., and King, J. A.: Evolution of gonadotropin-releasing hormone: Multiple usage of a peptide. News Physiol. Sci., 3:49, 1988.

Soules, M. F.: Problems in Reproductive Endocrinology and Infertility. New York, Elsevier Science Publishing Co., 1989.

Yen, S. S. C., and Jaffe, R.: Reproductive Endocrinology: Physiology, Pathophysiology and Clinical Management, 2nd Ed. Philadelphia, W. B. Saunders Co., 1986.

Index

Note: Page numbers in *italics* refer to illustrations; page numbers followed by (t) refer to tables.